Continental Tectonics

Geological Society Special Publications

*Series Editors*

A. J. HARTLEY

R. E. HOLDSWORTH

A. C. MORTON

M. S. STOKER

GEOLOGICAL SOCIETY SPECIAL PUBLICATION NO. 164

# Continental Tectonics

EDITED BY

## CONALL MAC NIOCAILL
University of Oxford, UK

## PAUL D. RYAN
National University of Ireland, Galway

1999

Published by

The Geological Society

London

# THE GEOLOGICAL SOCIETY

The Geological Society of London was founded in 1807 and is the oldest geological society in the world. It received its Royal Charter in 1825 for the purpose of 'investigating the mineral structure of the Earth' and is now Britain's national society for geology.

Both a learned society and a professional body, the Geological Society is recognized by the Department of Trade and Industry (DTI) as the chartering authority for geoscience, able to award Chartered Geologist status upon appropriately qualified Fellows. The Society has a membership of 8600, of whom about 1500 live outside the UK.

Fellowship of the Society is open to those holding a recognized honours degree in geology or cognate subject and who have at least two years' relevant postgraduate experience, or who have not less than six years' relevant experience in geology or a cognate subject. A Fellow with a minimum of five years' relevant postgraduate experience in the practice of geology may apply for chartered status. Successful applicants are entitled to use the designatory postnominal CGeol (Chartered Geologist). Fellows of the Society may use the letters FGS. Other grades of membership are available to members not yet qualifying for Fellowship.

The Society has its own publishing house based in Bath, UK. It produces the Society's international journals, books and maps, and is the European distributor for publications of the American Association of Petroleum Geologists, (AAPG), the Society for Sedimentary Geology (SEPM) and the Geological Society of America (GSA). Members of the Society can buy books at considerable discounts. The publishing House has an online bookshop (*http://bookshop.geolsoc.org.uk*).

Further information on Society membership may be obtained from the Membership Services Manager, The Geological Society, Burlington House, Piccadilly, London W1V 0JU, UK. (Email: *enquiries@geolsoc.org.uk*: tel: +44 (0)171 434 9944).

The Society's Web Site can be found at *http://www.geolsoc.org.uk/*. The Society is a Registered Charity, number 210161.

Published by The Geological Society from:
The Geological Society Publishing House
Unit 7, Brassmill Enterprise Centre
Brassmill Lane
Bath BA1 3JN, UK

(*Orders*: Tel. +44 (0)1225 445046
     Fax +44 (0)1225 442836)
Online bookshop: *http://bookshop.geolsoc.org.uk*

First published 1999

The publishers make no representation, express or implied, with regard to the accuracy of the information contained in this book and cannot accept any legal responsibility for any errors or omissions that may be made.

**British Library Cataloguing in Publication Data**
A catalogue record for this book is available from the British Library.

ISBN 1-86239-051-7

Typeset by Aarontype Ltd, Bristol, UK

Printed by Cambridge University Press, Cambridge, UK

**Distributors**

*USA*
  AAPG Bookstore
  PO Box 979
  Tulsa
  OK 74101-0979
  USA
*Orders*: Tel. +1 918 584-2555
     Fax +1 918 560-2652
     Email *bookstore@aapg.org*

*Australia*
  Australian Mineral Foundation Bookshop
  63 Conyngham Street
  Glenside
  South Australia 5065
  Australia
*Orders*: Tel. +61 88 379-0444
     Fax +61 88 379-4634
     Email *bookshop@amf.com.au*

*India*
  Affiliated East-West Press PVT Ltd
  G-1/16 Ansari Road, Daryaganj,
  New Delhi 110 002
  India
*Orders*: Tel. +91 11 327-9113
     Fax +91 11 326-0538

*Japan*
  Kanda Book Trading Co.
  Cityhouse Tama 204
  Tsurumaki 1-3-10
  Tama-shi
  Tokyo 206-0034
  Japan
*Orders*: Tel. +81 (0)423 57-7650
     Fax +81 (0)423 57-7651

# Contents

# Preface

It was a great surprise to the many friends and colleagues of John Frederick Dewey to learn that this enthusiastic tectonicist was 60 early in the summer of 1997. It was decided that this event could not pass without some formal acknowledgment of John's remarkable contribution to our understanding of fundamental tectonic processes. Plans for a major international symposium were mooted and I asked John what form he would like this meeting to take. He replied that he would most value a small meeting in Oxford with all his past and present graduate students (46 at that time), colleagues and friends to discuss current developments in Continental Tectonics followed by a field trip to his beloved West of Ireland. It is a tribute to John that 79 people (nearly 70% of those invited) were able to drop what they were doing at fairly short notice and book international flights to attend. This informal meeting took place in the Department of Earth Sciences, Oxford University on 15th to 17th September 1997. No book of abstracts was officially published, however, each session ended with 20 minutes discussion which was sometimes extremely robust. John attended every talk, listened intently, drew on his encylopaedic knowledge to ask searching questions and catalyse the livelier discussion sessions. We all left the meeting knowing that this had been a special event.

The meeting began with **J. F. Dewey** talking on the Caledonides of western Ireland: 40 years on! He emphasized how our improved understanding of the field geology had dictated a new understanding of the underlying tectonic processes. The next two talks by **J. K. Plant, P. Stone & J. R. Mendum**; and **M. Mange, J. F. Dewey, D. Wright & P. D. Ryan** illustrated how careful regional geochemical and heavy mineral provenance studies could be used to constrain tectonic processes. **G. Oliver** reviewed Pan African Mobile Belt theory. **P. D. Ryan & J. F. Dewey**; **M. Krabbendam**; **J. Wheeler**; **A. Wain**; and **M. Krabbendam & J.F. Dewey** discussed the role of continental eclogites in controlling lithospheric buoyancy during continental convergence and the subsequent mechanisms of their exhumation. The meeting then considered structural controls and mechanisms of tectonic exhumation with talks by: **J. Argent & V. Pease**; **V. Pease**; **G. Karner**; **K. C. Burke**; **N. Kusznir**; and **I. W. D. Dalziel** (University of Texas) discussed the role of mantle plumes during continental rifting. Processes at oceanic ridges and subduction zones and the problems in tectonically characterizing ophiolites were reviewed by: **J. Casey**, who gave 2 papers; **P. Ballance**; and **D. Bradley**. The meeting did not forget the commercial aspects of tectonics or our industrial sponsors in the next session in which talks by: **S. Lake**; **W. Mohriak**; **N. W. Driscoll, J. B. Diebold & G. Karner**; **J. A. Karson**; and **A. Khan** dealt with the processes of continental rifting and petroleum formation. Faulted rocks and methods of studying them were the theme of the next session with contributions from: **J. A. Cartwright**; **D. Sanderson**; and **S. Treagus & J. Treagus**. The final day began with a review of modern concepts in granite emplacement by: **K. McCaffrey**; **D. Hutton**; **F. Barker**; and **J. Reavy**, followed by a discussion of Appalachian geology given by: **N. Rast**, **D. E. Rast & F. R. Ettensohn**; **W. S. Kidd**; **D. Bradley**; and **A. Bobyarchick, M. Searle, D Waters, M. Dransfield, B. J. Stephenson, C. Walker** and **J. D. Walker**; **R. Butler & N. Harris** contributed to the next session which dealt with the evolution of the Alps and the Himalayas and emphasized new evidence for the rapid nature of exhumation processes. **L. Lonergan** described a new method for constraining tectonic denudation histories in mountain belts. **W. S. McKerrow** presented the latest Phanerozoic World Maps. **B. Yardley** gave the final paper which emphasized the influence of metamorphism and fluids on crustal rheology. Only one person could attempt a synthesis of all the diverse aspects of continental tectonics covered at the meeting: **J. F. Dewey** gave the closing review.

John Dewey graduated from the University of London with a First Class Honours in 1958 and then read for a PhD at Imperial College, London on the Ordovician and Silurian of South Mayo, western Ireland. His PhD was awarded in 1960, followed by MAs from both Cambridge and Oxford. John attained his ScD from Cambridge in 1987 and his DSc from Oxford in 1988. He lectured in Manchester, Cambridge, the State University of New York at Albany and was appointed to the Chair at Durham in 1982 and as Professor of Geology at Oxford in 1986. He has received many awards for his work including the Lyell (1983) and Wollaston Medals (1999) of the Geological Society, the Penrose Medal of the Geological Society of America (1992) and the Arthur Holmes Medal of the European Union of Geosciences (1993). John was elected a Fellow of the Royal Society in 1985. He was Founding Editor and Editor in Chief of Tectonics (1981–1984) and Basin Research (1989–present), and has served on the editorial boards of *Geology*, *Journal of Geology*, *Tectonophysics*, *Journal of Structural Geology*, *Transactions of the Royal Society of Edinburgh*, *Journal of South East Asian Earth Science* and *Geologica Balcanica*. John has recently been awarded honorary Doctorates by Memorial University, Newfoundland and the National University of Ireland, Maynooth. These awards recognized his contribution to world geoscience, and also the time he has spent in training and enthusing students (John has trained some 55 students whose field areas have literally covered the globe) and teaching them to cast their intellectual nets far wider than their small field areas. He has also played an active role in representing the science of Geology at national and international level and has always been at pains to emphasize the fundamental importance of fieldwork. Long may he continue this essential campaign!

John's scientific contribution began with papers on the Caledonides of western Ireland and kink bands before his Nature paper (Dewey 1969), which was the first attempt to apply the then new global tectonics to an ancient orogen. This was followed by a series of seminal papers, many with John M. Bird (see Dewey & Bird 1970) that related plate tectonics, which so well described the tectonics of the oceans, to continental tectonics. These were so successful that later generations have referred to the 2D, static plate tectonic cartoon sections as 'Deweygrams'. This is a little unfair as John has shown how complex, dynamic and three dimensional real world tectonics are (Dewey 1975). There then followed a series of papers, several with Kevin Burke, describing many of the world's principal orogens in plate tectonic terms and showing how to recognize plume generated triple junctions (Burke & Dewey 1973). John, from field studies in Newfoundland and his long-term association with the Lamont Doherty Observatory, became interested in the evolution of fracture zones (Casey & Dewey 1984), the obduction of ophiolites (Dewey 1976) and the recognition of ancient sutures (Dewey 1977). Articles followed, many with PhD students, relating detailed field observations often in the Tethysides or Newfoundland, to plate tectonic processes. These, perhaps, culminated in John's William Smith lecture to the Geological Society on 'Plate Tectonics and the Evolution of the British Isles' (Dewey 1982). John is not afraid to reject an old model in the light of new field evidence, but is very wary of rejecting field evidence in the light of a new model! Nature published a different plate tectonic model (Dewey & Shackleton 1984) for the Caledonides, which still stands as our template for understanding this orogen. John has an interest in the tectonics of continental break-up and sedimentary basins (Dewey 1988a). His studies in the Himalayas, the Alps and the Andes led to his seminal work on orogenic collapse (Dewey 1988b). I had the pleasure of introducing John to the Scandinavian Caledonides in 1989. This trip resulted in a series of papers arguing that metamorphic phase changes buffer topography during continental convergence (Dewey et al. 1993) and a new explanation for the Wilson Cycle (Ryan & Dewey 1997). John still continues with his life-long interest in the Caledonides but seasons it with a fascination in all aspects of continental tectonics, English music, water colour painting and cricket! I think John, as a scientist, was best summed up by his much loved and sadly missed friend Keith Cox (RIP) talking to an overawed graduate student; 'You could be just like John Dewey' opined Keith. The student glowed at this prospect and asked; 'Do you really think so?'. 'Yes', said Keith, 'Go home, young man, and learn the geology of the world off by heart'.

We would like to express the thanks of all those who attended this meeting to our industrial sponsors British Petroleum PLC and Amerada Hess Ltd. The money was partly used to help young researchers attend the meeting. Our sponsors' financial support and attendance was an excellent example of how good basic science can benefit both the academic and commercial branches of geoscience.

We would also like to thank the following for their contributions to refereeing papers in this volume: Torgeir Anderson (Universitet i Oslo); Ricardo Astini (Cordoba); Stephanie Baldwin (University of Cambridge); Andrew Brock (National University of Ireland, Galway); Ramon Carbonell (Institut Jaume Almera, Barcelona); Peter Clift (Woods Hole Oceanographic Institute); Keith Cox† (University of Oxford); John Dewey (University of Oxford); Richard England (BIRPS); Robert Hall (Royal Holloway); Laurent Jolivet (Université Pierre et Marie Curie, Paris); Lorcan Kennan (University of Oxford); Maarten Krabbendam (Monash University); Lidia Lonergan (Imperial College, London); Brian O'Reilly (Dublin Institute for advanced studies); John Platt (University of London); Dave Sanderson (Imperial College, London); Chris Stillman (Trinity College, Dublin); Philip Stone (British Geological Survey); Rob Strachan (Oxford Brookes); Cees van Staal (Geological Survey of Canada); Dave Waters (University of Oxford); Nigel Woodcock (University of Cambridge) and several anonymous reviewers.

We would also like to especially thank Mike Searle and Barry Wood (Oxford) for their assistance in revising the paper by Ziad Beydoun†, who passed away during the production of this volume.

Finally we wish to express our gratitude to Angharad Hills, Bob Holdsworth, Helen Knapp, and Alan Roberts for shepherding this volume through to completion!

Paul D. Ryan
Galway, May 1999

---

† Deceased.

## References

BURKE, K. C. A. & DEWEY, J. F. 1973. Plume-generated triple junctions: key indicators in applying plate tectonics to old rocks. *Journal of Geology*, **81**, 406–433.

CASEY, J. P. & DEWEY, J. F. 1984. Initiation of subduction zones along transform and accreting plate boundaries, triple junction evolution and spreadi,ng centres – implications for ophiolitic geology and obduction. *In*: GASS, I. G., LIPPARD, S. J. & SHELTON, A. W. (eds) *Ophiolites and Oceanic Lithosphere*. Geologlcal Society, London, Special Publications, **413**, 269–290.

DEWEY, J. F. 1969. The evolution of the Appalachian/Caledonian orogen. *Nature*, **222**, 124–129.

—— & BIRD, J. M. 1970. Mountain belts and the new global tectonics. *Journal of Geophysical Research*, **75**, 2625–2647.

——1975. Finite plate evolution: some implications for the evolution of rock masses at plate margins. *American Journal of Science*, **275-A**, 260–284.

——1976. Ophiolite obduction. *Tectonophysics*, **31**, 93–120.

——1977. Suture zone complexities: a review. *Tectonophysics*, **40**, 53–67.

——1982. Plate tectonics and the evolution of the British Isles (1981 William Smith Lecture, Geological Society of London). *Journal of the Geological Society, London*, **139**, 371–412.

——1988a. Lithospheric stress, deformation and tectonic cycles: the disruption of Pangea and the closure of Tethys. *In*: AUDLEY-CHARLES, M. G. & HALLAM, A. (eds) *Gondwana & Tethys*. Geological Society, London, Special Publications, **37**, 23–40.

——1988b. The Extensional collapse of orogens. *Tectonics*, **7**, 1123–1139.

——, RYAN, P. D. & ANDERSEN, T. B. 1993. Orogenic uplift and collapse, crustal thickness, fabrics and phase changes: the role of eclogites. *In*: PRICHARD, H. M., ALABASTER, T., HARRIS, N. B. W. & NEARY, C. R. (eds) *Magmatic Processes and Plate Tectonics*. Geological Society, London, Special Publications, **76**, 325–343.

—— & SHACKLETON, R. M. 1984. A model for the evolution of the Grampian Tract in the early Caledonides and Appalachians. *Nature*, **312**, 115–121.

RYAN, P. D. & DEWEY, J. F. 1997. Continental eclogites and the Wilson Cycle. *Journal of the Geological Society, London*, **154**, 371–412.

# Continental Tectonics: an introduction

P. D. RYAN,[1] & C. MAC NIOCAILL[2]

[1] *Department of Geology, National University of Ireland, Galway, Ireland*
*(e-mail: ryan@alisanos.nuigalway.ie)*
[2] *Department of Earth Sciences, University of Oxford, Parks Road, OX1 3PR, UK*
*(e-mail: conallm@earth.ox.ac.uk)*

The theory of plate tectonics has proven to be very successful in describing the behaviour of the oceanic plates; however the study of continental tectonics is complicated by the fact that continental lithosphere does not behave as a rigid plate. The narrow zones of seismicity that define plate boundaries in oceanic lithosphere become diffuse in continental lithosphere (Isacks *et al.* 1968). Continental crust is thicker, more buoyant with respect to the asthenosphere, of different composition and rheology, has higher heat productivity, and is up to 20 times older than oceanic crust. An understanding of continental tectonics requires that these effects upon rigid plate behaviour to be taken into account. A further, complicating, factor is that the principles of plate tectonics were derived from a remotely sensed geophysical dataset, acquired in the ocean basins over some 20 years, and held in relatively few laboratories. Continental tectonics requires the reconciliation of 150 year's worth of worldwide field data involving all branches of geology with these principles. The growth of the continents throughout time indicates that little continental lithosphere is subducted and, therefore, the records of past (pre-Jurassic) plate geometries are stored in the continental crust. The greater mean age of continental crust also means that any single continental rock mass is likely to have been affected by several cycles of creation and destruction of oceanic lithosphere (the so-called Wilson cycle). The transposed multiple fabrics associated with each cycle can only be elucidated by careful field study, if the critical evidence is still preserved. As a result, continental tectonics is a complex discipline. It involves the integration of extensive geological field and laboratory data sets with an understanding of plate tectonic principles, their comparison with modern and past analogues and often their integration in a numerical simulation.

This volume derives from the conference held on 15th–17th September 1997 in Oxford University to celebrate John F. Dewey's 60th birthday. This meeting addressed some of the current developments in the field of continental tectonics, many of which have benefited from John's insights, provided over the past 38 years. We first review some current topics in continental tectonics, including how the papers in this volume relate to them, and the contributions themselves are then presented, arranged in a pseudo Wilson Cycle, as follows: processes in closing oceans; collisional orogeny; orogenic collapse; and finally mechanisms of continental rifting.

## Some current controversies in continental tectonics

The exact dating of ancient orogenic events is often problematic. It relies upon precise, often difficult, correlations between sedimentary or volcanic events in the upper plate with plutonic and metamorphic events in the lower plate. The recent controversy concerning the age (estimates ranging from Neoproterozoic to Ordovician) of the Grampian event in Scotland and Ireland is an example of the difficulties that can be encountered (Soper *et al.* 1999). **Dewey & Mange** show how detailed field study can be linked to modern tectonic analogues to resolve such difficulties.

The relationship between metamorphism and deformation are dealt with by several articles. **Jolivet et al.** and **Forster & Lister** discuss the role of eclogite facies phase transformations, which can lower the buoyancy of the lithosphere (Dewey *et al.* 1993), in the lower crust of the Aegean region. Barrovian metamorphism is generally taken to have occurred because of increased radiogenic heating of a thickened crust (England & Thompson 1984; Thompson & England 1984). This mechanism is consistent with the tectonic history reported by **Searle et al.** for the Himalayan continent–continent collision, but it is inconsistent with the rapid arc–continent collisions, as described by **Dewey &**

*From*: MAC NIOCAILL, C. & RYAN, P. D. (eds) 1999. *Continental Tectonics*. Geological Society, London, Special Publications, **164**, 1–6. 1-86239-051-7/99/$15.00 © The Geological Society of London 1999.

**Mange**, which also produce Barrovian assemblages. **Dewey & Mange** also address the controversy concerning the nature of ophiolites. Do they represent slices of mid-ocean ridge crust and mantle (Moores & Vine 1971) or were they formed in supra–subduction zone environments (Dewey & Bird, 1971; Miyashiro 1973)? **Dewey & Mange** argue that a supra–subduction zone origin followed by emplacement during an arc collision event accounts for the complex nature of ophiolite suites and the short interval between their formation and obduction.

Terrane analysis (Coney *et al.* 1980) has shown how laterally mobile fault-bounded fragments of the continental crust can be. The varied nature of continental geology, however, means that a fault separating two provinces of differing geology may not have large strike slip displacement along it (Sengör & Dewey 1990; Stewart *et al.* 1999). **Plant *et al.*** adopt a novel approach to terrane analysis showing that tectonic lineaments do not always coincide with terrane boundaries.

Shallow-dipping detachment faults are widely recognized in regions where the continental crust has undergone extensional exhumation. Whereas modern syntheses (e.g. Jackson & White 1989) indicate that seismically-active, map-scale extensional faults tend to have dips of 30–60°, it remains controversial whether these regional scale faults initiated at low angles. Three contributions **Pease & Argent**, **Pease *et al.*** and **Karner *et al.*** all argue for the existence of tectonically-active low-angle detachments. A shallow dipping detachment goes some of the way to resolving the fact that whereas some detachments in the Basin and Range have up to 70 km of dip–slip motion, the highest grade metamorphic rocks exposed contain kyanite (Hamilton 1988).

Several tectonic models have been proposed for the formation of sedimentary basins in continental crust: pure shear (McKenzie 1978): simple shear (Wernicke 1985); flexural cantilever (Kuznsir *et al.* 1991); or flexural (**Karner & Driscoll**). There is also a vigorous debate as to whether elevated mantle temperatures are required to generate extensive basaltic magmatism on continental margins with low $\beta$ factors. Although this appears to be the case in Kenya (**Khan *et al.***) and east Greenland (**Karson & Brooks**), **Karner & Driscoll** argue that melting can also be caused by extension in the lower plate.

Finally, the reactivation of suitably orientated older structures by new tectonic processes is well known (Holdsworth *et al.* 1997) but, perhaps, not well understood. The fundamental problem is that lithospheric strength lies within the mantle, not the upper continental crust (Molnar 1992).

Two examples of reactivation along incipient plate boundaries are given in this volume. **Karson & Brooks** describe how a Precambrian fabric may have partly-controlled ridge segmentation in the northern Atlantic and **Beydoun** shows how a transform plate boundary has had its geometry modified by a pre-existing collisional structure.

## Tectonic processes within arc complexes

Ancient tectonic reconstructions rely upon the identification of rock assemblages at characterize plate boundaries. For example, the recognition of ancient arcs and their polarity is critical to the understanding of orogens. **Ballance** reviews the history of the southwest pacific Neogene arcs and concludes that initial complexity involving three arc systems, two of NE trend and one of NW trend, that existed for 10 Ma, underwent simplification due to the Miocene SE motion of the Pacific–Australasian pole. The proximity of the pole in the Oligocene led to amagmatic subduction of the Pacific Plate and the opening of a back-arc basin. Late Miocene plate reorganization was associated with ophiolite obduction in the North Island of New Zealand, the nucleation of a dextral fracture zone along the earlier, NW trending arc and the dextral translation and rotation of the earlier NW trending fore–arc to its present NE trend. This level of interpretation, made possible by the preservation of Neogene oceanic lithosphere, would be impossible in an ancient orogen. Furthermore, subsequent ocean closure could lead to removal of much of the critical positional information (Van Staal *et al.* 1998).

**Jolivet *et al.*** use data from the Japan Sea, the Tyrrhenian Sea and the Aegean Sea to review the mechanisms by which back-arc basins open. They conclude that the principal extension mechanism is slab–pull inducing subduction roll-back (Dewey 1980); however, in none of the basins reviewed this is the only mechanism active. Retreat of the Pacific subduction zone produced extension in the Japan Sea orthogonal to the trench and cannot explain the significant dextral strike–slip component observed. **Jolivet *et al.*** attribute this to far–field stresses caused by the India–Asia collision and the formation of a trans-Asian strike–slip fault system. Such a view would seem to support the analysis of Tapponnier *et al.* (1982) for the expulsion of material laterally from this collision zone, rather than that of England and Molnar (1990). The Tyrrhenian Sea opened in crust thickened during Alpine collision by the retreat of the Adriatic continental slab. **Jolivet *et al.*** suggest that eclogitisation of this slab provided the strong slab pull forces required. The extension of thickened lithosphere led to the exhumation of basement metamorphic cores. A similar history is proposed for the Aegean Sea with a stronger influence of orogenic collapse over a larger region and a geometric control imposed by the Arabia–Eurasia collision.

## Collisional orogenic processes

**Searle et al.** document the India–Asia collision, to form the Himalayas. They describe a large, isoclinal, SW-verging fold in the metamorphic isograds of the High Himalayas. They envisage the isograds, which cut folded rocks, as initially being horizontal during a crustal thickening phase from c. 50–30 Ma. The SW-verging Main Central Thrust and an extensional detachment, the Zanskar Shear Zone, in its hanging wall initiated between 30 and 21 Ma. There then followed a period (21–18.5 Ma) of very rapid exhumation (6–10 mm/a) associated with tectonic unroofing and rapid mechanical erosion of the footwall. Subsequent exhumation has been much slower (0.4 mm/a). Thrusting normally buries, not exhumes, rocks because the associated crustal thickening causes footwall subsidence. However, normal fault motion along the Zanskar Shear Zone unloaded the hanging wall of the Main Central Thrust and allowed very rapid exhumation rates to be associated with Miocene thrusting.

**Dewey & Mange** present a detailed study based upon high-resolution heavy mineral analysis of the Lower Palaeozoic strata of South Mayo, Ireland. This analysis provides support for earlier models (Dewey & Shackleton 1984; Ryan & Dewey 1990) which suggested that the Grampian orogeny of the Laurentian margin was the result of arc–continent collision in the Llanvirn. In particular, their study confirms the short-lived nature of this event (<10 Ma?) with emergence of metamorphic rocks of the lower plate occurring before cessation of sedimentation within a fore–arc basin in the upper plate. They suggest that Miocene Papua-New Guinea orogen, involving arc–collision with the NW Australian shelf followed by subduction flip, provides a good analogue for the Grampian event. They also show that the associated emplacement of ophiolites can lead to rapid crustal thickening and Barrovian metamorphism in the lower plate. **Dewey & Mange** suggest that the complex nature of ophiolite suites, plus the short time between their formation and emplacement are consistent with a supra–subduction

zone origin and obduction during an arc collision–subduction-flip event.

**Plant et al.** present multi-element regional geochemical maps for the northern British Caledonides based upon a density of one sample per 1.5 km$^2$. This systematic analysis is incorporated into existing terrane models for the orogen, which have, perhaps, been more influenced by the existence of major lineaments than the geological contrasts across them. Interestingly, a terrane boundary marking a putative suture along the Great Glen fault has no geochemical signature, calling into question the significance of this boundary (**Plant et al.**; Stewart et al. 1999). The distribution of sedimentary exhalative and volcanic massive sulphide and gold mineralization deposits clearly marks zones of long lived lithospheric extension and mafic vulcanism linked to the opening of the Iapetus Ocean. This method is a powerful tool for discriminating between flysch deposits from opposing margins deposited in a narrowing ocean which may not be distinguished using palaeontological or palaeomagnetic methods.

**Rast et al.** describe seismites, fluidization structures formed as a result of proximity to an M5 or greater earthquake, in the early Palaeozoic carbonate shelf of the Appalachians which underwent an Ordovician (Taconic) arc collision event (Van Staal et al. 1998). The wide areal and temporal extent of these structures raise the tantalizing possibility of mapping palaeo-seismicity on ancient continental margins. More importantly, the existence of considerable syn-sedimentary deformation in a 'stable' carbonate platform shows how careful we must be in interpreting early, isolated fabrics in higher grade metamorphic rocks as being the products of orogenesis. Earliest (D1) fabrics are often only observed in certain lithologies, or as inclusion trails in metamorphic minerals, but no large scale inversion of stratigraphy is associated with these structures. These early fabrics may actually be reflecting syn-sedimentary structures.

**Forster & Lister** describe a mafic boudin that preserves a complete (Eocene?) metamorphic history for the basement of the Cyclades. Three phases of high pressure metamorphism are recognized and related to the structural chronology of the region. Early blueschist assemblages are overprinted by an eclogite facies and then a further blueschist facies event. They argue that country rocks experienced the same metamorphic history. It seems eminently reasonable that the country rocks should have experienced the same metamorphic conditions as the rocks intruding them. **Forster & Lister** argue that blueschist facies assemblages recorded outside of the mafic boudins represent the later event and, therefore, mark the onset of collapse and exhumation rather than the initiation of burial metamorphism. Another fundamental point is that such parageneses rarely survive exhumation and usually only do so in mafic boudins which have been impermeable to fluids. Is it then possible that much of the lower crustal history of an orogen may only be preserved in a few boudins?

## Continental extensional tectonics

The mechanism of the tectonic exhumation of metamorphic core complexes bounded by flat lying detachments of large areal extent is controversial (Wernicke 1995). Were these faults initiated as flat lying structures or were they steep while active and then rotated, during isostatic relaxation of the footwall (the 'rolling hinge model': Buck 1988)? Two papers by **Pease & Argent** and **Pease et al.** describe the NW Sacramento Mountains region within the Colorado River extension corridor. Igneous rocks emplaced syn- and post-Miocene exhumation (19–14 Ma) allow precise dating of the sequence of events. Structural analysis (**Pease & Argent**) shows that the shallow west-dipping detachment, the Sacramento Mountains Detachment Fault, is the youngest. This is not consistent with the rolling hinge model, which predicts that the steeper faults should be younger. Thermochronologic and thermobarometric data (**Pease et al.**) support rapid cooling (38–53°C/Ma) and exhumation rates of 2 mm/a for a granodiorite complex intruded during shearing. The estimated geothermal gradients and cooling histories indicate an initial shallow dip angle of $c. 25$ for the extensional detachment within the footwall, a conclusion consistent with the structural data. It is interesting to speculate what role the emplacement of the granodiorite may have had upon the rheological behavior and geometry of the detachment fault.

## Continental rifting

The history of a continental transform fault, the Levant Fracture System, is reviewed by **Beydoun**. This fault links the Red Sea spreading centre in the south to the Tauride collisional orogen in the north and was initiated by the anticlockwise rotation of the Arabian plate away from the African plate during the Neogene. The fault consists of three portions, N–S left-stepping strike slip faults to the north and south of Lebanon and a central, braided fault system in

Lebanon. Miocene motion along the N and S segments (62 km sinistral) was accommodated by the central braided system until the development of a restraining bend, the Yammouneh Fault, along the pre-existing Palmyride fold belt in the Lebanon. Subsequent, Pliocene motion (45 km sinistral) has probably passed south of Beirut into the Mediterranean. This may be associated with the westward escape of the Anatolian wedge (Dewey et al. 1986). This fault is generally regarded as forming the boundary of the Arabian plate, which therefore, had different positions in the Miocene and the Pliocene.

The deep structure of the Kenya Rift based on the various KRISP experiments is reviewed by **Khan et al.** The rift has thick crust (38–44 km) at the margins with normal uppermost mantle P wave ($P_n$) velocities of 8.0 km/s. Under the rift, $P_n$ is 7.5–7.8 km/s throughout its length. This anomalously low value is attributed to the presence of hot mantle with 3–5% of partial melt that was probably emplaced at 20–30 Ma. The structure of the rift varies from south to north. In the north the crust is thinned to 20 km, whilst it is about 35 km in the south. This is due to higher $\beta$ factors and a thinner basal crustal layer ($V_p = 6.8$ km/s) containing significant underplated mafic igneous rocks in the north. The rifting varies in age from early Tertiary in the north to late Miocene in the south. Unlike the rifting model presented above for the Exmouth Plateau (**Karner & Driscoll**), the seismic evidence suggests that rifting is associated with lithospheric thinning due to the active uprising of anomalously hot mantle, as predicted by Burke & Dewey (1973).

Models concerning the nature of the continental lithosphere during extension are reviewed by **Karner & Driscoll**. They describe Permian to early Cretaceous (Valanginian) rifting of the Exmouth Plateau, western Australian margin and present a flexural model for the deformation of this lithosphere. The post-Valanginian 'thermal type' subsidence is much greater than would be predicted from the late-Cretaceous extension measured ($\beta = 1.15$–1.22) in the 'upper (brittle) plate' and is delayed by 30 Ma. This is attributed to the development of a flat lying detachment with ramp-flat geometry, that thinned the 'lower plate' (lower crust and upper mantle) during Tithonian–Valangian breaching of the southwestern Australian margin and the emplacement of oceanic lithosphere. This lower plate extension allowed generation of extensive magmatism and associated seaward dipping reflectors, which are recorded on this margin, without requiring the presence of a plume.

**Karson & Brooks** give an account of the Tertiary East Greenland Volcanic Rifted Margin. Voluminous basaltic vulcanism and plutonism are associated with the development of an asymmetrical crustal flexure that accommodated major subsidence along the ocean–continent boundary. Individual rift segments defined by structural discontinuities are recognized on three scales. First order segments occur between major triple-rift junctions, whilst second order segments (c. 100 km in length) are partly controlled by pre-existing Precambrian structures. The smallest segments, 10 km in length, correspond to smaller accommodation zones. The pattern of first and second order segmentation is similar to that recorded in mid-oceanic ridges, reflected in the spacing of transform faults. This raises the important question as to whether this segmentation is controlled by the spacing of magmatic centres or whether it is partly inherited from pre-existing structural fabrics at the time of continental rifting.

It is clear that, while plate tectonics has proven to be a spectacularly successful unifying framework for examining and understanding the tectonic histories of the plates that blanket our planet, much more work remains to be done in elucidating the complex history of that portion of the planet we live on; the continental crust. In addition, much remains to be understood about the dynamic evolution of continental crust and its deformation, the links between crustal dynamics and deeper mantle processes, and the evolution of these processes through time. The record of this is in the rocks. We hope that this wide ranging collection will further stimulate integrative research into the challenging field of continental tectonics.

We thank John Dewey and Bob Holdsworth for their thoughtful comments on this manuscript.

# References

BUCK, W. R. 1988. Flexural rotation of normal faults. *Tectonics*, **7**, 959–975.

BURKE, K. C. A. & DEWEY, J. F. 1973. Plume-generated triple junctions: key indicators in applying plate tectonics to old rocks. *Journal of Geology*, **81**, 406–433

CONEY, P. J., JONES, P. L. & MONGER, J. W. H. 1980. Cordilleran suspect terranes. *Nature*, **288**, 329–333.

DEWEY, J. F. 1980. Episodocity, sequence and style at convergent plate boundaries. *In*: STRANGEWAY, D. W. (ed.) *The Continental Crust and its Mineral Deposits*. Geological Association of Canada, Special Paper, **20**, 553–573.

—— & BIRD, J. M. 1971. The origin and emplacement of the ophiolite suite: Appalachian ophiolites in Newfoundland. *Journal of Geophysical Research*, **76**, 3170–3206.

—— & SHACKLETON, R. M. 1984. A model for the evolution of the Grampian Tract in the early Caledonides and Appalachians. *Nature*, **312**, 115–121.

——, HEMPTON, M., KIDD, W. S. F., SAROGLU, F. & SENGÖR, A. M. C. 1986. Shortening of continental lithosphere: Neotectonics of Eastern Anatolia – a young continental collision zone. *In*: COWARD, M. P. & REIS, A. C. (eds) *Collision Tectonics*. Geological Society, London Special Publications, **19**, 3–36.

——, RYAN, P. D. & ANDERSEN, T. B. 1993. Orogenic uplift and collapse, crustal thickness, fabrics and phase changes: the role of eclogites. *In*: PRICHARD, H. M., ALABASTER, T., HARRIS, N. B. W. & NEARY, C. R. (eds) *Magmatic processes and plate tectonics*. Geological Society, London Special Publications, **76**, 325–343.

ENGLAND, P. C. & MOLNAR, P. 1990. Right-lateral shear and rotation as the explanation for strike-slip faulting in Tibet. *Nature*, **344**, 140–142.

—— & THOMPSON, A. B. 1984. Pressure-temperature-time-paths of regional metamorphism I: heat transfer during the evolution of regions of thickened continental crust. *Journal of Petrology*, **25**, 894–928.

HAMILTON, W. 1988. Detachment faulting in the Death Valley region, California and Nevada. *Bulletin of the United States Geological Survey*, **1790**, 51–85

HOLDSWORTH, R. E., BUTLER, C. A. & ROBERTS, A. M. 1997. The recognition of reactivation during continental deformation. *Journal of the Geological Society, London*, **154**, 73–78.

ISACKS, B., OLIVER, J. & SYKES, L. R. 1968. Seismology and the new global tectonics. *Journal of Geophysical Research*, **73**, 5855–5899.

JACKSON, J. A. & WHITE, N. J. 1989. Normal faulting in the upper crust: observations from regions of active extension. *Journal of Structural Geology*, **11**, 15–36.

KUSZNIR, N. J., MARSDEN, G. & EGAN, S. S. 1991. A flexural cantilever simple shear/pre-shear model of continental lithosphere extension: application to the Jeanne D'Arc Basin and the Viking Graben. *In*: ROBERTS, A. M., YIELDING, G. & FREEMAN, B. (eds) *The Geometry of Normal Faults*. Geological Society, London, Special Publications, **56**, 41–60.

MCKENZIE, D. P. 1978. Some remarks on the development of sedimentary basins. *Earth and Planetary Science Letters*, **40**, 25–32.

MIYASHIRO, A. 1973. The Troodos ophiolite complex was probably formed in an island arc. *Earth and Planetary Science Letters*, **19**, 218–224.

MOLNAR, P. 1992. Mountain building – crust in mantle overdrive. *Nature*, **358**, 105–106.

MOORES, E. M. & VINE, F. J. 1971. The Troodos Massif, Cyprus and other ophiolites as oceanic crust: evaluation and implications. *Philosophical Transactions of the Royal Society, London*, **A268**, 433–466.

RYAN, P. D. & DEWEY, J. F. 1990. A geological and tectonic cross-section of the Caledonides of western Ireland. *Journal of the Geological Society, London*, **148**, 173–180.

SENGÖR, A. M. C. & DEWEY, J. F. 1990. Terranology: vice or virtue? *Philosophical Transactions of the Royal Society, London*, **A331**, 887–901.

SOPER, N. J., RYAN, P. D. & DEWEY, J. F. 1999. The Grampian orogeny in Scotland and Ireland. *Journal of the Geological Society, London*, **157**, in press.

STEWART, M., STRACHAN, R. A. & HOLDSWORTH, R. E. 1999. Structure and early kinematic history of the Great Glen Fault Zone, Scotland. *Tectonics*, **18**, 326–342.

TAPPONNIER, P., PELTZER, G., LE DAIN, A. Y., ARMIJO, R. & COBBING, P. 1982. Propagating extrusion tectonics in Asia: new insights from simple experiments with plasticine. *Geology*, **10**, 611–616.

THOMPSON, A. B. & ENGLAND, P. C. 1984. Pressure-temperature-time paths of regional metamorphism II: their inferences and interpretations using mineral assemblages in metamorphic rocks. *Journal of Petrology*, **25**, 929–955.

VAN STAAL, C. R., DEWEY, J. F., MAC NIOCAILL, C. & MCKERROW, W. S. 1998. The Cambrian-Silurian evolution of the northern Appalachians and the British Caledonides: history of a complex, southwestern Pacific-type segment of Iapetus. *In*: BLUNDELL, D. J. & SCOTT, A. C. (eds) *Lyell: the past is the key to the present*. Geological Society, London, Special Publications, **143**, 199–242.

WERNICKE, B. 1985. Uniform-sense normal simple shear of the continental lithosphere. *Canadian Journal of Earth Sciences*, **22**, 108–125.

WERNICKE, B. 1995. Low-angle normal faults and seismicity: a review. *Journal of Geophysical Research*, **100**, 20 159–20 174.

# Simplification of the Southwest Pacific Neogene arcs: inherited complexity and control by a retreating pole of rotation

PETER F. BALLANCE

*Department of Geology, The University of Auckland, Private Bag 92019, Auckland, New Zealand*

**Abstract:** Neogene arc activity in the Southwest Pacific began simultaneously at 25 Ma on three differently oriented sectors. Two sectors (Norfolk–Three Kings and Colville Ridges), aligned north and north-northeast, were near-orthogonal to subduction, while the intervening Northland–Reinga sector (northwest aligned) was strongly sinistral-oblique and produced twin parallel arcs of differing compositions. This complexity was inherited from a complex late Eocene–Oligocene margin that was convergent but amagmatic, and was due primarily to its proximity to the Pacific–Australia pole of rotation. The inception of arc magmatism at 25 Ma was triggered by a 20° increase in convergence angle between the north moving Australia Plate and the northwest moving Pacific plate (from 44° to 65°), and an increase in convergence rate from $c. 20$ to $30–40 \, \mathrm{mm \, yr^{-1}}$. Between 25 and 15 Ma three subduction zones were required. The Pacific Plate was subducted at the Colville Arc, and a South Fiji Basin Plate was subducted at the Norfolk–Three Kings Arc in a manner analogous to the present-day Philippine Sea Plate located between the Mariana Arc and the Japan Trench. During this interval the Norfolk Basin opened as a back-arc basin, while the Three Kings Arc moved eastwards between two fracture zones. Eastward movement of the Three Kings Arc ($c. 350 \, \mathrm{km}$) was sufficient to drive a magma-producing subduction rate of $c. 35 \, \mathrm{mm \, yr^{-1}}$, independently of Pacific Plate convergence which was therefore taken up at the Colville Subduction Zone. Sinistral-oblique subduction beneath the Northland–Reinga sector was also driven by Pacific Plate convergence of $c. 30 \, \mathrm{mm \, yr^{-1}}$. At 15 Ma the Colville Ridge extended 400 km southwestwards across North Island, New Zealand, to become a Colville–Coromandel–Taranaki Arc, and the two other arc segments died; this was the main simplification event, and it was probably a response to the progressive southeasterly retreat of the pole of Pacific–Australia rotation located to the south of New Zealand and the consequent increase in convergence rate. Between 15 and 5 Ma there was stability and no back-arc basin formed. A big change at 5 Ma was probably driven by further movement of the pole, this time to the southwest, and an increase in convergence rate across New Zealand. In response, the arc split to form the still-active Lau-Havre Back-arc Basin and the active Tonga–Kermadec–Taupo Arc, and the New Zealand mountains formed. Factors remaining to be resolved are the time of collision of the oceanic Hikurangi Plateau, which has entered the New Zealand Subduction Zone, and the timing of dextral displacement(s) of the forearc to its present location along the eastern margin of the North Island. Arc simplification was accompanied by further fragmentation and dispersal of New Zealand continental crust which had already been fragmented by late Cretaceous extension. The Vening Meinesz Fracture Zone, 900 km long, originated as an Eocene near-transform sinistral boundary, and had a complex Cenozoic record of strike-slip and transpressional activation and reactivation, with both sinistral and dextral sense.

Volcanic-magmatic arcs are important in the reconstruction of past tectonic scenarios because of their consistent spatial association with downgoing lithospheric slabs. They are typically located above the 90–100 km depth contour on the slab (Gill 1981) and can be expected to have an associated forearc basin (Karig 1970).

The history of the New Zealand portion of the volcanic Tonga–Kermadec–New Zealand Arc (Figs 1 and 2) has been enigmatic because of the apparent northwesterly alignment of the early–late Miocene arc centres in northern New Zealand, in contrast to the north-northeasterly alignment of the active arc (Fig. 3). Early attempts at reconstructing southwest Pacific Cenozoic arc history were hampered by lack of knowledge of the local offshore geology: approximately three-quarters of the New Zealand minicontinent is under water (Fig. 1). The attempts fell into two groups: those that proposed a

*From*: MAC NIOCAILL, C. & RYAN, P. D. (eds) 1999. *Continental Tectonics*. Geological Society, London, Special Publications, **164**, 7–19. 1-86239-051-7/99/$15.00 © The Geological Society of London 1999.

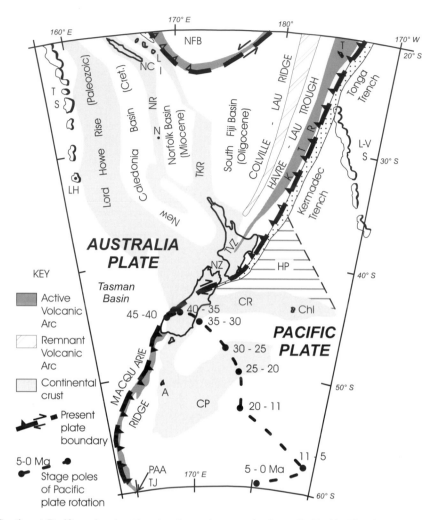

**Fig. 1.** Southwest Pacific regional setting showing active arcs, the Australia–Pacific Plate boundary and the migration of the Pacific Plate pole of rotation (after Sutherland (1995)). A, Auckland Island; ChI, Chatham Islands; CP, Campbell Plateau; CR, Chatham Rise; HP, Hikurangi Plateau; KTR, Kermadec–Tonga Ridge; LH, Lord Howe Island; LI, Loyalty Islands; L-VS, Louisville Seamount Chain; N , Norfolk Island; NC, New Caledonia; NFB, North Fiji Basin; NR, Norfolk Ridge; NZ, New Zealand; PAATJ, Pacific–Australia–Antarctica triple junction; T, Tonga; TKR, Three Kings Ridge; TS, Tasmantid Seamount Chain; TVZ, Taupo Volcanic Zone.

radical realignment of subduction in the late Neogene (e.g. Ballance *et al.* 1982; Cole 1986); and those that proposed that subduction had always had its present alignment, the north-westerly trend of the centres on land being due to trench roll-back and subduction retreat (Brothers 1986; Kamp 1984).

Recent research on the offshore extensions of northern New Zealand (Norfolk and Three Kings Ridges), and on the Colville Ridge remnant arc and Kermadec Ridge active arc, has revealed a complex pattern of arc segments in the latest Oligocene and early Miocene (Ballance *et al.* 1999; Herzer *et al.* 1997; Mortimer *et al.* 1998) (Fig. 3). It is now apparent that a major reorganization and simplification of subduction took place in the mid-Miocene (Herzer 1995). The knowledge base has been further extended by a reassessment of the Cenozoic poles of Pacific–Australia Plate rotation (Sutherland 1995) and a detailed investigation of part of the Vening Meinesz Fracture Zone, which played an important role in Cenozoic tectonic development of the region (Herzer & Mascle 1996).

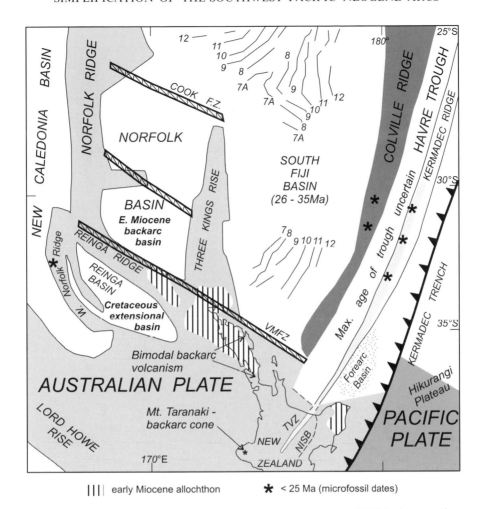

**Fig. 2.** Northern New Zealand, and adjacent bathymetric features, showing South Fiji Basin magnetic anomalies. Offshore data from sources cited in the text. Light stipple, continental crust; NISB, North Island Shear Belt; TVZ, Taupo Volcanic Zone; VMFZ, Vening Meinesz Fracture Zone.

This paper outlines the complex early Miocene arc scenario and attributes the complexity to inheritance from a complex Eocene–Oligocene precursor plate boundary. The ensuing arc simplification is attributed to the progressive southeasterly retreat of the pole of Pacific–Australia rotation, and the consequent increase in subduction rate and southward transfer of strongly oblique convergence. Simultaneously, however, the already fragmented New Zealand continental crust was further fragmented by back-arc spreading and developing strike-slip tectonism.

## Pre-Miocene plate organization

While there has long been robust evidence of Neogene subduction in the region, in the form of arc remnants and forearc geology (Pettinga 1982; Lewis & Pettinga 1993), and as required by global tectonic considerations (Weissel *et al.* 1977; Stock & Molnar 1982), the matter of subduction in the mid-Cenozoic is much more speculative. Various analyses of global plate development have concluded that a convergent margin is required in the southwest Pacific, beginning at 43 Ma (Molnar *et al.* 1975; Weissel *et al.* 1977; Stock & Molnar 1982). To date, few indications of convergence between 43 and 25 Ma have been apparent in central and northern New Zealand (Ballance 1993*b*), though there is a well-established spreading and transform boundary in southwestern and western New Zealand (Norris *et al.* 1978; Kamp 1986; Norris & Turnbull, 1993; Lamarche *et al.* 1997).

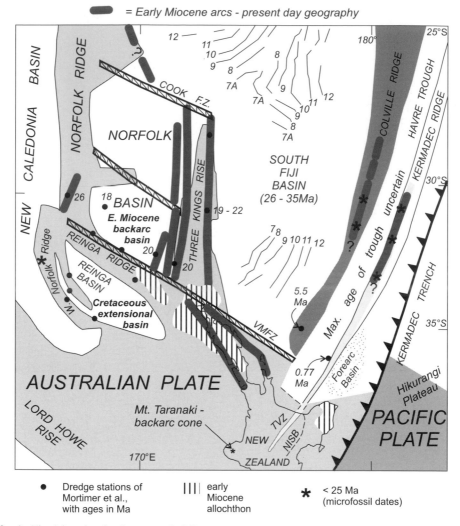

**Fig. 3.**  As Fig. 2 but showing known early Miocene arc segments.

While there is still no evidence of subduction volcanism in the interval 43–26 Ma, various considerations suggest that a complex convergent margin may have passed through northern New Zealand (Ballance *et al.* 1995; Smith *et al.* 1995) (Figs 4 and 5). The complexity was dictated in large part by the close proximity of the pole of Pacific–Australia rotation, located in present-day South Island, New Zealand (Sutherland 1995) (Fig. 1). A detailed consideration of the boundary is given by Ballance (in prep.). The particular aspect of interest here is the possible existence of a late Eocene convergent margin in two sections, a west-northwest striking sinistral transform alongside Reinga Ridge–

Northland, and a north striking amagmatic ('mute') subducting slab alongside the Three Kings Ridge (Fig. 4).

This situation is thought to have evolved to the late Oligocene situation, in which the South Fiji Basin (SFB) was fully open, by means of eastwards migration (roll-back) of the north striking subduction interface, which thus became the Colville–Lau Subduction Zone. In this scenario the slab was still amagmatic, but the SFB opened as a normal back-arc basin by spreading from at least two centres between *c.* 35 and 26 Ma (Davey 1982) (Fig. 2). The downgoing lithosphere may have been an extension of the present Pacific Plate lithosphere, which is old

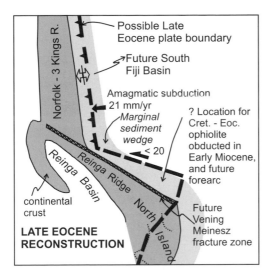

**Fig. 4.** A late Eocene reconstruction of northern New Zealand and the joined Norfolk–Three Kings Ridges, showing the location of the inferred convergent/sinistral transform Australia–Pacific Plate boundary, and the positions of the future South Fiji Basin and Vening Meinesz Fracture Zone.

and dense, and would therefore have been prone to roll-back; or it may have been younger, Cretaceous–Palaeogene, oceanic lithosphere which was the provenance for the Northland ophiolites (Malpas *et al.* 1992) and which has now disappeared. The importance of the Oligocene

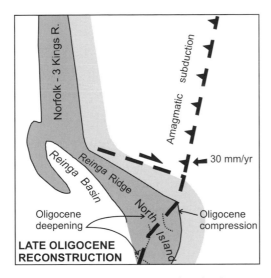

**Fig. 5.** A late Oligocene reconstruction showing eastward retreat of the Colville amagmatic convergence zone, opening of the South Fiji Basin and dextral transform motion across the precursor Vening Meinesz Fracture Zone.

developments is that they set the scene for the simultaneous, region-wide appearance of early Miocene arcs.

### Brief introduction to the Vening Meinesz Fracture Zone (VMFZ)

It is presumed that opening of the SFB utilized the Eocene transform, this time in a dextral fashion. This transform is thought to be the precursor of the present-day VMFZ. It evolved into the VMFZ during removal of the intervening crust by obduction, subduction and strike-slip, at various times since the Oligocene, as discussed below.

### The first convergence event at *c.* 25 Ma

If the scenario outlined above is correct, at 25 Ma the amagmatic subduction interface was located east of the present-day Colville–Lau Ridge. The Northland–Reinga Ridge transform sector of the Eocene–Oligocene plate boundary was still in existence as a steep crustal boundary, though not on the line of the present VMFZ. It is necessary that it was located northeast of the VMFZ, in order to allow the preservation of the Northland ophiolites and associated passive margin sediments which were incorporated in the Northland and East Coast Allochthons.

The increase of 20° in Australia–Pacific convergence angle at 25 Ma (Table 1), combined with southeasterly drift of the pole of rotation which had led to an increase in convergence rate from *c.* 20 to 30 mm yr$^{-1}$, appears to have brought

**Table 1.** *Convergence angles between Australia and Pacific Plates, measured from the Tasmantid seamount chains (McDougall & Duncan 1988) and the Louisville seamount chain (Lonsdale 1988)*

| Time of bends in chain (Ma) | Australian plate trajectory (°) | Pacific plate trajectory (°) | Convergence angle (°) |
|---|---|---|---|
| 0 | | | |
| | 018 | 300 | 78 |
| 5 | | | |
| | 018 | 285 | 93 |
| 10 | | | |
| | 010 | 295 | 75 |
| 18 | | | |
| | 010 | 305 | 65 |
| 25 | | | |
| | 000 | 315 | 45 |
| 43 | | | |
| | 032 | 340 | 52 |

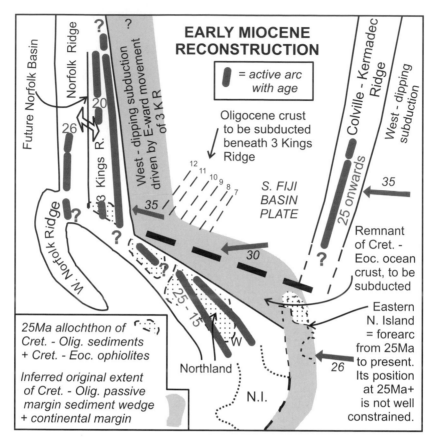

**Fig. 6.** Early Miocene reconstruction following opening of the South Fiji Basin and inception of arc volcanism, showing inferred relationship of three plates, three independent arc segments and the portion of South Fiji Basin anomalies subducted beneath the Three Kings Ridge during the early Miocene.

about a sudden change, from cryptic subduction (essentially a passive-margin situation) in northern New Zealand to strong convergence with obduction of passive-margin sediments and instant arc volcanism.

## Initial obduction event

The initial response to the new regime was the obduction, from the present-day north or northeast, of the Northland and East Coast Allochthons. The two allochthons are presently separated by the Bay of Plenty (Fig. 2) but were either contiguous at the time of emplacement in the earliest Miocene (Fig. 6) (Hayward 1993; Isaac *et al.* 1994; Mazengarb & Harris 1994), or comprised separate bodies squeezed out of transpressive flower structures (Herzer & Mascle 1996). Both contain sediment bodies that had formerly comprised a passive-margin sediment wedge along the Pacific margin of New Zealand (mid-Cretaceous–late Oligocene) (Ballance 1993*b*). Ophiolite bodies of various size are present throughout the allochthons (Parker *et al.* 1989) but the most extensive ophiolite nappes are structurally higher than the sediment nappes (Cassidy & Locke 1987), and prior to obduction are normally assumed to have lain outboard of and beneath the sediment wedge (Isaac *et al.* 1994).

It is assumed that the ophiolite nappes, which are of mid-ocean ridge basalt (MORB) composition with some younger seamount basalts (Malpas *et al.* 1992), were derived from a residual piece of Cretaceous–Palaeogene ocean crust located between the New Zealand continental crust and the SFB Oligocene age crust. That crust has now disappeared, presumably by early Miocene subduction beneath Northland (see below).

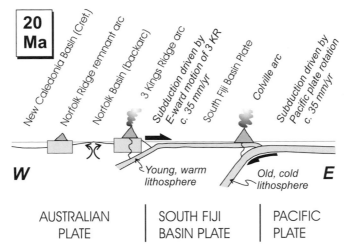

**Fig. 7.** Cartoon cross section across the South Fiji Basin plate at 20 Ma showing two parallel subduction zones.

## The instant-arc scenario

The combination of dredge samples, dates and magnetic anomalies (Adams *et al.* 1994; Herzer 1995; Ballance *et al.* 1999; Mortimer *et al.* 1998) suggests that between 26 and 19 Ma there were three independent arc segments (i.e. linear alignments of arc volcanoes) in the region (Figs 3 and 6). These were located on the present-day Norfolk–Three Kings Ridge, Northland Peninsula and Colville Ridge. The three subduction zones are presumed to have originated as a pre-existing slab (Colville), a steep transform crustal boundary (Northland–Reinga) and perhaps a dipping interface remnant from the Eocene subduction zone (Norfolk–Three Kings).

Thus, a subduction scenario is required in the early Miocene in order to generate contemporaneous arc segments on three different lineaments. Figures 6 and 7 show a possible early Miocene reconstruction on which Pacific Plate convergence directions and rates are plotted using Sutherland's (1995) poles and angular momentum data (Table 2), combined with *c.* 350 km of opening of the Norfolk Back-arc Basin in 10 Ma (an effective subduction rate of 35 mm yr$^{-1}$). The scenario implied is akin to the present-day western Pacific Ocean, where spreading from the mid-Pacific Ridge is driving two subduction

**Table 2.** *Convergence rates and Pacific Plate trajectories calculated from Sutherland's 1995 poles (see also Figs 4–8.)*

| Age (Ma) | Central Three Kings Ridge | Northland | Southern Colville Ridge |
|---|---|---|---|
| 40 | 21 mm yr$^{-1}$, near-orthogonal | <20 mm yr$^{-1}$, sinistral near-transform | – |
| 30 | – | ?c. 100 mm yr$^{-1}$, dextral transform | *c.* 30 mm yr$^{-1}$ plus roll-back, near-orthogonal |
| 25 | *c.* 35 mm yr$^{-1}$, near-orthogonal, driven by eastward movement of Three Kings Ridge | 30 mm yr$^{-1}$ strongly oblique sinistral | <35 mm yr$^{-1}$, slightly oblique dextral |
| | | **Coromandel** | |
| 10 | – | *c.* 45 mm yr$^{-1}$ moderately oblique dextral | 50 mm yr$^{-1}$ moderately oblique dextral |
| 5 | – | 47 mm yr$^{-1}$ moderately oblique dextral | 51 mm yr$^{-1}$ moderately oblique dextral |
| | | **East Cape** | **S. Kermadec** |
| 0 | – | 43 mm yr$^{-1}$ moderately oblique dextral | 51 mm yr$^{-1}$ near-orthogonal |

zones, Pacific Plate lithosphere beneath Philippine Sea Plate lithosphere at the Izu–Bonin–Mariana Trench, and both Philippine Sea and Pacific Plate lithosphere beneath Eurasia Plate at the Japan Trench and its southern extensions. In our case, subduction of Pacific Plate lithosphere was beneath the Colville Arc, while the SFB was the analogue of the Philippine Sea Plate. Thus, a SFB Plate was subducted at the Three Kings Ridge. Both subduction regimes were near-orthogonal and the Three Kings Arc may have continued northwards along the Loyalty Islands Ridge (Fig. 1).

The southern margin of the SFB Plate was coincident with the dextral transform produced during opening of the basin (precursor to the VMFZ), but it now became an obliquely subducting ($c. 45°$) sinistral margin beneath the Northland–Reinga Ridge.

## Driving forces for three subduction zones

The question arises as to the source of the driving forces for three independent subduction zones. Estimated Pacific Plate convergence rates range from $46\,mm\,yr^{-1}$ in the north to $<30\,mm\,yr^{-1}$ in the south (Table 2). This convergence may have been fully taken up at the Colville and Northland subduction zones, or it may have been partitioned between those two and the Three Kings. By analogy with the anomalies preserved in the northern part of the SFB (Fig. 2), it is normally assumed that western counterparts to the half-set of preserved magnetic anomalies in the southern SFB have been subducted westwards beneath the Three Kings Ridge (Davey 1982). Thus, the Norfolk Back-arc Basin ($c. 350\,km$ wide) opened simultaneously with apparent subduction of the western half of the SFB ($c. 480\,km$ wide). For this to occur without a contribution from the Pacific spreading ridge, trench roll-back of $c. 350\,km$ would be required at the Three Kings Ridge, with a slab angle of $c. 40°$. Since the SFB crust was no more than 10 Ma old, this angle might be unrealistically high. Therefore, some contribution from Pacific Plate spreading might have been required, but it cannot be quantified.

Trench roll-back at the Three Kings Ridge could not have accounted for the strongly oblique sinistral subduction beneath Northland. It would definitely have required Pacific Plate motion, thus suggesting that the transform created by the opening of the SFB (Oligocene), the precursor to the VMFZ, was functioning again during the early Miocene to separate the Colville–Three Kings subduction system from the Northland system. Oblique subduction

beneath Northland eventually removed all remnants of the Cretaceous–Palaeogene Northland ophiolite ocean crust, in the process bringing the SFB trench/transform closer to the fore-arc (the whereabouts of the early Miocene forearc are discussed below) and eventually to the VMFZ.

## Form of the early Miocene arc segments in Northland

Being exposed partly on land, the twin Northland arc segments have been intensively studied. There are two parallel, northwest trending bands of volcanic centres (Figs 2 and 4). That lying to the northeast comprises three discrete centres. They were largely subaerial, located on a basement of metagreywacke, and dominated by andesite and dacite (Smith et al. 1989). The southwestern band is broader and more diffuse, comprising six major and many minor centres (Herzer 1995). They were erupted through a basement of various volcanic terranes (Mortimer et al. 1999), were largely submarine, and were dominated by basaltic andesites and basalts (Wright & Black 1981; Smith et al. 1989). The two bands are 50–70 km apart. The southwestern band contributed large quantities of volcanic sediment to the bathyal-depth Waitemata Basin which lay between the two arc segments, as did the recently obducted Northland Allochthon (Ballance 1974; Hayward 1993).

There is no systematic age variation along the two volcanic belts and they are broadly contemporaneous; the age range is between $c. 25$ and 15 Ma (Hayward 1993; Adams et al. 1994; Herzer 1995). Both belts are calc-alkaline in character but they display no clear relationship to a single subducting slab (Smith et al. 1989). The strongly oblique nature of subduction implied on Fig. 6 may have been a contributing factor to the apparent complexity.

## Form of the Norfolk–Three Kings Ridge Arc segments

Dredge data is widely scattered, but magnetic anomalies suggest a single arc on the Norfolk Ridge and several north trending alignments of volcanic centres on the Three Kings Ridge (Mortimer et al. 1998) (Fig. 6). The sparse existing age data do not allow any conclusions to be drawn as to systematic age variation across the ridge (Mortimer et al. 1998), although the discussion above suggests that trench roll-back was very likely and, therefore, further age data may be expected to indicate eastward younging.

## Form of the Colville segment

The Colville Ridge is a remnant arc and is therefore incomplete, but like the Norfolk Ridge remnant arc it appears to comprise a single line of centres (Wright 1997; Ballance *et al.* 1999). It is continuous to the north with the Lau remnant arc, and the two together comprised a north-northeast trending arc covering >15° latitude. During the period from 25 to *c.* 5 Ma there was no back-arc basin associated with this arc.

## Simplification of the arc in the late Miocene

At *c.* 15 Ma, activity ceased on both volcanic belts in Northland and simultaneously the bathyal-depth Waitemata Basin was uplifted above sea level without any compressive tectonism (Spörli 1989). There is no evidence of activity on the Norfolk–Three Kings Ridges after 19 Ma, but an 18 Ma date in the Norfolk Basin (Fig. 3) suggests that activity may have continued to a later date, which by analogy with Northland may have been 15 Ma.

Late Miocene arc-type volcanism (15–5 Ma) is known in four locations (Fig. 8): Colville Ridge (Ballance *et al.* 1999a); Coromandel Peninsula (Adams *et al.,* 1994); west side of the Hauraki Rift (Kiwitahi Volcanics) (Adams *et al.* 1994); offshore western North Island between Auckland and the Taranaki Peninsula (Bergman *et al.*

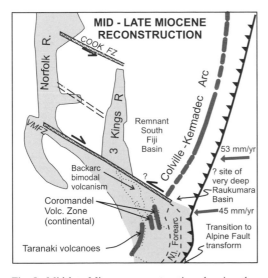

**Fig. 8.** Mid-late Miocene reconstruction showing the newly opened Norfolk Back-arc Basin, the extended Colville–Coromandel–Taranaki Arc and the Vening Meinesz Fracture Zone, now established in its present form.

1991; Herzer 1995). Herzer (1995) first linked all four sectors into a single, continuous north-northeast trending arc, thereby putting into perspective the confusion engendered previously by the apparent northwesterly alignment of centres in the two central sectors (Kamp 1984; Brothers 1986; Ballance 1988). It now seems that the two central sectors form en echelon segments oriented *c.* 45° to the arc, but approximately orthogonal to the local subduction direction (Fig. 8). The Kiwitahi segment displays a marked southeastward younging, possibly caused by subduction retreat. Age progression within the Coromandel segment is not clear because the base of the sequence is not exposed over much of its length.

A number of back-arc, within-plate, basaltic volcanic fields were established in Northland during this period and this type of activity has continued to the present (e.g. Auckland Volcanic Field) (Smith *et al.* 1993).

Thus, the biggest change in organization of the southwest Pacific Arc took place at *c.* 15 Ma, with the replacement of several pre-existing subduction segments (Three Kings Ridge, Northland) by an extension of the segment (Colville) located at the main locus of Pacific Plate subduction. The Colville Arc extended southwestward by *c.* 400 km. This simplification is attributed to the progressive southeasterly retreat of the Pacific Plate pole (Fig. 1), causing a regional 'smoothing' of convergence vectors and a steady increase in convergence rates. Also at about this time, the convergence angle between the plates again increased (Table 1) from 65 to 75°. The complexity in plate boundaries inherited from the Eocene–Oligocene phase of convergence was no longer stable.

## Further reorganization – the period 5 Ma to present

Ten million years of arc stability (15–5 Ma) was followed by another major reorganization. This time the main change was arc rifting and the appearance of the Lau–Havre Back-arc Basin. The change occurred earliest in the north (Lau) (Taylor *et al.* 1996), and is extending progressively southward into the North Island of New Zealand by way of a widening of the Bay of Plenty and Taupo Volcanic Zone (TVZ) (Wright 1993; Gamble *et al.* 1995).

The nature and cause of the relocation of arc volcanism within North Island, New Zealand, from the Coromandel–Taranaki trend to the parallel TVZ, a step of perhaps 50 km (the apparent distance has since increased to 80–100 km as a consequence of the southward extension of the back-arc basin), are unclear at present. There was an apparent hiatus in arc volcanism within the North Island between 4 and 2 Ma (Adams *et al.* 1994; Wilson *et al.*, 1984). Within the TVZ itself, volcanism has extended progressively southward over the past 2 Ma.

Over the same interval, 5 Ma to present, the major subsiding depocentre of the Wanganui Basin, which lies directly on-strike and south of the TVZ, has been migrating southward in parallel (Stern *et al.* 1993).

The main causative factor in this most recent change in the system has probably been an increase in convergence rate and compression across New Zealand, related to a further shift in the pole of rotation, this time in a southwesterly direction (Fig. 1) (Sutherland 1995, 1996). The convergence angle between the two plates, which had increased to over 90° at 10 Ma (Table 1), decreased again at 5 Ma to 78°. This may have been an additional factor contributing to the change. An independent analysis using hotspot tracks throughout the Pacific basin also predicted an increase of 44 mm yr$^{-1}$ in convergence rate across New Zealand at *c.* 4 Ma (Pollitz 1986). Related changes within New Zealand include the rise of the Southern Alps at *c.* 4 Ma and the North Island axial mountains at *c.* 1 Ma.

## The uncertain role of the Hikurangi Plateau

The Hikurangi Plateau (Figs 1 and 2) represents a wild card in this discussion. It is an oceanic plateau of Cretaceous age, presently located between the Chatham Rise and the Kermadec–Hikurangi Subduction Zone (Mortimer & Parkinson 1996). It has clearly entered the subduction zone, but to what extent and at what time have yet to be determined. The collision of an oceanic plateau has the potential to cause profound change in a subduction system.

## The role of the forearc

The entire eastern seaboard of the North Island of New Zealand comprises a well-developed forearc, which continues northwards off continental crust as the Raukumara Forearc Basin (Fig. 2) (Ballance 1993*a*; Lewis & Pettinga 1993; Davey *et al.* 1997). There is, however, a significant mismatch with the adjacent active arc (the TVZ). Unequivocal forearc features date back to early Miocene (Pettinga 1982; Delteil *et al.* 1996), whereas the TVZ dates back only 2 Ma, and the parallel and older Colville–Coromandel–Taranaki Arc only to 15 Ma. On

the other hand, there is no recognizable forearc adjacent to the twin early Miocene Northland arcs. A widespread feature of forearc basinal sediments is some degree of clockwise rotation (Walcott 1987).

It is suggested, therefore, that during the early Miocene the forearc was located closer to Northland than it is now and was oriented more north-northwesterly. Such a location would bring the East Coast Allochthon into closer contiguity with its analogue the Northland Allochthon and would provide the missing early Miocene forearc. Close constraints on its position are not available.

The time of major relocation of the forearc, by dextral movement and clockwise rotation along the North Island Shear Belt (Fig. 2), would logically be at the 15 Ma reorganization, when sinistral subduction beneath Northland was replaced by moderately oblique dextral subduction beneath the Colville–Coromandel–Taranaki Arc, but there is no constraint on this in the known history of the forearc. The Shear Belt is still an active dextral transcurrent feature, which links ultimately with the Marlborough and Alpine Faults of the South Island, so that dextral displacement of the forearc relative to older arcs is still proceeding. The displacement has been accentuated by the opening of the Havre Trough and Bay of Plenty in the last 2–3 Ma. The age of the Havre Trough is not well constrained.

## Discussion

The history of the southwest Pacific Neogene arcs – initial complexity for 10 Ma, followed by abrupt simplification at 15 Ma as all activity transferred to an extended Colville Arc – is a good example of the influence of external features on the evolution of a tectonic system. The initial complexity was dictated by inheritance from a complex Eocene–Oligocene convergent, but amagmatic, boundary, the complexity being due in large part to the close proximity of the pole of Pacific Plate rotation. The main simplifying factor seems to have been the progressive southeasterly retreat of the pole giving rise to more uniform, faster and more orthogonal convergence.

Other external factors which may have influenced the more recent evolution were the reorganization of Pacific Plate movement 4–5 Ma ago that caused the pole to shift in a different direction (to the southwest) and caused an increase in overall convergence rate in New Zealand, although the convergence angle, as indicated by hotspot chain trends (Table 1), actually decreased by c. 15°. These changes coincided with the apparent hiatus in arc activity from 4 to 2 Ma, the outstepping of the arc by 50 km or more, the inception of the Lau–Havre Back-arc Basin and the rise of the Southern Alps. There had not been an active back-arc basin, or mountains, for the preceeding 15 Ma. There is also the possibility that the oceanic Hikurangi Plateau collided at this time and contributed to events – at present there is no constraint on the timing of that event.

The inception of arc activity 25–26 Ma ago provides an example of the delicate balance that may prevail in a subduction zone. If the interpretation of plate convergence between 43 and 25 Ma is correct, slow, oblique convergence had emplaced a down-going slab along the interface between continental and oceanic crust, and a marginal basin had even opened (SFB) behind the retreating slab (Colville–Lau Lineament), but there had been no arc magmatism. An increase in conver-gence angle between plates at 25 Ma (from 45° to 65°; Table 1), accompanied by a small increase in convergence rate (from 20 to 30 mm yr$^{-1}$; Table 2), was sufficient to cause immediate obduction of the Northland and East Coast Allochthons and immediate inception of volcanism. The subduction zone had been primed, as it were, for a long time and required only a small change in convergence factors to pull the magmatic trigger.

Simultaneously, with simplification of the arc system, the continental crust of New Zealand was made more complex by fragmentation due to opening of back-arc basins and to dextral strike-slip displacement of the forearc. The latter gave rise to a major age mismatch between the forearc (25 Ma old) and the presently adjacent active arc (TVZ, 2 Ma old), combined with the absence of a forearc adjacent to the older Northland arcs (it is not clear if there is a forearc preserved east of the Three Kings Ridge). Fragmentation continues, as the opening of the TVZ unzips the North Island down the middle, and the forearc continues its dextral movement southwards.

The VMFZ presents a case history of a complex structural lineament that has been subject to repeated reactivation. It is presently an inactive, major strike-slip/transform lineament extending >900 km from the western margin of the Norfolk Ridge to the Colville Ridge (Herzer & Mascle 1996; Herzer et al. 1997) (Fig. 2). The analysis presented in this paper suggests that a precursor sinistral transform originated to the north of the present VMFZ, as a jog in an otherwise convergent margin initiated at c. 43 Ma. It was reactivated as a dextral transform during Oligocene opening of the SFB and has since experienced: (1) early Miocene partitioning with dextral opening of the Norfolk Basin in the west, and sinistral movement in the east if the Colville Ridge was translated westward; during this interval remnant oceanic crust separating the transform from Northland continental crust was subducted, bringing the transform close to the early Miocene forearc; (2) removal, by dextral strike-slip, of the early Miocene forearc at c. 15 Ma, producing the present VMFZ. There has been no further movement.

This analysis also suggests that the VMFZ and its precursors have never extended east of the Colville Ridge, as has been suggested by previous authors (e.g. van der Linden 1967).

I thank Robert Hall and Paul Ryan for their comments on this paper.

# References

ADAMS, C. J., GRAHAM, I. J., SEWARD, D. & SKINNER, D. N. B. 1994. Geochronological and geochemical evolution of late Cenozoic volcanism in the Coromandel Peninsula, New Zealand. New Zealand Journal of Geology and Geophysics, **37**, 359–379.

BALLANCE, P. F. 1974. An inter-arc flysch basin in northern New Zealand: Waitemata Group (Upper Oligocene to Lower Miocene). Journal of Geology, **82**, 439–471.

——1988. Late Cenozoic time-lines and calc-alkaline volcanic arcs in northern New Zealand – further discussion. Journal of the Royal Society of New Zealand, **18**, 347–358.

——1993a. The New Zealand Neogene forearc basins. In: BALLANCE, P. F. (ed.) South Pacific Sedimentary Basins. Sedimentary Basins of the World 2, Elsevier, 177–193.

——1993b. The paleo-Pacific, post-subduction, passive margin thermal relaxation sequence (Late Cretaceous–Paleogene) of the drifting New Zealand continent. In: BALLANCE, P. F. (ed.) South Pacific Sedimentary Basins. Sedimentary Basins of the World 2, Elsevier, 93–110.

——, PETTINGA, J. R. & WEBB, C. 1982. A model of the Cenozoic evolution of northern New Zealand and adjacent areas of the southwest Pacific. Tectonophysics, **87**, 37–48.

——, SMITH, I. E. M. & ISAAC, M. J. 1995. The role of a minor change in regional convergence vectors in causing almost immediate inception of arc volcanism in the early Miocene of northern New Zealand: was the slab already in position? (Abstract). Geological Society of New Zealand Miscellaneous Publication, **81A**, 75.

——, ABLAEV, A. G., PUSHCHIN, I. K. et al. 1999. Morphology and history of the Kermadec trench-arc-backarc basin-remnant arc system at 30 to

32°S: geophysical profiles, microfossil and K–Ar data. *Marine Geology*, **159**, 35–62.

BERGMAN, S. C., TALBOT, J. P. & THOMPSON, P. R. 1991. The Kora Miocene submarine andesite stratovolcano hydrocarbon reservoir, northern Taranaki Basin, New Zealand. *In*: COMMERCE, M. O. (ed.) *1991 New Zealand Oil Exploration Conference Proceedings*. Ministry of Commerce, Wellington, 178–206.

BROTHERS, R. N. 1986. Upper Tertiary and Quaternary volcanism and subduction zone regression, North Island, New Zealand. *Journal of the Royal Society of New Zealand*, **16**, 275–298.

CASSIDY, J. & LOCKE, C. A. 1987. Thin ophiolites of North Island, New Zealand. *Tectonophysics*, **139**, 315–319.

COLE, J. W. 1986. Distribution and tectonic setting of Late Cenozoic volcanism in New Zealand. *Bulletin Royal Society of New Zealand*, **23**, 7–20.

DAVEY, F. J. 1982. The structure of the South Fiji Basin. *Tectonophysics*, **87**, 185–241.

——, HENRYS, S. & LODOLO, E. 1997. A seismic crustal section across the East Cape convergent margin, New Zealand. *Tectonophysics*, **269**, 199–215.

DELTEIL, J., MORGANS, H. E. G., RAINE, J. I., FIELD, B. D. & CUTTEN, H. N. C. 1996. Early Miocene thin-skinned tectonics and wrench faulting in the Pongaroa district, Hikurangi margin, North Island, New Zealand. *New Zealand Journal of Geology & Geophysics*, **39**, 271–282.

GAMBLE, J. A., WRIGHT, I. C., WOODHEAD, J. D. & McCULLOCH, M. T. 1995. Arc and back-arc geochemistry in the southern Kermadec arc –Ngatoro Basin and offshore Taupo Volcanic Zone, SW Pacific. *In*: SMELLIE, J. L. (ed.) *Volcanism Associated with Extension at Consuming Plate Margins*, Geological Society, London, Special Publications, **81**, 193–212.

GILL, J. B. 1981. *Orogenic Andesites and Plate Tectonics*. Springer.

HAYWARD, B. W. 1993. The tempestuous 10 million year life of a double arc and intra-arc basin – New Zealand's Northland Basin in the Early Miocene. *In*: BALLANCE, P. F. (ed.) *South Pacific Sedimentary Basins. Sedimentary Basins of the World 2*. Elsevier, 113–142.

HERZER, R. H. 1995. Seismic stratigraphy of a buried volcanic arc, Northland, New Zealand, and implications for Neogene subduction. *Marine and Petroleum Geology*, **12**, 511–533.

—— & MASCLE, J. 1996. Anatomy of a continent-backarc transform — the Vening Meinesz Fracture Zone northwest of New Zealand. *Marine Geophysical Researches*, **18**, 401–427.

——, CHAPRONIERE, G. C. H., EDWARDS, A. R. et al. 1997. Seismic stratigraphy and structural history of the Reinga Basin and its margins, southern Norfolk Ridge system. *New Zealand Journal of Geology and Geophysics*, **40**, 425–451.

ISAAC, M. J., HERZER, R. H., BROOK, F. J. & HAYWARD, B. W. 1994. *Cretaceous and Cenozoic Sedimentary Basins of New Zealand*. Institute of Geological and Nuclear Sciences Monograph B, New Zealand.

KAMP, P. J. J. 1984. Neogene and Quaternary extent and geometry of the subducted Pacific plate beneath North Island, New Zealand: implications for Kaikoura tectonics. *Tectonophysics*, **108**, 241–266.

——1986. The mid-Cenozoic Challenger Rift System of western New Zealand and its implications for the age of Alpine Fault inception. *Bulletin of the Geological Society of America*, **97**, 255–281.

KARIG, D. E. 1970. Ridges and basins of the Tonga–Kermadec island arc system. *Journal of Geophysical Research*, **75**, 239–254.

LAMARCHE, G., COLLOT, J.-Y., WOOD, R. A., SOSSON, M., SUTHERLAND, R. & DELTEIL, J. 1997. The Oligocene–Miocene Pacific–Australia plate boundary, south of New Zealand: evolution from oceanic spreading to strike-slip faulting. *Earth and Planetary Science Letters*, **148**, 129–139.

LEWIS, K. B. & PETTINGA, J. R. 1993. The emerging, imbricate frontal wedge of the Hikurangi Margin. *In*: BALLANCE, P. F. (ed.) *South Pacific Sedimentary Basins. Sedimentary Basins of the World 2*, Elsevier, 225–250.

LONSDALE, P. 1988. Geography and history of the Louisville hotspot chain in the southwest Pacific. *Journal of Geophysical Research*, **93**, 3078–3104.

MALPAS, J., SPÖRLI, K. B., BLACK, P. M. & SMITH, I. E. M. 1992. Northland ophiolite, New Zealand, and implications for plate-tectonic evolution of the southwest Pacific. *Geology*, **20**, 149–152.

MAZENGARB, C. & HARRIS, D. H. M. 1994. Cretaceous stratigraphic and structural relations of Raukumara Peninsula, New Zealand; stratigraphic patterns associated with the migration of a thrust system. *Annales Tectonicae*, **8**, 100–118.

McDOUGALL, I. & DUNCAN, R. A. 1988. Age progressive volcanism in the Tasmantid seamounts. *Earth & Planetary Science Letters*, **89**, 207–217.

MOLNAR, P., ATWATER, T., MAMMERYCX, J. & SMITH, S. M. 1975. Magnetic anomalies, bathymetry, and the tectonic evolution of the South Pacific since the late Cretaceous. *Geophysical Journal of the Royal Astronomical Society*, **40**, 383–420.

MORTIMER, N. & PARKINSON, D. 1996. Hikurangi Plateau: a Cretaceous large igneous province in the southwest Pacific Ocean. *Journal of Geophysical Research*, **101**, 687–696.

——, HERZER, R. H., GANS, P. B., PARKINSON, D. L. & SEWARD, D. 1998. Basement geology from Three Kings Ridge to West Norfolk Ridge, southwest Pacific Ocean. *Marine Geology*, **148**, 135–162.

NORRIS, R. J. & TURNBULL, I. M. 1993. Cenozoic basins adjacent to an evolving transform plate boundary, southwest New Zealand. *In*: BALLANCE, P. F. (ed.) *Southwest Pacific Sedimentary Basins. Sedimentary Basins of the World 2*. Elsevier, 251–270.

——, CARTER, R. M. & TURNBULL, I. M. 1978. Cainozoic sedimentation in basins adjacent to a major continental transform boundary in southern New Zealand. *Journal of the Geological Society of London*, **135**, 191–205.

PARKER, R. J., BALLANCE, P. F. & SPÖRLI, K. B. 1989. Small 'Tangihua' volcanic masses at low levels in the Northland Allochthon, New Zealand: tectonic significance. *Royal Society of New Zealand Bulletin*, **26**, 127–136.

PETTINGA, J. R. 1982. Upper Cenozoic structural history, coastal southern Hawke's Bay, New Zealand. *New Zealand Journal of Geology & Geophysics*, **25**, 149–191.

POLLITZ, F. P. 1986. Pliocene change in Pacific-plate motion. *Nature*, **320**, 738–741.

SMITH, I. E. M., BALLANCE, P. F. & ISAAC, M. J. 1995. The role of a minor change in regional convergence vectors in causing almost immediate inception of arc volcanism in the early Miocene of northern New Zealand (Abstract). *Abstracts, IUGG XXI General Assembly*, Boulder, Colorado, B420.

——, RUDDOCK, R. S. & DAY, R. A. 1989. Miocene arc-type volcanic-plutonic complexes of the Northland Peninsula, New Zealand. *In*: SPÖRLI, K. B. & KEAR, D. (eds) *Geology of Northland: Accretion, Allochthons and Arcs at the Edge of the New Zealand Micro-continent*. Royal Society of New Zealand Bulletin, **26**, 205–213.

——, OKADA, T., ITAYA, T. & BLACK, P. M. 1993. Age relationships and tectonic implications of late Cenozoic basaltic volcanism in Northland, New Zealand. *New Zealand Journal of Geology and Geophysics*, **36**, 385–393.

SPÖRLI, K. B. 1989. Exceptional structural complexity in turbidite deposits of the piggy-back Waitemata Basin, Miocene, Auckland/Northland, New Zealand. *In*: SPÖRLI, K. B. & KEAR, D. (eds) *Geology of Northland: Accretion, Allochthons and Arcs at the Edge of the New Zealand Micro-continent*. Royal Society of New Zealand Bulletin, **26**, 183–194.

STERN, T. A., QUINLAN, G. M. & HOLT, W. E. 1993. Crustal dynamics associated with the formation of Wanganui Basin, New Zealand. *In*: BALLANCE, P. F. (ed.) *South Pacific Sedimentary Basins. Sedimentary Basins of the World 2*. Elsevier, 213–223.

STOCK, J. & MOLNAR, P. 1982. Uncertainties in the relative positions of the Australia, Antarctica, Lord Howe and Pacific plates since the late Cretaceous. *Journal of Geophysical Research*, **87**, 4697–4714.

SUTHERLAND, R. 1995. The Australia–Pacific boundary and Cenozoic plate motions in the SW Pacific: some constraints from Geosat data. *Tectonics*, **14**, 819–831.

——1996. Transpressional development of the Australia–Pacific boundary through southern South Island, New Zealand: constraints from Miocene–

Pliocene sediments, Waiho-1 borehole, South Westland. *New Zealand Journal of Geology and Geophysics*, **39**, 251–264.

TAPPIN, D. R. 1993. The Tonga frontal-arc basin. *In*: BALLANCE, P. F. (ed.) *South Pacific Sedimentary Basins. Sedimentary Basins of the World 2*. Elsevier, 157–176

—— & BALLANCE, P. F. 1994. Contributions to the sedimentary geology of 'Eua Island, Kingdom of Tonga: reworking in an oceanic forearc. *In*: STEVENSON, A. J., HERZER, R. H. & BALLANCE, P. F. (eds) *Geology and Submarine Resources of the Tonga–Lau–Fiji Region*. SOPAC Technical Bulletin, **8**, 1–20.

TAYLOR, B., ZELLMER, K., MARTINEZ, F. & GOODLIFFE, A. 1996. Sea-floor spreading in the Lau back-arc basin. *Earth and Planetary Science Letters*, **144**, 35–40.

VAN DER HILST, R. 1995. Complex morphology of subducted lithosphere in the mantle beneath the Tonga trench. *Nature*, **374**, 154–157.

VAN DER LINDEN, W. J. M. 1967. Structural relationships in the Tasman Sea and south-west Pacific Ocean. *New Zealand Journal of Geology & Geophysics*, **10**, 1280–1301.

WALCOTT, R. I. 1987. Geodetic strain and the deformational history of the North Island of New Zealand during the late Cenozoic. *Philosophical Transactions of the Royal Society of London, Series A*, **321**, 163–181.

WEISSEL, J. K., HAYES, D. E. & HERRON, E. M. 1977. Plate tectonic synthesis: the displacements between Australia, New Zealand, and Antarctica since the late Cretaceous. *Marine Geology*, **25**, 231–277.

WILSON, C. J. N., ROGAN, A. M. & SMITH, I. E. M. 1984. Caldera volcanoes of the Taupo Volcanic Zone, New Zealand. *Journal of Geophysical Research*, **89**, 8463–8484.

WRIGHT, A. C. & BLACK, P. M. 1981. Petrology and geochemistry of Waitakere Group, North Auckland, New Zealand. *New Zealand Journal of Geology and Geophysics*, **24**, 155–165.

WRIGHT, I. C. 1993. Pre-spread rifting and heterogeneous volcanism in the southern Havre Trough back-arc basin. *Marine Geology*, **113**, 179–200.

——1997. Morphology and evolution of the remnant Colville and active Kermadec arc ridges south of 33°30′S. *Marine Geophysical Researches*, **20**, 177–193.

# The kinematics of back-arc basins, examples from the Tyrrhenian, Aegean and Japan Seas

LAURENT JOLIVET,[1,3] CLAUDIO FACCENNA,[2] NICOLA D'AGOSTINO,[2] MARC FOURNIER[3] & DAN WORRALL[4]

[1] *Département des Sciences de la Terre, Université de Cergy-Pontoise, 8 Le Campus, 95011 Cergy-Pontoise Cedex, France, URA CNRS 1759*
[2] *Dipartimento di Scienze Geologiche, University of Rome Tre, Largo R. Murialdo 1, Roma 00154, Italy*
[3] *Département de Géotectonique, Université Pierre et Marie Curie, T 26–0 E1, 4 Place jussieu, 75252 Paris cedex 05, France, URA CNRS 1759*
[4] *Shell International Exploration and Production BV, PO Box 162, 2501 AN The Hague, The Netherlands*

**Abstract:** Three classical examples of marginal basins are explored to show the respective contributions of body forces and far-field stresses to the extension mechanism. Extensional stresses can be provided by: (1) the slab-pull force, which induces a retreat of the slab and which originates in the density contrast between the subducting slab (oceanic or continental) and the asthenosphere; (2) by lateral density contrasts within the crust (due to crustal thickening), which induce crustal spreading; and (3) by far-field stresses due to intra-continental deformation (continent–continent collision, for example). Slab pull is probably the most efficient extensional force to provide but its effects are modulated by the contributions of the two other forces. The Japan Sea opened along the eastern margin of the Eurasian continent because extensional boundary conditions were provided by the retreating Pacific subduction zone. The opening stopped as soon as the central Japan triple junction was established in its present position which resulted in a more efficient coupling between the Pacific and Eurasian Plates through the Philippine Sea Plate. The geometry of opening was further controlled by large-scale dextral strike-slip faults that run oblique to the subduction zone along >2500 km, and which are far-field effects of the India–Asia collision. The Northern Tyrrhenian Sea opened because of the retreat of the Adriatic continental slab. The strong slab-pull force is probably due to phase changes within the subducting lower crust. Crustal delamination leads to a warm lower crust which localizes the extensional strain. This extending domain migrated with time from west to east as the delamination and slab retreat proceeded. Upper crustal units incorporated in the Apennines accretionary wedge were later exhumed in the collapsing back-arc domain where their deformation and *P–T* history can now be observed. A similar history with an outward migration can be proposed for the Aegean Sea with, however, a stronger influence of crustal collapse over a larger domain. Here too continental collision (the Arabia–Eurasia collision) controlled the geometry of opening through the westward propagation of the North Anatolian Fault.

When back-arc basins form within the active margin of an intra-oceanic subduction they often present a clear pattern of magnetic anomalies. This is the case for some of the western Pacific back-arc basins which developed along the eastern border of the Philippine Sea Plate or the Autralian Plate, such as the Lau Basin (Parson & Hawkins 1994) or the Shikoku–Parece Vela Basin (Mrozowski & Hayes 1979; Chamot-Rooke *et al.* 1987) (Fig. 1). In these cases, the kinematics of opening can be derived from the

geometry of magnetic anomalies. The average velocity of opening is *c.* 2–3 cm yr$^{-1}$, although it can be much faster in some extreme cases, e.g. the northern Lau Basin (Bevis *et al.* 1995). Back-arc basins located on continental margins have much more complicated geometries and the pattern of magnetic anomalies is usually not helpful. With the notable exception of the South China Sea (Taylor & Hayes 1983; Briais *et al.* 1993), magnetic anomalies are usually very poor, e.g. in the Japan (Tamaki *et al.* 1992) or the Tyrrhenian

*From*: MAC NIOCAILL, C. & RYAN, P. D. (eds) 1999. *Continental Tectonics*. Geological Society, London, Special Publications, **164**, 21–53. 1-86239-051-7/99/$15.00 © The Geological Society of London 1999.

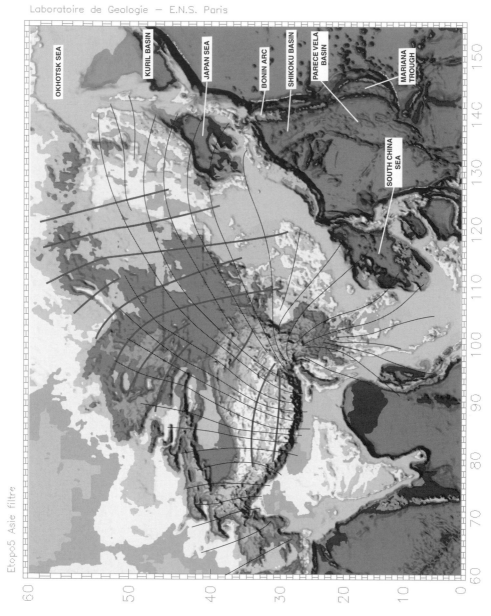

**Fig. 1.** Topographic map of Asia and the western Pacific region (etopo5) and the present-day stress field deduced from earthquake focal mechanisms. Thin lines, $\sigma_{Hmax}$; thick lines, $\sigma_{Hmin}$.

Seas where they have not been identified (Sartori 1990). In some cases, oceanic crust is either too recent to produce significant anomalies (the Okinawa Basin) or non-existent (the Aegean Sea).

To reconstruct the kinematics of such basins it is necessary to work out the regional geological history, particularly the history of deformation of the margins during the opening. Although the geometrical constraints are likely to be more uncertain than in the case of a clear magnetic anomaly pattern, the tectonic history of the margin usually allows the proposal of a reasonable tectonic evolution and a discussion of the tectonic evolution in terms of the forces involved.

The formation of back-arc basins represents one of the major problems in plate tectonics and several mechanisms have been proposed to explain their evolution (Taylor & Karner 1983). The geometry and the rate of extension is thought to be controlled by the roll-back of the hinge of the subducting slab (Dewey 1980). From a dynamical point of view, this process is possibly connected with the density contrast of the oceanic lithosphere sinking down into the asthenosphere (Forsyth & Uyeda 1975; Uyeda & Kanamori 1979). In this case, the length of the slab should be directly related to the velocity of retreat in the mantle (Zhong & Gurnis 1995; Faccenna et al. 1996). The presence of lateral (Dvorkin et al. 1993; Russo & Silver 1994) or global (Ricard et al. 1991) mantle flow can also accelerate this process. The way the overriding plate reacts to the retreat of the slab hinge is still enigmatic [for a review see Taylor & Karner (1983)]. Nevertheless, it has been suggested that extensional stress could be more easily transmitted from the sinking to the upper plate when the coupling between the two plates is low (Bott et al. 1989; Shemenda 1994; Scholz & Campos 1995; Hassani et al. 1997).

Besides the slab roll-back force, which is always present, continental margins are subjected to a stress regime which is controlled by local body forces and far-field boundary forces. Described here are three examples of back-arc basins where other external forces contribute with the slab-pull force producing large-scale lithospheric extension in the upper plate. Chosen for examination are the Japan, the Aegean and the Tyrrhenian Seas, examples of back-arc basins where gravitational collapse and continental collision have contributed with the slab-pull forces to the extensional process.

The Japan Sea is an example where the continental crust was not thickened before the opening and where far-field stresses, due to intracontinental deformation (India–Asia collision), control the geometry of opening (Jolivet

et al. 1990a, 1994c; Kimura & Tamaki 1986; Tamaki 1988).

The Mediterranean back-arc basins (Tyrrhenian and Aegean Seas) (Fig. 2) opened during the Neogene in an overall context of collision between the African and European Plates (Dercourt et al. 1986; Dewey et al. 1989b). Back-arc extension is active together with collapse of thick crustal domains (Reutter et al. 1980; Le Pichon 1981; Dewey 1988; Serri et al. 1993; Gautier & Brun 1994a). Extension initiated after the formation of a thick Alpine–Hellenic crustal wedge. While in the Aegean and Southern Tyrrhenian Seas, subduction involved Mesozoic oceanic lithosphere (Lallemant et al. 1994), in the Northern Tyrrhenian Sea a thinned continental lithosphere underthrusts the Northern Apennines (Dewey et al. 1989a). Moreover, in the Tyrrhenian area, subduction developed perpendicular to the African motion suggesting that sinking of the Adria–Ionian lithosphere is simply 'passive' (Patacca & Scandone 1989). Despite the continental nature of the subducting lithosphere in the north, slab roll-back and back-arc extension were active during the Neogene and are still active today (Malinverno & Ryan 1986). The Mediterranean area is therefore a favourable site to investigate the contributions of near- and far-field stresses to the back-arc opening process.

## The Japan Sea

Tertiary extension along the eastern margin of Asia has been attributed to two possible causes: (1) classical back-arc extension, thus a direct effect of the Pacific subduction (Uyeda & Kanamori 1979; Taylor & Hayes 1983); or (2) India–Asia collision (Tapponnier et al. 1982; Kimura & Tamaki 1986; Jolivet et al. 1990a). How subduction induces back-arc opening is not completely understood but a low mechanical coupling between the two plates is required. This low coupling can be due to retreat of an old, thermally mature, slab (Uyeda & Kanamori 1979) or to a decrease of the convergence velocity (Northrup et al. 1995). The Eocene–Miocene period was probably a period of slow convergence between the Pacific and Eurasia Plates (Northrup et al. 1995). The period of maximum extension (formation of back-arc basins between 30 and 15 Ma) is, however, not coincident with the period of slowest convergence (Eocene). Thus, this problem is not yet totally understood. The Japan Sea is probably the best studied back-arc basin, and it can therefore be used as a case example of the interaction between subduction and collision in shaping such basins.

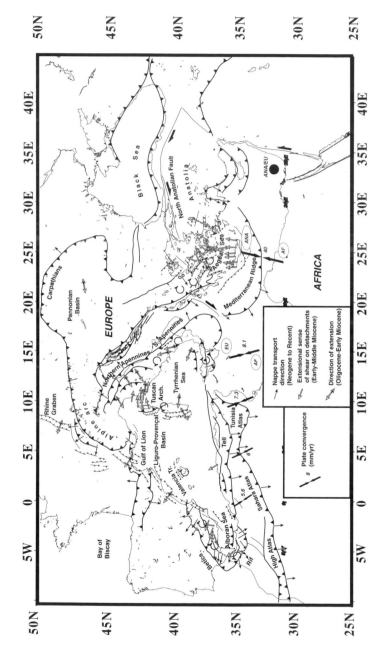

**Fig. 2.** General tectonic map of the Mediterranean region. Main thrust fronts, Miocene kinematic directions in the Mediterranean region from the Neogene to the Present (Frizon de Lamotte *et al.* 1991, 1995; Gautier & Brun 1994*b*; Jolivet *et al.* 1994*a*, *b*, 1996, 1999; Vissers *et al.* 1995; Jolivet & Patriat 1999).

## Geological Structure

The Japan Sea is situated immediately to the north of the Eurasia–Philippine Sea–Pacific Plates triple junction (Figs 1, 3 and 4). Extension and formation of oceanic crust occurred during the Early and Middle Miocene, after which time the whole region entered a stage of east–west compression 10 ma ago, ultimately leading to the incipient closure of the basin which started 2 Ma ago (Tamaki *et al.* 1992). The Pacific Plate underthrusts both the Eurasian and Philippine Sea Plates at a velocity of *c.* 10 cm yr⁻¹ in the Japan and Bonin Trenches, while the Philippine Sea Plate subducts below the Eurasian Plate at a much slower velocity, between 2 and 4 cm yr⁻¹ in the Nankai Trench (De Mets *et al.* 1990; De Mets 1992). Owing to the nearby position of the Philippine Sea Plate/Eurasia (PHSP–EU) rotation pole, the convergence rate of those two plates significantly increases from east to west along the Nankai Trench (Ranken *et al.* 1984; Huchon 1985; Seno *et al.* 1993). Vectors of the Pacific–Eurasia (PAC–EU) relative motion are almost perpendicular to the Japan Trench, while a significant obliquity is observed in the Nankai Trench for the PHSP–EU motion. This obliquity is partitioned between the nearly perpendicular convergence in the Nankai Trench and the dextral motion along the major tectonic boundary of southwest Japan, the Median Tectonic Line (MTL) (Fig. 4).

Deep waters are distributed within three deep basins (Fig. 4). The Japan Basin is floored by oceanic crust (Tamaki 1988). Its triangular shape is the consequence of a propagation of oceanic rifting from east to west (Tamaki *et al.* 1992). The Yamato and Tsushima Basins are instead floored with attenuated continental crust injected by basic magmas (Tokuyama *et al.* 1987).

A large rifted continental block, the Yamato Bank, remains isolated between the Yamato and Japan Basins. It shows a large aborted northeast–southwest trending rift in its middle part, and is bounded by large normal faults on its northwest and southeast sides. Normal faults with similar trends are found along the northern margin of the Japan Arc. They form dextral en echelon grabens along the western margin of Tohoku (Northeast Honshu) and Hokkaido. North-northwest–south-southeast trending dextral strike-slip faults transfer the motion from one graben to the next (Jolivet *et al.* 1991*b*).

The same trend of normal faults is found in the Tartary Strait between Siberia and Sakhalin, where they interfere with large north–south trending dextral strike-slip faults (Jolivet *et al.* 1992; Fournier *et al.* 1994; Worrall *et al.* 1996).

A large dextral pull-apart basin has been imaged on seismic reflection data in the middle of the strait [South Tartar Basin (STB); Fig. 4].

Extensional basins are also found along the MTL. They are filled with Middle Miocene sediments. Their geometry and the analysis of brittle deformation within them suggest that the MTL was a left-lateral, strike-slip fault with a large extensional component during the Middle Miocene (Fournier *et al.* 1995).

Major dextral shear zones bound the Japan Sea to the east and west. The Yangsan and Tsushima Faults were dextral during most of the Cenozoic (Sillitoe 1977). The Pohang and Tsushima (Ulleung) Basins formed during the Early and Middle Miocene as dextral pull-aparts (Lee & Pouclet 1988; Yoon *et al.* 1997), and are still active (Jun 1990). The Sakhalin–Hokkaido Shear Zone runs from central Japan to northern Sakhalin and further north to the mainland of Siberia for >2500 km (Jolivet *et al.* 1994*c*; Worrall *et al.* 1996). It is transtensional in the south with en echelon grabens described above. It is transpressional in the north along the Hidaka Ductile Shear Zone in Hokkaido (Kimura *et al.* 1983; Jolivet & Miyashita 1985), and several large faults and en echelon folds and thrusts in Sakhalin, such as the Tym-Poronaysk Fault which is known along >600 km (Rozhdestvenskiy 1982, 1986; Fournier *et al.* 1994).

## Tectonic timing

Palaeomagnetic data were first used to obtain the timing of opening of the Japan Sea (Otofuji & Matsuda 1983; Otofuji *et al.* 1985, 1991). Large-scale clockwise rotation of southwest Japan indicated fast opening during a very short time span, <1 Ma, some 15 Ma ago. More recent analysis seem to indicate a longer time span and some diffential rotation between several blocks in southwest Japan (Jolivet *et al.* 1995; Ishikawa 1997). Analysis of the tectonic history of the margins, the sedimentation in the basin and the depth–age relations, has suggested an opening from the Late Oligocene to the Middle Miocene (Tamaki 1986; Ingle *et al.* 1990; Tamaki *et al.* 1992). Drilling of the basement during two ODP cruises (127 and 128) confirmed those conclusions. The end of opening occurred *c.* 10 Ma ago, when extensional deformation was replaced by east–west compression. Some 2 Ma ago, thrusting started along the eastern margin of the Japan Sea.

It is difficult to reconcile what is known about the relative kinematics of the Pacific and Eurasia Plates with the dextral shear. In any case, it

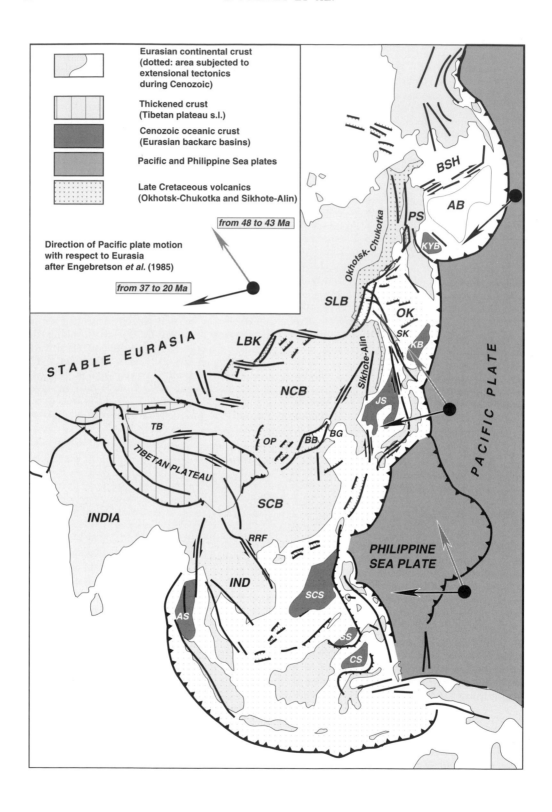

cannot simply be a matter of shear partitioning behind an oblique subduction zone as the dextral strike-slip system extends far inside the Asian continent obliquely to the Japan–Kuril Trenches. This leads to examination of the influence of the India–Asia collision.

## The influence of the India–Asia collision

The distance that collision-related deformation has propagated within the Asian continent is still an open question. A class of models emphasizes the crustal thickening responsible for rising of the Tibetan Plateau (England & Molnar 1990), whereas others put forward strike-slip faults which have propagated progressively to the north as far as the Stanovoy Ranges (Tapponnier et al. 1982; Davy & Cobbold 1988). A quantitative study of the amount of sediments deposited within and around Asia during the collision process suggests that between 30 and 60% of the collision process was not accommodated by crustal thickening and formation of significant relief. Lateral migration of continental blocks and extension might thus have accommodated a significant part of the convergence (Métivier 1996). The opening of the Japan Sea has been envisaged as a consequence of the India–Asia collision, also partly controlled by subduction roll-back extensional forces (Jolivet et al. 1994c). A recently published tectonic map of the Sea of Okhotsk (Worrall et al. 1996) extends further to the north collision-related tectonic features, and confirms the relation between back-arc opening in the Japan–Okhotsk Sea region and continental collision, and the importance of the extensional boundary condition to the east. Most studies of the deformation of the Asian continent were focused on its southern part near the collision zone. Since Tapponnier & Molnar's (1976, 1977) pioneer work little attention has been payed to far-field effects of the collision. Studies of active deformation and quantification of instantaneous motion in Asia recently enriched the debate for or against extrusion tectonics in the immediate vicinity of the indenter (Avouac & Tapponnier 1993; Molnar & Gipson 1996). Several papers, however,

proposed relating the opening of the Japan Sea to the India–Asia collision (Kimura & Tamaki 1986; Jolivet et al. 1990a, 1994c). Analogue models, scaled for gravity (Davy & Cobbold 1988; Fournier 1994), have shown that indentation tectonics can be produced far from the collision and dextral shear zones parallel to the eastern 'free boundary'. An additional component of extension (introduced in the experiments by gravitational spreading) induces the formation of basins controlled by those dextral shear zones (Fournier 1994). Those experiments show that the opening of the Japan Sea, partly controlled by strike-slip faults produced by indentation tectonics, is a physically feasable mechanism. Numerical models have so far been unable to reproduce the fault pattern observed in Asia. Any model describing the deformation of the Asian continent should include the formation of the major back-arc basins and major strike-slip faults which are first order facts, besides the spectacular crustal thickening in Tibet.

*The Pamir–Baikal–Okhotsk Shear Zone.* A simplified version of a new tectonic map of the Sea of Okhotsk (Worrall et al. 1996), together with the Japan Sea region (after Jolivet et al. (1994)), shows the entire dextral strike-slip system from southwest Japan to the northern Sea of Okhotsk where it meets a large-scale left-lateral system (Fig. 4). A striking feature is the 500 km long Shantar–Liziansky Basin along the northwestern margin of the Sea of Okhotsk, where basin-forming extension was active from the Eocene to the early Miocene. Extension is transferred at the southern and northern ends of the graben to strike-slip displacement along the left-lateral Stanovoy Fault (Tapponnier et al. 1982), which connects the Shantar–Liziansky Basin to the Baikal Rift. A steeply dipping shear zone of inferred left-lateral strike-slip origin, parallel to the northern margin of the Sea of Okhotsk, connects the Shantar–Liziansky Basin to Tertiary faults in the Pustarets and Penzhina Basins. Seismic data show that normal faults in the Shantar–Liziansky Basin accommodated a minimum of 15–20% extension (Worrall et al. 1996). The apparent offset of the Late Cretaceous

---

**Fig. 3.** Tectonic map of Asia from the India–Asia collision zone to the Bering Strait [modified after Worrall et al. (1996)] and the directions of Pacific–Eurasia relative motion (Engebretson et al. 1985). AS, Andaman Sea; BB, Bohai Basin; BG, Bohai Gulf; AB, Aleutian Basin; BSH, Bering Shelf; CJCZ, central Japan collision zone; CS, Celebes Sea; DD, Derugin Deep; IP, Izu Peninsula; ISTL, Itoigawa–Shizuoka Tectonic Line; JS, Japan Sea; KYB, Komandorsky Basin; KB, Kuril Basin; LB, Liziansky Basin; LBK, Lake Baikal; MTL, Median Tectonic Line; NCB, north China; OK, Sea of Okhotsk; OP, Ordos Plateau; PS, Pustartets Basin; RRF, Red River Fault; SAF, Sikhote Alin Fault; SB, Shantar Basin; SCB, South China Block; SCS, South China Sea; SK, Sakhalin; SLB, Shantar Liziansky Basin; SS, Sulu Sea; STB, South Tartar Basin; TB, Tarim Basin; TB, Tsushima Basin; TTL, Tanakura Tectonic Line; YB, Yamato Basin; YBK, Yamato Bank; YF, Yangsan Fault.

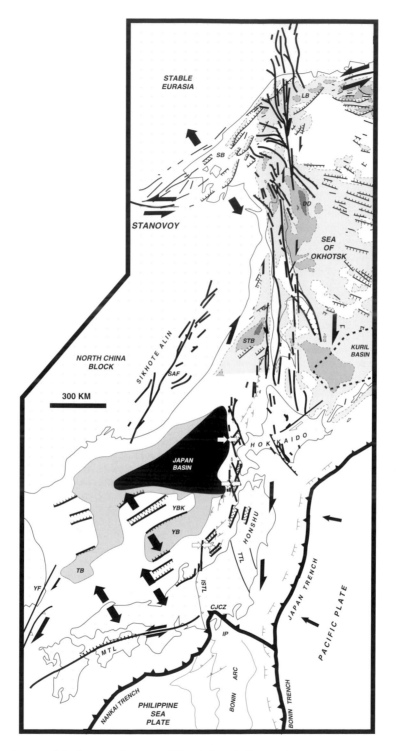

**Fig. 4.** Tectonic map of the Japan and western Okhotsk Seas (Jolivet *et al*. 1994*c*; Worrall *et al*. 1996). White convergent arrows along the eastern margin of Japan Sea indicate direction of recent shortening. For abbreviations see legend of Fig. 3.

Okhotsk–Chukotka and Sikhote Alin volcanic belts (Fujita & Newberry 1982; Parfenov *et al.* 1993) likewise suggests left-lateral shear along the northern edge of the Sea of Okhotsk. This succession of extensional basins and strike-slip faults is kinematically compatible with the left-lateral shear zone that runs from the Pamir Range to Lake Baikal (Davy & Cobbold 1988).

*Right-lateral Shear Zones along the Pacific Margin.* North trending dextral shear zones controlled the opening of intracontinental or back-arc basins as dextral pull-aparts (the Bohai Basin, the Japan Sea) during the Tertiary (Kimura *et al.* 1983; Lallemand & Jolivet 1985; Jolivet *et al.* 1994c). This situation prevailed in Japan until the active east–west compression took over in the late Miocene at 10 Ma. The Sakhalin–Hokkaido Shear Zone extends into the northern Sea of Okhotsk, where it cuts through the Shantar–Liziansky Basin and dies out further north. Transtensional dextral motion until the middle Miocene changed to transpressional motion from the late Miocene to the Present in the north (Worrall *et al.* 1996). The dextral shear zone is thus traceable for a distance of >2500 km. Parallel to the Pacific margin in the south, it cuts right through the continent further north. On the basis of a comparison with analogue experiments (Davy & Cobbold 1988), it has been proposed that the left- and right-lateral shear zones both accommodated the internal deformation of Asia due to collision (Jolivet *et al.* 1990a). The new data in the Okhotsk Sea reinforce the comparison with those experiments, and some other more recent studies (Fournier 1994; Jolivet *et al.* 1994c) which show the contemporaneous formation of north–south dextral shear zones with an extensional component along the eastern border of Asia, and northeast–southeast sinistral ones linking the northwest corner of the indenter (western Himalayan syntaxis) to the northeast corner of the indented lithosphere (Okhotsk Sea region).

*Timing of Deformation in Northeast Asia.* The extrusion model for the India–Asia collision was originally based upon plasticine experiments (Tapponnier *et al.* 1982) which suggested the progressive extrusion of continental blocks along left-lateral shear zones, with a northward propagation of deformation from the Red River Fault to the Stanovoy Ranges. Three lines of evidence now seem to show that the Pamir–Baikal–Stanovoy Shear Zone might instead be an early product of the India collision. Firstly, the age of the inception of the Baikal Rift is clearly pre-Neogene (Kashik & Mazilov 1994; Logatchev & Zorin 1987), and arguably Eocene; see reviews in Logatchev (1993) and Houdry-Lémont (1994) (Table 1). Two stages of Baikal Rift development are generally recognized. An early stage of slow extension, starting between Late Cretaceous and Late Oligocene is followed by a second stage of more rapid subsidence and extension in the Pliocene–Quaternary. Secondly, dextral shear in Sakhalin and in the Tatar Basin may have started as early as the Eocene, and exposed and subsurface rift grabens in northern Sakhalin contain Eocene non-marine infill above a sharp basal unconformity (Kosygin & Sergeyev 1992; Worrall *et al.* 1996). Finally, subsurface correlations suggest that the early fill in Shantar–Liziansky is also Eocene in age (Worrall *et al.* 1996). While not absolutely conclusive, available evidence now suggests that the Pamir–Baikal–Okhotsk Shear Zone had its inception in Eocene times, approximately coincident with the India collision, and is thus coincident with the earliest, not the latest, development of collision-related shear in northeastern Asia.

*Evolving Kinematic Boundary Conditions.* The evolution of boundary conditions can be viewed by using stress-field variations in Asia (Fig. 5). The present-day stress field can be roughly deduced from focal mechanisms of earthquakes (Zoback 1992). Figure 1 shows the directions of $\sigma_{Hmax}$ and $\sigma_{Hmax}$ and $\sigma_{Hmin}$ deduced from focal mechanisms. The strike of $\sigma_1$ is generally northeastward in northeast Asia, except near Japan where it changes to become perpendicular to the subduction zone. In most of northeast Asia, a strike-slip regime is dominant, whereas in Japan, a compressional regime is recorded. The 25 Ma map (Fig. 5) was constructed by using the orientation of large-scale structures, such as rifts of this age as well as more detailed studies of palaeostresses (Fournier *et al.* 1995). The geometry of the stress field is similar to the present one except near Japan. Interpretation of the 45 Ma stage was obtained from the strike of large-scale structures only. The result is consequently less reliable and can be used only for comparison. An older stage (pre-45 Ma) could also be proposed that would show a totally different scheme in northeast Asia, where most of the faults parallel to the eastern border were left-lateral. Thus, major changes appear at 45–43 Ma if the hypothesis for the formation of the large-scale left-lateral Pamir–Baikal–Okhotsk Shear Zone *c.* 10 Ma, when east–west compression began in Japan, is followed:

- *From 45 to 10 Ma.* The geometry of deformation in the early stages of collision (45–10 Ma) is very similar to that predicted by

**Table 1.** *Tectonic history of Asia and its eastern boundaries*

| Age (Ma) | Pamir–Baikal–Okhotsk shear zone | Himalaya–Tibet | South China Sea | Japan Sea Sakhalin | Northern Sea of Okhotsk | Bering Shelf |
|---|---|---|---|---|---|---|
| 45 | Left-lateral motion and extension–First episode of rifting | Inception of collision | Rifting | Inception of Pacific subduction | Extension and left-lateral motion | Right-lateral motion and basin formation |
| 32 | Rifting and sinistral motion continues | Crustal thickening | First oceanic crust | Rifting and dextral motion | → | → |
| 25 | → | Crustal exhumation and first exhumation | Spreading | First oceanic crust, dextral motion | → | → |
| 20 | → | South Tibetan detachment; fast exhumation | Kinematic change; inception of subduction / End of spreading | Spreading and extension, dextral motion / End of spreading | → | → |
| 12 | → | → | Subduction | End of dextral motion, inversion in Tsushima | End of extension | Subsidence and dextral deformation slow |
| 8 | → | Lithospheric root detachment and uplift? | → | Inception of east-west compression | Continued slow regional subsidence | Minor deformation |
| 2 | Fast extension on the Baikal rift; second stage of rifting | Continuous thickening and extension on the plateau | → | Inception of Japan Sea subduction | Faster extrusion of Okhotsk block? | Deformation ceases |

analogue models scaled for gravity (Davy & Cobbold 1988; Jolivet *et al.* 1990*a*). As in other types of models (Tapponnier *et al.* 1982; Dewey *et al.* 1989*a*, England & Molnar 1990; Huchon *et al.* 1994; Rangin *et al.* 1995), the eastern boundary is free of stress other than lithostatic.

Such conditions require a frontal subduction of old lithosphere east of Asia (Uyeda & Kanamori 1979). This situation was in fact established only after the abrupt change in the absolute motion of the Pacific Plate at 43 Ma. This major kinematic change in plate motion was recently challenged (Norton 1995), based on a Farallon–Pacific–Antarctica–Africa–North America–Asia Plate circuit. Norton's original observations are that the 43 Ma elbow exists only along the Hawai–Emperor Chain and that no major tectonic event ever occurred on the margins of the continental plates surrounding the Pacific Ocean. Those two assertions can, however, be disputed: (1) the Louiville Ridge in the southwest Pacific shows the same elbow, as well as most other hotspot lines (Fleitout & Moriceau 1992); (2) a number of events along the Pacific margins could be associated with the 43 Ma event, such as the possible capture of the Philippine Sea Plate (Hilde *et al.* 1977), the inversion of strike-slip sense along the Bering Shelf (Worrall 1991) or the absence of Eocene sedimentation in northeast Japan, as well as the cessation of activity of major left-lateral faults along the margin of northeast Asia (Otsuki & Ehiro 1978). Palaeomagnetic data obtained on the Emperor seamounts suggest that the hotspot might have moved at several cm yr$^{-1}$ during the early Tertiary (Tarduno & Cottrell 1997). The data still allow a significant northward motion of the Pacific Plate before the Hawaiian–Emperor bend. This discussion is placed under the hypothesis that the 43 Ma event actually happened.

The Pacific–Eurasia relative motion was oblique on the plate boundary before 43 Ma

(Engebretson *et al.* 1985). This motion was recorded on land by numerous left-lateral faults (Otsuki & Ehiro 1978). Effects of the oblique subduction were felt far inside the Asian continent, suggesting a strong mechanical coupling between the two plates. It is only when frontal subduction was established that the stress-free boundary

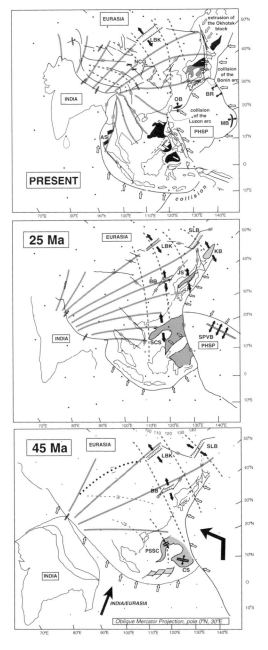

**Fig. 5.** Possible tectonic evolution of Asia from the first collision to the Present with emphasis on the deformation in Far East and the stress-field variations [principal stress directions in the Present-day configuration were inferred from focal mechanisms taken from Dziewonski *et al.* (1987), Parfenoy *et al.* (1987), Fournier *et al.* (1994) Huchon *et al.* (1994)]. Thick lines represent $\sigma_{Hmax}$. See Fig. 3 for abbreviations; SPVB, Shikoku-Parece Vela Basin; PHSP, Philippine Sea Plate; MB, Mariana Basin; PSSC, Proto South China Sea; BR, Bonin Rift; OB, Okinawa Basin.

conditions were met. It can be speculated that at this stage, as soon as these conditions were established, a left-lateral shear zone propagated inside Asia at least as far as the Sea of Okhotsk and perhaps to the margin of the Arctic Ocean.

• *From 10 Ma to the Present.* Compression has been recorded in the marginal basins since the middle Miocene (in the south) or late Miocene (in the north) (Tamaki *et al.* 1992). In Japan, it can be due to the progressive collision of the Bonin Arc, since the late middle or early late Miocene (Taira *et al.* 1989), and to a faster extrusion of the Okhotsk block in the Quaternary (Riegel *et al.* 1993), which might have produced a more efficient compressional component along the dextral Sakhalin–Hokkaido Shear Zone. The most reasonable explanation is that the arrival of the Philippine Sea Plate between the Pacific and Eurasia Plates below central Japan has simply coupled the two larger plates, inducing east–west compression parallel to the Pacific Plate motion in the eastern margin of Asia. Although a large part of the Philippine Sea Plate is thermally mature (in its southwestern part), its northern part is young (<12 Ma for the youngest) and was still younger and buoyant 10 Ma ago. The Pacific Plate is much older all along the western Pacific subduction zone. The arrival of the buoyant Philippine Sea Plate between the Eurasian and Pacific Plates might have considerably changed the mechanical coupling.

## Conclusions

Before the present-day tectonic setting, characterized by crustal thickening in Tibet, strike-slip and extensional tectonics dominated in Asia. The left-lateral shear zone that links the western Himalayan syntaxis to the Baikal and Stanovoy Ranges can now be extended through the Sea of Okhotsk and perhaps as far as the margins of the Arctic Ocean. The western margin of the Sea of Okhotsk is characterized by the 2500 km long dextral Sakhalin–Hokkaido Shear Zone, running from central Japan to the northern margin of the Sea of Okhotsk. The deformation related to collision reached the Baikal Graben and the Sea of Okhotsk, and perhaps the Bering Shelf, by Eocene times. The early propagation of a left-lateral shear zone across far eastern Russia might have been favoured by the creation of a stress-free, or extensional boundary, condition east of Asia after the change in the absolute

motion of the Pacific Plate at 43 Ma. The distribution of sinistral and dextral shear zones in northeast Asia is similar to that predicted by analogue models of collision scaled for gravity.

When compared to more simple back-arc basins, such as the Shikoku–Parece Vela Basin, the Japan Sea shows similarities in terms of timing. Both basins opened at the same time, a period when extension prevailed along the entire western Pacific margin. Basin geometries are instead very different and the deformation history of the Japan Basin cannot be explained solely by extension due to slab roll-back. Models involving asthenospheric injection (Tatsumi *et al.* 1989) are alternative explanations for back-arc opening which cannot take into account the asymmetry of the basin and the existence of large-scale strike-slip faults. Intracontinental deformation only can explain the observed strike-slip component. The simplest model, described here, involves the India–Asia collision which modifies the stress regime in the back-arc region and leads to the observed strike-slip history. Extension is fundamentally a consequence of subduction, slab roll-back and/or asthenospheric injection, and strike-slip a consequence of collision.

## Northern Tyrrhenian Sea

### *Geological setting*

The Tyrrhenian Sea formed during the last 15 Ma in the back-arc region of the Apennines–Calabrian subduction system (Kastens *et al.* 1988; Kastens & Mascle 1990; Sartori 1990) (Figs 2, 6 and 7). Extension affected a continental crust partly thickened during the Alpine Orogeny, as opposed to the nearby Liguro–Provençal Basin where no such thickening has occurred (Burrus 1984; Faccenna *et al.* 1997). A continuum of extension is recorded from the Gulf of Lion and Provence margin in the early Oligocene or even late Eocene, to the present extension in the Apennines (Chamot-Rooke *et al.* 1999; Jolivet *et al.* 1998). Oceanic crust was formed in the Liguro–Provençal Basin during the Early Miocene when Corsica and Sardinia rotated counter-clockwise about a pole located in the Ligurian Sea (Montigny *et al.* 1981; Vigliotti & Kent 1990; Mauffret *et al.* 1995).

Extension migrated eastward and reached Corsica by the Early Miocene, the Corsica Basin in the Middle Miocene, Elba Island in the Late Miocene, and Tuscany and the Apennines in the Quaternary (Jolivet *et al.* 1990*b*, 1998) (Fig. 8). This continuous evolution is

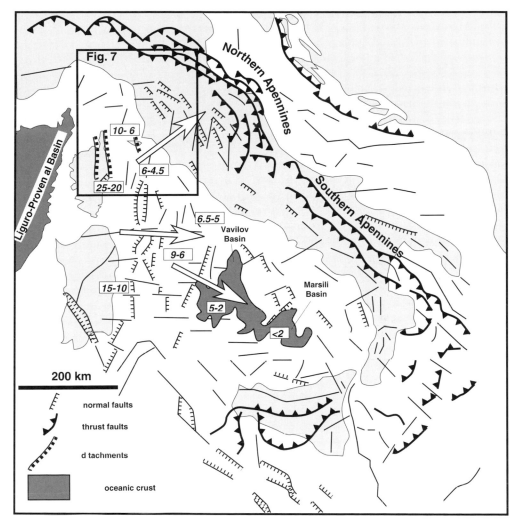

**Fig. 6.** Tectonic map of the Tyrrhenian Sea and the Apennines. Numbers in square boxes are the ages of extension in Ma (Kastens & Mascle 1990; Jolivet *et al.* 1999).

recorded in the ages of synrift deposits, radiometric ages of exhumed metamorphic rocks and of magmatic events (Elter *et al.* 1975; Serri *et al.* 1993; Bartole 1995; Brunet *et al.* 1997; Chamot-Rooke *et al.* 1997; Jolivet *et al.* 1998). A contemporaneous migration of the magmatic arc is recorded along the same direction from the western margin of Corsica to the western Apennines (Serri *et al.* 1993). During the same period, the Apennines belt was formed as a large accretionary complex at the expense of the sedimentary cover of the Adriatic Plate (Lavecchia & Stoppa 1989; Patacca *et al.* 1990; D'Offizi *et al.* 1994).

Despite this simple kinematic history, the geometry and the mode of extension is different between the two basins: a 'narrow' and 'single-rift' style of extension of the Liguro-Provençal basin contrasts with the 'wide' and 'multi-rift' style of the Tyrrhenian Basin. It has been proposed (Faccenna *et al.* 1997) that these two styles of extension depend upon the pre-rift rheology linked with its geological heritage: the Tyrrhenian extension, in fact, occurred on thickened and weakened Alpine crust whereas the Liguro-Provençal Basin developed on a cold and resistant Hercynian crust. Oceanic crust formed only in the southern Tyrrhenian Sea, north of

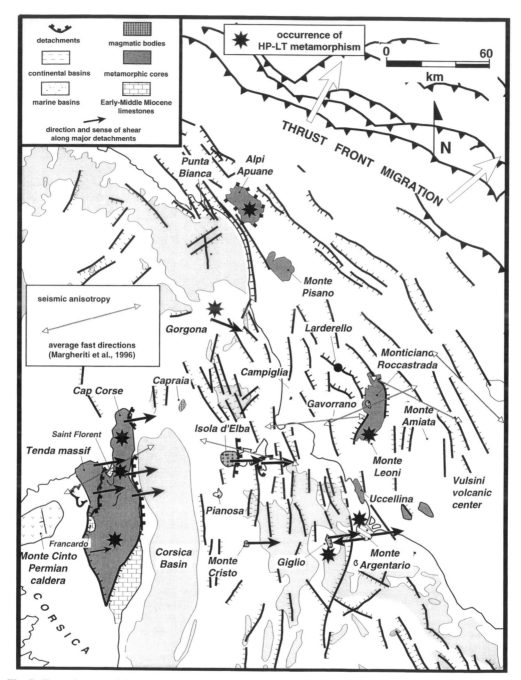

**Fig. 7.** Tectonic map of the Northern Tyrrhenian Sea [modified after Jolivet *et al.* (1998)].

Sicily, within the Marsili and Vavilov Basins from 5 Ma to the Present (Kastens *et al.* 1988). This fast opening in the southern regions is coeval with fast rotations about vertical axes in the southern Apennines and Calabrian Arcs (Scheepers *et al.* 1993).

Extension in the Northern Tyrrhenian Sea has been active for most of the Neogene (Fig. 8). As opposed to its southern counterpart, no large-scale rotation of crustal blocks is recorded in the palaeomagnetic record (Mattei *et al.* 1996). Only an east–west extension is observed

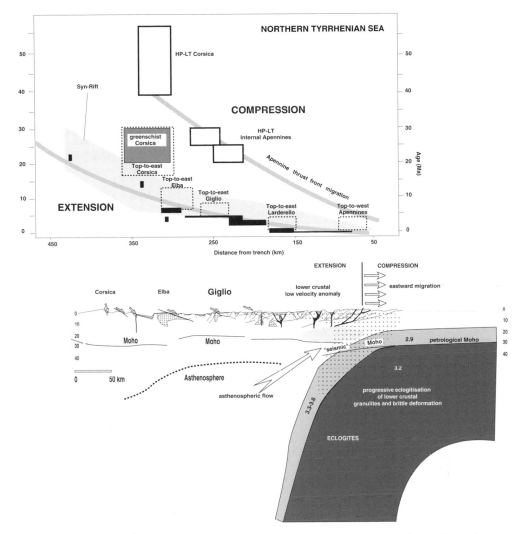

**Fig. 8.** Cross-section through the Apennines and Tyrrhenian Sea, and the migration of extension and magmatism. (**a**) Diagram showing the migration of magmatism (black boxes), synrift deposits (grey area), HP–LT metamorphism (white boxes) and activity of extensional shear zones (dotted boxes). (**b**) Lithospheric-scale cross-section (Wigger 1984; Della Vedova *et al.* 1991; Ponziani *et al.* 1995), seismicity (Selvaggi & Amato 1992), geological structure (Zitellini *et al.* 1986; Lavecchia *et al.* 1987), age of syn-rift deposits (Bossio *et al.* 1993; Martini & Sagri 1993; Bartole 1995), radiometric ages of magmatism (Serri *et al.* 1993) and radiometric ages of metamorphic rocks (Brunet *et al.* 1997).

from Corsica to the Apennines (Jolivet *et al.* 1994*b*, 1998). The direction of extension is attested by the overall geometry of grabens, and by the shear direction recorded in exhumed metamorphic core complexes of Corsica and the Tuscan Archipelago (Jolivet *et al.* 1998). A clear migration of the locus of extension is evidenced from the Early Miocene in Corsica to the Present in the Apennines. Migration of exten-

sion is coeval with a similar migration of the compressional front to the east and a concommittant migration of the volcanic arc always along an east–west direction. The large-scale geometry is quite simple and can be considered as 2D.

It has been suggested that the difference in style and amount of extension in these two regions corresponds to a difference in style of

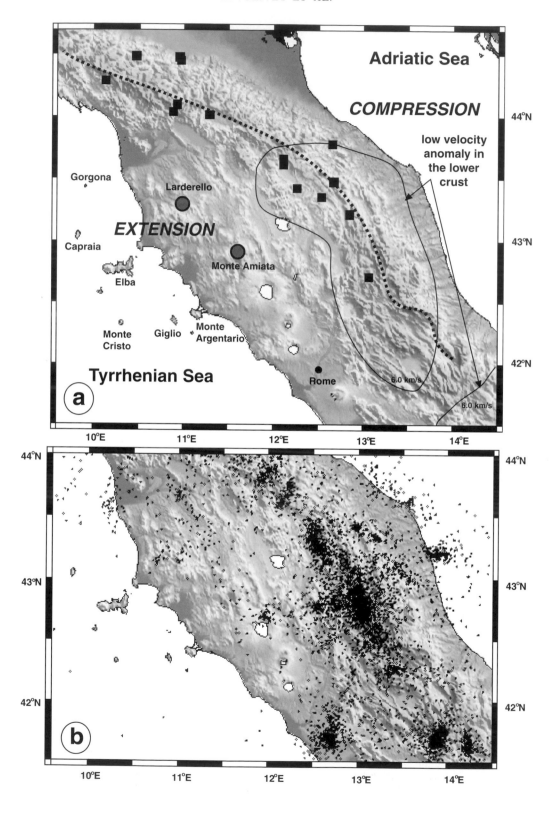

subduction (Faccenna *et al.* 1997). From the Oligocene to the Present, only continental lithosphere has been subducting below the Northern Apennines (Coli *et al.* 1991), whereas Ionian oceanic lithosphere is still subducting below the Calabrian Arc (Selvaggi & Chiarabba 1995).

Slab roll-back (Malinverno & Ryan 1986; Royden 1993; Keller *et al.* 1994) is considered the most plausible mechanism for extension in this region. Modelling of the shape of the slab shows that topographic load cannot be responsible for the observed geometry and gravimetric signal, and that subsurface loads at mantle depth (slab) are the acting driving forces for the recent period (Royden 1993). On the other hand, it has been shown that exhumation of high-pressure (HP) and low-temperature (LT) metamorphic rocks was active in the frontal domain until the middle Miocene and that HP–LT metamorphism had also migrated eastward with time (Jolivet *et al.* 1998). Extension was thus set on a thick crust until at least the Middle Miocene. Topographic loads and body forces leading to crustal collapse were thus active in the early history of the Tyrrhenian Sea. It can be shown that both compression in the south (African collision) and crustal collapse are necessary to initiate the subduction process (Faccenna *et al.* 1996). As subduction and extension proceed, the slab-pull component increases as the length of the subducted slab increases.

However, from the Oligocene to the present only continental lithosphere has subducted below the Northern Apennines (Coli *et al.* 1991). The geodynamic situation is thus very different from a classical Mariana-type subduction. Furthermore, although subduction and compression are still active along the trench in the Adriatic Sea, and a drastic shortening observed in the sediments offscrapped from the Adriatic crust, no significant crustal thickening is recorded at present (Chiarraba & Amato 1996). Thus, the continental crust has to be partly subducted (D'Offizi *et al.* 1994; Pialli *et al.* 1995). Eclogitization of the subducted crust could have reduced its buoyancy and permitted its subduction (Pialli *et al.* 1995). The recent discovery of intermediate earthquakes (down to 90 km) below the Apennines, as well as tomographic images, confirm this conclusion (Selvaggi & Amato 1992) (Fig. 9). This deep seismic activity somehow apparently contradicts the presence of a low-velocity anomaly in the lower crust below the Northern Apennines

(Chiarraba & Amato 1996) (Fig. 9) and a severe attenuation of Pn and Sn waves in the upper mantle (Mele *et al.* 1997).

## Compression v. extension and subduction of continental lithosphere

Summarized here are some of the most important aspects of the geodynamics of this region.

- Before the Early Oligocene, extension was active only on the future northern margin of the Gulf of Lion (Burrus 1984), more easterly regions were subjected to compression. The nappe stack and the associated HP–LT metamorphism of Alpine Corsica is evidence of this essentially compressional stage (Mattauer *et al.* 1981; Caron 1994). From the Late Oligocene onward the Apennines thrust front started to move eastward and basins formed behind by extension. The most spectacular result of this extension is the rifting and spreading of the Liguro-Provençal Basin and concommittant counter-clockwise rotation of Corsica and Sardinia in the Late Oligocene and Early Miocene (Burrus 1984). Extension started in Alpine Corsica some 30 Ma ago (Jolivet *et al.* 1991a, 1998; Daniel *et al.* 1996). It was accommodated by shallow east dipping extensional shear zones at the brittle–ductile transition and normal faults above. Evidence for extension is found further east with a similar geometry and a progressive younging toward the Apennines: Middle Miocene in the Corsica Basin, Late Miocene in Elba and Monte Cristo, Pliocene in Giglio (Rossetti *et al.* 1999), Late Pliocene and Early Pleistocene in Tuscany. At present, extension is active in the Apennines (Frepoli & Amato 1997). During the whole Neogene the direction of extension observed in exhumed metamorphic cores remained unchanged, trending approximately east–west. Recent data on the seismic anisotropy along a transect from Corsica to the Apennines shows that the whole Tyrrhenian lithosphere has been stretched along the same direction (Margheriti *et al.* 1996) (Fig. 7). These results also show a progressive change of the fast direction from approximately east–west in the extended domain to northwest–southeast in the Apennines. This evolution shows the

**Fig. 9.** Topography and seismicity in the Northern Apennines and Tuscany. (**a**) Deep earthquakes (Selvaggi & Amato 1992) (black squares) and a low-velocity anomaly in the lower crust (Chiarraba & Amato 1996). The dotted line represents the drainage divide in the Apennines. (**b**) All earthquake epicentres.

progressive deformation of the mantle and lower crust from the front of the belt to the back-arc domain.

- The thrust front has migrated eastward with a somewhat lower velocity (Patacca & Scandone 1989). It can be followed on a map by the migration of the HP–LT metamorphic rocks and by the migration of preserved thrusts in the Apennines. Along the entire transect from Corsica to the Apennines there is little evidence that the continental basement has been involved in the crustal thickening process. Crustal thrust sheets are included in the Schistes Lustrés Nappe of Alpine Corsica (Mattauer *et al.* 1981) but further east the Palaeozoic basement crops out only in the Apuan Alps, with no indication of deep crustal lithologies (Carmignani & Kligfield 1990). In all metamorphic complexes observed in the Tuscan archipelago and in Tuscany, HP–LT metamorphics rocks are metasediments of Permo-Triassic age (Verrucano facies) without any trace of a continental basement. The frontal Apennines are made of imbricated thrust sheets of sedimentary origin scrapped off the subducting Adriatic crust (D'Offizi *et al.* 1994). Nowhere is the Adriatic continental crust involved in the shortening process as seen at the surface. The upper crust might still be involved at depth below the Apennines but at least a part of the crust, and certainly the lower crust, has to be subducted in the asthenosphere.

- The transition between compression and extension coincides today with the highest elevation all along the Apennines belt (D'Agostino *et al.* 1998). The example of the Gran Sasso Range is very characteristic of this behaviour. The Gran Sasso basal thrust is reworked at shallow depths by a shallow west dipping normal fault which crops out on the internal flank of the mountain range. Steeper west dipping normal faults, which affect the entire brittle crust, are then found at close proximity further west. Major short-wavelength (30 km) topographic gradients are related to active normal faults. This geometry is different to that observed in the Himalayan–Tibetan belt or the Andes, where the transition between compression and extension coincides with a given topographic contour (Molnar & Lyon-Caen 1988). Regions situated above 3000 m show active extension, while compression is recorded below. This might suggest a general sudden collapse of the entire belt (Platt & England 1994). The geometry observed in the Apennines instead suggests that a wave of extension progressively lowers the altitude of the belt being formed further east. This is compatible with the eastward migration described above for the Neogene period.

- The present day Apennines form a 100 km wide topographic bulge which culminates at Gran Sasso where most of the shallow seismic energy is released. This domain is supported by a crust not thicker than 40 km. It also corresponds to a low-velocity anomaly in the lower crust with P-wave velocities $<6.0 \, \text{km s}^{-1}$ (Chiarraba & Amato 1996). The lateral extent of this anomaly is very close to that of the topographic bulge. To obtain these low velocities, high temperatures must exist in the lower crust below the Apennines at present. This anomaly is also located east of the most intensely extended domain in the Northern Apennines.

The location of this anomaly is somehow in contradiction with the location of the recent volcanic arc further to the west. Volcanic rocks recently erupted have a clear mantle signature with almost no crustal contamination (Serri *et al.* 1993). During the Late Miocene and the Pliocene a larger crustal component was present in magmatic rocks. The granitoids of Elba, Monte Cristo and Giglio contain cordierite, and are thought to derive from crustal anatexis. This suggests that the position of the volcanic arc is controlled by the geometry of the slab at depth and the partial melting domain in the lower crust is due to some other mechanism (see below). The absence of a high heat flow at the surface, above the low-velocity anomaly, also suggests that the anomaly is quite recent. It can be speculated that it was previously further west and that it has migrated eastward to the present position only recently. The Quaternary age of the main phase of uplift, where extension was already established, and the geographical distribution of thermal anomalies in the lower crust and upper mantle, suggest a thermal origin for the main topographic buldge of the Apennines.

Despite a hot lower crust, deep earthquakes are found at mantle depths down to 90 km below the belt (Selvaggi & Amato 1992). Four of those earthquakes provided first-motion focal mechanisms: three are extensional and one is compressional. A question then arises: in which material are those deep earthquakes produced? Tomographic data suggest that the slab is continuous below the Apennines, down at least to

250 km (Amato *et al.* 1993). Why then would the earthquakes be recorded only down to 90 km if the subducted continental mantle is continuous down to deeper depths?

- As shown on Fig. 8, the eastward migration of extension and compression proceeds at two different rates, extension migrating faster than compression. The distance between the volcanic domain and the trench thus decreases with time. This decrease can suggest a progressive verticalization of the slab. This is evidence for a major role played by slab pull in the trench retreat process.

## Eclogitization of the lower crust

The recent discovery of pseudotachylites associated to eclogitization of granulites in western Norway north of Bergen (Austrheim 1987; Austrheim & Boundy 1994; Austrheim *et al.* 1997), suggest that brittle deformation might occur at great depths during metamorphic phase transformation. This brittle deformation is seen only during the first stages of eclogitization of intermediate to basic granulites. While the transformation proceeds, lower crustal material becomes less resistant and more ductile. Only ductile deformation is seen when the transformation is complete. It is uncertain whether these deformations are the result of sudden volume change or whether they correspond to a more regional strain field imposed by the geometry of plate convergence. The existence of eclogites below the Moho, below the Tibetan Plateau, has also been postulated (Le Pichon *et al.* 1997; Sapin & Hirn 1997), in spite of the presence of a hot and weak lower crust.

Lower crustal material brought to eclogite conditions above 13–14 kbar pressure ($P$) [at 550°C temperature ($T$)], can attain high densities, similar to what is expected for mantle rocks (Bousquet *et al.* 1997). The density increase is controlled by the appearance of garnets >500°C and by the transformation of plagioclase to pyroxene. It has been proposed for the Caledonides of western Norway (Dewey *et al.* 1993; Andersen 1999), the Himalayas (Henry *et al.* 1997; Le Pichon *et al.* 1997) and the Alps (Bousquet *et al.* 1997), that a part of the continental crust might be hidden below the Moho which would then mark the phase transition between amphibolite facies and eclogites.

One can postulate that the deep earthquakes seen below the Apennines are within the subducted continental crust and that they show the progressive equilibration of lower crustal material to the $P$–$T$ conditions of the eclogite facies. This hypothesis can be paralleled to that proposed by Pialli *et al.* (1995).

## Lower crustal delamination

This hypothesis has several important implications (Fig. 10).

- It offers the possibility of a delamination of the lower crust together with the subcontinental Adriatic mantle (D'Offizi *et al.* 1994; Pialli *et al.* 1995). The delaminated lid has a higher density than the surrounding asthenosphere and it sinks under its own weight leading to roll-back of the slab.
- The delaminated crust is replaced by asthenosphere flowing upward. The remaining crust is thus put in direct contact with the hot astenosphere and partial melting ensues. The low-velocity anomaly would then be a consequence of this delamination controlling the position of the topographic buldge.
- Partially molten lower crust leads to the diapiric ascent of anatectic granitoids. It also acts as a velocity discontinuity which will localize extensional strain in the lower crust and control shear senses along the brittle–ductile transition (Jolivet *et al.* 1998). During slab roll-back the molten lower crust also migrates eastwards, while the crust molten in previous stages cools back and becomes more resistant. Extension then follows the migration.
- Upper crustal rocks underthrust below the Apennines tend to rise up again once the lower crust has delaminated between the frontal thrust and the first extensional fault (Chemenda *et al.* 1995).

The acceleration of the eastward migration with time is evidence that the slab-pull component is more and more efficient with time, and overcomes the extensional forces due to collapse of the thickened crust. This is also compatible with the progressive steepening of the slab, suggested by the faster migration of extension rather than compression. This smaller contribution of body forces can also be seen in the delay between the inception of extension at a given point along the transect and the extrusion of magmatic rocks at the surface: it suggests that crustal thickness has decreased with time. The maximum pressure recorded in HP–LT metamorphic cores also decreases eastward, which also suggests a thinner crust (Jolivet *et al.* 1998). This is further compatible with the observation that the anatectic component in magmas decreases both with time and eastwards (Serri *et al.* 1993). Last but not least, the exhumation

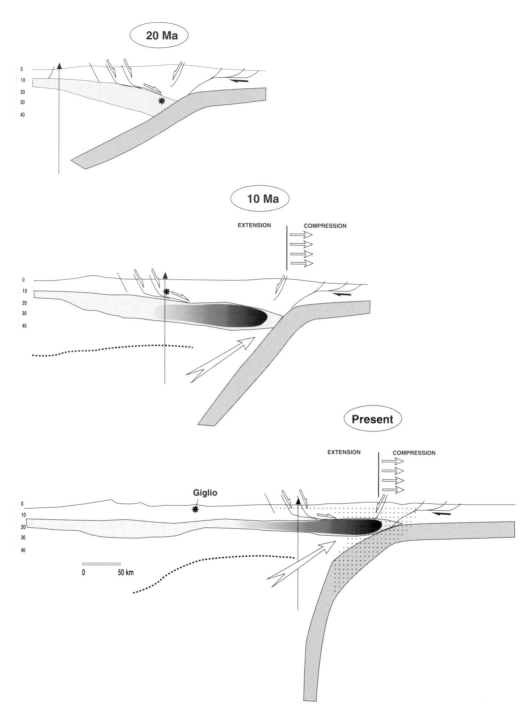

**Fig. 10.** Schematic evolution of the Corsica–Apennines transect.

of well-preserved HP–LT parageneses in the Tuscan archipelago during the Miocene suggests that the crust was thick enough to prevent reheating before the rocks were exhumed.

## Conclusion

Several mechanisms have been proposed to explain the evolution of the Northern Tyrrhenian–Apennine system. It has been suggested that the roll-back of the hinge of the slab, driven by subsurface loads or a slab-pull force, is able to explain the gravimetric signal of the area and the formation of deep foredeep basins (Malinverno & Ryan 1986; Royden 1993). An alternative model proposed by other Carmignani et al. (1994) suggests that the collapse of a thickened, post-orogenic crust can contribute to produce extension in the hinterland and compression in the foreland. Topographic loads and body forces leading to crustal collapse were indeed active in the early history of the Tyrrhenian Sea, and can contribute together to the fast retreat of the hinge of the slab (Faccenna et al. 1996). However, they cannot alone explain the progressive evolution and migration of the extension–compressional system, especially in most recent times when crustal thickening did not occurr.

Some authors (Reutter et al. 1980; Lavecchia & Stoppa 1989; D'Offizi et al. 1994; Keller et al. 1994; Pialli et al. 1995) proposed that the delamination of the lower crust–upper mantle from the upper crust can better explain the architecture of the Apenninic wedge, allowing a passive subduction of the continental lithosphere. In agreement with this model (D'Offizi et al. 1994; Pialli et al. 1995), it is suggested here that phase transformation and eclogitization of the lower crust is a fundamental requisite for the evolution of the Northern Tyrrhenian–Northern Apennine area, leading to the delamination of the continental crust and stable subduction of the lower crust–upper mantle during the last 30 Ma. The eclogitization of the lower crust is a possible mechanism to explain the limited occurrence of deep earthquakes in the high-velocity zone present below the Apennines. As already proposed for other mountain belts (Andersen & Jamtveit 1990; Dewey et al. 1993; Le Pichon et al. 1997), this hypothesis suggests that most of the eclogitized lower continental crust is subducted. This does not conflict with the rule that continental crust is generally buoyant and is not subducted. Eclogites are rather rare rocks on the surface of continents and eclogitized granulites are even less frequent. Old portions of buoyant continental crust are not made of eclogitized granulites but of granulites or high temperature gneiss with some partial melting, remains of post-orogenic collapse of ancient mountain belts.

The occurrence of deep earthquakes below the Northern Apennines is interpreted as the result of brittle deformation during the progressive equilibration of the underthrusted Adriatic continental lower crust to the P–T conditions of the eclogite facies. The delamination of the eclogitized lower crust can provide an additional force to drive slab roll-back and extension in the upper plate. In this hypothesis, the Moho seen below the Northern Apennines would be intracrustal, separating amphibolites from eclogites.

An immediate comparison with the Alboran Sea can be seen (Platt & Vissers 1989; Seber et al. 1996). In both cases, the driving mechanism for back-arc extension is the roll-back of a delaminated lid of sinking lithosphere. As argued by previous authors, this mechanism is the only one that can explain the formation of arcs without simple kinematic relations with the Africa–Eurasia convergence. The phase changes invoked here below the Apennines can only enhance the efficiency of a backward motion of the subducting plate. It also has the advantage of solving the problem of the fate of the lower crust which is not found in the accreted units forming the Apennines.

This example shows that slab roll-back is feasible even with the subduction of a continental lithosphere. In the case of the Tyrrhenian Sea, as opposed to the Aegean or Japan Seas, there is no real convergence between the two plates and subduction is only the consequence of passive sinking of the Adriatic lithosphere in the mantle, at least for recent periods. Extension within the upper plate is mainly a consequence of slab roll-back and delamination, its detailed geometry is controlled by that of the pre-existing crustal wedge which is now collapsing. The active faulting in the Apennines is then driven by two concurrent mechanisms: (1) extension induced by the slab roll-back; (2) gravitational collapse of the topographic bulge located above the low velocity anomaly. The relative importance of the two mechanisms depends on the crustal structures undergoing extension: thicker crust will reflect a greater component of the gravitational collapse while extension in the thinner crust would be driven by slab roll-back only.

## The Aegean Sea

The tectonic history of the Aegean Sea is partly similar to that of the Tyrrhenian Sea as it

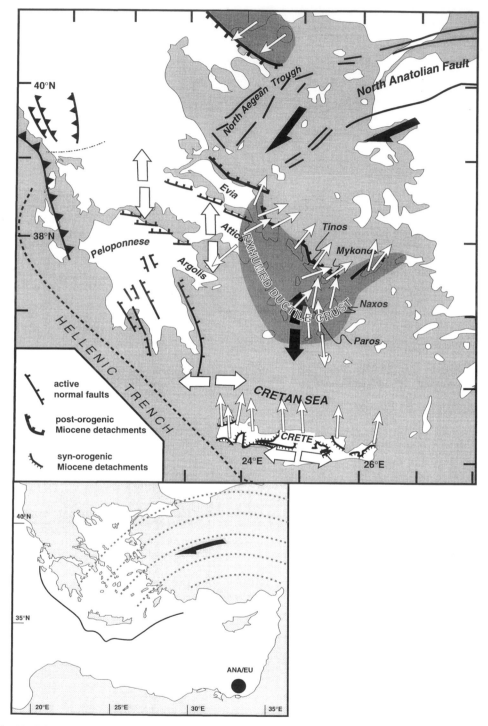

**Fig. 11.** Schematic tectonic map of the Aegean Sea showing the exhumed ductile crust in the Aegean region and the direction and sense of ductile shear (Buick 1991; Faure *et al.* 1991; Dinter & Royden 1993; Sokoutis *et al.* 1993; Gautier & Brun 1994*a*, Jolivet *et al.* 1996). The insert shows the rigid extrusion of the Anatolian block documented by geodetic displacements as well as the calculated rotation pole (Le Pichon *et al.* 1994).

involves the collapse of a thickened crust and the outward migration of the subduction front. However, two major differences are to be noted: (1) extension is more widely distributed in the Aegean Sea, which suggests a larger component of gravitational collapse, and (2) the recent geometry of extension is partly controlled by dextral strike-slip faults which guide the westward extrusion of the Anatolian block.

## Geological setting

The Aegean Sea (Figs 2, 11 and 12) formed in the back-arc region of the Hellenic Trench during the Late Oligocene until Present (Le Pichon 1981; Le Pichon & Angelier 1981; Gautier et al. 1993). Subduction of the African Plate beneath the Anatolian block proceeds at a fast rate of c. 4–5 cm yr$^{-1}$ (Le Pichon et al. 1994). Extension has taken place on a previously thickened continental crust, the deep parts of which are now exhumed at the surface (Lister et al. 1984). Subduction of oceanic crust started at least 40 Ma ago, as attested by seismic tomography data (Spakman 1990). Transition from continental collision to oceanic subduction might have released the compressional stresses and favoured gravitational collapse of the thickened crust.

Active extension is principally localized along the outer Hellenic Arc (Crete), in the Corinth Gulf region and around Volos (McKenzie 1972, 1978; Taymaz et al. 1991; Hatzfeld et al. 1993; Armijo et al. 1996; Rigo et al. 1996), leading to an intense seismicity. The central part of the extended domain, the Cyclades Islands, is less seismic though it was subjected to high extensional strain during the Miocene.

Large finite extension is recognized in regions of thin continental crust in the North Aegean Trough and the Cretan Sea where the deepest waters occur. It is also important in the Cyclades islands where Cordilleran-type metamorphic core complexes have recently been studied (Lister et al. 1984; Buick 1991; Faure et al. 1991; Gautier et al. 1993; Gautier & Brun 1994a; Jolivet et al. 1994a; Jolivet & Patriat 1999). The exhumation of metamorphic rocks is partly the consequence of the Aegean extension, even though a large part of it could be contemporaneous with compressional tectonics (Avigad et al. 1997; Jolivet & Patriat 1999). The exhumation of the eclogites and blueschists in the Cyclades Islands is probably older than the formation of the Aegean Sea. The Aegean extension probably exhumed the amphibolites and greenschist facies rocks cropping out in the core of metamorphic core complexes such as those of Naxos and Paros.

Exhumation of HP–LT metamorphic rocks occurred in Crete in the Early–Middle Miocene, while HT–LP rocks were exhumed further north in the Cyclades Islands (Fassoulas et al. 1994; Jolivet et al. 1996). Exhumation in Crete is interpreted as the result of extension in the upper parts of a thick accretionnary complex near the thrust front.

Extension is further controlled by the North Anatolian Fault, a dextral strike-slip fault which runs along the northern border of Anatolia from eastern Turkey to the Aegean Sea (McKenzie 1978; Le Pichon et al. 1994; Armijo et al. 1996). It connects to the Aegean Trench through a number of east–west grabens, among which the Gulf of Corinth is the most active (King & Ellis 1990; Rigo et al. 1996). Current displacements reveal an almost rigid rotation of the Anatolian block, including the Aegean region, about a pole located in the southeast Mediterranean Sea (Le Pichon et al. 1994) (Fig. 11). Internal deformation of the Aegean domain is at present a second order phenomenon in terms of velocities relative to Eurasia. It could have been a first order phenomenon before the initiation of the north Anatolian Fault in the middle Miocene.

## Geometry of extension

Extension was more widely distributed during the Miocene than at Present (Gautier & Brun 1994a). It is recognized in the whole Aegean region as well as in western Turkey in the Menderes Massif (Hetzel et al. 1995a, b). The progressive localization with time is probably due to the formation of the North Anatolian Fault and its southward propagation (Armijo et al. 1996).

However, active extension in the Corinth Gulf and fossil extension seen in the Cyclades Islands show a very similar geometry with shallow, north dipping ductile shear zones (Jolivet et al. 1994a, Jolivet & Patriat 1999; Patriat et al. 1999). Active north dipping normal faults that control sediment deposition in the Corinth Gulf are planar down to the depth of the brittle–ductile transition, e.g. 8–10 km (King et al. 1985; Jackson & White 1989). Seismogenic low-angle, north dipping normal faults have been recognized in the Corinth Gulf (Bernard et al. 1997). They are then relayed by shallow, north dipping extensional shear zones which produce microearthquakes (King et al. 1985; Rigo et al. 1996). Geophysical studies show that the deep crust in the same region is rich in fluid circulation (Chouliaras et al. 1997).

Exhumed ductile shear zones of Miocene age in Tinos or Paros show exactly the same features

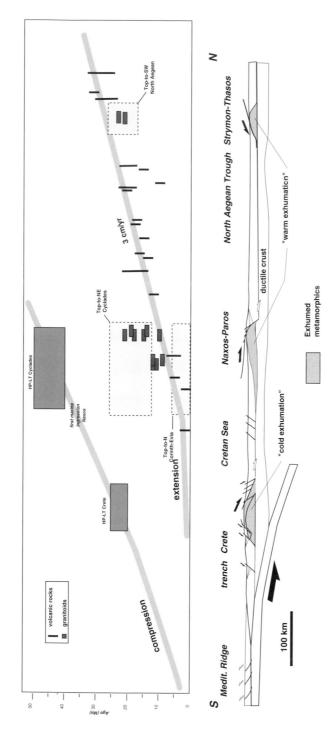

**Fig. 12.** Cross-section through the Aegean domain showing the major core complexes and the migration of metamorphic and magmatic events. (Altherr *et al.* 1982; Seidel *et al.* 1982; Fytikas *et al.* 1984; Kyriakopoulos *et al.* 1988; Wijbrans *et al.* 1993; Jolivet *et al.* 1996, 1999; Wawrzenitz and Krohe 1997).

(Gautier *et al.* 1993; Patriat 1996; Jolivet & Patriat 1999). Top-to-the-north shear criteria are concentrated along a prominent north dipping shear zone where deformation evolves with time from ductile to brittle. The ductile deformation is contemporaneous with the crystallization of greenschist parageneses that rework earlier (Eocene) HP–LT ones. The shear zones are further characterized by intense fluid circulations best seen in cataclastic levels immediately beneath the brittle fault.

Similar north dipping detachments are observed in most islands of the northern Cyclades Islands, including Mykonos, Paros and Naxos (Avigad & Garfunkel 1989; Buick 1991; Faure *et al.* 1991; Gautier & Brun 1994*b*). The metamorphic conditions active during extension in those three islands was much hotter than in Tinos and Andros, reaching the amphibolite facies and anatexy. The island of Ios shows a complex interaction between north and south directed shear zones (Vandenberg & Lister 1996). In the northern Aegean region, the southern Rhodopian Massif (Strymon detachment) (Dinter & Royden 1993) and Thassos Island (Sokoutis *et al.* 1993; Wawrzenitz & Krohe 1999)) also show extensional shear zones with top-to-the-northeast sense of shear.

Recent palaeomagnetic studies of the Aegean granitoids suggest that some of the stretching lineations shown in Fig. 11 have been rotated since the Miocene by some 25° or so, either clockwise (Tinos and Mykonos) or counter-clockwise (Naxos) (Morris & Anderson 1996; Avigad *et al.* 1999). Though it might be too early to have a complete picture of the pre-rotation lineation pattern, it remains that from the central Aegean to Crete all senses of shear point to the north as opposed to those of the Northern Aegean domain which point to the south.

While post-orogenic extension was active in the Cyclades Islands, coeval with high temperature evolution, synorogenic extension occurred in Crete in a much colder environment (Jolivet *et al.* 1996) (Fig. 12). Data from the islands of Syros and Sifnos suggest that a similar history occurred in the Eocene when HP–LT rocks were being exhumed in the Cyclades below north dipping detachments in a cold environment, and HT–LP rocks were exhumed further north in the Rhodopian Massif.

### Gravitational collapse v. slab roll-back

A simple model involving the progressive migration of the subduction front from north to south can be proposed with synorogenic extension above the accretionary complex accompanying post-orogenic extension in the back-arc region (Jolivet *et al.* 1994*b*; Jolivet & Patriat 1999). The migration of the magmatism documents the roll-back of the subducting slab (Fytikas *et al.* 1984) (Fig. 12).

Rocks were first buried within the accretionary complex near the trench where they recrystallized in the blueschist or eclogite facies. They are then soon exhumed at the surface by the activity of large-scale detachments that dip toward the back-arc region and its low topography. Those which are exhumed soon enough will preserve their HP–LT parageneses, while those which remain at depth longer will see a more or less complete high temperature overprint. During the southward migration of the slab, the accretionary complex grows at the expense of sediments carried by the subducting slab as in the case of the Apennines. The once frontal zones are then progressively transferred to the back-arc domain where the heat flow is higher. The thick crust is then heated and its resistance decreases. It thus collapses and HT–LP metamorphic rocks are exhumed. This process went on from the Eocene–Present more or less continuously.

As shown in Fig. 12, extension is coeval with the emplacement of granitoids. As opposed to the volcanic rocks which migrated regularly from north to south at *c.* $3 \, \text{cm} \, \text{yr}^{-1}$, granitoids mostly intruded the Cyclades Islands and are grouped in a rather narrow domain which corresponds to the zone thickened during the Eocene.

### Conclusions

As in the Tyrrhenian Sea, crustal collapse after initial thickening controls the geometry of extension. Extensional boundary conditions are provided by the retreat of the subducting African slab. Far field stresses due to the Arabia–Eurasia collision also played their part for the recent period as extension became localized and asymmetric as a consequence of the southward propagation of the North Anatolian Fault.

### Discussion

Back-arc extension in all the three examples described above suggests that a large component of slab roll-back was involved, and that the geometry of extension and the distribution of strain within the extended domain is further controlled by long-distance effects of continental

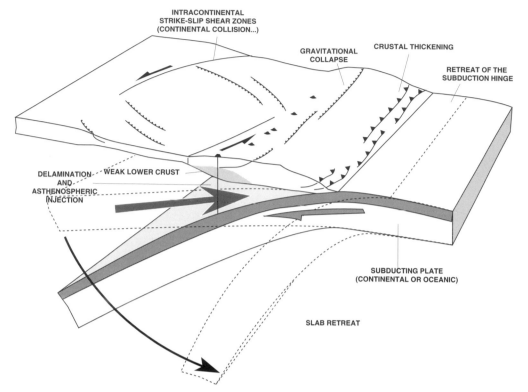

**Fig. 13.** Schematic diagram showing the various phenomena involved in the formation of marginal basins.

collision and/or body forces during collapse of mountain belts (Fig. 13).

## Slab roll-back

The Japan Sea opened during the Miocene above a retreating Pacific subduction, which also promoted the opening of the Shikoku-Parece Vela Basin along the eastern margin of the Philippine Sea Plate during the same period. The end of opening in the Late Miocene and the recent incipient closure of the Japan Sea affects only the Eurasian margin. Extension still prevails above the Bonin (Taylor 1992) and Mariana Trenches at present. Slab roll-back is still active but the degree of mechanical coupling between the Eurasian and Pacific Plates was modified in the Late Miocene by the formation of the central Japan triple junction. The introduction of the young Philippine Sea Plate between the two larger plates, and the collision of the Bonin Arc with central Japan, might have coupled their displacements and induced compression in the upper plate strictly parallel to the convergence vector.

The Northern Tyrrhenian Sea extension is presently active in the Apennines as the con-

tinuation of a long-lived process which started as early as the Oligocene. Slab roll-back has been active during this entire period, even though there has been little east–west convergence between the European and the Adriatic Plates. Passive sinking of the Adriatic continental lithosphere is still active at present. Eclogitization of the lower continental crust might be one factor which increases the density of the lithosphere and permits sinking and roll-back.

The Aegean Sea started to form (30 Ma ago) soon after the inception of an oceanic subduction which replaced continental collision (40–45 Ma ago). The old oceanic lithosphere of the eastern Mediterranean has been subducted below the Anatolian Plate at a fast rate (4–5 cm yr$^{-1}$) since the Late Miocene at least. Before the formation of the North Anatolian Fault the rate of convergence Africa–Eurasia was probably lower.

## Crustal collapse

The collapse of a thick and warm continental crust leads to widely distributed extension (Wernicke 1992). The Mediterranean back-arc basins make no exception to that rule. The striking difference between the Liguro-Provencal Basin, where extension is localized and oceanic crust formed in rather early stages and the Northern Tyrrhenian Sea, where a Basin-and-Range topography is observed, is due in large part, to the pre-extension crustal thickness (Faccenna *et al.* 1997; Jolivet *et al.* 1998). In the Tyrrhenian Sea oceanic crust formed in the south only in the latest stages of extension after extreme thinning of the continental crust. Post-orogenic extension lead to the exhumation of metamorphic rocks equilibrated in HT–LP conditions. Exhumation of blueschists and eclogites occurs within the accretionary complex.

Extension is even more widely distributed in the Aegean Sea and core complexes more spectacular than in the Tyrrhenian Sea, although the tectonic histories of the regions are quite similar. Crustal collapse was probably initiated after the transition from continental collision to oceanic subduction some 40 Ma ago, leading to a sudden crustal collapse of the entire thickened domain while the subduction started its southward retreat.

## Continental collision

Long-distance effects of continental collision are recorded both in the Japan and Aegean Seas. The recent propagation of the North Anatolian Fault in the Aegean Sea has led to a localization of strain along the fault and at its southwestern termination. The formation is this fault is due to the extrusion of the Anatolian block as a consequence of the Arabia–Eurasia collision (McKenzie 1972).

The geometry of opening of the Japan Sea can be explained as a consequence of large-scale strike-slip shear zones due to the indentation of India into Asia. The asymmetry of opening from the beginning of rifting to the end of spreading is a consequence of the India–Asia collision. This is possible only because a low-stress boundary condition to the east (Pacific subduction) allowed the propagation of strike-slip faults up to the margin. After the formation of the central Japan triple junction, deformation in the upper plate became controlled by the Pacific–Eurasia relative movements and east–west compression ensued.

We are indebted to Lydia Lonergan and John Platt who provided useful comments to prepare a better manuscript. A special thanks is due to Jean Paul Cadet and Renato Funiciello who considerably helped several of us during many years.

# References

ALTHERR, R., KREUZER, H., WENDT, I. *et al.* 1982. A Late Oligocene/Early Miocene high temperature belt in the anti-cycladic crystalline complex (SE Pelagonian, Greece). *Geologische Jachbuch*, **23**, 97–164.

AMATO, A., ALESSANDRINI, B., CIMINI, G., FREPOLI, A. & SELVAGGI, G. 1993. Active and remanant subducted slabs beneath Italy: evidence from seismic tomography and seismicity. *Annali di Geofisica*, **XXXVI**, 201–214.

ANDERSEN, T. B. 1998. Extensional tectonics in the Caledonides of southern Norway, an overview. *Tectonophysics*, **285**, 333–352.

—— & JAMTVEIT, B. 1990. Uplift of deep crust during orogenic extensional collapse: a model based on field studies in the Sogn–Sunnfjord region of Western Norway. *Tectonics*, **9**, 1097–1111.

ARMIJO, R., MEYER, B., KING, G. C. P., RIGO, A. & PAPANASTASSIOU, D. 1996. Quaternary evolution of the Corinth Rift and its implications for the Late Cenozoic evolution of the Aegean. *Geophysical Journal International*, **126**, 11–53.

AUSTRHEIM, H. 1987. Eclogitization of lower crustal granulites by fluid migration through shear zones. *Earth and Planetary Science Letters*, **81**, 221–232.

—— & BOUNDY, T. 1994. Pseudotachylites generated during seismic faulting and eclogitization of the deep crust. *Science*, **265**, 82–83.

——, ERAMBERT, M. & ENGVIK, A. K. 1997. Processing of crust in the root of the Caledonian continental collision zone: the role of eclogitization. *Tectonophysics*, **273**, 129–154.

AVIGAD, D. & GARFUNKEL, Z. 1989. Low-angle faults above and below a blueschist belt: Tinos Island, Cyclades, Greece. *Terra Nova*, **1**, 182–187.

——, BAER, G. & HEIMANN, A. 1999. Block rotations and continental extension in the Central Aegean Sea: paleomagnetic and structural evidence from Tinos and Mykonos. *Earth and Planetary Science Letters*, in press.

——, GARFUNKEL, Z., JOLIVET, L. & AZAÑÓN, J. M. 1997. Back-arc extension and denudation of Mediterranean eclogites. *Tectonics*, **16**, 924–941.

AVOUAC, J. P. & TAPPONNIER, P. 1993. Kinematic model of active deformation in central Asia. *Geophysics Research Letters*, **20**, 895–898.

BARTOLE, R. 1995. The North Tyrrhenian–Northern Apennines post-collisional system: constrain for a geodynamic model. *Terranova*, **7**, 7–30.

BERNARD, P., BRIOLE, P., MEYER, B. *et al.* 1997. The Ms = 6.2, June 15, 1995 Aigion earthquake (Greece): Evidence for low angle normal faulting in the Corinth rift. *Journal of Seismology*, **1**, 131–150.

BEVIS, M., TAYLOR, F. W., SHUTZ, B. E. *et al.* 1995. Geodetic observations of very rapid convergence and back-arc extension at the Tonga arc. *Nature*, **374**, 249–251.

BOSSIO, A., COSTANTINI, A., LAZZAROTTO, A. *et al.* 1993. Rassegna delle conoscenze sulla stratigrafia del Neautoctono Toscano. *Memoires de la Societe Geologique Italiana*, **49**, 17–98.

BOTT, M. H. P., WAGHORN, G. D. & WHITTAKER, A. 1989. Plate boundary forces at subduction zones and trench–arc compression. *Tectonophysics*, **170**, 1–15.

BOUSQUET, R., GOFFÉ, B., HENRY, P., LE PICHON, X. & CHOPIN, C. 1997. Kinematic, thermal and petrological model of the Central Alps: Lepontine metamorphism in the Upper Crust and eclogitisation of the lower crust. *Tectonophysics*, **273**, 105–128.

BRIAIS, A., PATRIAT, P. & TAPPONNIER, P. 1993. Updated interpretation of magnetic anomalies and seafloor spreading stages in the South China Sea: implications for the tertiary tectonics of SE Asia. *Journal of Geophysical Research*, **98**, 6229–6328.

BRUNET, C., MONIÉ, P. & JOLIVET, L. 1997. Geodynamic evolution of Alpine Corsica based on new $^{40}$Ar/$^{39}$Ar data. *Terra Nova, Special Issue EUG*, 493.

BUICK, I. S. 1991. Mylonite fabric development on Naxos, Greece. *Journal of Structural Geology*, **13**, 643–655.

BURRUS, J. 1984. Contribution to a geodynamic synthesis of the Provençal basin (north-western Mediterranean). *Marine Geology*, **55**, 247–269.

CARMIGNANI, L. & KLIGFIELD, R. 1990. Crustal extension in the northern Apennines: the transition from compression to extension in the Alpi Apuane core complex. *Tectonics*, **9**, 1275–1305.

——, DECANDIA, F. A., FANTOZZI, P. L., LAZZAROTTO, A., LIOTTA, D. & MECCHERI, M. 1994. Tertiary extensional tectonics in Tuscany (Northern Apennines, Italy). *Tectonophysics*, **238**, 295–315.

CARON, J. M. 1994. Metamorphism and deformation in Alpine Corsica. *Schweizerische Mineralogische and Petrographische Mitteilungen*, **74**, 105–114.

CHAMOT-ROOKE, N., GAULIER, J. M. & JESTIN, F. 1999. Constraints on Moho depth and crustal thickness in the Liguro–Provençal basin from a 3D gravity inversion: geodynamic implications. *Journal of the Geological Society, London*, in press.

——, RENARD, V. & PICHON, X. L. 1987. Magnetic anomalies in the Shikoku Basin, a new interpretation. *Earth and Planetary Science Letters*, **83**, 214–223.

CHEMENDA, A. I., MATTAUER, M., MALAVIEILLE, J. & BOKUN, A. N. 1995. A mechanism for syn-collision rock exhumation and associated normal faulting: results from physical modelling. *Earth and Planetary Science Letters*, **1995**, 225–232.

CHIARRABA, C. & AMATO, A. 1996. Crustal velocity structure of the Apennines (Italy) from P-wave travel time tomography. *Annali di Geofisica*, **39**, 1133–1148.

CHOULIARAS, G., VAN NGOC PHAM, BOYER, D., BERNARD, P. & STAVRAKAKIS, G. N. 1997. Crustal structure of the Gulf of Corinth in Central Greece, determined from magnetotelluric sounding. *Annali di Geofisica*, **50**, 61–67.

COLI, M., NICOLICH, R., PRINCIPI, G. & TREVES, B. 1991. Crustal delamination of the Northern Apennines thrust belt. *Bollettino della Società Geologica Italiana*, **110**, 501–510.

D'AGOSTINO, N., CHAMOT-ROOKE, N., FUNICIELLO, R., JOLIVET, L. & SPERANZA, F. 1998. The role of

pre-existing thrust faults and topography on the styles of extension in the Gran Sasso range (Central Italy). *Tectonophysics*, **292**, 229–254.

D'OFFIZI, S., MINELLI, G. & PIALLI, G. 1994. Foredeeps and thrust systems in the northern Apennines. *Bolletino di Geofisica Teorica ed Applicata*, **36**, 141–144.

DANIEL, J. M., JOLIVET, L., GOFFÉ, B. & POINSSOT, C. 1996. Crustal-scale strain partitionning: footwall deformation below the Alpine Corsica Oligo-Miocene detachement. *Journal of Structural Geology*, **18**, 41–59.

DAVY, P. & COBBOLD, P. R. 1988. Indentation tectonics in nature and experiments. Experiments scaled for gravity. *Bulletin of the Geological Institutions of Upsalla*, **14**, 129–141.

DE METS, C. 1992. A test of present-day plate geometries for northeast Asia and Japan. *Journal of Geophysical Research*, **12**, 17627–17636.

——, GORDON, R. G., ARGUS, D. F. & STEIN, S. 1990. Current plate motions. *Geophysical Journal International*, **101**, 425–478.

DELLA VEDOVA, B., MARSON, I., PANZA, G. F. & SUHADOLC, P. 1991. Upper mantle properties of the Tuscan–Tyrrhenian area: a framework for its recent tectonic evolution. *Tectonophysics*, **195**, 311–318.

DERCOURT, J., ZONENSHAIN, L. P., RICOU, L. E. *et al.* 1986. Geological evolution of the Tethys belt from the Atlantic to the Pamir since the Lias. *Tectonophysics*, **123**, 241–315.

DEWEY, J. F. 1980. Episodicity, sequence, and style at convergent plate boundaries. *In*: STANGWAY, D. W. (ed.) *The Continental Crust and its Mineral Deposits*. Geological Association of Canada Special Paper, **20**, 553–573.

——1988. Extensional collapse of orogens. *Tectonics*, 7, 1123–1139.

——, CANDE, S. & PITMAN, W. C. I. 1989*a*. Tectonic evolution of the India–Eurasia collision zone. *Eclogae Geological Helvetrae*, **82**, 717–734.

——, RYAN, P. D. & ANDERSEN, T. B. 1993. Orogenic uplift and collapse, crustal thickness, fabrics and metamorphic phase changes: the role of eclogites. *In*: PRICHARD, H. M., ALABASTER, T., HARRIS, N. B. W. & NEARY, C. R. (eds) *Magmatic Processes and Plate Tectonics*. Geological Society, London, Special Publication, **76**, 325–343.

——, HELMAN, M. L., TORCO, E., HUTTON, D. H. W. & KNOTT, S. D. 1989*b*. Kinematics of the Western Mediterranean. *In*: COWARD, M. P., DIETRICH, D. & PARK, R. G. (eds) *Alpine Tectonics*. Geological Society, Special Publication, **45**, 265–283.

DINTER, D. A. & ROYDEN, L. 1993. Late Cenozoic extension in northeastern Greece: Strymon valley detachment system and Rhodope metamorphic core complex. *Geology*, **21**, 45–48.

DVORKIN, J., NUR, A., MAVKO, G. & BEN-AVRAHAM, Z. 1993. Narrow subducting slab and the origin of back-arc basins. *Tectonophysics*, **227**, 63–79.

DZIEWONSKI, A. M., EKSTROM, G., FRANZEN, J. E. & WOODHOUSE, J. H. 1987. Global seismicity of 1979: centroid-moment tensor solutions for 524 earthquakes. *Physics of the Earth and Planetary Interiors*, **48**, 18–46.

ELTER, P., GIGLIA, G., TONGIORGI, M. & TREVISAN, L. 1975. Tensional and contractional areas in the recent (Tortonian to Present) evolution of the Northern Apennines. *Bollettino Geofisica Teoricata Applicata*, **17**, 3–18.

ENGEBRETSON, D. C., COX, A. & GORDON, R. G. 1985. Relative motions between oceanic and continental plates in the Pacific basin. *Geological Society of America Special Paper*, **206**, 59.

ENGLAND, P. & MOLNAR, P. 1990. Right-lateral shear and rotation as the explanation for strike-slip faulting in eastern Tibet. *Nature*, **344**, 140–142.

FACCENNA, C., MATTEI, M., FUNICIELLO, R. & JOLIVET, L. 1997. Styles of back-arc extension in the Central Mediterranean. *Terra Nova*, **9**, 126–130.

——, DAVY, P., BRUN, J. P., FUNICIELLO, R., GIARDINI, D., MATTEI, M. & NALPAS, T. 1996. The dynamic of backarc basins: an experimental approach to the opening of the Tyrrhenian Sea. *Geophysical Journal International*, **126**, 781–795.

FASSOULAS, C., KILIAS, A. & MOUNTRAKIS, D. 1994. Postnappe stacking extension and exhumation of high-pressure/low-temperature rocks in the island of Crete, Greece. *Tectonics*, **13**, 127–138.

FAURE, M., BONNEAU, M. & PONS, J. 1991. Ductile deformation and syntectonic granite emplacement during the late Miocene extension of the Aegean (Greece). *Bulletin de la Societe Geologique de France*, **162**, 3–12.

FLEITOUT, L. & MORICEAU, C. 1992. Short-wavelength geoid, bathymetry and the convective pattern beneath the Pacific ocean. *Geophysical Research International*, **110**, 6–28.

FORSYTH, D. W. & UYEDA, S. 1975. On the relative importance of driving forces on plate motion. *Geophysical Journal of the Royal Astronomical Society*, **43**, 163–200.

FOURNIER, M. 1994. *Collision continentale et ouverture des bassins marginaux, l'exemple de la Mer du Japon.* Thèse de Doctoriat, Université Pierre et Marie Curie.

——, JOLIVET, L. & FABBRI, O. 1995. Neogene stress field in SW Japan and mechanism of deformation during the Japan Sea opening. *Journal of Geophysical Research*, **12**, 24 295–24 314.

——, ——, HUCHON, P., ROZHDESTVENSKY, V. S., SERGEYEV, K. F. & OSCORBIN, L. 1994. Neogene strike-slip faulting in Sakhalin, and the Japan Sea opening. *Journal of Geophysical Research*, **99**, 2701–2725.

FREPOLI, A. & AMATO, A. 1997. Contemporaneous extension and compression in the Northern Apennines from earthquakes plane solutions. *Geophysical Journal International*, **129**, 368–388.

FRIZON DE LAMOTTE, D., ANDRIEUX, J. & GUÉZOU, J. C. 1991. Cinématique des chevauchements Néogènes dans l'arc bético-Rifains, discussion sur les modèles géodynamiques. *Bulletin de la Societe Geologique de France*, **4**, 611–626.

——, POISSON, A., AUBOURG, C. & TEMIZ, H. 1995. Chevauchements post-tortoniens vers l'ouest puis vers le sud au coeur de l'angle d'Isparta (Taurus, Turquie). Conséquences géodynamiques. *Bulletin de la Societe Geologique de France*, **166**, 59–67.

FUJITA, K. & NEWBERRY, J. T. 1982. Tectonic evolution of northeastern Siberia and adjacent regions. *Tectonophysics*, **89**, 337–357.

FYTIKAS, M., INNOCENTI, F., MANETTI, P., MAZZUOLI, R., PECCERILLO, A. & VILLARI, L. 1984. Tertiary to Quaternary evolution of volcanism in the Aegean region. *In*: DIXON, J. R. & ROBERTSON, A. H. F. (eds) *The Geological Evolution of the Eastern Mediterranean*. Geological Society, London, Special Publications, **17**, 687–699.

GAUTIER, P. & BRUN, J. P. 1994a. Crustal-scale geometry and kinematics of late-orogenic extension in the central Aegean (Cyclades and Evvia island). *Tectonophysics*, **238**, 399–424.

—— & ——1994b. Ductile crust exhumation and extensional detachments in the central Aegean (Cyclades and Evvia islands). *Geodinamica Acta*, **7**, 57–85.

——, —— & JOLIVET, L. 1993. Structure and kinematics of upper Cenozoic extensional detachement on Naxos and Paros (Cyclades Islands, Greece). *Tectonics*, **12**, 1180–1194.

HASSANI, R., JONGMANS, D. & CHÉRY, J. 1997. Study of plate deformation and stress in subduction process using two-dimensional numerical models. *Journal of Geophysical Research*, **102**, 17 951–17 966.

HATZFELD, D., BESNARD, M., MAKROPOULOS, K. & HATZIDIMITRIOU, P. 1993. Microearthquake seismicity and fault-plane solutions in the southern Aegean and its geodynamic implications. *Geophysical Journal International*, **115**, 799–818.

HENRY, P., LE PICHON, X. & GOFFÉ, B. 1997. Kinematic, thermal and petrological model of the Himalayas: constraints related to metamorphism within the underthrust Indian crust. *Tectonophysics*, **273**, 31–56.

HETZEL, R., PASSCHIER, C. W., RING, U. & DORA, O. O. 1995a. Bivergent extension in orogenic belts: the Menderes massif (southwestern Turkey). *Geology*, **23**, 455–458.

——, RING, U., AKAL, A. & TROESCH, M. 1995b. Miocene NNE-directed extensional unroofing in the Menderes massif, southwestern Turkey. *Journal of the Geological Society, London*, **152**, 639–654.

HILDE, T. W. C., UYEDA, S. & KROENKE, L. 1977. Evolution of the western Pacific and its margins. *Tectonophysics*, **38**, 145–165.

HOUDRY-LÉMONT, F. 1994. *Mécanismes de l'extension continentale dans le rift Nord-Baikal, Sibérie.* Thèse de Doctorat, Université Pierre et Marie Curie.

HUCHON, P. 1985. *Géodynamique de la zone de collision d'Izu et du point triple du Japon Central.* Thèse de Doctorat, Université Pierre et Marie Curie.

——, LE PICHON, X. & RANGIN, C. 1994. Indochina peninsula and the collision of India and Eurasia. *Geology*, **22**, 27–30.

INGLE, C. J., SUYEHIRO, K. & BREYMANN, M. V. 1990. *Initial Reports of the ODP.* College Station, TX.

ISHIKAWA, N. 1997. Differential rotations of north Kyushu island related to middle Miocene clockwise rotation of SW Japan. *Journal of Geophysical Research*, **102**, 17 729–17 745.

JACKSON, J. A. & WHITE, N. J. 1989. Normal faulting in the upper continental crust: observations from regions of active extension. *Journal of Structural Geology*, **11**, 15–36.

JOLIVET, L. & MIYASHITA, S. 1985. The Hidaka Shear Zone (Hokkaido, Japan): genesis during a right-lateral strike slip movement. *Tectonics*, **4**, 289–302.

—— & PATRIAT, M. 1999. Ductile extension and the formation of the Aegean Sea. *Journal of the Geological Society, London, special publication*, in press.

——, DANIEL, J. M. & FOURNIER, M. 1991*a*. Geometry and kinematics of ductile extension in alpine Corsica. *Earth and Planetary Science Letters*, **104**, 278–291.

——, DAVY, P. & COBBOLD, P. 1990*a*. Right-lateral shear along the northwest Pacific margin and the India–Eurasia collision. *Tectonics*, **9**, 1409–1419.

——, SHIBUYA, H. & FOURNIER, M. 1995. Paleomagnetic rotations and the Japan Sea opening. *In*: TAYLOR, B. & NATLAND, J. (eds) *Active Margins and Marginal Basins of the Western Pacific*. Geophysical Monograph, **88**, 355–369.

——, TAMAKI, K. & FOURNIER, M. 1994*c*. Japan Sea, opening history and mechanism, a synthesis. *Journal of Geophysical Research*, **99**, 22 237–22 259.

——, DANIEL, J. M., TRUFFERT, C. & GOFFÉ, B. 1994*b*. Exhumation of deep crustal metamorphic rocks and crustal extension in back-arc regions. *Lithos*, **33**, 3–30.

——, BRUN, J. P., GAUTIER, P., LALLEMANT, S. & PATRIAT, M. 1994*a*. 3-D kinematics of extension in the Aegean from the Early Miocene to the Present, insight from the ductile crust. *Bulletin de la Societe Geologique de France*, **165**, 195–209.

——, DUBOIS, R., FOURNIER, M., GOFFÉ, B., MICHARD, A. & JOURDAN, C. 1990*b*. Ductile extension in Alpine Corsica. *Geology*, **18**, 1007–1010.

——, FOURNIER, M., HUCHON, P., ROZHDESTVENSKIY, V. S., SERGEYEV, S. & OSCORBIN, L. S. 1992. Cenozoic intracontinental dextral motion in the Okhotsk-Japan Sea region. *Tectonics*, **11**, 968–977.

——, GOFFÉ, B., MONIÉ, P., TRUFFERT-LUXEY, C., PATRIAT, M. & BONNEAU, M. 1996. Miocene detachment in Crete and exhumation *P–T–t* paths of high pressure metamorphic rocks. *Tectonics*, **15**, 1129–1153.

——, HUCHON, P., BRUN, J. P., CHAMOT-ROOKE, N., LE PICHON, X. & THOMAS, J. C. 1991*b*. Arc deformation and marginal basin opening, Japan Sea as a case study. *Journal of Geophysical Research*, **96**, 4367–4384.

——, FACCENNA, C., GOFFÉ, B. *et al.* 1998. Mid-crustal shear zones in post-orogenic extension: the northern Tyrrhenian Sea case. *Journal of Geophysical Research*, **103**, 12 123–12 160.

JUN, M. S. 1990. *Source parameters of shallow intraplate earthquakes in and around the Korean peninsula and their tectonic implication.* Acta Universitatis Upsaliensis, Comprehensive Summaries of Uppsala Dissertation from the Faculty of Science, Uppsala.

KASHIK, S. A. & MAZILOV, V. N. 1994. Main stages and paleogeography of Cenozoic sedimentation in the Baikal rift system (Eastern Siberia). *Bulletin des de Recherches Exploration–Production Elf-Aquitaine*, **18**, 453–462.

KASTENS, K. A. & MASCLE, J. *et al.* 1990. The geological evolution of the Tyrrhenian Sea: an introduction to the scientific results of ODP Leg 107. *In*: KASTENS, K. A., MASCLE, J. *et al.* (eds) *Proceedings of the ODP, Scientific Results*, **107**, 3–26.

——, AUROUX, C. *et al.* 1988. ODP Leg 107 in the Tyrrhenian Sea: insight into passive margin and back-arc basin evolution. *Geological Society of America Bulletin*, **100**, 1140–1156.

KELLER, J. V. A., MINELLI, G. & PIALLI, G. 1994. Anatomy of late orogenic extension: the Northern Apennines case. *Tectonophysics*, **238**, 275–294.

KIMURA, G. & TAMAKI, K. 1986. Collision, rotation and back arc spreading: the case of the Okhotsk and Japan seas. *Tectonics*, **5**, 389–401.

——, MIYASHITA, S. & MIYASAKA, S. 1983. Collision tectonics in Hokkaido and Sakhalin. *In*: HASHIMOTO, M. & UYEDA, S. (eds) *Accretion Tectonics in the Circum Pacific Regions*. Terrapub, Tokyo, 117–128.

KING, G. & ELLIS, M. 1990. The origin of large local uplift in extensional regions. *Nature*, **348**, 689–693.

——, OUYANG, Z., PAPADIMITRIOU, P. *et al.* 1985. The evolution of the Gulf of Corinth (Greece) an aftershock study of the 1981 earthquake. *Geophysical Journal of the Astronomical Society*, **80**, 677–693.

KOSYGIN, Y. A. & SERGEYEV, K. F. 1992. *Geological map of Sakhalin, scale 1:500 000*. Vostlgeology, Yuzhno-Sakhalinsk, Russia.

KYRIAKOPOULOS, K., PEZZINO, A. & DEL MORO, A. 1988. Rb–Sr chronological, petrological and structural study of the Kavala Plutonic complex (N. Greece). *Bulletin of the Geological Society of Greece*, **23**, 545–560.

LALLEMAND, S. & JOLIVET, L. 1985. Japan Sea: a pull apart basin. *Earth and Planetary Science Letters*, **76**, 375–389.

LALLEMANT, S., TRUFFERT, C., JOLIVET, L., HENRY, P., CHAMOT-ROOKE, N. & VOOGD, B. D. 1994. Spatial transition from compression to extension in the western Mediterranean Ridge accretionary complex. *Tectonophysics*, **234**, 33–52.

LAVECCHIA, G. & STOPPA, F. 1989. Il 'rifting' tirrenico: delaminazione della litosfera continentale e magmatogenesi. *Bollettino dell Societa Geologica Italiana*, **108**, 219–235.

——, MINELLI, G. & PIALLI, P. 1987. Contractional and extensional tectonics along the transect Trasimeno Lake-Pesaro (Central Italy). *In*: BORIANI, A., BONAFEDE, M., PICCARDO, G. B. & VAI, G. B. (eds) *The lithosphere in Italy*. Advances in Earth Science Research, **80**, 139–142.

LE PICHON, X. 1981. Land-locked oceanic basins and continental collision, the eastern Mediterranean as a case example. Zurich. *In*: HSUE, K. J. (ed.) *Mountain Building Processes*. Academic Press, London, 201–211.

—— & ANGELIER, J. 1981. The Aegean Sea. *Philosophical Transactions of the Royal Society of London*, **300**, 357–372.

——, HENRY, P. & GOFFÉ, B. 1997. Uplift of Tibet: from eclogites to granulites – implications for the Andean Plateau and the Variscan Belt. *Tectonophysics*, **273**, 57–76.

——, CHAMOT-ROOKE, N., LALLEMANT, S. L., NOOMEN, R. & VEIS, G. 1994. Geodetic determination of the kinematics of Central Greece with respect to Europe: implications for eastern Mediterranean tectonics. *Journal of Geophysical Research*, **100**, 12 675–12 690.

LEE, J. S. & POUCLET, A. 1988. Le volcanisme néogène de Pohang (SE Corée), nouvelles contraintes géochronologiques pour l'ouverture de la mer du Japon. *Compte-Rendus de l'Académie des Sciences, Paris*, **307**, 1405–1411.

LISTER, G. S., BANGA, G. & FEENSTRA, A. 1984. Metamorphic core complexes of cordilleran type in the Cyclades, Aegean Sea, Greece. *Geology*, **12**, 221–225.

LOGATCHEV, N. A. 1993. History and geodynamics of the Lake Baikal rift in the context of the eastern Siberia rift system: a review. *Bulletin des Centres de Recherches Exploration – Production Elf-Aquitaine*, **17**, 353–370.

—— & ZORIN, Y. A. 1987. Evidences and causes of the two-stage development of the Baikal rift. *Tectonophysics*, **143**, 225–234.

MCKENZIE, D. 1972. Active tectonics in the Mediterranean region. *Geophysical Journal of the Royal Astronomical Society*, **30**, 109–185.

MCKENZIE, D. 1978. Active tectonics of the Alpine–Himalayan belt: the Aegean Sea and surrounding regions. *Geophysical Journal of the Royal Astronomical Society*, **55**, 217–254.

MALINVERNO, A. & RYAN, W. 1986. Extension in the Tyrrhenian sea and shortening in the Apennines as result of arc migration driven by sinking of the lithosphere. *Tectonics*, **5**, 227–245.

MARGHERITI, L., NOSTRO, C., COCCO, M. & AMATO, A. 1996. Seismic anisotropy beneath the Northern Apennines (Italy) and its tectonic implications. *Geophysical Research Letters*, **23**, 2721–2724.

MARTINI, I. P. & SAGRI, M. 1993. Tectono-sedimentary characteristics of Late Miocene–quaternary extensional basins of the Northern Apennines, Italy. *Earth Science Review*, **34**, 197–233.

MATTAUER, M., FAURE, M. & MALAVIEILLE, J. 1981. Transverse lineation and large scale structures related to Alpine obduction in Corsica. *Journal of Structural Geology*, **3**, 401–409.

MATTEI, M., KISSEL, C., SAGNOTTI, L., FUNICIELLO, R. & FACCENNA, C. 1996. Lack of Late Miocene to Present rotation of the northern Tyrrhenian margin (Italy): a constraint on geodynamic evolution. *In*: MORRIS, A. & TARLING, D. H. (eds) *Paleomagnetism and Tectonics of the Mediterranean Region*. Geological Society, London, Special Publication, **105**, 141–146.

MAUFFRET, A., PASCAL, G., MAILLARD, A. & GORINI, C. 1995. Tectonics and deep structure of the north-western Mediterranean basin. *Marine and Petroleum Geology*, **12**, 645–666.

MELE, G., ROVELLI, A., SEBER, D. & BARAZANGI, M. 1997. Shear wave attenuation in the lithosphere beneath Italy and surrounding regions: tectonic implications. *Journal of Geophysical Research*, **102**, 11 863–11 875.

MÉTIVIER, F. 1996. *Volumes sédimentaires et bilans de masse en Asie pendant le Cénozoïque*. Thèse de Doctorat, Université Denis Diderot.

MOLNAR, P. & GIPSON, J. M. 1996. A bound on the rheology of continental lithosphere using very long baseline interferometry: the velocity of south China with respect to Eurasia. *Journal of Geophysical Research*, **101**, 545–553.

—— & LYON-CAEN, H. 1988. Some simple physical aspects of the support, structure, and evolution of mountain belts. *Geological Society of America Special Paper*, **218**, 179–207.

MONTIGNY, R., EDEL, J. B. & THUIZAT, R. 1981. Oligo-Miocene rotation of Sardinia: K–Ar ages and paleomagnetism data of Tertiary volcanics. *Earth and Planetary Science Letters*, **54**, 261–271.

MORRIS, A. & ANDERSON, A. 1996. First paleomagnetic results from the Cyclaic Massif, Greece, and their implications for Miocene extension directions and tectonic models in the Aegean. *Earth and Planetary Science Letters*, **142**, 397–408.

MROZOWSKI, C. L. & HAYES, D. E. 1979. The evolution of the Parece Vela basin, eastern Philippine sea. *Earth and Planetary Science Letters*, **46**, 49–67.

NORTHRUP, C. J., ROYDEN, L. H. & BURCHFIEL, B. C. 1995. Motion of the Pacific plate relative to Eurasia and its potential relation to Cenozoic extension along the eastern margin of Eurasia. *Geology*, **23**, 719–722.

NORTON, I. O. 1995. Plate motions in the North Pacific: the 43 Ma non-event. *Tectonics*, **14**, 1080–1094.

OTOFUJI, Y. & MATSUDA, T. 1983. Paleomagnetic evidence for the clockwise rotation of Southwest Japan. *Earth and Planetary Science Letters*, **62**, 349–359.

——, HAYASHIDA, A. & TORII, M. 1985. When did the Japan sea open? paleomagnetic evidence from southwest Japan. *In*: NASU, N., UYEDA, S., KUSHIRO, I., KOBAYASHI, K. & HAGAMI, H. (eds) *Formation of Active Ocean Margin*. TERRAPUB, Tokyo, 551–556.

——, ITAYA, T. & MATSUDA, T. 1991. Rapid rotation of southwest Japan – paleomagnetism and K–Ar ages of Miocene volcanic rocks of southwest Japan. *Geophysical Journal International*, **105**, 397–405.

OTSUKI, K. & EHIRO, M. 1978. Major strike slip faults and their bearing on the spreading of the Japan Sea. *Journal Physical Earth, Supplement Issue*, **26**, 537–555.

PARFENOV, L. M., KOZ'MIN, B. M., IMAYEV, V. S. & SAVOSTIN., L. A. 1987. The tectonic character of the Olekma–Stanovoy seismic zone. *Geotectonics*, **21**, 560–572.

——, NATAPOV, L. M., SOKOLOV, S. D. & TSUKANOV, N. V. 1993. Terrane analysis and accretion in North-East Asia. *The Island Arc*, **2**, 35–54.

PARSON, L. M. & HAWKINS, J. W. 1994. Two-stage ridge propagation and the geological history of the Lau backarc basin. *In*: HAWKINS, J. PARSON, L. & ALLAN, J. (eds) *Proceedings of the ODP, Scientific Results*, **135**, 819–828.

PATACCA, E. & SCANDONE, S. 1989. Post-Tortonian mountain building in the Apennines. The role of the passive sinking of a relic lithospheric slab. *In*: BORIANI, A. BONAFEDE, M., PICCARDO, G. B. & VAI, G. B. (eds) *The Lithosphere in Italy. Advances in Science Research*. Accademia Nazionale dei Lincei, Rome, 157–176.

——, SARTORI, R. & SCANDONE, P. 1990. Tyrrhenian basin and Apenninic arcs: kinematic relations since late Tortonian times. *Memoire della Societa Geologica Italiana*, **45**, 425–451.

PATRIAT, M. 1996. *Etude de la transition cassant-ductile en extension, application au transect Olympe–Naxos, Grèce*. Thèse de Doctorat, Université Pierre et Marie Curie.

——, JOLIVET, L. & GOFFÉ, B. 1999. Evolution from thickened crust to metamorphic core complex during post-orogenic extension: direct observations from the Olympos–Naxos transect (Greece). *Tectonics*, in press.

PIALLI, G., ALVAREZ, W. & MINELLI, G. 1995. Geodinamica dell'Appenino settentrionale e sue ripercusioni nella evoluzione tettonica miocenica. *Studi Geologici Camerti, Volume Speciale*, **1995/1**, 523–536.

PLATT, J. P. & ENGLAND, P. 1994. Convective removal of lithosphere beneath moutain belt: thermal and mechanical consequences. *American Journal of Science*, **294**, 307–336.

—— & VISSERS, R. L. M. 1989. Extensional collapse of thickened continental lithosphere: A working hypothesis for the Alboran Sea and Gibraltar arc. *Geology*, **17**, 540–543.

PONZIANI, F., DE FRANCO, R., MINELLI, G., BIELLA, G., FEDERICO, C. & PIALLI, G. 1995. Crustal shortening and duplication of the Moho in the Northern Apennines: view from seismic refraction data. *Tectonophysics*, **252**, 391–418.

RANGIN, C., HUCHON, P., BELLON, H. *et al.* 1995. Cenozoic tectonics of central and South Vietnam. *Tectonophysics*, **251**, 179–196.

RANKEN, B., CARDWELL, R. K. & KARIG, D. E. 1984. Kinematics of the Philippine sea plate. *Tectonics*, **3**, 555–575.

REUTTER, K. J., GIESE, P. & CLOSS, H. 1980. Lithospheric split in the descending plate: observation from the Northern Apennines. *Tectonophysics*, **64**, T1–T9.

RICARD, Y., DOGLIONI, C. & SABADINI, R. 1991. Differential rotation between lithosphere and mantle: a consequence of lateral mantle viscosity variations. *Journal Geophysical Research*, **96**, 8407–8415.

RIEGEL, S. A., FUJITA, K., KOZ'MIN, B. M., IMAEV, V. S. & COOK, D. B. 1993. Extrusion tectonics of the Okhotsk plate, northeast Asia. *Geophysical Research. Letters*, **20**, 607–610.

RIGO, A., LYON-CAEN, H., ARMIJO, R. *et al.* 1996. A microseismicity study in the western part of the Gulf of Corinth (Greece): implications for large-scale normal faulting mechanisms. *Geophysical Journal International*, **126**, 663–688.

ROSSETTI, F., FACCENNA, C., JOLIVET, L., TECCE, F., FUNICIELLO, R. & BRUNET, C. 1999. Syn- versus post-orogenic extension in the Tyrrhenian Sea, the case study of Giglio Island (Northern Tyrrhenian Sea, Italy). *Tectonics*, in press.

ROYDEN, L. H. 1993. Evolution of retreating subduction boundaries formed during continental collision. *Tectonics*, **12**, 629–638.

ROZHDESTVENSKIY, V. S. 1982. The role of wrench faults in the structure of Sakhalin. *Geotectonics*, **16**, 323–332.

——1986. Evolution of the Sakhalin fold system. *Tectonophysics*, **127**, 331–339.

RUSSO, R. M. & SILVER, P. G. 1994. Trench-parallel flow beneath the Nazca plate from seismic anisotropy. *Science*, **263**, 1105–1111.

SAPIN, M. & HIRN, A. 1997. Seismic structure and evidence for eclogitization during the Himalayan convergence. *Tectonophysics*, **273**, 1–16.

SARTORI, R. 1990. The main results of ODP Leg 107 in the frame of neogene to Recent geology of peri-Tyrrhenian areas. *In*: KASTENS, K. A. MASCLE, J. *et al.* (eds) *Proceedings of ODP, Scientific Results*, **107**, 715–730.

SCHEEPERS, P. J. J., LANGEREIS, C. G. & HILGEN, F. J. 1993. Counter-clockwise rotations in the southern Apennines during the Pleistocene: paleomagnetic evidences from the Matera area. *Tectonophysics*, **225**, 379–410.

SCHOLZ, C. H. & CAMPOS, J. 1995. On the mechanism of seismic decoupling and back-arc spreading at subduction zones. *Journal of Geophysical Research*, **100**, 22 103–22 115.

SEBER, D., BARAZANGI, M., IBENBRAHIM, A. & DEMNATI, A. 1996. Geophysical evidence for lithospheric delamination beneath the Alboran Sea and Rif-Betic mountains. *Nature*, **379**, 785–790.

SEIDEL, E., KREUZER, H. & HARRE, W. 1982. The late Oligocene/early Miocene high pressure in the external hellenides. *Geologische Jachbuch*, **E23**, 165–206.

SELVAGGI, G. & AMATO, A. 1992. Intermediate-depth earthquakes in the Northern Apennines (Italy): evidence for a still active subduction? *Geophysics Researcj Letters*, **19**, 2127–2130.

—— & CHIARABBA, C. 1995. Seismicity and P-wave velocity image of the Southern Tyrrhenian subduction zone. *Geophysical Journal International*, **121**, 818–826.

SENO, T., STEIN, S. & GRIPP, A. E. 1993. A model for the motion of the Philippine Sea plate consistent with NUVEL-1 and geological data. *Journal of Geophysical Research*, **98**, 17 941–17 948.

SERRI, G., INNOCENTI, F. & MANETTI, P. 1993. Geochemical and petrological evidence of the subduction of delaminated Adriatic continental lithosphere in the genesis of the Neogene–Quaternary magmatsim of central Italy. *Tectonophysics*, **223**, 117–147.

SHEMENDA, A. I. 1994. *Subduction, Insight from Physical Modeling. Modern Approaches in Geophysics*. Kluwer.

SILLITOE, R. H. 1977. Metallogeny of an Andean type continental margin in South Korea, implications for opening of the Japan Sea. *In*: TALWANI, M. & PITMAN, W. C. I. (eds) *Island Arcs, Deep Sea Trenches and Back Arc Basins*. AGU, Maurice Ewing Series, **1**, 303–310.

SOKOUTIS, D., BRUN, J. P., DRIESSCHE, J. V. D. & PAVLIDES, S. 1993. A major Oligo-Miocene detachment in southern Rhodope controlling north Aegean extension. *Journal of the Geological Society, London*, **150**, 243–246.

SPAKMAN, W. 1990. Tomographic images of the upper mantle below central Europe and the Mediterranean. *Terra Nova*, **2**, 542–553.

TAIRA, A., TOKUYAMA, H. & SOH, W. 1989. Accretion tectonics and evolution of Japan. *In*: BEN AVRAHAM, Z. (ed.) *The Evolution of the Pacific Ocean Margin*. Oxford University Press, 100–123.

TAMAKI, K. 1986. Age estimation of the Japan sea on the basis of stratigraphy, basement depth and heat flow data. *Journal of Geomagnetics and Geoelectrics*, **38**, 427–446.

——1988. Geological structure of the Japan sea and its tectonic implications. *Bulletin of the Geological Survey of Japan*, **39**, 269–365.

——, SUYEHIRO, K., ALLAN, J., INGLE, J. C. & PISCIOTTO, K. 1992. Tectonic synthesis and implications of Japan Sea ODP drilling. *Proceedings of ODP, Scientific Results*, **127/128**, 1333–1350.

TAPPONNIER, P. & MOLNAR, P. 1976. Slip line field theory and large-scale continental tectonics. *Nature*, **264**, 319–324.

——1977. Active faulting and tectonics in China. *Journal of Geophysical Research*, **82**, 2905–2930.

——, PELTZER, G., DAIN, A. Y. L., ARMIJO, R. & COBBOLD, P. 1982. Propagating extrusion tectonics in Asia: new insights from simple experiments with plasticine. *Geology*, **10**, 611–616.

TARDUNO, J. A. & COTTRELL, R. D. 1997. Paleomagnetic evidence for motion of the Hawaiian hotspot during formation of the Emperor seamounts. *Earth Planetary Science Letters*, **153**, 171–180.

TATSUMI, Y., OTOFUJI, Y. I., MATSUDA, T. & NOHDA, S. 1989. Opening of the Sea of Japan back-arc basin by asthenospheric injection. *Tectonophysics*, **166**, 317–329.

TAYLOR, B. 1992. Rifting and the volcanic-tectonic evolution of the Izu–Bonin–Mariana arc. *In*: TAYLOR, B. & FUJIOKA, K. (eds) *Proceedings of the ODP Scientific Results*, **126**, 627–651.

—— & HAYES, D. E. 1983. Origin and history of the South China Sea basin. *In*: HAYES, D. E. (ed.) *The Tectonic and Geologic Evolution of Southeast Asian Seas and Islands, Part 2*. AGU, Geophysical Monograph, **27**, 23–56.

—— & KARNER, G. D. 1983. On the evolution of marginal basins. *Review of Geophysics and Space Physics*, **21**, 1727–1741.

TAYMAZ, T., JACKSON, J. & MCKENZIE, D. 1991. Active tectonics of the north and central Aegean Sea. *Geophysical Journal International*, **106**, 433–490.

TOKUYAMA, H., SUYEMASU, M., TAMAKI, K. *et al.* 1987. Report on DELP cruises in the Japan Sea, part III: seismic reflection studies in the Yamato basin and the Yamato Rise area. *Bulletin Earthquake Research Institute, University of Tokyo*, **62**, 367–390.

UYEDA, S. & KANAMORI, H. 1979. Backarc opening and the mode of subduction. *Journal of Geophysical Research*, **84**, 1049–1106.

VANDENBERG, L. C. & LISTER, G. S. 1996. Structural analysis of basement tectonics from the Aegean metamorphic core complex of Ios, Cyclades, Greece. *Journal of Structural Geology*, **18**, 1437–1454.

VIGLIOTTI, L. & KENT, D. V. 1990. Paleomagnetic results of Tertiary sediments from Corsica: evidence for post-Eocene rotation. *Physics of the Earth and Planetary International*, **62**, 97–108.

VISSERS, R. L. M., PLATT, J. P. & VAN DER WAL, D. 1995. Late orogenic extension of the Betic Cordillera and the Alboran domain: a lithospheric view. *Tectonics*, **14**, 786–803.

WAWRZENITZ, N. & KROHE, A. 1999. Exhumation and doming of the Thasos metamorphic core complex (S'Rhodope, Greece): structural and geochronological constraints. *In*: JOLIVET, L. & GAJAIS, D. (eds) *Extensional Tectonics and Exhumation of Metamorphic Rocks in Mountain Belts*. Tectonophysics Special Volume, **285**, 301–332.

WERNICKE, B. 1992. Cenozoic extensional tectonics of the, U.S. cordillera. *In*: BURCHFIEL, B. C., LIPMAN, P. W. & ZOBACK, M. L. (eds) *The Cordilleran Orogen: Conterminous, U.S.* Geological Society of America, Boulder, Colorado, **G-3**, 553–581.

WIGGER, J. 1984. Die Krustenstruktur des NordApennines und angereurender Gebeite mit besonder Bernecksichtigung der geotermischen Anomalie der Toskana. *Berliner Geowissenschaftliche Abbendlungen*, **9**, 1–87.

WIJBRANS, J. R., VAN WEES, J. D., STEPHENSON, R. A. & CLOETHINGH, S. A. P. L. 1993. Pressure–temperature–time evolution of the high-pressure metamorphic complex of Sifnos, Greece. *Geology*, **21**, 443–446.

WORRALL, D. M. 1991. Tectonic history of the Bering Sea and the evolution of Tertiary strike-slip basins of the Bering shelf. *Geological Society of America, Special Paper*, **257**, 120.

——, KRUGLYAK, V., KUNST, F. & KUSNETSOV, V. 1996. Tertiary tectonics of the Sea of Okhotsk, Russia: far-field effects of the India-Asia collision. *Tectonics*, **15**, 813–826.

YOON, S. H., PARK, S. J. & CHOUGH, S. K. 1997. Western boundary fault systems of Ulleung Back-arc basin: further evidence of pull-apart opening. *Geoscience Journal*, **1**, 75–88.

ZHONG, S. & GURNIS, M. 1995. Mantle convection with plates and mobile, faulted margins. *Science*, **267**, 838–843.

ZITELLINI, N., TRINCARDI, F., MARANI, M. & FABBRI, A. 1986. Neogene tectonics of the Northern Tyrrhenian sea. *Giornale di Geologia*, **48/1-2**, 25–40.

ZOBACK, M. L. 1992. First- and second-order patterns of stress in the lithosphere: the world stress map project. *Journal of Geophysial Research*, **97**, 11 703–11 728.

# Petrography of Ordovician and Silurian sediments in the western Irish Caledonides: tracers of a short-lived Ordovician continent–arc collision orogeny and the evolution of the Laurentian Appalachian–Caledonian margin

JOHN DEWEY & MARIA MANGE

*Department of Earth Sciences, University of Oxford, Parks Road, Oxford OX1 3PR, UK*

**Abstract:** Ordovician orogeny affected the Laurentian margin of the Appalachian–Caledonian Belt from the Southern Appalachians to the British Isles and is dated, stratigraphically, as post-uppermost Lower Cambrian and pre-upper Llandovery. Geochronological data favour a short Grampian orogeny from *c.* 470 to 460 Ma during the late Arenig–mid-Llanvirn, which is supported by the late Arenig–earliest Llanvirn termination of the Laurentian rifted margin carbonate shelf. The likeliest plate tectonic model for the Grampian Orogeny is of a continent-facing oceanic (Laurentia) arc, created as an infant mafic arc during the Middle Cambrian, evolving into an Early Ordovician intermediate-silicic arc with suprasubduction-zone ophiolites. This arc collided with the Laurentian margin and forearc ophiolites were obducted across the margin creating an orogen and accreting the arc to the margin. Continued plate convergence was accommodated by a flip in subduction polarity that terminated late orogenic retrocharriage, and led to the rapid final stages of uplift and unroofing of the orogen and to subduction accretion in the new mid-Ordovician trench.

Presented in this paper are definitive new detrital heavy mineral evidence that supports and enhances this model of a short-lived orogenic event involving ophiolite obduction, and the rapid development and unroofing of a Grampian Barrovian metamorphic complex over *c.* 10 Ma. High-resolution heavy mineral analysis (HRHMA) has been undertaken on most of the Ordovician and Silurian stratigraphical units in the South Mayo Trough in western Ireland. It is shown that ophiolite unroofing began during the Arenig and that, by the early Llanvirn, a Barrovian complex was being eroded. The Killadangan 'Formation', of the Clew Bay Complex is confirmed as part of a pre-Grampian early Ordovician accretionary prism by both HRHMA and by quartz–felspar–lithic (QFL) analysis, which show it to have been derived from Precambrian rocks of the Laurentian margin. QFL analysis also shows that the Ordovician provenance of the South Mayo Trough sequence evolves from an undissected arc through a dissected arc to a recycled orogenic detrital pattern, except for the earliest sediments (Letterbrock), which were derived from the transitional continental sediments of the Killadangan accretionary prism, which, in turn, were derived from Laurentia. The problem of timing the Grampian deformation and metamorphism of the Dalradian is now regarded as being solved. It was an Arenig–Llanvirn event lasting *c.* 10 Ma, which occurred during and is recorded, faithfully, by the detrital heavy mineral assemblage in the conformable Ordovician sequence of the South Mayo Trough. The Miocene evolution of New Guinea is strikingly similar to the Arenig–Llanvirn evolution of the Laurentian margin and was, analogously, the short-lived result of the collision of an arc with the north Australian margin followed by subduction polarity flip.

The British Caledonides have been divided (Dewey 1969) into a northern 'orthotectonic' zone, between the Moine Thrust and the Highland Boundary Fault (Fig. 1a), with very complex structures and a southern paratectonic zone consisting mainly of low-grade or anchimetamorphic Ordovician and Silurian rocks with steeply dipping slaty cleavages in sinistral polyphase transpression zones (Soper *et al.* 1992). Most deformation in the paratectonic zone occurred, uncontentiously, during Late Silurian–Early Devonian times and by diachronous Silurian subduction accretion in the Southern Uplands, but the precise age of deformation and metamorphism in the orthotectonic zone has been argued keenly and there is no general agreement about either its timing or the tectonic model responsible for it. The orthotectonic zone forms the northwest boundary of the Appalachian–Caledonian Belt from the British

*From*: MAC NIOCAILL, C. & RYAN, P. D. (eds) 1999. *Continental Tectonics*. Geological Society, London, Special Publications, **164**, 55–107. 1-86239-051-7/99/$15.00 © The Geological Society of London 1999.

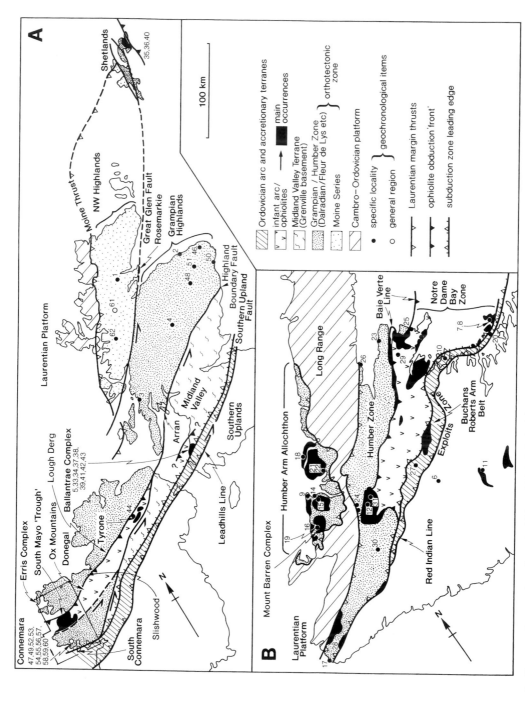

**Fig. 1.** Outline tectonic map of northwest British and Irish Caledonides (**a**) and the northwest Newfoundland Appalachians. Numbered localities refer to geochronological items recorded in Appendix 3.

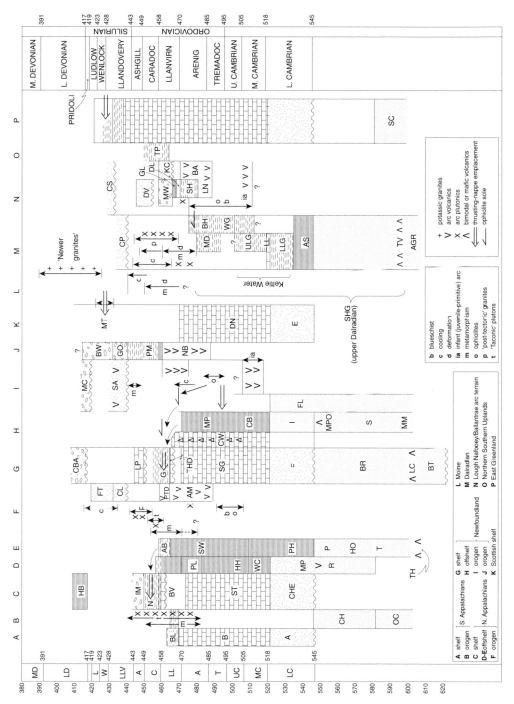

**Fig. 2.** Outline tectonostratigraphies of the Appalachians and Caledonides. Data from: Williams & Stevens (1969); Bird & Dewey (1971); Harris *et al.* (1975); Hatcher & Odom (1980); Hurst & McKerrow (1981); Dewey & Shackleton (1984); Drake *et al.* (1989); Graham *et al.* (1989); Rankin *et al.* (1989); Surlyk (1991) and Van Staal *et al.* (1999). Key to abbreviation in Appendix 4.

Isles (Fig. 1a) through Newfoundland (Fig. 1b) to the Southern Appalachians, in which the deformed rocks are variously termed Moine/Dalradian (British Isles), Fleur de Lys (Newfoundland), Pinnacle/Hoosac/Tyson (Western New England) and Chilhowee/Ocoee (Southern Appalachians) (Fig. 2). Ordovician orogeny in the zone is referred to as Grampian (British Isles), Humberian (Newfoundland) and Taconian (Western New England). These strata are of late Proterozoic (Riphean–Vendian) age and accumulated in continental rifts overlain by Cambrian–Early Ordovician continental margin sediments. In the Appalachians, the age and cause of the principal deformation in the orthotectonic zone is generally agreed to have been a Middle Ordovician (Fig. 2) collision of the rifted Laurentian margin with a continent-facing arc(s) (Karson & Dewey 1978; Rowley & Kidd 1981; Casey & Dewey 1984; Karabinos *et al.* 1998). This general agreement results from the relative clarity of relationships in the Appalachians where Ordovician deformation may be directly linked with the obduction of suprasubduction-zone ophiolites of a primitive oceanic arc, with subjacent continental margin thrust sheets, across the orthotectonic zone onto the Laurentian shelf (Fig. 1b), with an associated clear stratigraphic record (Fig. 2) (Van Staal *et al.* 1999). Even where ophiolites are absent, klippen of continental margin rocks lie in well-dated middle Ordovician foreland basins on the old Cambro-Ordovician shelf (e. g. Hamburg and Taconic Klippen). In the British Isles, the picture is less clear, principally because major Ordovician–Silurian sinistral faulting, probably with many hundreds of kilometres of relative motion, has partially obscured the original Ordovician tectonic pattern with repetition (Connemara Dalradian) and excision (along the Highland Boundary Fault Zone) (Fig. 1a) of tectonic elements (Dewey & Shackleton 1984; Hutton & Dewey 1986). One cannot look at Scotland and Ireland alone; they are part of a long orogen and other segments, such as Newfoundland, provide a clear template against which the heavily terraned British Isles can be judged and interpreted. Along-strike comparisons are important for the understanding of any orogenic belt with the *caveats* that tectonic patterns may change profoundly at old triple junctions, and that an apparent semi-linear long-distance continuation of zones and elements may be the result of intense flattening and straightening during terminal continent–continent collision. However, the Newfoundland orthotectonic zone data are so clear and relatively unequivocal that they can be used as a broad template for

the orthotectonic zone in the British Isles, because most of the Appalachian elements are present, although in a faintly or severely confused condition.

The British and Irish Caledonides occupy a key position between the Appalachian orogen of western North America to the south and the Caledonides of East Greenland and Scandinavia to the north. They are, probably, the most intensively researched part of any of the world's orogenic belts. Despite, perhaps because of, this and because the problems of this data-rich segment have been argued in geographic isolation, fundamental interpretative problems have arisen. The view persists that the British Isles orthotectonic zone is fundamentally different from that in the Appalachians in having one or more post-Grenville contractional orogenic events before the Ordovician Grampian Orogeny (Lambert 1969; Van Breemen *et al.* 1974; Brook *et al.* 1976, 1977; Piasecki & Van Breemen 1979; Powell *et al.* 1983; Barr *et al.* 1985; Bowes 1986; Friend *et al.* 1995; Rogers *et al.* 1995*a, b*; Noble *et al.* 1996, 1997). An extreme position is that the Moine/Dalradian block is a terrane with Precambrian Grampian orogeny wholly exotic to Laurentia (Bluck & Dempster 1991). This, we believe, cannot be correct for three reasons: (1) the Leny Limestone, an integral part of the Dalradian (Harris 1969; Tanner 1995), has a Laurentian fauna (Fletcher 1989); (2) Ordovician arcs with Laurentian faunal affinities (Williams 1976; Van Staal *et al.* 1998) and ophiolites are outboard of the Grampian terranes; (3) Dalradian rocks have an autochthonous relationship with Grenville basement in western Ireland (Winchester 1992). In contrast, we believe that the British and Irish orthotectonic zone data can be interpreted, fairly precisely, in the same way as those of Newfoundland, that the first contractional orogenic event was of Middle Ordovician age (Dewey *et al.* 1998) and was generated by the collision of the Laurentian margin of the Iapetus Ocean with a continent-facing primitive oceanic arc (Casey & Dewey 1984; Dewey & Shackleton 1984; Dewey & Ryan 1990). Ordovician deformation of the orthotectonic zone does not continue into East Greenland, where the first Phanerozoic contractional orogenic deformation occurred during the mid-Silurian. This event (Scandian) was pervasive in the Scandinavian Caledonides and East Greenland, generated a final phase of thrusting on the Moine Thrust and was the result of the collision of Baltica with Laurentia. It does not appear to have been a substantial orogenic event in the Appalachians, although it has been claimed, we believe incorrectly, that the principal orthotectonic zone contractional deformation in Newfoundland occurred during the Silurian (Cawood *et al.* 1994).

The geometry and stratigraphy of basins in and adjacent to orogenic belts is central to developing models for the tectonic evolution of those belts. Furthermore, the sedimentary petrography of their stratigraphic sequences, as expressed in clast size, distribution and composition, is a guide to general tectonic environment (Dickinson *et al.* 1983), provenance, and unroofing geometries and rates. Of particular importance are heavy mineral suites and their geochemistry (Mange & Maurer 1992), especially when combined with the quartz and feldspar types. In this paper, the stratigraphy, clast composition and heavy mineral suites of Ordovician and Silurian rocks in the Caledonides of western Ireland are described, in particular to address the problems of the age, rate and tectonic nature of Grampian orogenic event in Dalradian rocks on the margin of Laurentia, the rate of its unroofing, the palaeogeography and palaeotectonics of western Ireland, and the general evolution of the Appalachian–Caledonian orogenic belt.

The guiding principles, in this and any other analysis of a piece of regional geology, are as follows. First, a narrow, purely local geological approach to solving tectonic problems is doomed; one must look at a very broad regional scale to make analyses, contrasts and comparisons. Secondly, plate tectonic palaeogeography/palaeotectonics is a major key to frame and understand more local geological relationships. Thirdly, to understand palaeotectonics, especially of pre-Mesozoic age, the Lyellian principle is followed, i.e. one must use modern tectonic analogues and analyse and understand modern and recent tectonic patterns according to basic plate tectonic principles, at least during the Phanerozoic. *Ad hoc* non-actualistic models have no relevance for understanding old orogenic belts. Above all, models and data must be consistent and make sense. Fourthly, it is important not to take geochronological, palaeomagnetic and, indeed, any published geological data at face value. Many Rb/Sr 'pseudochrons' and some K–Ar dates are demonstrable nonsense, have dogged tectonic analyses for far too long and should be discarded. No geochronological system or method is perfectly established and unequivocal. Cited as of little value is the 490 Ma date for the Cashel Intrusion in Connemara (Jagger *et al.* 1988), the 475 Ma date for the Slieve-Gamph Intrusion in Mayo (Pankhurst *et al.* 1976), the Ordovician ages for the Donegal Granite Suite (Leggo *et al.* 1969) and

the 510–500 Ma ages for the slates of the Dalradian margins (Dewey & Pankhurst 1970) in so far as they are supposed to date specific, defined geological events. These numbers are saying something but the message is unclear. The established relationships of basic geology and stratigraphy are the only starting points in tectonic analysis. In the ensuing analysis, we have used, almost exclusively, zircon, monazite and titanite (sphene) ages with some K–Ar and $^{40}Ar/^{39}Ar$, and very few Rb/Sr numbers (catalogued in Appendix 3). We take the Popperian, rather than the Baconian, scientific methodology that tectonic thinking, logic and model building is the central and critical component of understanding regional tectonics, provided, of course, that such model building conforms with clearly established geological data. The big achievement of plate tectonics has been the use of better understood modern analogues and processes in building models for the past. Put another way, science does not advance by merely assimilating facts, so-called facts or pseudofacts but, rather, by paradigm building controlled by established facts and models that can be falsified. In geological arguments fraught with opinion and ideas, like the problem of the age and origin of the deformation and metamorphism of the Dalradian, the Popperian methodology is critical. Consequently, we, unashamedly, make judgements about the quality of published data in a broader framework of tectonic understanding, continuity and rationalism. Especially, we do not take any timescale [e.g. Tucker & McKerrow (1995)], into which geochronological and stratigraphic data is slotted, as established. Timescales evolve and improve by continual iteration.

## Outline of the western Irish Caledonides

In Connemara and west Mayo, Dalradian and fossiliferous Ordovician and Silurian strata are unconformably overlain by Lower Carboniferous fluviatile red sediments beneath marine limestones (Figs 3 and 4). North of Clew Bay, a Dalradian metamorphic complex unconformably overlies a Grenville basement and is in tectonic contact, to the south, with the Clew Bay Complex in Achill Island and in the Ox Mountains. In Connemara, a Dalradian metamorphic complex ('Connemara Schists') is overlain unconformably to the north by Silurian strata and intruded, to the south, by a late Silurian–Devonian granite plutonic complex (Galway Granite). Between the Mayo and Connemara Dalradian 'blocks' lies the South Mayo Trough (Dewey 1963), which consists of a central zone

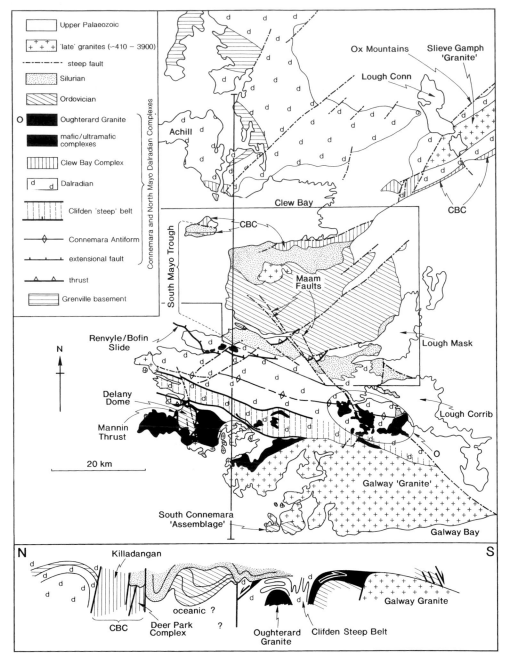

**Fig. 3.** Outline geological/tectonic map of west Galway and Southwest Mayo with north–south section. Simplified after Leake & Tanner (1984); Graham *et al.* (1989); Long *et al.* (1995).

of Ordovician rocks with strips of Silurian strata to the north and south. The northern (Croagh Patrick) and southern (Killary/Kilbride) Silurian sequences rest unconformably upon the lowest parts of the Ordovician sequence so that, clearly, the Ordovician rocks lay in a broad synclinorium prior to Silurian deposition. The Silurian also rests unconformably upon both the Connemara Dalradian and the Clew Bay Complex, thus obscuring their relationship with the

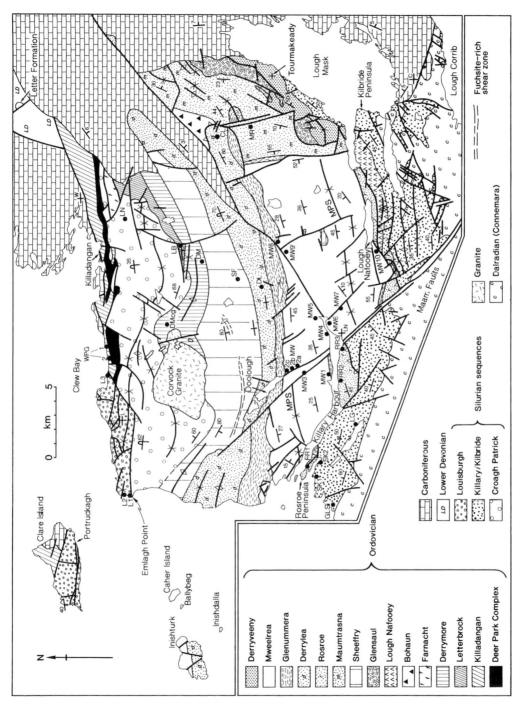

**Fig. 4.** Simplified geological map of the South Mayo Trough. Simplified and modified from Graham *et al.* (1989). Large labelled dots indicate localities and formations for which samples were collected for high-resolution heavy mineral analysis.

Ordovician but, clearly, major faults must exist below the Silurian.

The Connemara Dalradian terrain may be considered, conveniently, in four zones (Fig. 3). First, a northern low-grade zone of Southern Highland Group Dalradian strata, with syntectonic mafic/ultramafic intrusions (Dawros, Currywongaun, Doughruagh) separated by a low-angle extensional slide (Renvyle–Bofin Slide) from, secondly, the main Dalradian outcrop, a zone of uppermost Appin Group, Argyll Group and lowermost Southern Highland Group *c*. 5 km thick, arranged in north verging recumbent folds, affected by a Barrovian metamorphism with, variably, sillimanite, kyanite, andalusite, staurolite (Yardley 1976); these rocks were folded into the Connemara antiform and intruded by the late to post-tectonic Oughterard Granite suite. Thirdly, to the south, fold axial surfaces steepen into the Clifden Steep Belt, which contains syntectonic mafic intrusions (Cashel/Lough Wheelaun) similar to those of the northern zone. Fourthly, to the south, the steep dips of the Clifden Steep Belt roll over into gentler northward dips in a zone of sillimanite-grade semipelites injected by huge volumes of syntectonic gabbros and ultramafics (Errisbeg Complex) and quartz diorites (Leake 1986 1989), which was thrust (Mannin Thrust) south-southeastwards across silicic volcanics, probably of early Ordovician age, exposed in the Delany Dome (Leake *et al.* 1983).

The Dalradian rocks of west Mayo and the Ox Mountains are a Barrovian metamorphic assemblage in a region of dextral transpression with higher pressure phengite- and pyrope-bearing assemblages to the south near the contact with the Clew Bay Complex (Harris 1995; Harris & Harris 1996). The Clew Bay Complex consists of a dismembered ophiolite complex (Deer Park Complex) to the south (Dewey & Ryan 1990; Ryan & Dewey 1991) and the low-grade/anchimetamorphic Killandangan Formation/Ballytoohy Series (Appendix 1) to the north, along the southern shores of Clew Bay, on Clare Island and in the southern Ox Mountains, and a disrupted assemblage of albite schists, crossite-bearing mafic rocks, graphitic schists and ultramafic melanges. The contact between the Clew Bay Complex and the Dalradian in Claggan Bay in South Achill is sharp and consists of intensely sheared and interposed flaggy quartzites in black schists.

The distribution of the Ordovician rocks of the South Mayo Trough is shown in Fig. 4, their age and correlation in Fig. 5, and their lithology in Appendix 1. Figure 6 illustrates the correlation, with thickness and facies changes, of the Ordovician of Central Murrisk on the north limb of the Mweelrea/Partry Syncline with the Ordovician of Rosroe, Leenane, Lough Nafooey and Tourmakeady on it's southern limb.

The Silurian rocks are arranged in three discrete sequences (Figs 4 and 7), whose lithologies are summarized in Appendix 2. The Llandovery Croagh Patrick succession and the upper Llandovery–Wenlock Killary Harbour/Kilbride succession were deposited unconformably across Ordovician, Dalradian and Clew Bay Complex, and were deformed with the Ordovician in a Late Silurian low-grade sinistral transpression zone, whereas the ?Ludlow/Pridoli Louisburgh succession was deposited in a strike-slip pull-apart basin during Late Silurian transpression of the South Mayo Trough.

The relationship between the Dalradian of Connemara and Mayo, and the fossiliferous Ordovician strata of Central Murrisk has been a difficult and contentious problem, because a contact is not exposed, yet it is central to understanding the nature and timing of the Grampian Orogeny. Dewey (1961) considered the Dalradian deformation and metamorphism to have been pre-Arenig, principally because the Ordovician of the South Mayo Trough is anchimetamorphic to low grade, with a moderately simple structural style and sequence with steep cleavages, whereas the adjacent Dalradian has Barrovian metamorphic assemblages and a more complicated polyphase structural history involving early recumbent fold nappes. However, another particular reason is that the upper part of the Derrylea Group contains detrital staurolite (Dewey 1961) and is part of a conformable stratigraphical sequence that passes down into Arenig rocks (Figs 5 and 6). It was Dewey's (1961) view that the detrital staurolite was derived from the Connemara Dalradian and, therefore, that the age of deformation and metamorphism of the Connemara Dalradian must have been earlier than the oldest Tremadoc–Arenig part of the South Mayo Trough Ordovician sequence. This view was challenged by Phillips *et al.* (1976) who argued, now known correctly, a mid-Ordovician age for the Grampian event, which was first suggested by Kennedy (1955 1958) for the Scottish Highlands.

Dewey & Shackleton (1984) related the Grampian event to the northward obduction of a nappe of ophiolitic and accretionary prism rocks across the Laurentian continental margin, and Dewey & Ryan (1990) developed a model in which the Ordovician strata of the South Mayo Trough in Central and Eastern Murrisk (north limb of the Mweelrea/Partry Syncline) accumulated in a fore-arc basin to the intra-oceanic arc

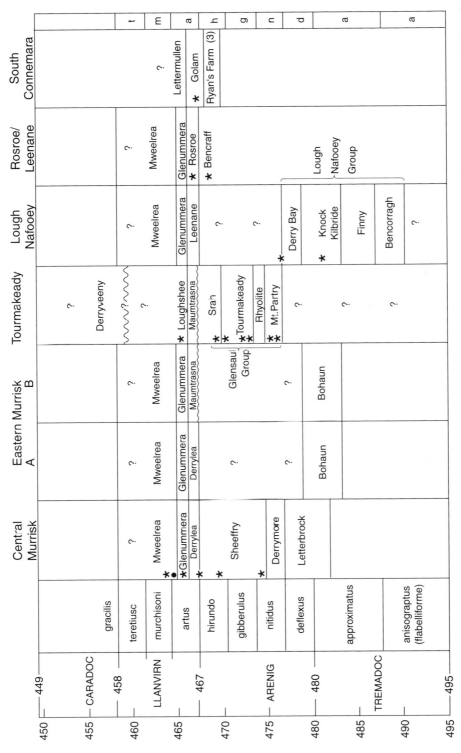

**Fig 5.** Correlation chart of the Ordovician stratigraphy of the South Mayo Trough. Stars indicate fossil control; timescale after Tucker & McKerrow (1995). Data from: McKerrow & Campbell (1960); Stanton (1960); Dewey (1963); Dewey *et al.* (1970); McManus (1972); Archer (1977); Ryan & Archer (1977); Graham (1987); Harper *et al.* (1988); Graham *et al.* (1989) and Long *et al.* (1995). See Appendix 1 for lithologies.

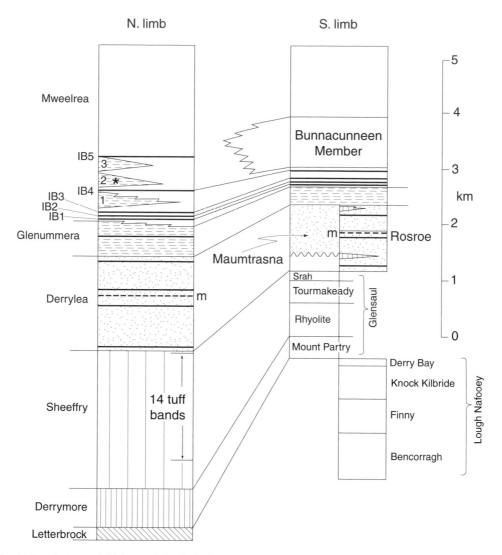

**Fig. 6.** Correlation and thickness of the Ordovician strata of the north and south limbs of the Mweelrea–Partry Syncline. Thick lines, intermediate to silicic tuff bands and silicic ignimbrites (IB 1–5); dashed thick line (m), horizon in Derrylea and Rosroe, the formations at which the metamorphic detrital 'flood' occurs; thin dashed lines, mudrocks and siltstones.

represented by the Lough Nafooey and Glensaul Groups (Figs 5 and 6). The Clew Bay Complex was seen by Dewey & Ryan (1990) as a forearc accretionary prism (Ballytoohy/Killadangan/South Achill melanges) with an ophiolitic back stop (Deer Park Complex). It was argued by Dewey & Shackleton (1984), Hutton & Dewey (1986) and Dewey & Ryan (1990) that the Connemara Dalradian massif was sinistrally transported by hundreds of kilometres, as a terrane, prior to the early Silurian, thus causing repetition of the Dalradian/collided arc relationship.

## Heavy mineral analysis of the South Mayo Trough

In this paper, the detrital mineralogy of the Ordovician of the South Mayo Trough is used to characterize the origin, evolution, timing, duration and unroofing of the Grampian Orogen in the orthotectonic zone. Understanding the sediment composition of basin-fill sequences, attached to a past orogenic hinterland, is essential for the reconstruction of orogenic evolution. Because tectonic movements produce

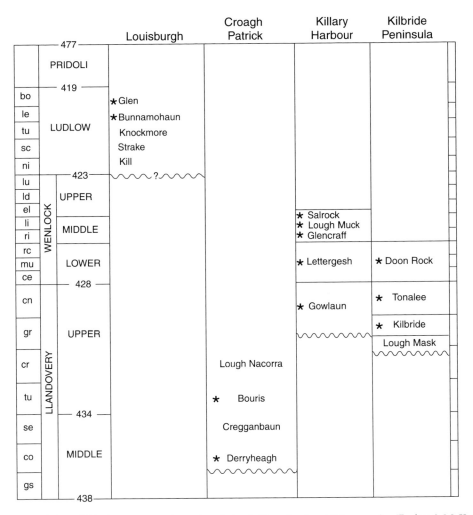

**Fig. 7.** Correlation of Silurian sequences in western Ireland. Stars, fossil control, ages after Tucker & McKerrow (1995). Data from: McKerrow & Campbell (1960); Dewey (1963); Phillips (1974); Graham *et al.* (1989) and Maguire & Graham (1996). See Appendix 2 for lithologies.

changes in sediment source areas, influence relief and erosion, and define sediment conduits and basin configuration, detritus transported from a progressively changing hinterland carries signatures of tectonic driving mechanisms. Therefore, the mineralogy of the resulting sediments can be regarded as an archive, preserving the history of an extinct or displaced hinterland. The recognition of index minerals of specific source rocks in the sediments is of great significance because they help constrain the time of uplift and erosion of their parent lithologies. Therefore, the study of heavy minerals is central to the reconstruction of the evolution of any dynamic hinterland. Many keynote works have

demonstrated the potential of heavy minerals in provenance studies (van Andel 1950; Füchtbauer 1967; Morton 1985; Allen & Mange-Rajetsky 1992). Several species crystallize in restricted parageneses so that their detrital occurrence points clearly to their respective source lithologies. Those of foreland-basin sediments are frequently used to reconstruct the unroofing history of the source domains (Mange-Rajetsky & Oberhänsli 1982; Winkler & Bernoulli 1986; Lonergan & Mange-Rajetsky 1994). They are important, also, in highlighting regionally significant lithological markers.

During a sedimentary cycle, the original heavy mineral composition of a sediment may undergo

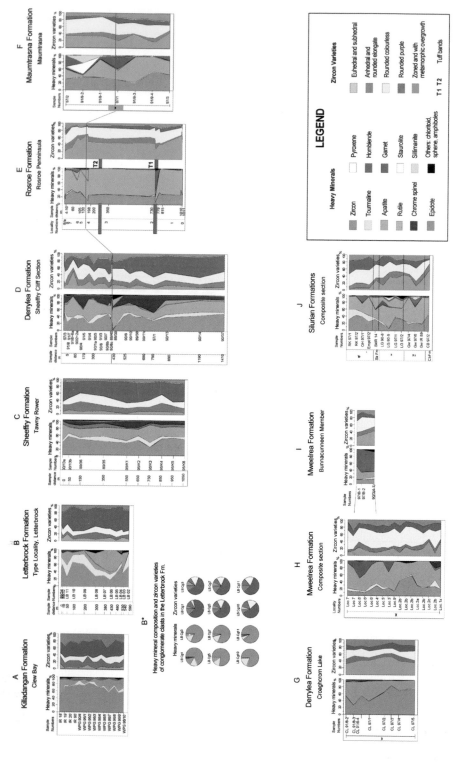

**Fig. 8.** Heavy minerals and zircon varieties: A, Killadangan Formation; B, Letterbrock Formation; C, Sheeffry Formation; D, Derrylea Formation; E, Rosroe Formation; F, Maumtrasna Formation; G, Derrylea Formation at Croaghcrom Lough; H, Mweelrea Formation; I, Bunnacunneen Member; J, Silurian sequences.

considerable changes, caused by various modifying factors in the sedimentary environment. The most influential of these are: (1) hydraulic processes during transport, caused by the differing densities of the individual species, resulting in preferential sorting according to size, shape and density; (2) post-depositional dissolution, because of the low resistance of the majority of heavy minerals to both acidic or alkaline conditions. Their behaviour during diagenesis is in accord with their chemical stability and is manifested by the progressive decline of individual heavy mineral species with increasing depth of burial. Advanced geological age also contributes to the impoverishment of heavy mineral suites (Pettijohn 1941). Ultimately, the sediments reach the stage of high mineralogical maturity, where the heavy mineral fraction consists of only ultrastable detrital minerals such as zircon, tourmaline, rutile and apatite (Hubert 1962; Morton 1984 1985 1986; Morton & Johnson 1993).

Conventional, species-level, heavy mineral studies carried out on low-diversity assemblages often prove inconclusive. An innovative approach, designed to alleviate this problem and to overcome the effects of hydraulic sorting, is termed high-resolution heavy mineral analysis (HRHMA; Mange-Rajetsky 1995; Lihou & Mange-Rajetzky 1996). It is based on the recognition that the majority of rock-forming and accessory minerals form in a diversity of size and habit, and are represented by several chemical, structural, colour and optical varieties, controlled principally by specific conditions during crystallization. Because a wide range of lithologies provide detritus to siliciclastic sediments, their heavy mineral assemblages are complex and an individual heavy mineral species may comprise several kinds of varieties, each preserving a different genetic and/or sedimentological history. HRHMA focuses especially on the ultrastable species: zircon, tourmaline and apatite. They are ubiquitous in detrital sediments, occur in significant quantities and are represented by several varieties. Commonly, when unstable heavy minerals are absent, specific varieties can provide clues to their former presence. During HRHMA, the varieties of a single species are studied and, therefore, density controlled sorting during transport is likely to have been unimportant.

Variables include grain morphology, colour, internal structure, etc. The measurement of grain shape (roundness and sphericity) is a standard sediment petrological procedure and is an important part of the description of sedimentary textures (Powers 1953 1982; Pryor 1971;

Pettijohn *et al.* 1973). Colour varieties, especially those of zircon, can be provenance-diagnostic; e.g. Mackie (1923) and Tomita (1954) successfully traced sediments to respective parent lithologies by using purple zircon varieties. Structural types of zircon (showing either euhedral zoning or overgrowth, etc.) are significant because they are linked to specific parageneses (Speer 1980).

The number of samples selected from each formation ranged from 10 to 40 depending on its thickness. Sample preparation was carried out using the technique described by Mange & Maurer (1992). For the heavy mineral separation, both bromoform (d. 2.89) and di-iodomethane (d. 3.20) were used. Di-iodomethane was used to obtain sufficient zircon from the generally chlorite-dominated samples for the varietal study and to detect the presence of provenance-diagnostic species that occur in low number. Heavy mineral residues were mounted in liquid Canada balsam for the microscopy: 200 grains were counted for the first phase of the analysis which was carried out at the species level. This provided information on the overall heavy mineral compositions. The second phase of the analysis involved the study of the varietal types of zircon and, when sufficient grains were present, apatite and tourmaline. The total of 70–100 grains of each species were counted, allocating the varieties in discrete categories; then data were recalculated and plotted as number percentages (Fig. 8a–j).

The low-grade sediments with abundant chlorite are highly indurated with a generally fine grain size. Zircon and apatite are the most common species. Other species include tourmaline, rutile, chrome spinel, epidote, sphene, various amphiboles and pyroxenes, garnet, staurolite, chloritoid and various authigenic phases. Zircon is represented by several distinctive varieties, which were recorded in the following categories (Fig. 9):

- sharp euhedral crystals, derived dominantly from volcanics and intrusive granites;
- subhedral grains from granitoids, gneisses, migmatites and some mafic rocks;
- anhedral fragments with irregular morphology;
- elongated, rounded morphology with a generally complex derivation, which can be reworked sedimentary or metasedimentary;
- rounded to well-rounded, colourless grains with polycyclic histories, including sedimentary or metasedimentary parentages;
- rounded to well-rounded, pink and purple grains, always polycyclic and, ultimately,

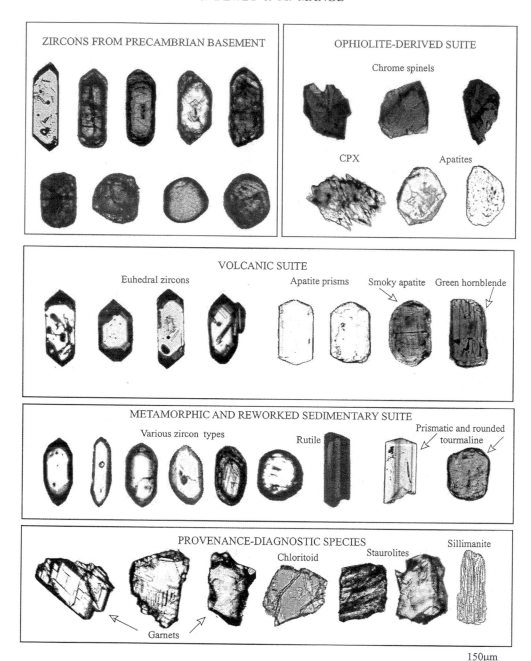

**Fig. 9.** Forms of selected heavy mineral species in the Ordivician and Silurian strata of western Ireland.

from the Precambrian of the Laurentian Shield;

● zircons with metamorphic overgrowth derived from high-grade metamorphic parageneses, ortho- and paragneisses, migmatites, contact metamorphic rocks, etc.: some show signatures of more than one phase of metamorphism and may also have been through several sedimentary cycles.

Tourmaline occurs in small numbers, either as prismatic or rounded grains with green or brown

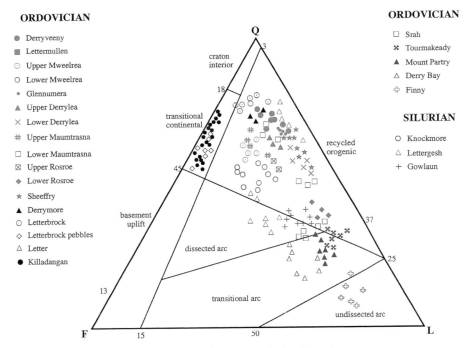

**ORDOVICIAN**

- ● Derryveeny
- ■ Lettermullen
- ◎ Upper Mweelrea
- ○ Lower Mweelrea
- · Glennumera
- ▲ Upper Derrylea
- × Lower Derrylea
- # Upper Maumtrasna
- □ Lower Maumtrasna
- ⊠ Upper Rosroe
- ◆ Lower Rosroe
- ★ Sheeffry
- ▲ Derrymore
- ○ Letterbrock
- ◇ Letterbrock pebbles
- △ Letter
- ● Killadangan

**ORDOVICIAN**

- □ Srah
- ✳ Tourmakeady
- ▲ Mount Partry
- △ Derry Bay
- ✢ Finny

**SILURIAN**

- ○ Knockmore
- △ Lettergesh
- + Gowlaun

**Fig. 10.** Quartz–felspar–lithic diagram for Ordovician and Silurian formations.

colours, and convey only limited local information (Fig. 9). Apatite, with few exceptions, is present in all samples and is most abundant in the volcaniclastic Rosroe and Maumtrasna Formations, and in the predominantly fluviatile Mweelrea Formation. Apatite is sensitive to acidic conditions, and studied grains are commonly etched and corroded in response to acidic leaching. Its morphological types mirror the nature of the particular depositional setting, e.g. angularity is high in the predominantly volcaniclastic Rosroe Formation, contrasting with the large number of rounded to well-rounded grains in the Mweelrea Formation (Fig. 9).

In addition to the HRHMA presented, the lithology (of formations studied with pebble contents where present), thickness, facies and provenance are documented in Appendices 1 (Ordovician) and 2 (Silurian). Also, carried out and presented are quartz–feldspar–lithic (QFL) analyses of most of the formations studied (Fig. 10). QFL analysis is a powerful tool in understanding the tectonic associations of sedimentary packages (Dickinson *et al.* 1983), especially in discriminating between lithic-poor, mature, cratonic-derived sediments and lithic-rich, immature arc-derived sequences.

## Heavy mineral suites

### Killadangan 'Formation'

Strongly contrasting opinions have existed about the age and origin of this 'formation', Dewey & Ryan (1990) considered it to have been part of an Ordovician subduction-accretion wedge whereas Williams *et al.* (1994) argued for a Silurian (Wenlock) age on the basis of trilete spores recovered from dark shales. The probably partly correlative Ballytoohy 'Series' on Clare Island (Phillips 1973) has an oceanic affinity with cherts and umbers which have yielded faunas with possible ages from middle Cambrian (*Protospongia hicksi*; Rushton & Phillips 1973) to Early or Middle Ordovician (euconodonts and chitinozoans; Harper *et al.* 1989; Williams *et al.* 1994, 1996). Two facies of the Ballytoohy 'Series', the Benilra and Oonaghcarragaun, are indistinguishable from the Killadangan 'Formation'; the Siorr chert/umber/sandstone facies is very similar to the Letter Formation (Graham & Smith 1981) northeast of Westport (Fig. 4). It is believed that the Ballytoohy 'Series' (Phillips 1973), Killadangan 'Formation' and Letter 'Formation' (Graham & Smith 1981) are various oceanic chert/shale/seamount and trench turbidite components of a broadly correlative subduction–accretion prism to a north facing early–mid-Ordovician arc (Dewey & Ryan 1990). The cherts of the assemblage are probably pre-Caradocian because a major change occurred, through the World's Ordovician oceans, from Llanvirn chert sequences to Caradocian black, carbonaceous, graptolitic mudstones (Leggett 1978). QFL ratios of Killadangan sandstones (Fig. 8) indicate a mature transitional continental origin. This indicates that the Early Ordovician arc was fairly close to the Laurentian margin and that its trench was fed by Laurentia derived turbidites. The only lithic components are grey chert and white polycrystalline clasts of slightly strained vein quartz, which are believed to be rapidly recycled components of a deforming and dewatering accretionary prism. Heavy mineral analysis indicates a uniform, low-diversity, zircon-rich assemblage (Fig. 8a) with apatite, rutile, minor tourmaline and traces of epidote, garnet and amphiboles. The zircon population is dominated by deep-purple, rounded grains, some with an opaque core and some with overgrowths. Chrome spinel is virtually absent. The provenance is from a metamorphic parentage of gneisses and migmatites with some granitoids (euhedral–subhedral zircons) with minor metasedimentary and mafic rocks. The abundance of purple zircons indicates mostly Pre-Caledonian basement sources, probably from the Laurentian continent.

The interpretation of the origin and age of the Killadangan 'Formation' and related parts of the Clew Bay Complex north of the ophiolitic Deer Park Complex, is that they were assembled in an oceanic trench north of a north facing oceanic arc (Dewey & Ryan 1990), the sediments sourced primarily from the Laurentian continent. For reasons of petrography and provenance, it is not accepted that the Silurian age for the Killadangan 'Formation' proposed by Williams et al. (1994) is correct. The petrography and heavy mineral suites are wholly dissimilar from the dissected arc and recycled orogenic nature of other Silurian sequences in western Ireland (Figs 8 and 10). For this reason (and reasons elaborated in 'Letterbrock Formation'), it is concluded that the Killadangan assemblage is of Early–Middle Ordovician age. Five further lines of evidence oppose a Silurian age for the Killandangan:

- all the certain Silurian sequences in western Ireland contain silicic to andesitic tuffs with much volcaniclastic detritus in most sandstones, indicating a suprasubduction zone, high-K, calc-alkaline origin on continental crust (Menuge et al. 1995), whereas no such material has been discovered in the Killadangan;
- derivation of the Killadangan turbidites from a transitional continental province untouched by vulcanism is supported by whole-rock geochemistry (Harkin et al. 1996);
- conglomerates of the Llandovery Cregganbaun Formation rest unconformably upon the Killadangan on the south side of Clew Bay (Johnston & Phillips 1995; authors unpublished mapping);
- rutilated blue quartz is abundant in the Killadangan, which invites comparison with the Dalradian Southern Highland Group greywackes of Scotland;
- the absence of macrofauna, especially graptolites, in Killandangan shales, in spite of assidious searches, is, at least, suspicious.

## Letterbrock Formation

The Letterbrock Formation (Figs 5 and 8b) is subdivided into a lower and upper section, indicated by two mineralogically contrasting intervals. The lower 400 m portion is characterized by the abundance of rounded epidote, indicating a predominantly detrital origin. Amphiboles and pyroxenes are also present in small amounts. In the upper 300 m interval, epidote and associated species are absent and chrome spinel appears. One sample, close to the top of the sequence, is rich in apatite, which, with the large number of acicular, unabraded, euhedral zircons, may signal a volcanic influx. Tourmaline content is unusually high (Fig. 8b), with abundant newly formed sharp prismatic varieties.

Rounded, deep-purple zircons dominate the zircon population, similar to the Killadangan 'Formation'. Clasts from conglomerate horizons were examined for their heavy mineral content to obtain information on provenance, and to ascertain whether there is any provenance link between the Letterbrock Formation and the Killadangan 'Formation', i.e. whether detritus from the Killadangan was recycled into the Letterbrock. Fifteen well-rounded sandstone pebbles were analysed, which revealed zircon-rich assemblages, with low-apatite content, and with the dominance of rounded, deep-purple zircons among the zircon varieties. These all show a nearly identical composition to that of the Killadangan sandstones (Fig. 8). Similarly to the Letterbrock sandstone samples, newly formed prismatic tourmaline is common in the clasts. Letterbrock sandstone clasts (Fig. 10) are indistinguishable from Killadangan sandstones from which they were probably derived. The Letterbrock sandstone matrix is very similar but slightly more quartz and lithic rich (Fig. 10). It is concluded that the Killadangan is older than the Letterbrock and was providing sandstone clasts together with ophiolitic trondjemite, dolerite and gabbro clasts from the Deer Park Complex.

Letterbrock provenance evolved in two distinct phases. The Lower Letterbrock Formation was sourced from mafic rocks (indicated by amphiboles, pyroxenes and abundant epidote) from Killadangan-type sandstones and dark shales, from pre-Caledonian Laurentian basement lithologies. By the deposition of the Upper Letterbrock Formation, a marked change took place in palaeogeography. Detritus from the mafic source no longer reached the depocentre. At the same time, with the emergence of the ophiolitic basement, a significant new source appeared and this is signalled by the presence of chrome spinel. Occasionally, debris from the volcanic arc reached the locus of Letterbrock deposition.

## Sheeffry Formation

The heavy mineral distribution (Fig. 8c) is remarkably uniform, characterized by the ubiquitous presence of chrome spinel, following the trend that started in the underlying Upper Letterbrock. In some uppermost Sheeffry turbidites, chrome spinel can be up to 45% of the heavy mineral fraction, with substantial amounts of associated serpentinous detritus, in places as serpentinite pebbles, and arenites have densities as high as 3.0. Kinahan *et al.* (1876) noted that sediments (Sheeffry Formation) on Inishturk are commonly talcose and 'sheared'. Recorded here is that many of the cleaved and sheared upper Sheeffry turbidites have a talcose 'soapy' feel. In several parts of the middle and upper Sheeffry Formation, between just north of Doolough and eastwards into the Sheeffry Mountains, shear zones (Fig. 4) are localized in ultramafic-rich, dark green dense greywackes in which fragmented lineated chromite grains are mantled in horizontally lineated sheaths of fuchsite. Rounded purple zircons dominate the zircon suites. In addition to the rounded forms, euhedral volcanic zircons occur in moderate quantities. It is noteworthy that zircons with overgrowth are present in very small amounts. These varieties are purple and are

most abundant in the Killadangan sandstone, derived from pre-Caledonian basement lithologies. In the Letterbrock, their proportions are lower and show a further decline in the Sheeffry Formation. Provenance suggests that influx from the exposed ophiolitic basement continued, with recycled detritus from older formations. A first-cycle component from volcanics is also detectable, whereas contributions from the pre-Caledonian Laurentian basement diminished. Overall, the Sheeffry Formation is more volcaniclastic than the Letterbrock Formation.

## Derrylea Formation

The lower 500 m of the sampled section is mineralogically uniform (Fig. 8d) and strongly resembles the underlying Sheeffry Formation. In the overlying 460 m thick middle section, the proportion of chrome spinel is markedly higher, with the progressive increase of colourless rounded zircons and by the decline of the purple varieties. Purple zircons with overgrowth are present in small amounts. However, towards the top of the section, colourless zircons with overgrowth appear in larger numbers, highlighted by peaks in the zircon pattern shown in Fig. 8d. The upper section starts at sample No. 39/89; this horizon, with the first appearance of staurolite (Dewey 1961) and garnet, and containing traces of chloritoid (Fig. 9), records the emergence of markedly different source lithologies. Detritus from new sources is also constrained by changes in the zircon population. The rounded purple zircons became diluted by the colourless varieties and colourless zircons with overgrowths are common as are sharp euhedral zircons, which are especially frequent in the uppermost beds. Chrome spinel quantities remain generally high. White mica is rare in the lower section but is abundant in the upper interval, with large, field-visible quantities in No. 39/89, a prominent thick coarse turbidite.

The source areas of the Sheeffry Formation continued to provide detritus to the lowermost section, whereas the ultramafic part of the ophiolitic basement contributed significantly larger amounts of material to the middle section. Based on the appearance of index minerals of regionally metamorphosed rocks, the source can be traced to regionally metamorphosed lithologies with Dalradian affinities, indicating the beginning of their unroofing. An intensive unroofing phase of the volcanic arc is signalled by the abundance of euhedral zircons in the pale lateral turbidites of the uppermost beds.

## Rosroe Formation

The Rosroe Formation consists mainly of proximal turbidite fans, roughly correlative with the Derrylea Formation (Figs 4 and 5), and was densely (40) sampled, but only representative samples from each locality are shown in the plots. Up to locality 6, the heavy mineral spectrum is uniform (Fig. 8e) and is characterized by the abundance of apatite with the sporadic presence of epidote, pyroxenes and green hornblende, but no chrome spinel. Apatite morphologies show high angularity with prismatic grains and gently rounded prisms being the most common (Fig. 9). The predominantly volcaniclastic nature of the formation is reflected by the abundance of sharp euhedral, colourless zircons and the small number of detrital assemblages, such as rutile and tourmaline, rounded purple zircons and zircons with overgrowth. At locality 6, garnet appears in small amounts and, then, is continuously present in the section. Changes can also be detected in the zircon distribution with a moderate increase in the rounded, purple and overgrowth category, occurring with higher tourmaline and rutile proportions.

The provenance is essentially arc derived, predominantly first-cycle sediments with a minor input from pre-existing sediments and, probably, low-grade metasediments. The presence of garnet at locality 6, signals the first arrival of detritus from metamorphic sources reflecting lithological changes in the hinterland. This top interval is probably correlative with the upper section of the Derrylea Formation. The transition from the Lower to the Upper Rosroe, and from the correlative Lower to Upper Derrylea, shows a QFL change towards quartz dominance resulting from the unroofing of Dalradian rocks. At Croaghcrom Lake (Fig. 8g), the uniform, zircon-dominated assemblages of the lower part change to apatite-rich suites. Traces of hornblende, pyroxene and garnet are present at the top, and euhedral zircon is abundant, with small pebbles of low-grade metasedimentary rocks.

## Maumtrasna Formation

The Maumtrasna Formation is roughly correlative with the Derrylea and Rosroe Formations (Graham 1987). Sandstone beds and small sandstone packages within the conglomerates were sampled (Fig. 8f). The lower beds contain simple heavy mineral suites with abundant apatite, similar to the Rosroe Formation, and chrome spinel is almost absent. The increase in diversity with stratigraphy is significant; about half-way up the section, garnet becomes abundant, accompanied by hornblende. This is followed by a pyroxene- and sillimanite-rich horizon. The samples near the top are rich in blue–green hornblende. The zircon spectrum is dominated by abundant sharp euhedral grains. The influence of the volcanic arc is demonstrated by the high frequency of sharp euhedral zircons, hornblende and pyroxenes. The arrival of index metamorphic minerals in the Maumtrasna section coincides with the appearance of similar suites in the Derrylea and Rosroe Formations, and define a laterally correlative horizon that documents a regionally significant provenance change, brought about by the emergence of a metamorphic terrain, probably by the stripping of a superjacent ophiolite nappe.

## Mweelrea Formation and Bunnacunneen Member

Localities 1x and 2b (Fig. 8h) are at the transition between the underlying Glenummera Formation and the basal Mweelrea, and are the only samples that contain chrome spinel. The assemblages are rich in apatite, showing dominantly rounded to well-rounded morphologies. Garnet is common in samples at locality 3 and becomes common in the overlying sandstones. The provenance change is also shown by the increase in abundance of tourmaline. Euhedral zircon is common and rounded colourless zircons are very common. The Bunnacunneen Member is rich in both garnet and euhedral zircons (Fig. 8i).

The provenance of the Glennumera–Mweelrea transitional zone appears to be the last part of the succession that received material from an ophiolitic source because there is a notable absence of chrome spinel in the overlying sequences. Source rocks included ignimbrites, recycled sediments and metamorphic lithologies, indicated by chlorite–muscovite schists and semi-pelites, and the presence of garnet in large amounts and traces of staurolite and colourless zircons. The Mweelrea continues the QFL trend (Fig. 10) from arc to progressively more quartz-rich recycled orogenic sources.

## Silurian Formations

Silurian sediments were analysed to enhance the understanding of provenance evolution and sediment dispersal in the South Mayo Trough. The following formations were included: the Cregganbaun Formation, the Gowlaun Member of the Lettergesh Formation, the Lettergesh Formation, the Salrock Formation and the Louisburgh Group. Heavy mineral distributions are plotted in a composite section (Fig. 8j).

Each formation shows distinctive heavy mineral compositions. The Cregganbaun Formation, with the dominance of euhedral zircons and apatite, shows strong first-cycle volcaniclastic affinities. The Gowlaun Member contains garnet, staurolite and pyroxenes. The Lettergesh and Salrock Formations show typical volcaniclastic assemblages with the addition of staurolite, garnet and traces of rounded chrome spinel.

The heavy mineral signatures of the Louisburgh Group are in marked contrast with the above formations and are typified by the dominance of polycyclic assemblages with abundant rounded tourmaline. Euhedral zircons are rare and the increased number of rounded purple zircons suggest reworking of detritus from earlier Ordovician formations. The continued presence of a metamorphic source is shown by the occurrence of first-cycle garnet and staurolite. QFL ratios (Fig. 8) indicate a transition between dissected arc and recycled orogenic sources, accurately reflecting the contemporaneous volcanism (Menuge et al. 1995) and local Dalradian metamorphic-source lands. It is again emphasized that this contrasts strongly with the petrography of the transitional–continental Killadangan Formation (Fig. 10) and, we believe, militates strongly against a Silurian age for this sequence.

## Evolution of the South Mayo Trough

The interpretation of the South Mayo Trough as a north facing forearc to an Ordovician oceanic arc (Dewey & Ryan 1990), whose collision with the Laurentian rifted margin was responsible for the Grampian Orogeny (Dewey & Shackleton 1984), is followed. The Dewey & Shackleton (1984) model involved subduction flip during or after the Grampian Orogeny so that, by Caradocian times, the Iapetus Ocean floor was subducting beneath the Laurentian margin to which an arc had been accreted. The sediments of the South Mayo Trough, uniquely in the British and Irish Caledonides, preserve a clear record of these tectonic events, whose signals and timing can be decoded by the distribution of provenance-index heavy mineral species and diagnostic varietal types, supported by QFL analysis and pebble–boulder lithologies where present.

The broad pattern of stratigraphic trends of chrome spinel, metamorphic minerals and colourless and purple zircons are shown in Fig. 11. Chrome spinel first appears in the Upper

Letterbrock Formation, indicating the likely uplift and erosion of the mantle component of ophiolitic basement by about the middle Arenig. Following the disappearance of detritus from the mafic parentages of upper ophiolitic basalts, dolerites, gabbro and trondjemite. The ophiolitic basement continued to shed material until the end of Glennumera Formation deposition. The source of the ophiolitic detritus is uncertain beyond that it was derived from the north; it may have been from an ophiolitic backstop, the Deer Park Complex (Dewey & Ryan 1990), or from a more widespread ophiolite nappe obducted onto and causing much of the deformation of the Dalradian, or from both. The lower and middle part of the Letterbrock Formation is similar to the Killadangan in its heavy mineral content and its QFL ratios (Fig. 10). Sandstone clasts in the Letterbrock Formation are indistinguishable from Killadangan sandstones in their heavy minerals (Figs 8, 10, 12–14) and QFL ratios (Fig. 10), and it is concluded that the Killadangan 'Formation' is part of an early Ordovician pre-collisional accretionary wedge, uplifted and recycled back into the Letterbrock sediments of the forearc basin (Figs 14 and 15).

Garnet, staurolite, sillimanite and some amphiboles, believed to be from a Dalradian metamorphic source, first appear in the upper parts of the Derrylea, and the laterally equivalent Rosroe and Maumstrana Formations. Their appearance is abrupt in all three formations and is accompanied by a 'flood' of detrital muscovite, reflecting the widespread dispersal of detritus from a common metamorphic source and linking the three formations by a laterally correlatable horizon (Figs 8 and 14). This horizon of provenance change, critically, provides a time constraint on the beginning of unroofing of metamorphic rocks at the source. Further support is drawn from the trends of coloured zircon varieties; a marked decrease of the purple zircons coincides with the first occurrence of staurolite and garnet in the Derrylea Formation. The decline of purple zircon was accompanied by the influx of colourless zircons from the new lithologies. Detrital staurolite was recorded by Dewey (1961) from thin section work but was not 'rediscovered' until the present work began, emphasizing that high-resolution heavy mineral analysis is the only effective way to make quantitative and unequivocal statements about detrital heavy minerals.

The heavy mineral composition of the early Llanvirn northern and southern limb sequences are remarkably different, indicating contrasting provenances. Comparison of the Derrylea Formation of the northern limb with the Rosroe

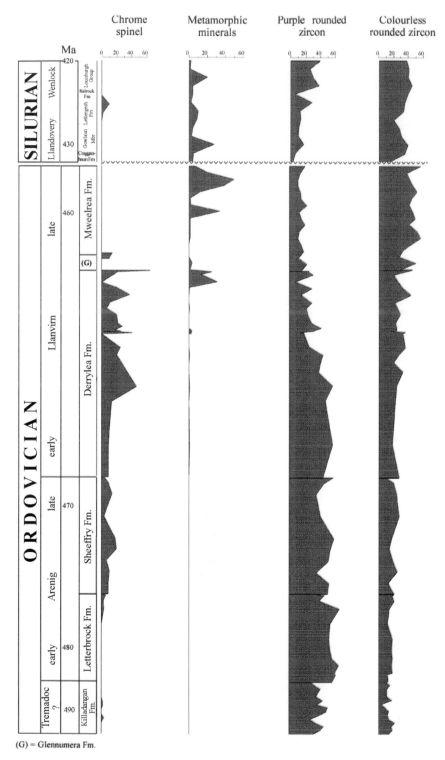

**Fig. 11.** Stratigraphic trends of heavy minerals and zircon varieties through the Ordovician and Silurian sequences.

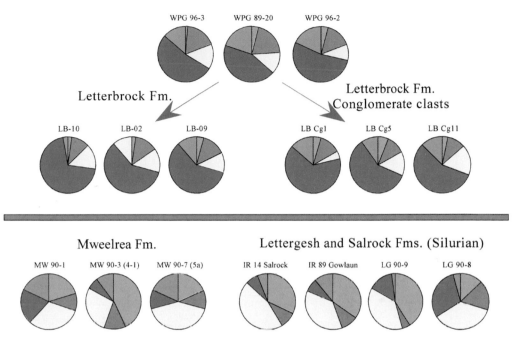

**Fig. 12.** Zircon varietal compositions of the Killadangan, Letterbrock, Mweelrea, Lettergesh and Salrock Formations.

and Maumtrasna Formations of the southern limb is particularly instructive. The axial turbidites (Dewey 1962) of the Derrylea Formation are characterized by chrome spinel and abundant purple zircon, especially in the lower and middle sections. Reworked detritus is represented by rounded tourmaline and rutile throughout. Provenance from ophiolitic basement and continental sources is followed by the addition of first-cycle metamorphics above the muscovite-rich marker horizon, complemented by arc-derived detritus towards the top. The pale, southerly derived, lateral turbidites (Dewey 1962) become much less common in the upper part of the Derrylea, contain no metamorphic or ophiolitic heavy minerals, and are dominated by arc-derived clear and smoky quartz and volcaniclastic detritus.

In the Rosroe Formation, on the southern limb, chrome spinel is absent and the negligible amount of reworked detritus indicates that a volcanic hinterland dominated the source area to the south. The uplift and denudation of this southerly arc source is reflected in the northward progradation of proximal fans with rounded boulders of hypabyssal and shallow plutonics (tonalites, monzonites, granodiorites, quartz, felspar porphyrite and felsites). However, in the upper part of the Rosroe and Maumtrasna, with the unroofing of the new metamorphic terrain, detritus from a Dalradian metamorphic source reached this part of the depocentre as a result of changes in basin configuration. Coinciding with the appearance of garnet in the Rosroe Formation is a noticeable increase in tourmaline and rutile percentages, and also in metamorphic zircons supporting a provenance change.

Prior to the unroofing of the Dalradian metamorphic source during the early part of the *artus* zone in the early Llanvirn (Fig. 5), the palaeogeographic picture of the South Mayo Trough, during the Arenig and earliest Llanvirn, is clearly

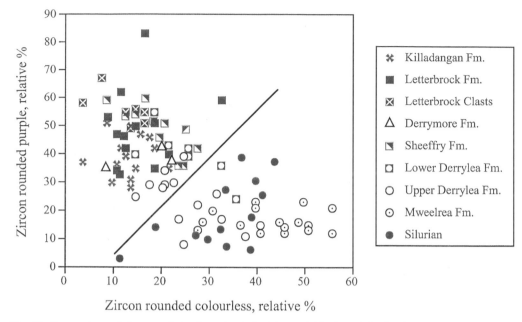

**Fig. 13.** Comparison of zircon varieties from the Ordovician and Silurian of the South Mayo Trough.

one of an arc to the south with the progressive unroofing of an ophiolite complex to the north (Dewey & Shackleton 1984), with the interleaving of sediments from the two sources indicated by their heavy mineral content. The unroofing of an ophiolite is also indicated by the work of Wrafter & Graham (1989), who showed that, in the north limb succession, high early Ti and Fe (mafic source) progressively diminish accompanied by an increase in Cr, Ni and Mg (ultramafic source); in the upper part of the Derrylea Formation, a marked drop in Cr, Ni and Mg values accompanies the appearance and denudation of a subophiolite northerly Dalradian metamorphic source. The appearance of metamorphic detritus in the upper Rosroe Formation presents an interesting tectonic problem. As far as is known, the whole of the lower Rosroe Formation was derived from a southerly volcanic plutonic source (Archer 1977) with very minor amounts of foliated silicic material, which, it is believed, has an igneous protolith. It might be argued that the appearance of metamorphic detritus in the upper Rosroe Formation reflects the unroofing of a southerly Dalradian basement upon which the arc was developed and that all the metamorphic detritus was southerly derived, spreading northwards into the South Mayo Trough. Two lines of evidence militate against this. First, the arc was clearly intra-oceanic (Clift & Ryan 1993), indicated by

its evolution and geochemistry from high-magnesian early units progressively through andesites of the Lough Nafooey Group to the more silicic explosive volcanics of the Glensaul Group. Secondly, the upper Derrylea metamorphic detritus was not derived from the south; it is contained in the darker green, axial turbidites, whereas the interbedded, southerly derived, pale, lateral turbidites contain only arc-derived detritus.

The middle Llanvirn Mweelrea Formation contains a flood of metamorphic heavy minerals (Fig. 8), including colourless rounded zircons in fluviatile sandstones deposited by west flowing rivers in a humid constricted South Mayo Trough (Pudsey 1984), passing westwards into shallow-marine shale and siltstone slate bands (Fig. 6). The absence of chrome spinel indicates that the stripping of an ophiolite cover from the Dalradian was complete. A further diminution of purple zircons was accompanied by an increase in euhedral zircons (Figs 12 and 13), indicating continued vulcanism, witnessed by five ignimbrite horizons (Fig. 6). The Derryveeny Formation of uncertain age (post-Glensaul Group and pre-Upper Llandovery), contains conglomerates with boulders of unmistakable Dalradian (probably Connemara) lithologies, including foliated corundum-bearing migmatites, Oughterard Granite, quartzites, two-mica schists and foliated gabbros. This suggests that

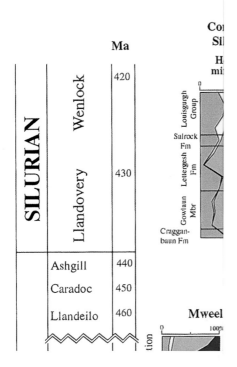

P

Co
Sil

H
mi

SILURIAN

Wenlock

Llandovery

| Ma |
|---|
| 420 |
| 430 |

Ashgill    440
Caradoc    450
Llandeilo    460

Louisgurgh Group

Salrock Fm

Lettergesh Fm

Gowlaun Mbr

Cragganbaun Fm

tion

Mweel

0            100%

the Connemara Terrane was in place relative to the Ordovician South Mayo Trough by, or during Derryveeny times (Dewey & Ryan 1990).

QFL studies (Fig. 10) strongly support the evolution of the South Mayo Trough from undissected arc through transitional and dissected arc into recycled orogenic fields, with a general drift from lithic-rich to quartz-rich assemblages. The exception is the Killadangan to Letterbrock pathway passing from a transitional continental (Laurentian) field into a feldspar-diminished, and slightly lithic-enhanced, field during the early Arenig.

By Early Silurian times, the Ordovician South Mayo Trough had been folded into a giant synclinorium and covered by a blanket of marine sediments (Dewey & Ryan 1990). The Silurian unconformity is complicated and, at least, locally developed on, and roughly parallel with, an extensional detachment (Williams & Rice 1989), suggesting rapid, perhaps catastrophic, marine transgression. The Silurian sequences of Croagh Patrick and Killary Harbour/Kilbride are dissimilar in age (Fig. 7) and facies (Appendix 2). The former was deposited in a shallow-marine sandstone 'shelf' whereas the latter was deposited in a rapidly deepening and filling suprasubduction zone basin (Williams & Harper 1991; Menuge et al. 1995), prior to their sinistral transpressive deformation. Again, distinctly, the younger Louisburgh Silurian accumulated in a sinistral strike-slip pull-apart basin within the Clew Bay Fault Zone during Late Silurian transpressive strain. The Killary Harbour/Kilbride Silurian lies astride the dissected arc/recycled orogenic QFL fields (Fig. 10), reflecting it's substantial volcaniclastic component, shallow subvolcanic diorite sills and in a diverse assemblage of metamorphic minerals. The Louisburgh Silurian, lying further into the recycled orogenic field, has a much smaller volcaniclastic component and an increased purple zircon component, probably derived from the adjacent Killadangan.

## Age and duration of the Grampian Orogeny

The Dalradian stratigraphic sequence of Connemara, between 5 and 6 km thick, can be correlated with the middle part of the Scottish Dalradian sequence, especially the Connemara Marble/Cleggan Tillite/Bennabeola Quartzite with the Islay Limestone/Portaskaig Tillite/Jura Quartzite (Fig. 16). The Grampian regional polyphase deformation and Barrovian metamorphic sequence is dated as part-uppermost Lower Cambrian by the Laurentian fauna of the Leny Limestone (Fletcher 1989), which is in clear stratigraphic and structural continuity within the Dalradian along the Highland Border near Callander (Harris 1969; Tanner 1995; Harris et al. 1999). A lower minimum age of Early Ordovician may supercede the Leny Limestone evidence if the chitinozoan data of Downie et al. (1971) on the Macduff Slates at the very top of the Scottish Dalradian is validated (Molyneux 1999). The Ben Vuirich Granite, once thought to post-date the earliest phases of Grampian deformation is now regarded as predating all the regional Grampian deformation (Tanner & Leslie 1994; Tanner 1996; personal observations) and probably formed during rifting events associated with the Tayvallich Volcanics (Fig. 16) that immediately predated opening of the Iapetus Ocean (Anderton 1980). The Durine Formation carbonates, at the top of the Durness Laurentian margin shelf succession, is of uppermost Arenig, possibly very earliest Llanvirn age (Higgins 1967), and provides an important maximum age for the Grampian destruction of the Iapetan Laurentian shelf, although Grampian orogenic events may have begun earlier in a more oceanic direction (Dewey 1982).

A certain direct minimum age for the Grampian Orogeny is given by the Llandovery–Dalradian unconformity in north Connemara (McKerrow & Campbell 1960), a terrane-linking unconformity that ties the Dalradian to slightly deformed Ordovician rocks. The Derryveeny Formation (Graham et al. 1991), of uncertain post-mid Llanvirn–pre-upper Llandovery age (Fig. 17), contains certain Dalradian detritus. If we are correct in our belief that the upper Derrylea metamorphic detritus was sourced by Dalradian rocks of the Grampian metamorphic complex, their Barrovian metamorphism and a large part, if not all, of the Grampian deformation had been achieved by the early Llanvirn (Fig. 18). Caradoc greywackes (subduction–accretion prism) in the Southern Uplands contain detritus from arc and metamorphic (probably Dalradian) sources (Elders 1987). Hutchison & Oliver (1999) have argued, from zoning and composition, a Dalradian source for detrital garnets in Caradocian sandstones in the northern belt of the Southern Uplands. Bracketing ages for the Grampain Orogeny are also given by the Upper Llanvirn unconformity on Middle Arenig at Ballantrae (Stone 1996) and the Caradoc unconformity on Arenig along the Highland Border (Curry et al. 1984).

Dewey (1961) argued for a pre-Arenig age for the deformation and metamorphism of the Dalradian on the basis of the quite dissimilar

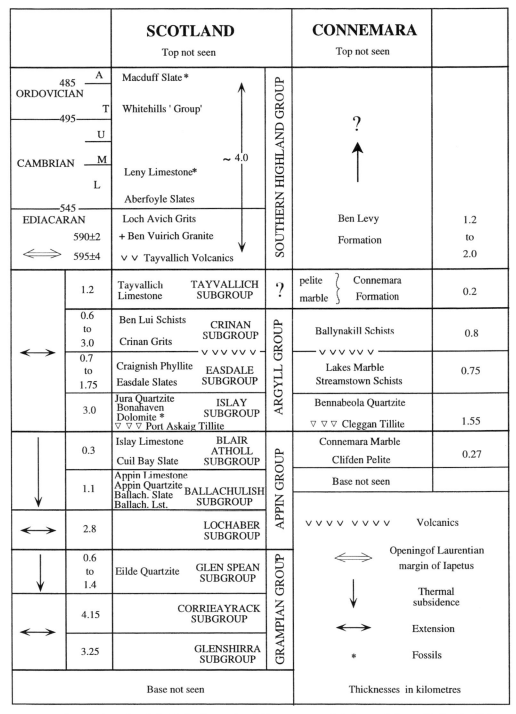

**Fig. 16.** Outline stratigraphy of the Dalradian rocks of Connemara and their correlation with the Scottish Dalradian. Data from: Sutton & Watson (1955); Leake (1963, 1986, 1989); Harris & Fettes (1972); Harris *et al.* (1975, 1978); Brasier & McIlroy (1977); Leake *et al.* (1983); Leake & Tanner (1984) and Hambrey *et al.* (1991).

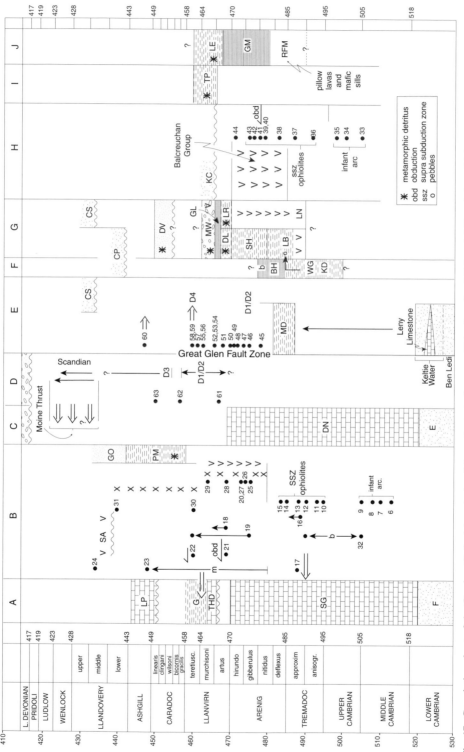

**Fig. 17.** Correlation chart of the Cambro-Ordovician strata of the orthotectonic zone of Newfoundland and the British and Irish Caledonides [(**c**)–(**j**)].
A, Newfoundland shelf; B, Newfoundland origin; C, Scottish shelf; D, Moine; E, Dalradian; F, Clew Bay Zone; G, South Mayo Trough; H, Ballantrae and Shetlands;
I, northern Southern Uplands; J, South Connemara. Key to abbreviations in Appendix 4.

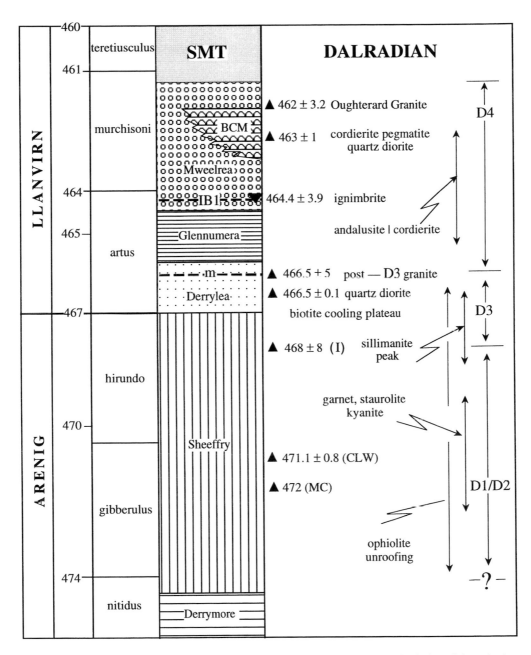

**Fig. 18.** Ordovician stratigraphy and events in the South Mayo Trough related to the timing of the polyphase deformation and metamorphism of the Dalradian. The 464.4 ± 3.9 Ma for Ignimbrite Band 1 (IB1), in the lower Mweelrea, is the only direct isotopic evidence for the age of the Ordovician sequence and is a concordant U–Pb age on the abraded tips of large acicular zircon crystals measured by Dr Stephen Noble at the NERC Isotope Geosciences Laboratory. Timescale after Tucker & McKerrow (1995); geochronological items in Appendix 3. BCM, Bunnacunneen Member; CLW, Cashel Lough Wheelaun Intrusion; I, Insch; IB1, Ignimbrite Band 1; MC, Morven Cabrach; SMT, South Mayo Trough.

deformation and metamorphism of the Dalradian and Ordovician rocks of western Ireland, and because the Derrylea Formation, containing detrital staurolite supposedly from a Dalradian source, forms part of a conformable sequence, the lower part of which is Arenig (Fig. 5). Dewey (1961) argued that the existence of detritus from a Dalradian metamorphic source in the Ordovician sequence means that the metamorphism must have predated the whole of the conformable sequence. It is now known from, e.g. foreland basins, that this view is incorrect. Dewey (1961 1962 1963) viewed the South Mayo Trough as a post-Grampian, fault-bounded trough (Dewey et al. 1970; Ryan & Archer 1977) fed by metamorphic Dalradian sources to the north and south, by a volcanic arc astride the Connemara 'Cordillera' to the south, and by the mafic and ultramafic rocks of the Deer Park Complex (Dewey 1962 1963). This rather primitive pre-plate tectonic 'model' has been superceded by more cogent models involving the collision of a north facing arc with the Laurentian margin (Mitchell 1978, 1981, 1989; Casey & Dewey 1984; Dewey & Shackleton 1984; Dewey & Ryan 1990), although it is emphasized that the principal arguments in this paper are about the timing and duration of the Grampian Orogeny rather than it's precise palaeogeography, palaeotectonics and plate tectonic setting.

The Ordovician South Mayo Trough was probably a forearc sequence (Fig. 15), very similar to those of the present Luzon in the Philippines and to the Cretaceous Great Valley (Dewey & Ryan 1990), during the deposition of which, it is suggested, the Grampian deformation and metamorphism occurred, and which, uniquely in the Appalachian–Caledonian belt, faithfully records the nature and precise timing of the Grampian Orogeny. This is illustrated in Fig. 18 in which the stratigraphy of the South Mayo Trough, with its biostratigraphic and isotopic ages, has been plotted v. well-documented isotopic ages of structural, igneous and metamorphic events in the Dalradian of Connemara and two isotopic ages from the Scottish Dalradian. The newest and most reliable ages for the Connemara and Scottish mafic–ultramafic suites are (Appendix 3) 471 (Friedrich et al. 1997) and 472 Ma (Rogers et al. 1995a, b), respectively. These calc-alkaline to tholeiitic suites were injected synchronously with the first major north vergent nappe-forming event in Connemara (Wellings 1998), essentially the first major deformation phase in the Dalradian (so-called D2). Earlier fabrics (so-called D1) are recorded in garnet porphyroblasts but it is believed that these are immediately earlier,

kinematically linked, small-scale parts of 'D2' and are, therefore, referred to as D2(D1), the first structural event in the Dalradian polyphase sequence, probably generated beneath northward obducted fore-arc suprasubduction-zone ophiolite sheets (Van Staal et al. 1999) and/or accretionary prisms. The syn D2(D1) calc-alkaline mafic–ultramafic suite of Connemara was probably developed above a diachronously flipping subduction zone (Van Staal et al. 1999). Wellings (1998) has argued a rapid transition from D2(D1) nappe emplacement to D3 north vergent recumbent folding, probably a direct kinematic continuation of D2(D1). A plateau of c. 467 Ma in the biotite 'cooling' ages (Elias et al. 1988) suggests a period of denudation and cooling during the early Derrylea, which led to the arrival of metamorphic detritus c. 1 Ma later. The quartz diorite 'sea' of Connemara, south of the Clifden Steep Belt, is dated at $466.5 \pm 0.6$ Ma (Friedrich et al. 1997), post-D3 granite and pegmatites are dated at $466 \pm 5$ Ma (Cliff et al. 1996) and the syn to post-D4 Oughterard Granite is dated at $462 \pm 3.2$ Ma (Friedrich et al. 1997). The D4 growth of the Connemara Antiform, rotation of the 'roots' of D2–D3 nappes in the Clifden Steep Belt and the retrocharriage south-southeastwards thrusting of Dalradian assemblages across (probably) Ordovician silicic volcanics in the Delany Dome, are probably related to subduction flip that led to the progressive subduction–accretion of the Middle Ordovician rocks of South Connemara (Dewey & Ryan 1990).

The Connemara geochronology data (Friedrich et al. 1997) indicate that the Grampian deformation of the Dalradian took c. 10 Ma from c. 471 to 462 Ma, from about the gibberulus zone of the Arenig to the late murchisoni zone of the Llanvirn, i.e. from about the middle part of the Sheeffry Formation to the upper part of the Mweelrea Formation (Fig. 18). It is suggested that arc–continent collision began in Derrymore times but substantial thrusting and thickening of the Laurentian margin was achieved by c. 470 Ma. Late Arenig uplift denudation and cooling of the Dalradian generated a 'biotite plateau' in the cooling ages and D4 retrocharriage was completed by the end of the murchisoni zone (Fig. 18). Ophiolite unroofing during Sheeffry and early Derrylea time led to the unroofing of Dalradian metamorphic rocks during the artus zone, upper Derrylea. The first flood of Dalradian metamorphic detritus in the upper Derrylea progressed to a massive flood during the deposition of the Mweelrea Formation (molasse) and the end of calc-alkaline volcanism with the eruption of the Mweelrea ignimbrites.

The Derryveeny Formation records the transpressional strike-slip docking of the Connemara Terrane with the South Mayo Trough.

The essential message of the Ordovician South Mayo Trough is that it records, in a conformable stratigraphic sequence, the rapid evolution of the Grampian Orogen during a 10 Ma history (Fig. 18) of arc–continent collision and subduction flip. Whereas the U–Pb zircon, monazite and titanite data constrain the Grampian orogenic event to a period of only *c*. 10 Ma during the Early–Middle Ordovician, the K–Ar and $^{40}Ar/^{39}Ar$ data for the Dalradian of the British and Irish Caledonides yield a spread from over 480 to 390 Ma (Fig. 19). These data have been interpreted (Dempster *et al.* 1995; Elias *et al.* 1988) as representing cooling of the Grampian metamorphic complex (Dewey & Pankhurst 1970) or as a late Caledonian overprint on Grampian ages (Fitch *et al.* 1964). The new geochronological and stratigraphic data make the K–Ar data and ideas irrelevant and equivocal.

A very short (<10 Ma) duration of the whole Dalradian polyphase deformation sequence and Barrovian metamorphism is supported by Wellings' (1998) work on the Currywongaun–Doughruagh disrupted mafic–ultramafic sheets. Wellings (1998) showed that the sheet was injected during the first phase of deformation [D2(D1)] but immediately prior (*c*. 0.5 Ma) to D3. Simple kinematic modelling of the Grampian Orogen also indicates a very short period for the Grampian thickening of the crust, irrespective of the orogenic mechanism responsible or the geometry of the structures. Taking the present Grampian Orogen as *c*. 200 km wide, a double-thickness orogenic crust as a culmination of Grampian shortening and a plate convergence component of 10 cm yr$^{-1}$ normal to the strike of the orogen, crustal thickening should have taken 2 Ma at a horizontal strain rate of *c*. 10$^{-15}$. Reducing the plate convergence rate to an unlikely low figure of 1 cm yr$^{-1}$ gives an outside maximum of 20 Ma, whereas a 5 cm yr$^{-1}$

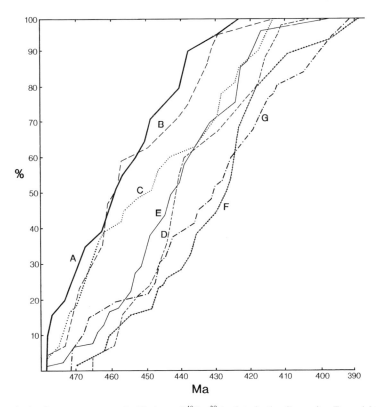

**Fig. 19.** Cumulative frequency curves for K–Ar and $^{40}Ar/^{39}Ar$ data in the Grampian Zone. (**a**) K–Ar hornblende, Connemara; Miller *et al.* (1991); (**b**) K–Ar muscovite, Connemara; Miller *et al.* (1991); (**c**) $^{40}Ar/^{39}Ar$ biotite, Connemara; Elias *et al.* (1988); (**d**) K–Ar biotite, Connemara; Miller *et al.* (1991); (**e**) K–Ar muscovite and whole rock, Scottish Highlands; Dewey & Pankhurst (1970); (**f**) K–Ar biotite, Scottish Highlands; Dewey & Pankhurst (1970); (**g**) K–Ar biotite, Taconics; Drake *et al.* (1989).

rate yields a 4 Ma duration. A minimum and maximum of 2 and 10 Ma, respectively, are suggested to span the likely range for the Grampian orogenic event of arc collision, ophiolite obduction, crustal thickening subduction flip and unrooting.

This short duration for the Grampian Orogeny during the Middle Ordovician raises the vexing question of how the associated Barrovian metamorphism was achieved. The orogenic crust cannot have heated up by conduction alone with a thickening lithosphere that included mantle thickening because that would have taken substantially longer than 2–10 Ma (England &

Thompson 1984; Thompson & England 1984). Advective lithospheric thinning of the kind modelled by England & Houseman (1989), that leads to orogenic collapse (Dewey 1988), is not a candidate for allowing access of asthenosphere to, and heating the base of, the deforming crust because the time interval between lithospheric thickening and advective thinning is again too long, typically c. 20 Ma. It is believed that an answer to the problem of rapidly heating the Grampian metamorphic pile is allied directly with the problem of the origin and injection of the mafic–ultramafic magmas of Currywongaun/ Doughruagh, Cashel/Lough Wheelaun above a

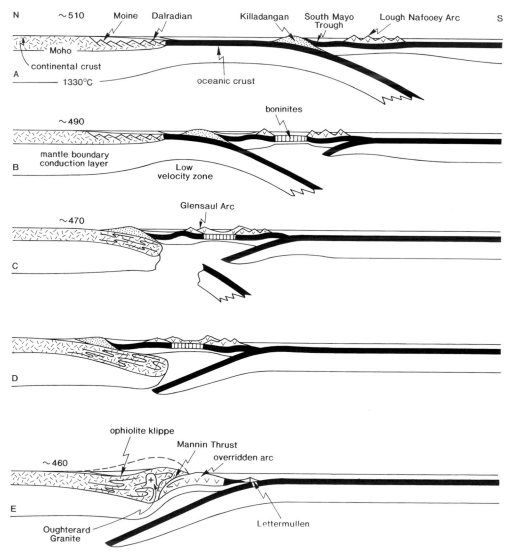

**Fig. 20.** Schematic sections illustrating the Ordovician tectonic evolution of the Laurentian margin.

subduction zone (Yardley & Senior 1982), and the Scottish bodies of Insch, Belhelvie and Haddo House into the thickening crust. South of the Clifden Steep Belt, the injection of vast amount of mafic–ultramafic magma was immediately followed by the D3 injection of a 'sea' of quartz-diorite gneiss, both with a calc-alkaline affinity (Leake 1989) with an associated and clearly derivative regional metamorphism to the sillimanite grade. Two sources of asthenosphere may have been individually or jointly responsible (Fig. 20) for the origin of the mafic–ultramafic suite, for the rapid heating of the metamorphic pile and for the quartz-diorite sea. First, the arc slab that collided with and overrode the Laurentian margin was probably quite thin because infant arc suprasubduction ophiolites of Tremadocian age formed just prior to their obduction and were, therefore, hot at the base, anything from *c.* 950 to 1300°C depending upon the thickness of the slab. The slab may have carried asthenosphere onto the subducting continental margin and, in any case, the horizontally collapsing continental edge was probably plunging into subarc asthenosphere (Fig. 20), which could have fed mafic magmas and heat laterally into the nappe pile. Secondly, if a subduction polarity flip model is appropriate, it is virtually certain that the flip involved opposing, polarity subduction zones that overlapped substantially (Van Staal *et al.* 1999) so that, in any transorogenic section, coeval older and younger subduction zones existed (Fig. 20). This leads to the likelihood that slab break-off, with the consequent renewed pulse of rising asthenosphere and the newly 'arriving' flipped subducted slab with its consequent corner convective flow influence, yields a further source of new suborogenic heating and calc-alkaline plutonism. This also explains the duration of calc-alkaline volcanism that spans almost the entire history of the Grampian Orogen in western Ireland from the pre-collisional infant oceanic arc of the Lough Nafooey Group, the early syncollisional Connemara mafic–ultramafic suite, the evolved syncollisional arc volcanism of the Glensaul and Sheeffry Groups, and the post-collisional silicic volcanism of the Mweelrea Formation, much of which overlaps with the Grampian deformation and metamorphism.

## The Nature of the Grampian Orogeny: the New Guinea analogue

The geological evidence for the Ordovician collision of an oceanic arc (Ryan *et al.* 1980) with the Laurentian margin and the consequent

obduction of suprasubduction zone ophiolites followed by subduction polarity flip is, we believe, clear. However, the clearly documented extreme brevity of the Grampian Orogeny also testifies to the likelihood of the model for its development. Orogeny occurs at convergent plate boundaries in, at least, the following eight ways, each with its peculiar tectonic/structural geometry and kinematics, metamorphic patterns, plutonism and basins.

- Continent–continent collision, following the completed subduction of a wide ocean as in the Himalayan–Tibetan system, produces wide zones of deformation evolving over periods of up to *c.* 50 Ma. Ophiolite obduction and blueschists appear not to be associated with this type of orogeny, which has rather long-wavelength foreland basins and flexural (peripheral) bulges.
- Continent–continent collision resulting from the contractional collapse of super-rifts/para-oceans, as in the European Alps, develop basement thrust sheet and fold nappe complexes without or with little earlier calc-alkaline arc magmatism because little or no oceanic lithosphere *sensu stricto* was subducted. Such orogens have blueschists and other evidence of the burial and exhumation of high-pressure metamorphic rocks.
- Andean-style orogeny, developed at convergent leading edge continental margins with intermittent bursts of silicic magmatism and rapid, large-magnitude shortening, confined principally to foreland thrust belts, and probably related to phases during which the continent moves trenchward in a mantle reference frame.
- New Zealand style orogeny that has developed the Southern Alps during the Late Cenozoic as a result of a narrow transpressional sliver of continent caught in an obliquely convergent plate boundary in which decollement occurs between a shortening wedge of continental crust and lithospheric mantle that is swallowed beneath the wedge.
- Intensely-localized orogeny occurs at major restraining, locking, bends of major intra-continental transform faults such as the Transverse Ranges of California on the San Andreas Fault and the Anti-Lebanon Mountains on the Dead Sea Transform.
- Complex, looping, strongly oroclinal orogens commonly develop during the closure of irregular 'remnant' oceans just prior to their extinction by terminal and complete continent–continent collision. This is particularly well illustrated by the young narrow

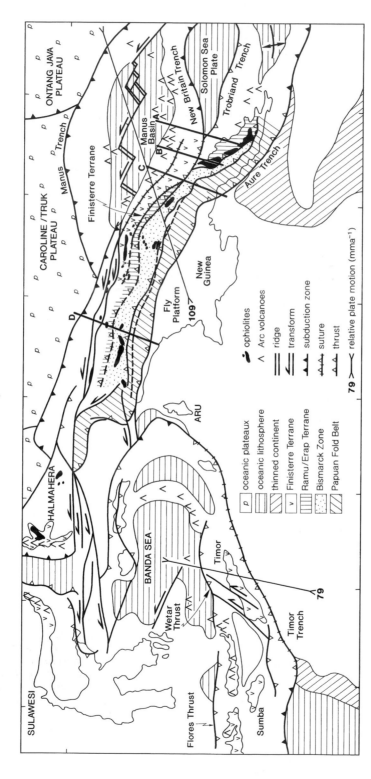

**Fig. 21.** Simplified tectonic map of the collisional interaction of northern Australia with Indonesia and Papua New Guinea. Data for this figure and Figs 22 and 23 from Geological World Atlas (1976); Hamilton (1979); Tectonic Map of the World (1985); De Smet *et al.* (1990); Harris (1991); Pigram & Symonds (1991); Silver *et al.* (1991); Abbott *et al.* 1994*a*,*b*); Abers & McCaffrey (1994); Pegler *et al.* (1995); Auzende *et al.* (1996); Crowhurst *et al.* (1996); Cullen (1996); Genrich *et al.* (1996); Hall (1996, 1997); Malaihollo & Hall (1996); McCaffrey (1996); McDowell *et al.* (1996); Richardson & Blundell (1996) and Snyder *et al.* (1996).

orogens marginal to the western Mediterranean, such as the Apennines, that appear to have formed by subduction roll-back of Jurassic remnant oceanic lithosphere leading to the 'sucking' and oroclinal bending of arcs to open young back-arc basins (Ligurian, Balearic and Tyrrhenian Seas), and to allow these arcs to collide with and accommodate themselves to the shape of irregular remnant-rifted margins (Dewey et al. 1988). The back-arc extension commonly leads to the exhumation of blueschists and other high-pressure metamorphic rocks of earlier formed orogens. Similarly, Aegean extension and bending of the Cretan fore-arc will probably contribute to the demise of the Eastern Mediterranean and the development of an orogenic belt along the Benghazi rifted margin of North Africa.

- Intra-oceanic arcs whose polarity is continentward facing are likely, eventually, to collide with a rifted continental margin; e.g. the Vanuatu Arc will most likely collide with the Lord Howe Rise and the Great Barrier Reef margin of Queensland, and the Luzon subduction zone will close the South China Basin much as the collision of Taiwan with China has developed a narrow orogen to isolate the South China Basin from the Okinawa Trough. The diachronous collision of the Sumba–Timor forearc (Fig. 21) is witnessing the development of a short-lived collisional Timorese orogen and subduction polarity flips along the Flores and Wetar Thrusts (Fig. 22a). However, it is believed that this is a poor analogue for the Grampian Orogeny because there is little post-flip oceanic lithosphere to be subducted and there is no syncollisional ophiolite obduction.

- In explaining the Grampian Orogeny, it is believed that extensional or transtensional continent-facing oceanic arcs that develop suprasubduction-zone ophiolites followed rapidly by collision with a rifted continental margin, and rapidly generated subophiolite orogens, such as the Oman whose brevity is explained by the ease with which continued plate convergence is accommodated by subduction polarity flip (McKenzie 1969). An evolving system of this kind is Papua New Guinea (Figs 21, 22b and 23), which is a very instructive tectonic system in the context of explaining the Grampian Orogeny. The north Australian shelf, the Fly Platform, collided with a continentward-facing arc (Tsumba Volcanics) during the early Miocene, with the obduction of an ophiolite sheet(s), subophiolite blueschist

metamorphism, a Barrovian metamorphic complex in the Bismarck Ranges, subduction polarity flip and the accretion of the younger arc of the Finisterre Ranges. The stratigraphic sequences of the Ramu and Erap Zones (Fig. 23) bear a striking resemblance to the Ordovician sequence of the South Mayo Trough. The particular point to be made here is that the continent–arc collision and ophiolite obduction between the Bismarck and Ramu Zones developed and unroofed a Barrovian metamorphic orogen; this was of similar proportions and duration to the Grampian Orogen in c. 5 Ma. It is suggested that this brevity is typical of orogens generated by arc–continent collisions. The Papua New Guinea Orogen is complicated by dual polarity subduction zones, the New Britain Trench and Trobriand Trough (Fig. 21), that converge and plunge conically westwards beneath the Orogen (Fig. 22b). The Trobriand Trend probably represents the flipped subduction zone following the Ramu–Bismarck collision that is in face-to-face collision with the New Britain Trench that allows collision with the Finisterre Arc (Figs 21 and 22b).

An important feature of orogens developed beneath obducted ophiolite sheets is that they do not develop substantial orographic expressions during the earliest phases of deformation and crustal thickening while the heavy ophiolite nappe is preserved to depress the orogen. Only when the ophiolite nappe is substantially eroded will mountains begin to grow. This is the most likely explanation of the progressive growth of the Grampian Orogen with ophiolite unroofing preceding the sudden appearance of metamorphic detritus.

Since Gass's (1968) ground-breaking work on the Troodos Complex, there has been a developing controversy about the origin and emplacement of the ophiolite suite from the point of view (Moores & Vine 1971) that they represent slices of mid-ocean ridge crust and mantle, to the opinion that they formed in intra-arc and back-arc basins (Dewey & Bird 1971) to the more recent view, initially promulgated by Miyashiro (1973) from geochemistry, that they represent a very complex assemblage of rock suites formed above a subduction zone, from forearc boninites to back-arc tholeiites that inject, distend and are overlain by assemblages of calc-alkaline volcanics (Falloon et al. 1992; Smellie & Stone 1992; Bloomer et al. 1995; Bedard et al. 1998; Van Staal et al. 1998). It is believed that the 'ophiolite problem' is now

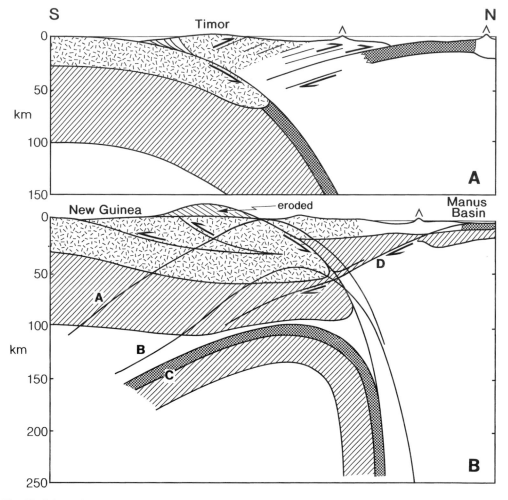

**Fig. 22.** Schematic true-scale sections across the northern margin of Australia and its collided arcs. (**a**) Section across Timor; (**b**) sections across Papua New Guinea (lines of four sections shown in Fig. 21). Data sources in caption to Fig. 21.

essentially solved in favour of the suprasubduction zone origin; not only does it explain the complex polyphase relationships between igneous and structural events, including clearly arc-related assemblages, but it accounts for the very short interval between origin and obduction (Dewey 1974), which is so typical of all large obducted ophiolite sheets. There may be rare fragments of oceanic crust and lithosphere, generated at spreading ridges and transform faults, caught up in orogenic belts, especially slices from near-rifted margins (Littleport Complex; Van Staal *et al.* 1989) or sliced from transforms (?Annieopsquotch Complex; Van Staal *et al.* 1999), but it is believed the overwhelming

fate of the oceans, with the exception of seamounts, is subduction.

## Evolution of the Laurentian margin

In this section, data from the Appalachians, and the British and Irish Caledonides, are used to develop a simple outline composite picture of the tectonic evolution of the Laurentian margin of Iapetus (Fig. 20). Although there are some differences, to be explored briefly below, between the Appalachians and Caledonides, the profound similarities allow confidence that along-strike broad tectonic homologies that are well-preserved in the Appalachians but disrupted

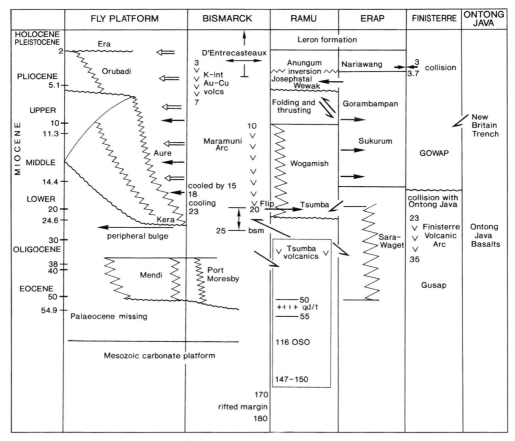

**Fig. 23.** Simplified chart illustrating the Cenozoic tectonostratigraphic and evolution of Papua New Guinea. Data from sources in caption to Fig. 21.

by substantial sinistral strike-slip motion in the Caledonides are being dealt with. The two striking similarities are: a series of late Proterozoic rift complexes and a Cambro-Ordovician carbonate platform marking the rifted margin of the Iapetus Ocean; and a short-lived Ordovician orogeny (Taconian, Humberian and Grampian) that destroyed the rifted margin. It has been claimed that Proterozoic (Knoydartian and Morarian) Orogeny (Lambert 1969) has affected the Moine and Dalradian sequences that were deformed by the Grampian Orogeny. Following Soper & Harris (1997), this is disputed for the following reasons.

- No regional unconformities have been found within the Moine and Dalradian sequences; it is difficult to see how orogeny could have occurred within and during the deposition of a conformable sequence.
- The *c.* 830 Ma event(s) is, spatially, vaguely defined in the Moine and the Monadliath

Mountains with no obvious fronts or edges.
- The well-defined Grampian (Ordovician) deformation sequence that is clearly the first event to affect the Moine of Sutherland (Friend & Strachan, pers. comm.), seems to pass imperceptibly southwards into the Morar region where similar or identical structures are claimed to be Proterozoic (Brook *et al.* 1976, 1977; Brewer *et al.* 1979).
- There are no obvious preserved molasse basins that record the supposed Late Proterozoic orogeny.
- Silicic igneous rocks are common in orogenic belts, especially late in their evolution following substantial crustal thickening, and there has developed the belief that granite plutons, pegmatites and migmatites always signify orogeny and crustal thickening. This view is incorrect; these rocks are also common in extensional rift valley regimes where they are accompanied, commonly, by alkaline/peralkaline riebeckite and aegirine-bearing

phases, such as the Carn Chuinneag Granite, and mafic plutons and dyke swarms.

- The flat-lying shear zones and migmatites with their Proterozoic zircon ages (Friend *et al.* 1995; Noble *et al.* 1996, 1997) in the Moine Series and Grampian Group, that are supposed to document Proterozoic contractional orogeny, are just as likely to have been generated during continental extension. Flat, penetrative ductile structures occur both in orogenic belts, where they are generally non-coaxial, and in extensional regimes where they are non-coaxial, beneath extensional detachments, passing down into more coaxial structures (Dewey *et al.* 1993). In metamorphic core complexes throughout the world, almost-undeformed strata in extensional basins above detachments are juxtaposed with 'basement' rocks that may vary even very locally, both laterally and vertically, from little-deformed upper crustal rocks with mafic/silicic dyke swarms to exceedingly deformed mid-crustal rocks with both deformed and undeformed silicic plutons. Hanging-wall, rift-basin strata may bear either a structural or an unconformable relationship with basement. It is suggested that the flat-lying fabrics of the Moine, if they are indeed of Proterozoic rather than Grampian age, are extensional fabrics possibly with internal extensional/unconformable contacts. Moreover, transtension can yield a substantial component of horizontal shortening normal to the stretching direction where the transport direction is at a low angle to the walls of the deforming zone, giving bulk constrictional fabrics.
- If Neoproterozoic orogenies occurred in the Scottish Highlands, they would be unique to the Laurentian margin. Bluck & Dempster (1991) have 'solved' this problem by arguing that Dalradian, and possibly Moine, 'blocks' are exotic and foreign to the margin of Laurentia with an exotic, Gondwana, provenance. Their view is predicated on the assertion that the Dalradian Block was undergoing compression at *c.* 590 Ma, whereas Laurentia was undergoing extension at that time. In turn, this is based upon the notion that the Ben Vuirich Granite post-dates an early and substantial phase of shortening in the Dalradian (Rogers *et al.* 1989), which both our and Tanner's (1996) observations have shown to be incorrect.

There are several reasons why neither the Moine nor the Dalradian Blocks can be exotic to Laurentia and have a Cadomian provenance.

- The Dalradian is on the Laurentian side of allochthonous arc and ophiolite terranes, which extend from the Southern Appalachians to the British Caledonides and whose collision with the Laurentian rifted margin was responsible for the Ordovician Grampian Orogeny. These ophiolite–arc sequences were initiated in the Middle Cambrian so that the Dalradian and Moine Block would have had to rift from the Cadomian margin of Iapetus across an ocean and collide with the Laurentian rifted margin all in the Early Cambrian.
- The upper Dalradian Leny Limestone contains *Pagetides*, a Laurentian tribolite.
- Although the Moine and Dalradian rift complexes are wider and better developed than elsewhere, such rift sequences extend along the whole Appalachian–Caledonian margin of Laurentia.
- There is no discernible orogenic event-unconformity or molasse sequence to mark any collision of Dalradian Moine Blocks with the Laurentian margin.
- There is no obvious suture along which Moine–Dalradian Blocks might have collided with Laurentia. One would have to appeal to a suture buried beneath the Scandian Moine Thrust. A transform terrane boundary, such as the Great Glen, is of no avail in docking Highland terranes because a transform does not consume oceanic lithosphere but merely redistributes blocks and fragments along the same orogenic margin.
- The continuity of the M and J deep reflectors (Hynes & Snyder 1995) beneath the Highlands suggest crustal continuity from the Laurentian foreland to the Highland Boundary Fault.
- Bluck & Dempster (1991) argue that the present distribution of Cambrian Ordovician carbonates of the Durness Laurentian margin platform, truncated by the Moine Thrust, are not sufficiently distal shelf edge and, therefore, that the Moine is exotic to Laurentia. The current authors find this argument hard to understand because the Durness carbonates could have extended well over the Moine and part of the Dalradian, and the original shelf geometry cannot be deduced. Also, there is an interesting and peculiar difference between the carbonate platform of the Southern and Northern Appalachians v. Newfoundland and the British Caledonides. In the Southern Appalachians, a full thermal subsidence curve is apparent with a steep Early Cambrian portion whereas, in Newfoundland and the British Caledonides, the

subsidence curve is much flatter in thinner platform sequences. This could mean that: (1) the earliest phases of thermal subsidence were above sea level in decaying rift shoulders; or (2) that outboard regions of the shelf have been displaced after the early Llanvirn; or (3) that outboard regions of the shelf were overridden substantially by marginal thrust sheets. Reason 1 is preferred because the shelf edge in Newfoundland can be defined precisely and the platform shares precisely the same subsidence pattern as the Scottish Durness platform.

- The Moine rests unconformably upon a Grenville basement of clearly Laurentian provenance.
- Dalradian rocks, in western Ireland, are thrust over the Grenville Erris Complex. Along this thrust, there is not the faintest hint of ophiolites or arc rocks that would allow its interpretation as a suture and it is clear that the Dalradian in western Ireland belongs to the Laurentian margin (Winchester 1992).

There is no compelling evidence for Proterozoic contractional orogeny and it is concluded, following Soper (1994), Soper & England (1995) and Soper & Harris (1997), that the Scottish Highlands were dominated by regional Riphean–Vendian lithospheric extension from immediately post-Grenville to latest Neoproterozoic times (Glover et al. 1995), principally in a complex of half-grabens that progressed mainly eastwards from the Laurentian foreland. Stretching was locally associated with magmatism, commonly alkaline and sometimes bimodal with mafic dyke swarms, and was periodic, alternating with phases of thermal subsidence. Locally, subhorizontal penetrative ductile fabrics were developed in the floors of rifts together with 'detachment' shear zones that probably have both structural and unconformable relationships with rift strata. Most fabrics were probably localized and not regionally pervasive. Garnet growth accompanied early extension whose fabrics are preserved as inclusion trails. Other sites of very localized contractional deformation in overlapping rift complexes might have been strike-slip and transpressional relays between rifts, transpressional jogs in these relays, and in propagating rift tips with their rotational strains. In this tectonic environment, differing bulk strains can be very patchy and localized, with quite rapid variations between horizontal extension and shortening.

The concept of a long, slow, but intermittent, extension of the continental lithosphere prior to

Late Riphean opening of Iapetus, inherent in our view of the Moine and Dalradian, is not a unusual concept in continental tectonics. In the Canning Basin, in northwest central Australia, complex diachronous rifting occurred through Devonian–Carboniferous times (Veevers 1984); in East Greenland, intermittent rifting from Baltica occurred from Carboniferous to Palaeocene times (Escher & Wyatt 1976) when continental rupture occurred. Also, individual yet substantial portions of a rifted margin may develop at substantially different times: e.g. the Central Atlantic Ocean opened between North America and Africa at c. 165 Ma, then the South Atlantic at c. 125 Ma, the Labrador Sea from c. 80 Ma and the North Atlantic at c. 55 Ma (Dewey et al. 1988). Thus, individual sections of a rifted margin may have protracted and complex histories, and long-rifted margins may develop at different times.

If the opening of Iapetus represents the termination of Neoproterozoic rifting, it's precise timing is a minor problem in relation to the available data. The Eriboll Quartzite (Early Cambrian) transgression at c. 545 Ma represents thermal subsidence following rift separation of Laurentia, probably from western South America (Dalziel 1994), whereas the Lighthouse Cove/Tayvallich Volcanics at 595 Ma, with the Long Range dykes at 605–615 Ma, most likely reflect the end of lithospheric stretching and the beginning of flood basalt volcanism that was the immediate harbinger of continental separation and the growth of new ocean floor. The southeastern Laurentian margin was a classic rifted continental margin of late Precambrian–Cambrian age with great along-strike continuity from Alabama to Newfoundland (Dewey 1969). In Newfoundland, a rapid facies change from a Cambro-Ordovician carbonate shelf, following a lower Cambrian quartzite transgression, occurs through shelf-edge carbonate breccias into offshelf continental-rise, finer grained clastics (Fig. 2). A similar facies change occurs along the western margin of the entire Southern and Northern Appalachians. In Britain, the shelf edge is not preserved; the Moine series overthrusts the Laurentian shelf along the Moine Thrust. The carbonate shelf records thermal subsidence following continental stretching, rifting and separation, probably from the pre-Andean western margin of South America, prior to 545 Ma. Continental stretching with the concurrent development of thick rift sequences occurred, with sequential periods of rifting and thermal recovery, from at least 900 Ma to latest Precambrian times (Fig. 3). Continental stretching was finally terminated by separation and the beginning of the oceanic growth of Iapetus during the period from *c.* 600 Ma to 545 Ma (Anderton 1980), characterized by extensional suites of mafic dykes, plateau basalts and alkaline granites [Tibbit Hill Volcanics (554 Ma; Kumarepeli *et al.* 1989); Lighthouse Cove Basalts, Lady Slipper Pluton (555 Ma; Cawood *et al.* 1996); Hare Hill Granite (608 Ma; Currie *et al.* 1992); Round Point Granite (602 Ma; Williams *et al.* 1985); Tayvallich Volcanics (595 Ma; Halliday *et al.* 1989); Ben Vuirich Granite (590 Ma; Rogers *et al.* 1989); Carn Chuinneag Granite (563 Ma; Pidgeon & Johnson 1974); Long Range dykes (615 Ma; Kamo *et al.* 1989)]. The similar (600 Ma) ages of alkaline to subalkaline plutons in Colonsay (Muir *et al.* 1994), $550 \pm 8$ Ma for the Cheticamp Pluton (Jamieson *et al.* 1986) and $551 \pm 1$ Ma (Lin *et al.* 1997) for metavolcanic sequences in the western Cape Breton Highlands, are probably also related to terminal continental stretching. The Lady Slipper and Carn Chuinneag intrusions may be related to persistent hotspot magmatism along a magmatic continental margin. In the northeastern Ox Mountains, western Ireland, garnetiferous psammites and metabasites of the Slishwood Division show isothermal decompression of eclogite, followed by isobaric cooling of kyanite granulites, followed again by amphibolite-facies decompression and retrogression with later mylonite zones (Sanders 1994). The eclogite-facies metamorphism and deformation was probably in Grenville basement (Daly 1996); the amphibolite-facies event was probably of Grampian (Ordovician) age (Sanders 1994), whereas the 605 Ma decompression may have been associated with immediately pre-Iapetus continental extension. One of the authors (JFD) has seen subhorizontal amphibolite-facies extensional fabrics wrapping eclogite boudins in the northeastern Slishwood and in the Lough Derg Inlier to the northeast; this may also date coeval extension. Similarly, the 576 Ma garnet metapyroxenite in the Ballantrae Complex (Hamilton *et al.* 1984) may reflect extreme continental stretching along the Iapetan margin similar to that along the Galicia rifted mar-in where lower continental crust and mantle are exhumed along detachment footwalls (Boillot *et al.* 1988).

During the Early to Middle Ordovician, the Laurentian rifted continental margin collided with a continentward-facing oceanic arc(s). This is especially clear in Newfoundland where the obduction of suprasubduction-zone ophiolites was immediately followed by subduction polarity flip and the establishment of the continental margin the Notre Dame Arc, characterized by

the tonalite 'sea' (Fig. 17, items 25–30), and the Humber and Notre Dame Bay Zones (Fig. 1) (Van Staal *et al.* 1999). Even within Newfoundland, the collision and flip was complex and diachronous, beginning in the Tremadocian (Fig. 17, item 17) in the southwest while suprasubduction-zone ophiolites, such as the Betts Cove (Bedard *et al.* 1998) and Bay of Islands Complex (Fig. 17, items 10–15), were still developing, perhaps the result of tectonic 'headlands' and 're-entrants' in the Laurentian rifted margin (Van Staal *et al.* 1999).

The suprasubduction-zone ophiolites extensionally fragmented a Middle–Upper Cambrian infant oceanic arc (Fig. 17, items 6–9, 33–95) which may have nucleated, at least partly, on an oceanic transform fault/fracture zone (Karson & Dewey 1978; Casey & Dewey 1984). The ophiolites were generated, as is also common for the Tethyan Cretaceous ophiolite complexes (Dewey 1974), just prior to their obduction and formed very close to the Laurentian margin (Trench & Torsvik 1992). They form not only the highest sheets (e. g. Bay of Islands Complex) of Ordovician thrust complexes but acted as accretionary prism back stops (Deer Park Complex), and the foundation of Arenig–Llanvirn fore-arcs and fore-arc basins (South Mayo Trough) (Dewey & Ryan 1990). Structural, igneous and metamorphic relationships in rocks of the infant arc (Lake Ambrose/Twillingate, early Ballantrae Complex), the ophiolites and the evolved pre-collisional arc (Lough Nafooey/Glensaul, Balcreuchan group) (Fig. 17) are varied and complicated, and comprise an assemblage with an already complex history when it collided with the Laurentian margin. The Mount Barren Complex (Fig. 1) probably formed in an oceanic transform/fracture zone system with complicated polyphase relationships between structural and igneous events (Karson & Dewey 1978), and are intruded by mafic rocks of the Bay of Islands Complex. Foliated low-grade amphibolite xenoliths in the infant arc Twillingate calc-alkaline pluton may represent fragments of the oceanic lithosphere upon which the infant arc was developed; similarly, the dismembered ophiolites of the Deer Park Complex contain foliated amphibolites. The extensional fragmentation of the infant arc that led to the suprasubduction ophiolites is spectacularly recorded by the mafic dyke complexes that cut the Twillingate Pluton and locally formed sheeted complexes such as the Moreton's Harbour (Fig. 17, item 20). The Ballantrae Complex records an evolving primitive arc with Arenig tholeiites and boninites developed in an extensional, probably intra-arc, setting on and within a Tremadoc and older calc-alkaline arc (Smellie & Stone 1992). The ultramafic components of Ballantrae suprasubduction-zone ophiolites have a calc-alkaline affinity (Smellie & Stone 1992) and complicated, partly diapiric, relationships with the complex (Dewey 1974), recalling the relationships described by Bloomer *et al.* (1995) and Fryer *et al.* (1995) for the Izu-Bonin primitive oceanic arc, where boninites, island arc tholeiites and diapiric forearc serpentinites fragment and disrupt the earlier parts of the arc.

The origin, development and collision of the evolved oceanic arc with its polyphase igneous and structural complexity was a progressive event beginning with the nucleation of the arc in the Middle Cambrian, subduction to form the infant and evolved arc, then collision leading to ophiolite obduction with high-temperature sole granulites and amphibolites (Appendix 3, items 21, 22 and 41) during the Arenig and Llanvirn. The collisional suture is well preserved in Newfoundland as the Baie Verte Line (Fig. 1b), in Ireland as the Clew Bay Zone and, in Scotland, as the Highland Border Zone (or Complex) in a strike-slip disrupted form. Within this suture zone are variably preserved rock assemblages that testify to the progressive sweeping-up (subduction-accretion) of deformed packages by the oceanic arc, a process that led directly, continuously and diachronously into the deformation of the Dalradian and Moine. These progressively-accreted packages include the Quebec blueschists (505–490 Ma, item 32), the blueschists and phengite/pyrope-rich garnet-bearing rocks, the albite schists/Ballytoohy/Killadangan assemblages of the Clew Bay Zone (Yardley *et al.* 1987; Harris 1995), and the progressive involvement by tectonic feathering (Shackleton 1986) of the distal Dalradian. This concept is expressed in the confusion over what constitute Dalradian v. exotic assemblages along the Highland Border in Scotland. The confusion and argument are semantic and artificial. The Highland Border Zone contains a spectrum of assembled scraped-up rocks from Proterozoic to Cambro-Ordovician Dalradian lithologies to Tremadoc–Arenig forearc rocks. For example, the suture zone contains amphibolites of perhaps at least five types and ages mangled in fairly indecipherable form including Early Cambrian Iapetus rifted margin mafics such as the Bute Amphibolite (540 Ma; Dempster & Bluck 1991), Cambrian transform amphibolites (Twillingate xenoliths and Deer Park amphibolites), seamounts, intra-arc shear zone amphibolites and obduction sole amphibolites. Essentially, the suture zone contains a range of unrelated and casually-related rock assemblages swept up

at the leading subduction edge of the arc and then accreted to the Dalradian sequence at the 'feathering' onset of deformation. It is suggested that the eclectic range of lithologies and ages, and their relationships along the Highland Border (Curry *et al.* 1982, 1984), are the result of this subduction tectonic sweeping-up process. Arc collision and ophiolite obduction led to the progressive Grampian Arenig–Llanvirn shortening of the Dalradian and Moine sequences of the Laurentian margin. During Arenig times, the high-level ophiolite sheet was eroded during its emplacement and, by the late *artus* Biozone, a subophiolite Barrovian D2–D3 metamorphic pile had been developed and begun to be eroded. The access of asthenosphere to the base of the deforming nappe pile allowed the injection of substantial volumes of mafic and ultramafic magmas, probably yielding the thermal source for the Barrovian metamorphism.

The vergence of the D2–D3 fold nappes of Connemara appears to have been towards the Laurentian continent, and the Grampian nappes and thrusts of the Moine have a northwesterly vergence. However, there is a problem with the vergence of the early Dalradian fold nappes in Scotland. Mendum & Thomas (1996) have argued for a northwestward D2 vergence whereas Krabbendam *et al.* (1997) asseverate a southeastward vergence which follows the classical view (Sturt 1961) of a dual-polarity Grampian Orogen with a *zwischengebirg* Loch Awe Syncline. A southeastward vergence is not obviously consistent with an arc–continent collision in its simplest form but, if correct, points to geometric and kinematic differences between the Taconic–Humber Orogen of Newfoundland and the Grampian Orogen in the British Isles. In Newfoundland, the sharp Early Ordovician shelf edge and continental rise, and their progressive destruction by arc collision and ophiolite obduction to form a rather narrow orogen, the Humber Zone, are very clear, whereas, in the British Isles, the Grampian Orogen is substantially wider (Fig. 1) with a more complex deformation history and geometry, a higher metamorphic grade and the shelf edge was destroyed by Scandian, mid-Silurian shortening across the Moine Thrust. It is suggested that these differences are a consequence of differences in the geometry of the Laurentian rifted margin of Iapetus, similar to the variations seen along present-day rifted margins which can vary from margin to margin and along-strike from quickly developed sharp, well-defined continent–ocean boundaries to more complex, wider, diffuse zones of lithospheric stretching and rifting developed over long periods, wide plateaux of attenuated continental lithosphere (Exmouth Plateau) or volcanic margins with aggradational basaltic sequences, like the Rockall Bank, developed at hotspot sites. Perhaps the ophiolite obduction, foreland basin and narrow orogenic style of Newfoundland reflects a sharp margin with single, continentward, structural polarity, whereas the wider, more structurally complex style of the Scottish Grampian Zone resulted from the bulldozing of a wide rift complex with more limited ophiolite obduction. Another substantial, perhaps related, difference, is that the Appalachian Laurentian foreland has a Grenville basement (*c.* 1000 Ma), whereas, in Scotland, the foreland basement is Lewisian (*c.* 2500–1700 Ma). The Grampain Orogen is underlain by Grenville rocks (Sanders *et al.* 1984) and the Moine Thrust Zone is probably a zone of repeated contraction and extension as the Grenville Thrust Front, extensional detachment at the beginning of Moine sedimentation, Grampian thrusting and extensional detachment, and, finally, Silurian thrusting (Kelley 1988; Kelley & Bluck 1989; Kelley *et al.* 1994).

Grampian contraction and crustal thickening on early recumbent structures led to horizontal shortening across steep structures (e.g. D3 in the Moine) and to back-rotation and retrocharriage reflected, for example, in the D4 growth of the Connemara Antiform, the development of the Clifden Steep Belt and the south-south-eastwards thrusting along the Mannin Thrust, which placed Dalradian metamorphic rocks back over part of the arc. The steepening of early structures along the Baie Verte Line, the Clew Bay Zone and the Highland Border are probably part of this event. It is also possible that the southward vergent D2 flat-belt structures in the Grampian Highlands (Krabbendam *et al.* 1997) were developed during early retrocharriage. This retrocharriage phase, at *c.* 460 Ma in Connemara, marks the end of orogenic growth by collisional tightening and the propagation of subduction polarity flip. This led to the diachronous growth of a new trench in which Iapetus oceanic lithosphere was subducted beneath the Laurentian margin, with its newly accreted arc. Along this trench, Ordovician subduction accretion is recorded by 'clipped-off' slices of oceanic crust (Annieopsquotch 'Ophiolite' Complex) and seamounts (mafic rocks of the South Connemara Series and of the northern belt of the Southern Uplands), which continued into the Silurian (Van Staal *et al.* 1999) with the development of the clastic accretionary prism of the Southern Uplands and led to collision with exotic arcs (Grangegeeth; Owen *et al.* 1992) and,

finally, to Late Silurian transpressional collision (Soper *et al.* 1992) that closed the last remnants of Iapetus.

Sinistral strike-slip was endemic along the British and Irish segment of the Laurentian margin from, at least, post-Mweelrea–Early Devonian times. The detachment of the Connemara Terrane from west of Ireland is recorded, probably, in the extensional detachment system immediately beneath the upper Llandovery unconformity (Williams & Rice 1989), and it's docking with the Laurentian margin by the Derryveeny conglomerates and the post-Mweelrea–pre-Upper Llandovery shortening of the Ordovician rocks of the South Mayo Trough. That very large sinistral motion may have occurred along some terrane boundaries is suggested by the apparent absence of Silurian deformation in the Dalradian, which is present in the Moine and recorded by mid-Silurian thrusting along the Moine Thrust between *c.* 430 and 420 Ma. This Scandian deformation was the result of the Laurentia–Baltica collision; it's absence in the Dalradian may be because the Great Glen Fault had enough post-Scandian sinistral motion to bring the Dalradian from a position along the Laurentian margin sufficiently to the southwest that escaped the Scandian collision.

We recognize, with gratitude, discussions on Appalachian–Caledonian geology in the field, office and meetings, over the years, with many geologists, especially Tony Harris, Donny Hutton, Bill Kidd, Stuart McKerrow, Paul Ryan, Robert Shackleton, Jack Soper, Cees Van Staal and Hank Williams. We thank David Wright for assistance with the heavy mineral work; Claire Carlton for her drafting expertise; Sally Thompson for her patient work on the manuscript; Peter Clift, Stuart McKerrow and Philip Stone for meticulous reviews; and Conall Mac Niocaill for his patient encouragement.

## Appendix 1 (Ordovician Strata)
## (for references see caption to Fig. 5)

*Ballytoohy 'Series'* (from base to top; Phillips 1973). *Carrickarollagh Formation*. Black mudstones, grey charts, umbers and sideritic mudstones, black graphitic slates, laminated mudstones, oolitic and hydrozoan limestones, green greywacke turbidites and mass flow deposits. *Siorr Formation*. Black, grey and green cherts, chocolate Mn–Fe umbers and sandstones (cherts contain *Protospongia hicksii*). *Tonalattarive Formation*. Spilite and carbonate. *Formation*. Thick sandstones and melanges containing blue quartz, spilite, quartzite and muscovite–plagioclase schist. *Oonaghcarragaun Formation.* Black pelites and sandstones. The whole series is probably a melange assemblage.

*Killadangan (Leckanvy) 'Formation'.* Green, graded, poorly sorted, medium–fine-grained sandstones and subarkoses, pebbly sandstones, siltstones, black graphitic slates, grey cherty argillites, dark calcareous mudstones, green and black contorted mudstones, thin carbonates, black mud–chip, chert, white quartz and vein quartz pebble beds, abundant blue rutilated quartz, quartz (both monocrystalline and polycrystalline), subrounded to subangular. Probably correlative with the Benilra and Oonaghcarragaun Formation of the Ballytoohy Series, and with the Ardvarney, Callow and Clooneygowan 'Formations' in the Ox Mountains, the Portruckagh 'Formation' on Clare Island and the Westport Grits. The 'formation' is a melange assemblage, probably developed in a subduction–accretion setting.

*Letter 'Formation'* (Graham & Smith 1981). Brown to green sandstones, brown manganiferous and sideritic mudstones, mafic volcanics. Probably correlative with the Siorr Formation of the Ballytoohy Series.

*Bohaun Formation (>1000 m).* Magnesian (boninitic) spilitized massive pillow lavas with jasper lenses and pillow breccias (albite–oligoclase, epidote and chlorite), nodular green cherts and jaspers, calcareous shales. Probably correlative with the Knock Kilbride Formation.

*Srah Formation (120 m).* Granule sandstones and polymict conglomerates (quartz–porphyry, black chert, tuff and spilite).

*Tourmakeady Formation (400 m).* Silicic tuffs, grits, mudrocks, shales, limestones and limestone breccias.

*Rhyolite Formation (600 m).* Massive flow-banded, flow-brecciated and columnar rhyolite.

*Mount Partry Formation (400 m).* Chert, argillite, andesite breccias and volcaniclastics.

*Derry Bay Formation (130 m).* Volcaniclastics, arkosic arenites, sandstones, siltstones, black pyritous slates and pebble beds (silicic volcanics, granite and spilite).

*Knock Kilbride Formation (390–600 m).* Mafic pillow lavas and hyaloclastites with middle member of epiclastic andesite breccias, tuffs and cherts.

*Finny Formation (390–590 m).* Spilitic pillow lavas, keratophyres, epiclastic andesite breccias, limestone breccias and cherts with air-fall volcanic blocks.

*Bencorragh Formation (>855 m).* Magnesian mafic pillow lavas, pillow breccias and hyaloclastites (plagioclase–phyric with no albite and microcline–albite groundmass). Volcanics of Lough Nafooey Group essentially submarine tholeiitic to calc-alkaline basalts and basaltic andesites.

*Letterbrock Formation (>250 m base not seen).* Blue-green slates and greywacke (massive lithic–felspathic) turbidites (both proximal and distal), common quartz, chlorite and chloritized rock fragments, frequent plagioclase, alkali felspar, actinolite epidote, chlorite, spilite and dole-mite, rare muscovite, sphene, zoizite and chert; clay–pebble conglomerates, conglomerates (white trondjemite, black-green chert, sandstone, shale–mudstone, jasper, spilite, dolerite, minor gabbro, silicic volcanics, low-grade meta-sandstones, vein quartz, white-pink quartzite, grey-green schist and albite schist). Provenance from the north.

*Derrymore Formation (700 m).* Thinly interbedded, mainly fine–medium-grained felspathic sandstone turbidites and grey-green to dark grey or black slates, rarely with fuchsite/chlorite/pyrite shear zones with secondary dolomite and siderite, a few thin orange-weathering tuffs, scarce thin pebble beds (jasper, spilite and albite–strained quartz). Provenance from the north, northeast and east.

*Sheeffry Formation (2500 m).* Green to grey-green turbidites (much clear volcanic quartz and chloritized spilite, frequent plagioclase, alkali felspar, chlorite, biotite and jasper, minor epidote and muscovite) and dark grey slates with white-weathering silicic tuffs and submarine lahars, common siderite–ankerite nodules, and fuchsite–brunerite shear zones. Provenance from the north, east and south, becoming thinner and more pelitic westwards.

*Derrylea Formation (1700 m).* Massive greywacke turbidites, with common ferroan dolomite/siderite/ankerite nodules, green slates and cream-weathering porcellanous tuffs. Dark green turbidites (chromite and chlorite rich) derived from the north and east, paler grey-green turbidites (with abundant clear and smoky quartz and alkali felspar) derived from the south; lower part of formation with detrital albite–oligoclase changing to andesine near the top. Turbidites contain common quartz, orthoclase, microcline and plagioclase, frequent muscovite, chlorite and chromite, and minor epidote, hornblende, zircon and sphene. Towards the top of formation, metamorphic detritus (staurolite and garnet metamorphic quartz) becomes important and minor pebble conglomerates in channels contain microcline–granite, fine-grained quartzite, slate, jasper, black chert, spilite, chlorite–muscovite–schist and serpentinite. The formation coarsens to the east in thicker beds.

*Maumstrasna Formation (900 m).* Coarsening-upwards sequence of alluvial and shallow-marine fanglomerates and coarse subarkosic sandstones with common quartz, alkali-feldspar, frequent plagioclase and, towards the top of the formation, muscovite and biotite. Pebbles and cobbles of quartz-rich granite, quartz–felspar porphyry, vein quartz, jasper, grey-green to purplish mudstone, silicic tuff, sandstone, biotite–muscovite, and hornblende granite and granodiorite. Towards the top of the formation, a minor metamorphosed pebble component appears of mica schist, psammite, quartzite and migmatite. Provenance principally from the southeast.

*Rosroe Formation (1200 m).* Deep-water marine proximal fans of coarse, thick, channel sandstones (common igneous quartz, alkali felspar and plagioclase, frequent chlorite and mafic rock fragments, rare muscovite and garnet in the upper part of the formation) and conglomerates, thin sandstones, interbedded thin sandstones and grey mudstones and grey/grey-green mudstones derived principally from the south, southeast and east; three green andesitic tuffs (50–50% andesine, 10–15% basaltic hornblende and 10% quartz) and several discontinuous units of dark grey biosparites, biomicrites and limestone breccia. Rounded boulder conglomerates dominate in the east and diminish westwards (massive and flow-banded quartz–felspar porphyry, coarse pinkish granite, pink-green rhyolite, spilite, jasper, green, black and turquoise chert, biotite ± hornblende and biotite ± muscovite granite, tonalite, mongonites and granodiorite; towards the top of the formation, psammitic schists, quartzites and schists appear).

*Glenummera (Lough Shee) Formation (600 m in north to 300 m in south).* Grey-green slates, mudrocks and plate cherty argillites with minor

thin 'dilute' distal and nepheloid sandstone turbidites with common metamorphic and igneous quartz, alkali-felspar and muscovite, frequent plagioclase, chlorite and biotite and rare chromite.

*Mweelrea Formation (>3200 m in the north to >2500 m in the south, top not seen).* Deposited in humid (Pudsey 1984) alluvial fans, fan deltas and braided rivers that flowed westwards; red, brown, pink, grey and brown-green litharenites (common igneous and metamorphic quartz and alkali felspar, frequent muscovite, plagioclase and chlorite, garnet and biotite). Pebble beds and conglomerates contain welded tuff, pale chert, white vein quartz, jasper, quartzite, psammitic schists, muscovite–biotite schist, graphitic schist, garnet schist, pink and white siliceous minimum-melt granite, dark green basalt, rhyolite, rhyodacite porphyry, foliated granite and alluvial fan conglomerates coarser, common and thicker to the south with boulders up to 1 m derived from the southeast and east. Three marine slate bands thickening to the northwest and six ignimbrite horizons.

*Derryveeny Formation.* Polymict conglomerates with minor sandstones. Boulders to 0.5 m of cordierite and garnet schist, quartzite, gneiss, migmatite, foliated and non-foliated granite, quartz–felspar porphyry, vein quartz, jasper, chert, biotite–muscovite garnet peraluminous granite, leucogranite, biotite granite, mylonite, mafic igneous rocks, tuff and felsite. Provenance from the southeast.

*Lettermullen Formation (>1035 m).* Turbidites with conglomerate lenses containing boulders and pebbles of quartz–plagioclase porphyry non-foliated granite, granite pegmatite, greenschist slates and phyllite quartzite, psammitic schist, foliated granite, muscovite–biotite schist, andesine–oligoclase gabbro, and andesine-oligoclase amphibolite.

*Golam Formation (350 m).* White-weathering bedded grey chert and argillitic chert with manganiferous cherty argillites at base.

## Appendix 2 (Silurian Strata) (for references see caption to Fig. 7)

*Derryheagh Formation (150 m).* Dark pelites in the north, shallow marine calcareous sandstones in the south.

*Cregganbaun Formation (1360 m).* Shallow-marine, cross-bedded quartzites with 250 m of conglomerate at base with boulders and pebbles of white quartzite, jasper, slate, vein quartz, tuff and fine-grained mafic rocks.

*Bouris Formation (1000 m).* Shallow-marine calcareous sandstones and pelites.

*Lough Nacorra Formation (350 m).* Shallow-marine, current-bedded massive subarkoses.

*Gowlaun Formation (400 m).* Upward-deepening sequence from non-marine basal breccia through shallow-marine sandstones and grey-green mudstones to turbidites and slide boulder conglomerates containing quartzite, greyish pink sandstone (Mweelrea), rhyolite, ignimbrite, andesite, schist, dolomite, jasper and green chert.

*Lettergesh Formation (730 m in east plus 1520 m in west).* Feldspathic greywacke turbidites with tuffs and granodiorite sills.

*Glencraff Formation (65 m).* Fine-grained arkosic greywackes and mudstones, tuffs near top. Provenance to northwest.

*Lough Muck Formation (2005–280 m).* Transition from greywackes and mudstones of the Glen-craff Formation to shallow-marine sediments of the Salrock Formation.

*Salrock Formation (>817 m).* Red and green shales and siltstones, thin sandstones and tuffs.

*Lough Mask Formation (170 m).* Red fluvial to tidal flat sandstones with a basal albite trackyte.

*Kilbride Formation (340 m).* Upward-deepening, shallow-marine sandstones.

*Doon Rock Formation (400 m) (top not seen).* Green-grey turbidites.

*Kill Formation (0–30 m).* Cross-bedded pebbly sandstone (metamorphic quartz, orthoclase, microcline, albite–andesine, muscovite, biotite, zircon, tourmaline, magnetite, hornblende and chert) in channels with calcreted red mudstones. Basal conglomerates with pebbles of vein quartz, jasper, psammitic schist, biotite–schist and amphibolite. Provenance from the north.

*Strake Formation (250 m).* Fluvial laminated red siltstones, and red and green sandstones and tuffs. Provenance from the west and southeast.

*Knockmore Formation (400 m).* Mainly fluvial sandstones derived from the east.

*Bunnamohaun Formation (300 m).* Non-marine mudrocks and subordinate sandstones.

*Glen Formation (550 m).* Non-marine coarse sandstones, polymict conglomerates and thin red mudrocks. Provenance mainly from the northeast.

## Appendix 3

1. Carn Chuinneag Granite $550 \pm 10$ (U/PbZr); Pidgeon & Johnson (1974)
2. Portsoy Granite $669 \pm 17$ (Rb/Sr); Pankhurst (1974)
3. Tayvallich Volcanics $595 \pm 4$ (U/PbZr); Halliday *et al.* (1989)
4. Ben Vuirich Granite $590 \pm 2$ (U/PbZr); Rogers *et al.* (1989)
5. Ballantrae eclogite $576 \pm 32$ (U/PbZr); Hamilton *et al.* (1984)
6. Lake Ambrose rhyolite $513 \pm 2$ (U/PbZr); Dunning *et al.* (1991)
7. Twillingate Pluton $510 ^{+17}_{-16}$ (U/PbZr); Williams *et al.* (1976)
8. Twillingate Pluton $507 ^{+3}_{-2}$ (U/PbZr); Elliott *et al.* (1991)
9. Littleport trondjemite $505 ^{+3}_{-2}$ (U/PbZr); Jenner *et al.* (1991)
10. Brighton Gabbro $495 \pm 5$ ($^{40}Ar/^{39}Ar$); Stukas & Reynolds (1974)
11. Pipestone Pond ophiolite $493 ^{+2.5}_{-1.9}$ (U/PbZr); Dunning & Krogh (1985)
12. Grand Lake trondjemite $490 \pm 4$ (U/PbZr); Cawood *et al.* (1996)
13. Betts Cove Ophiolite $488.6 ^{+2.6}_{-2.0}$ (U/PbZr); Dunning & Krogh (1985)
14. Bay of Islands Ophiolite Complex $485.7 ^{+1.9}_{-1.2}$ (U/PbZr); Dunning & Krogh (1985)
15. Bay of Islands Ophiolite Complex $484 \pm 5$ (U/PbZr); Jenner *et al.* (1991)
16. Mount Barren Complex (cooling) $489 \pm 6$ ($^{40}Ar/^{39}Ar$); McCaig & Rex (1986)
17. Cape Ray Igneous Complex $488 \pm 3$ (U/PbZr); Dubé *et al.* (1996)
18. Littleport Complex (cooling) $469 \pm 14$ (K/Ar); Archibald & Farrar (1976)
19. Lewis Hills shear zone (range) 475–460 ($^{40}Ar/^{39}Ar$); Idleman (1990)
20. Moreton's Harbour dykes $473 \pm 9$ ($^{40}Ar/^{39}Ar$); Williams *et al.* (1976)
21. Bay of Islands sole $469 \pm 5$ ($^{40}Ar/^{39}Ar$); Dallmeyer & Williams (1975)

22. Bay of Islands sole $460 \pm 5$ ($^{40}$Ar/$^{39}$Ar); Dallmeyer & Williams (1975)
23. Fleur de Lys garnets $448.6 \pm 2.3$ (Sm/Nd); Vance & O'Nions (1990)
24. Corner Brook Lake pegmatite $434 \, ^{+2}_{-3}$ (U/PbZr); Cawood et al. (1994)
25. Cape Brûle porphyry $475 \pm 10$ (U/PbZr); Mattinson (1977)
26. Coney Head Complex $474 \pm 2$ (U/PbZr); Dunning (1987)
27. Buchans Volcanics $473 \, ^{+3}_{-2}$ (U/PbZr); Dunning et al. (1987)
28. Cape Ray Tonalite $469 \pm 2$ (U/PbZr); Dubé et al. (1996)
29. Burlington Granodiorite $464 \pm 6$ (U/PbZr); Mattinson (1977)
30. Cormack's Lake Complex $460 \pm 10$ (U/PbZr); Currie et al. (1992)
31. Glover Island Granodiorite $440 \pm 2$ (U/PbZr); Cawood et al. (1996)
32. Taconian blueschists 505–490 ($^{40}$Ar/$^{39}$Ar); Laird et al. (1993)
33. Ballantrae metapyroxenite $576 \pm 11$ (Sm/Nd); Hamilton et al. (1984)
34. Ballantrae basalt $501 \pm 12$ (Sm/Nd); Thirlwall & Bluck (1984)
35. Shetland hornblende schist $498 \pm 2$ ($^{40}$Ar/$^{39}$Ar); Flinn et al. (1991)
36. Shetland trondjemite $492 \pm 3$ (U/PbZr); Spray & Dunning (1991)
37. Ballantrae gabbro $487 \pm 8$ (K/Ar); Bluck et al. (1980)
38. Ballantrae trondjemite $483 \pm 4$ (U/PbZr); Bluck et al. (1980)
39. Ballantrae gabbro 479 (K/Ar); Harris et al. (1965)
40. Shetland Hornblende schist $479 \pm 6$ (K/Ar); Spray (1988)
41. Ballantrae sole $478 \pm 4$ (K/Ar); Bluck et al. (1980)
42. Ballantrae basalt $476 \pm 14$ (Sm/Nd); Thirlwall & Bluck (1984)
43. Ballantrae gabbro $475 \pm 8$ (K/Ar); Harris et al. (1965)
44. Tyrone tonalite $471 \pm 24$ (U/PbZr); Hutton et al. (1988)
45. Connemara (D3) metasomatites $478 \pm 25$ (U/PbTi); Cliff et al. (1993)
46. Strichen Granite $475 \pm 5$ (U/PbMo); Pidgeon & Aftalion (1978)
47. Oughterard Granite $473 \pm$ ?(Rb/Sr); Tanner et al. (1997)
48. Morven Cabrach Gabbro 472 (Pb/PbZr); Rogers et al. (1995a)
49. Cashel–Lough Wheelaun metagabbro $471.1 \pm 0.8$ (U/PbZr); Friedrich et al. (1997)
50. Aberdeen Granite $470 \pm 1$ (U/PbMo); Kneller & Aftalion (1987)

51. Insch Gabbro $468 \pm 8$ (U/PbZr); Rogers et al. (1995a, b)
52. Connemara quartz–diorites $466.5 \pm 0.6$ (U/PbZr); Friedrich et al. (1997)
53. Connemara pegmatite $466 \pm 3$ (U/PbTi); Cliff et al. (1996)
54. Connemara post-D3 granite $466 \pm 5$ (U/PbZr); Cliff et al. (1996)
55. Connemara cordierite pegmatites $463 \pm 4$ (U/PbMo); Cliff et al. (1996)
56. Connemara quartz–diorite gneiss $463 \pm 4$ (U/PbZr); Cliff et al. (1996)
57. Oughterard Granite $462 \pm 3.2$ (U/PbXe); Friedrich et al. (1997)
58. Mannin Thrust $460 \pm 25$ (Rb/SrWR); (Leake et al. 1983)
59. Oughterard Granite $460 \pm 5$ (Rb/SrWR); Leggo et al. (1969)
60. Mannin Thrust $447 \pm 4$ (Rb/Sr); Tanner et al. (1989)
61. Moine metamorphism $467 \pm 20$ (Rb/Sr); Brewer et al. (1979)
62. Glen Dessary Syenite $456 \pm 5$ (U/PbZr); Van Breemen et al. (1979a)
63. Moine (Glenfinnan) pegmatites, $455 \pm 4$, $450 \pm 10$ (U/PbMo); Van Breemen et al. (1974)

## Appendix 4

A, Antietam; AB, Albee; AGR, Argyll Group; AM, Ammonoosuc; AS, Aberfoyle Slates; B, Beekmantown; BA, Ballantrae Complex; BE, Ballantrae Eclogite; BH, Ballytoohy; BL, Blountian; BR, Bradore; BT, Bateau; BV, Balmville; BW, Botwood; CB, Cooks Brook; CBA, Clam Bank; CH, Chilhowee; CHE, Cheshire; CL, Clough; CP, Croagh Patrick; CS, Connemara Silurian; CW, Cow Head; DL, Derrylea; DN, Durness; DV, Derryveeny; E, Eriboll; F, Forteau; FL, Fleur de Lys; FT, Fitch; G, Goose Tickle; GL, Glenummera; GM, Golam; GO, Goldson; HB, Helderberg; HH, Hatch Hill; HO, Hoosac; I, Irishtown; IM, Illinois Mountain; KC, Kirkland Conglomerate; KD, Killadangan; LB, Letterbrock; LC, Lighthouse Cove; LE, Lettermullen; LL, Leny Limestone; LLG, Lower Leny Grits; LN, Lough Nafooey; LP, Long Point; LR, Leenane–Rosroe; MC, Mic Mac; MD, Macduff Slates; MM, Mount Musgrave; MP, Mudd Pond; MPO, Maiden Point; MT, Moine Thrust; MW, Mweelrea; N, Normanskill; NB, New Bay; OC, Ocoee; P, Pinnacle; PH, Pinney Hollow; PL, Poultney; PM, Point Leamington; PTD, Partridge; R, Rensselaer; RFM, Ryan's Farm; S, Summerside; SA, Sops Arm; SC, Spiral Creek; SG, Saint George; SH,

Sheeffry; SHG, South Highland Group; ST, Stockbridge; SW, Stowe; T, Tyson; TH, Tibbit Hill; THD, Table Head; TP, Tappins; TV, Tayvallich Volcanics; ULG, Upper Leny Grits; WC, West Castleton; WG, Westport Grits.

# References

ABBOT, L. D., SILVER, E. A. & GALEWSKY, J. 1994a. Structural evolution of a modern arc–continental collision in Papua New Guinea. *Tectonics*, **13**, 1007–1034.

——, ——, THOMPSON, P. R., FILEWICZ, M. V., SCHNEIDER, C. & ABDOERRIAS, A. R. 1994b. Stratigraphic constraints on the development and timing of arc-continent collision in northern Papua New Guinea. *Journal of Sedimentary Research*, **364**, 169–189.

ABERS, G. A. & McCAFFREY, R. 1994. Active arc–continent collision: earthquakes, gravity anomalies and fault kinematics in the Huen–Finisterre collision zone, Papua New Guinea. *Tectonics*, **13**, 227–245.

ALLEN, P. A. & MANGE-RAJETZKY, M. A. 1992. Devonian–Carboniferous sedimentary evolution of the Clair area, offshore north-western UK: impact of changing provenance. *Marine and Petroleum Geology*, **9**, 29–52.

ANDERTON, R. 1980. Did Iapetus start to open during the Cambrian? *Nature*, **286**, 706–708.

ARCHER, J. B. 1977. Llanvirn stratigraphy of the Galway–Mayo border area, western Ireland. *Geological Journal*, **12**, 77–98.

ARCHIBALD, D. A. & FARRAR, E. 1976. K–Ar ages from the Bay of Islands ophiolite and the Little Port Complex, western Newfoundland and their geological implications. *Canadian Journal of Earth Sciences*, **13**, 520–529.

AUZENDE, J. M., KROENKE, L., COLLOT, J. Y., LAJOY, Y. & PELLETIER, B. 1996. Compressive tectonism along the eastern margins of Malaita Island (Solomon Islands). *Marine Geophysical Research*, **18**, 289–304.

BARR, D., ROBERTS, A. M., HIGHTON, A. J., PARSON, L. M. & HARRIS, A. L. 1985. Structural setting and geochronological significance of the West Highland Granitic Gneiss, a deformed early granite in the Proterozoic, Moine rocks of NW Scotland. *Journal of Geological Society, London*, **142**, 663–675.

BEDARD, J. H., LAUZIÈRE, K., TREMBLAY, A. & SANGSTER, A. 1998. Evidence for fore-arc sea-floor-spreading for the Betts Cove Ophiolite, Newfoundland: oceanic crust of boninitic affinity. *Tectonophysics*, **284**, 233–245.

BIRD, J. M. & DEWEY, J. F. 1970. Lithospheric plate-continental margin tectonics and the evolution of the Appalachian Orogen. *Geological Society of American Bulletin*, **81**, 1031–1060.

BLOOMER, S. W., TAYLOR, B., MacLEOD, C. J., STERN, R. J., FRYER, P., HAWKINS, J. W. & JOHNSON, L. 1995, Early arc volcanism and the ophiolite problem: a perspective from drilling in the Western Pacific. *American Geophysical Union of Geophysics Monologue*, **88**, 1–30.

BLUCK, B. J. & DEMPSTER, T. J. 1991. Exotic metamorphic terranes in the Caledonides: tectonic history of the Dalradian block, Scotland. *Geology*, **19**, 1133–1136.

——, HALLIDAY, A. N., AFTALION, M. & MAC-INTYRE, M. 1980. Age and origin of Ballantrae ophiolite and its significance to the Caledonian orogeny and Ordovician time scale. *Geology*, **9**, 492–495.

BOILLOT, G., GIRARDEAU, J. & KORNPROBST, J. 1988. Rifting of the Galicia margin: crustal thinning and emplacement of mantle rocks on the seafloor. *Proceedings of the Ocean Drilling Programme, Scientific Results*, **103**, 741–756.

BOWES, D. R. 1986. The absolute time-scale and the subdivision of the Precambrian rocks of Scotland. *Geologiska Foreningens i Stockholm Forhandl.*, **90**, 175–188.

BRASIER, M. D. & McILROY, D. 1997. *Neonereites uniserialis* from *c.* 600 years old rocks in western Scotland and the emergence of animals. *Journal of the Geological Society, London*, **155**, 132–139.

BREWER, M. S., BROOK, M. & POWELL, D. 1979. Dating of the tectonic-metamorphic history of the southwestern Moine, Scotland. Geological Society, London, Special Publications, **8**, 129–137.

BROOK, M., BREWER, M. S. & POWELL, D. 1976. Grenville age for the rocks in the Moine of north western Scotland. *Nature*, **260**, 515–517.

——, POWELL, D. & BREWER, M. S. 1977, Grenville events in the Moine rocks of the Northern Highlands, Scotland. *Journal of the Geological Society, London*, **133**, 489–496.

BROWN, P. E., MILLER, J. A., SOPER, N. J. & YORK, D. 1965. Potassium–argon age pattern of the British Caledonides. *Yorkshire Geological Society*, **35**, 103–138.

CASEY, J. F. & DEWEY, J. F. 1984. Initiation of subduction zones along transform and accretionary plate boundaries, triple junction evolution and fore-arc spreading centres – implications for ophiolite geology and obduction. *In*: GASS, I. G., LIPPARD, S. J. & SHELTON, A. W. (eds) *Ophiolites and Oceanic Lithosphere*. Geological Society, London, Special Publications, **13**, 269–290.

CAWOOD, P. A., DUNNING, G. R., LUX, D. & VAN GOOL, J. A. M. 1994. Timing of peak metamorphism and deformation along the Appalachian margin of Laurentia in Newfoundland: Silurian, not Ordovician. *Geology*, **22**, 399–402.

——, VAN GOOL, J. A. M. & DUNNING, G. R. 1996. Geological development of eastern Humber and western Dunnage zones: Corner Brook–Glover Island region, Newfoundland. *Canadian Journal of Earth Science*, **33**, 182–198.

CLIFF, R. A., YARDLEY, B. W. D. & BUSSY, F. R. 1993. U–Pb isotopic dating of fluid infiltration and metasomatism during Dalradian regional metamorphism in Connemara, western Ireland. *Journal of Metamorphic Geology*, **11**, 185–191.

——, —— & ——1996. U–Pb and Rb–Sr geochronology of magmatism and metamorphics in the Dalradian of Connemara, western Ireland. *Journal of the Geological Society, London*, **153**, 109–120.

CLIFT, P. D. & RYAN, P. D. 1993. Geochemical evolution of an Ordovician island arc, South Mayo, Ireland. *Journal of the Geological Society, London*, **150**, 1–14.

CROWHURST, P. V., HILL, K. C., FOSTER, D. A. & BENNETT, A. P. 1996. Thermochronological and geochemical constraints on the tectonic evolution of northern Papua New Guinea. *In*: HALL, R. & BLUNDELL, D. J. (eds) *Tectonic Evolution of Southeast Asia*. Geological Society, London, Special Publications, **106**, 523–537.

CULLEN, A. B. 1996. Ramu Basin, Papua New Guinea; a record of late Miocene terrane collision. *American Association of Petroleum Geology Bulletin*, **80**, 663–684.

CURRIE, K. L., VAN BREEMEN, O., HUNT, P. A. & VAN BERKEL, J. T. 1992. Age of high-grade gneisses south of Grand Lake, Newfoundland. *Atlantic Geology*, **28**, 153–161.

CURRY, G. B., INGHAM, J. K., BLUCK, B. J. & WILLIAMS, A. 1982. The significance of a reliable Ordovician age for some Highland Border rocks in Central Scotland. *Journal of the Geological Society, London*, **139**, 451–454.

——, BLUCK, B. J., BURTON, C. J., INGHAM, J. K., SIVETER, D. J. & WILLIAMS, A. 1984. Age, evolution and tectonic history of the Highland Border Complex, Scotland. *Roy. Soc. Edin. Trans.*, **75**, 113–134.

DALLMEYER, R. D. & WILLIAMS, H. 1975 $^{40}$Ar/$^{39}$Ar ages from the Bay of Islands metamorphic aureole: their bearing in the timing of Ordovician of ophiolite obduction. *Canadian Journal of Earth Sciences*, **12**, 1685–1690.

DALY, J. S. 1996. Pre-Caledonian history of the Annagh Gneiss Complex, northwestern Ireland, and correlations with Laurentia–Baltica. *Irish Journal of Earth Sciences*, **15**, 5–18.

DALZIEL, I. W. D. 1994. Precambrian Scotland as a Laurentia–Gondwana link; origin and significance of cratonic promontories. *Geology*, **22**, 589–592.

DAVIES, H. L. 1971. Peridotite–gabbro–basalt complex in eastern Papua: an overthrust plate of oceanic mantle and crust. *Australian Bureau Mineral Research Bulletin*, **128**.

DE SMET, M. E. M., FORTUIN, A. R., TROELSTRA, S. R., VAN MARLE, L. J., KARMINI, M., TJOKROSAPOETRO, S. & HADIWASASTRA, S. 1990. Detection of collision-related vertical movements in the Outer Banda Arc (Timor, Indonesia), using micropalaeontological data. *Journal of Southeast Asian Earth Sciences*, **4**, 337–356.

DEMPSTER, T. J. & BLUCK, B. J. 1991. The age and tectonic significance of the Bute amphibolite, Highland Border Complex, Scotland. *Geological Magazine*, **128**, 77–80.

——, HUDSON, N. F. C. & ROGERS, G. 1995. Metamorphism and cooling of the NE Dalradian. *Journal of the Geological Society, London*, **152**, 383–390.

DEWEY, J. F. 1961. A note concerning the age of the metamorphism of the Dalradian rocks of western Ireland. *Geological Magazine*, **98**, 399–404.

——1962, The provenance and emplacement of upper Arenigian turbidites in Co. Mayo, Eire. *Geological Magazine*, **99**, 238–252.

——1963. The Lower Palaeozoic stratigraphy of central Murrisk, Co. Mayo, Ireland and the evolution of the South Mayo Trough. *Journal of the Geological Society, London*, **119**, 313–344.

——1969. Evolution of the Appalachian/Caledonian orogen. *Nature*, **222**, 124–129.

——1971. A model for the Lower Palaeozoic evolution of the southern margin of the early Caledonides of Scotland and Ireland. *Scottish Journal of Geology*, **7**, 219–240.

——1974. Continental margin and ophiolitic obduction: Appalachian Caledonian systems. *In*: BURKE, C. A. & DRAKE, C. L. (eds) *Geology of Continental Margins*. Springer, New York, 933–950.

——1982. Plate tectonics and the evolution of the British Isles. *Journal of the Geological Society, London*, **139**, 371–412.

——1988. The extensional collapse of orogens. *Tectonics*, **7**, 1123–1139.

—— & BIRD, J. M. 1971. The origin and emplacement of the ophiolite suite; Appalachian ophiolites in Newfoundland. *Journal of Geophysical Research*, **76**, 3179–3206.

——, & PANKHURST, R. J. 1970. The evolution of the Scottish Caledonides in relation to their isotopic age pattern. *Transactions of the Royal Society of Edinburgh*, **68**, 361–369.

—— & RYAN, P. D. 1990. The Ordovician evolution of the South Mayo Trough, western Ireland. *Tectonics*, **9**, 887–902.

—— & SHACKLETON, R. M. 1984. A model for the evolution of the Grampian tract in the early Caledonides and Appalachians. *Nature*, **312**, 115–121.

——, MCKERROW, W. S. & MOORBATH, S. E. 1970. Relationships between uplift, sedimentation and isotopic ages during Ordovician times in western Ireland. *Scottish Journal of Geology*, **6**, 445–457.

——, RYAN, P. D. & ANDERSEN, T. B. 1993. Orogenic uplift and collapse, crustal thickness, fabrics and phase changes; the rôle of ecologites. *In*: PRICHARD, H. M., ALABASTER, T., HARRIS, N. B. W. & NEARLY, C. R. (eds) *Magmatic Processes and Plate Tectonics*. Geological Society, London, Special Publications, **76**, 325–343.

——, —— & SOPER, N. J. 1999. Ordovician orogeny in Scotland and Ireland. *Journal of the Geological Society, London*, in press.

——, HELMAN, M. L., TURCO, E., HUTTON, D. H. W. & KNOTT, S. D. 1988. Kinematics of the western Mediterranean. *In*: COWARD, M. P., DIETRICH, D. & PARK, R. G. (eds) *Alpine Tectonics*. Geological Society, London, Special Publications, **45**, 265–283.

DICKINSON, W. R., BEARD, S. L., BRAKENRIDGE, G. R. *et al.* 1983. Provenance of North American Phanerozoic sandstones in relation to tectonic setting.

*Geological Society of American Bulletin*, **94**, 222–235.

DOWNIE, C., LISTER, T. R., HARRIS, A. L. & FETTES, D. J. 1971. A palynological investigation of the Dalradian rocks of Scotland. *Institute of Geological Science, Report*, **71/9**, 1–30.

DRAKE, A. A., SINHA, A. K., LAIRD, JO & GUY, R. E. 1989. The Taconic Orogen. *In*: HATCHER, R. D., THOMAS, W. A. & VIELE, G. W. (eds) *The Appalachian–Ouachita Orogen in the United States*. Geological Society of America, 101–178.

DUBÉ, B., DUNNING, G. R., LAUZIÈRE, K. & RODDICK, J. C. 1996. New insights into the Appalachian Orogen from geology and geochronology along the Cape Ray fault zone, southwest Newfoundland. *Geological Society of America Bulletin*, **108**, 101–116.

DUNNING, G. R. 1987. U/Pb geochronology of the Coney Head Complex, Newfoundland. *Canadian Journal of Earth Sciences*, **24**, 1072–1078.

—— & KROGH, T. E. 1985. Geochronology of ophiolites of the Newfoundland Appalachian. *Canadian Journal of Earth Sciences*, **22**, 1659–1670.

——, KEAN, B. F., THURLOW, J. G. & SWINDEN, H. S. 1987. Geochronology of the Buchans, Roberts Arm, and Victoria Lake Groups and Mansfield Cove Complex, Newfoundland. *Canadian Journal of Earth Sciences*, **24**, 1175–1184.

——, SWINDEN, H. S., KEAN, B. F., EVANS, D. T. W. & JENNER, G. A. 1991. A Cambrian island arc in Iapetus: geochronology and geochemistry of the Lake Ambrose volcanic belt, Newfoundland Appalachians. *Geological Magazine*, **128**, 1–17.

——, O'BRIEN, S. J., COLMAN-SADD, S. P., BLACKWOOD, R. F., DICKSON, W. L., O'NEILL, P. P. & KROGH, T. E. 1990. Silurian orogeny in the Newfoundland Appalachians. *Journal of Geology*, **98**, 893–913.

ELDERS, C. F. 1987. The provenance of granite boulders in conglomerates of the Northern and Central Belts of the Southern Uplands of Scotland. *Journal of the Geological Society, London*, **144**, 853–864.

ELIAS, E. M., MCINTYRE, R. M. & LEAKE, B. E. 1988. The cooling history of Connemara, western Ireland, from K–Ar and Rb–Sr studies. *Journal of the Geological Society, London*, **143**, 649–660.

ELLIOTT, C. G., DUNNING, G. R. & WILLIAMS, P. F. 1991. New U/Pb zircon age constraint on the timing of deformation in north-central Newfoundland and implications for early Palaeozoic Appalachian orogenesis. *Geological Society of America Bulletin*, **103**, 125–135.

ENGLAND, P. C. & HOUSEMAN, C. A. 1989. Extensional collapse during continental convergence with applicaiton to the Tibetan Plateau. *Journal of Geophysical Research*, **94**, 17 561–17 579.

—— & THOMPSON, A. B. 1984. Pressure–temperature–time paths of regional metamorphism. I, Heat transfer during the evolution of regions of thickened continental crust. *Journal of Petrology*, **25**, 894–928.

ESCHER, A. & WATT, W. S. 1976. Summary of the Geology of Greenland. *In*: ESCHER, A. & WATT,

W. S. (eds) *Geology of Greenland*. Geological Survey of Greenland, 10–17.

FALLOON, T. J., MALAHOFF, A., ZONENSHEIN, L. P. & BOGDANOV, Y. 1992. Petrology and geochemistry of back-arc basin basalts from Low Basin spreading ridges at 15°, 18° and 19°S. *Mineralogy and Petrology*, **47**, 1–35.

FETTES, D. J. 1979. The structural and metamorphic state of the Dalradian rocks and their bearing on the age of emplacement of the basic sheets. *Scottish Journal of Geology*, **6**, 108–118.

FITCH, F. J., MILLER, J. A. & BROWN, P. E. 1964. Age of Caledonian orogeny and metamorphism in Britain. *Nature*, **203**, 275–278.

FLETCHER, T. P. 1989. *The Leny Limestone fauna – preliminary report*. British Geological Survey Technical Report HI/TPF/89/2.

FLINN, D., MILLER, J. A. & RODDOM, D. 1991. The age of the Norwick hornblende schists of Unst and Fetlar of the Shetland ophiolite. *Scottish Journal of Geology*, **27**, 11–19.

FRIEDRICH, A. M., HODGES, K. V. & BOWRING, S. A. 1997. *Geochronological constraints on the tectonic evolution of Connemara*. Geological Society, London, Tectonic Studies Group, Glasgow Meeting 1997, Abstracts with Programme.

FRIEND, C. R. L., KINNY, P. D., ROGERS, G., STRACHAN, R. A., PATTERSON, B. A. & WOLDSWORTH, R. E. 1995. *U–Pb zircon (SHRIMP) systematics of the Ardgour Gneiss: implication for the timing of Moine Orogenesis*. Geological Society, London, Tectonic Studies Group, Durham Meeting 1995, Abstracts with Programme.

FRYER, P., MOTTL, M., JOHNSON, L., HAGGERTY, PHIPPS, S. & MAEKAWA, H. 1995. Serpentine bodies in the forearcs of western Pacific convergent margins: origin of associated fluids. *American Geophysical Union, Geophysics Monologie*, **88**, 259–279.

FÜCHTBAUER, H. 1967. Die Sandsteine in der Molasse nördlich der Alpen. *Geologische Rundschau*, **56**, 266–300.

GASS, I. G. 1968. Is the Troodos Massif of Cyprus a fragment of Mesozoic ocean crust? *Nature*, **220**, 39 42.

GENRICH, J. F., BOCK, Y., McCAFFREY, R., CALAIS, E., STEVENS, C. V. & SUBARYA, C. 1996. Accretion of the southern Banda arc to the Australian plate margin determined by Global Positioning System measurements. *Tectonics*, **15**, 288–295.

GEOLOGICAL WORLD ATLAS (SHEETS 13–16) 1976. Commission for the Geological Map of the World and UNESCO, Paris.

GLOVER, B. W., KEY, R. M., MAY, F., CLARK, G. C., PHILLIPS, E. R. & CHACKSFIELD, B. C. 1995. A Neoproterozoic multi-phase rift sequence: the Grampian and Appin groups of the southwestern Monadhliath mountains of Scotland. *Journal of the Geological Society, London*, **152**, 391–406.

GRAHAM, J. R. 1987. The nature and field relations of the Ordovician Maumstrasna Formation, County Mayo, Ireland. *Geology Journal*, **22**, 347–369.

—— & SMITH, D. G. 1981. The age and significance of a small Lower Palaeozoic inlier in County Mayo. *Royal Dublin Society Earth Science Journal*, **4**, 1–5.

——, LEAKE, B. E. & RYAN, P. D. 1989. *The Geology of South Mayo, Western Ireland*. Scottish Academic Press.

——, WRAFTER, J. P., DALY, J. S. & MENUGE, J. F. 1991. A local source for the Ordovician Derryveeny formation, western Ireland: Implications for the Connemara Dalradian. *In*: MORTON, A. C., TODD, S. P. & HAUGHTON, P. D. W. (eds) *Developments in Sedimentary Proveance Studies*. Geological Society, London, Special Publications, **57**, 199–213.

HALL, R. 1996. Recontrusting SE Asia. *In*: HALL, R. & BLUNDELL, D. J. (eds) *Tectonic Evolution of Southeast Asia*. Geological Society, London, Special Publications, **106**, 153–184.

——1997. Cenozoic plate tectonic reconstructions of SE Asia. *In*: FRASER, A. J., MATTHEWS, S. J. & MURPHY, R. W. (eds) *Petroleum Geology of Southeast Asia*. Geological Society, London, Special Publications, **126**, 11–23.

HALLIDAY, A. N., GRAHAM, C. M., AFTALION, M. & DYMOKE, P. 1989. The deposition age of the Dalradian Supergroup: U–Pb and Sm–Nd isotopic studies of the Tayvallich Volcanics, Scotland. *Journal of the Geological Society, London*, **146**, 3–6.

HAMBREY, M. J., FAIRCHILD, I. J., GLOVER, B. W., STEWART, A. D., TREAGUS, J. E. & WINCHESTER, J. A. 1991. The Late Precambrian geology of the Scottish Highlands and Islands. *Geological Association Guide*, **44**.

HAMILTON, P. J., BLUCK, B. J. & HALLIDAY, A. N. 1984. Sm–Nd ages from the Ballantrae complex, S.W. Scotland. *Royal Society of Edinburgh Transactions*, **75**, 183–187.

HAMILTON, W. 1979. *Tectonics of the Indonesian region*. US Geological Survey, Professional Paper 1078.

HARKIN, J., WILLIAMS, D. M., MENUGE, J. F. & DALY, J. S. 1996. Turbidites from the Clew Bay Complex, Ireland: provenance based on petrography, geochemistry and crustal residence values. *Geology Journal*, **31**, 379–388.

HARPER, D. A. T., WILLIAMS, D. M. & ARMSTRONG, H. A. 1989. Stratigraphical correlations adjacent to the Highland Boundary Fault in the west of Ireland. *Journal of the Geological Society, London*, **146**, 381–384.

——, GRAHAM, J. R., OWEN, A. W. & DONOVAN, S. K. 1988. An Ordovician fauna from Lough Shee, Partry Mountains, Co. Mayo, Ireland. *Geology Journal*, **23**, 293–310.

HARRIS, A. L. 1969. The relationship of the Leny Limestone to the Dalradian. *Scottish Journal of Geology*, **5**, 187–190.

—— & FETTES, D. J. 1972. Stratigraphy and structure of Dalradian rocks at the Highland Border. *Scottish Journal of Geology*, **8**, 253–264.

——, —— & SOPER, N. J. 1999. New evidence that the Lower Cambrian Leny Limestone at Callander, Perthshire, belongs to the Dalradian Supergroup,

and a reassessment of the 'exotic' status of the Highland Border Complex – Discussion. *Geological Magazine*, in press.

——, SHACKLETON, R. M., WATSON, J., DOWNIE, C., HARLAND, W. B. & MOORBATH, S. E. 1975. A correlation of Precambrian rocks in the British Isles. Geological Society, London, Special Reports, **6**.

HARRIS, D. H. M. 1995. Caledonian transpressional terrane accretion along the Laurentian margin in Co. Mayo, Ireland. *Journal of the Geological Society, London*, **152**, 797–806.

—— & HARRIS, A. L. 1996. Caledonian deformation along the Laurentian margin in Scotland and Ireland. *Geological Association of Canada, Special Paper*, **41**, 81–91.

HARRIS, P. M., FARRAR, E., MACINTYRE, R. M. & YORK, D. 1965. Potassium–argon age measurements from the Ordovician system of Scotland. *Nature*, **205**, 352–353.

HARRIS, R. A. 1991. Temporal distribution of strain in the active Banda orogen: a reconciliation of rival hypotheses. *Journal of Southeast Asian Earth Sciences*, **6**, 373–386.

HATCHER, R. D. & ODOM, A. L. 1980. Timing of thrusting in the southern Appalachians, USA. A model for orogeny? *Journal of the Geological Society, London*, **137**, 321–327.

HIGGINS, A. C. 1967. The age of the Durine Member of the Durness Limestone formation of Durness. *Scottish Journal of Geology*, **3**, 381–388.

HUBERT, W. 1962. A zircon–tourmaline–rutile maturity index and the interdependence of the composition of heavy mineral assemblages with the gross composition and texture of sandstones. *Journal of Sedimentary Petrology*, **32**, 440–450.

HURST, J. M. & MCKERROW, W. S. 1981. The Caledonian nappes of eastern North Greenland. *Nature*, **290**, 772–774.

HUTCHISON, A. R. & OLIVER, G. J. H. 1999. Garnet provenance studies and their juxtaposition of Laurentian marginal terranes in Scotland during the Ordovician. *Journal of the Geological Society, London*, in press.

HUTTON, D. H. W. & DEWEY, J. F. 1986. Palaeozoic terrane accretion in the western Irish Caledonides. *Tectonics*, **5**, 1115–1124.

——, AFTALION, M. & HALLIDAY, A. N. 1988. An Ordovician ophiolite in County Tyrone, Ireland. *Nature*, **315**, 210–212.

HYNES, A., & SNYDER, D. D. 1995. Deep-crustal mineral assemblages and potential for crustal rocks below the Moho in the Scottish Caledonides. *Geophysical Journal International*, **123**, 1–17.

IDLEMAN, B. D. 1990. *Geology and $^{40}Ar/^{39}Ar$ geochronology of the Coastal Complex near Trout River and Lark Harbour, Western Newfoundland*. PhD Thesis, SUNY Albany.

JAGGER, M. D., MAX, M. D., AFTALION, M. & LEAKE, B. E. 1988. U–Pb zircon ages of basic rocks and gneisses intruded into the Dalradian of Cashel, Connemara, western Ireland. *Journal of the Geological Society, London*, **145**, 645–648.

JAMIESON, R. A., VAN BREEMEN, O., SULLIVAN, R. W. & CURRIE, K. L. 1986. The age of igneous and metamorphic events in the Western Cape Breton Highlands, Nova Scotia. *Canadian Journal of Earth Sciences*, **23**, 1891–1901.

JENNER, G. A., DUNNING, G. R., MALPAS, J., BROWN, M. & BRUCE, T. 1991. Bay of Islands and Little Post Complexes, revisited, age, geochemical and isotopic evidence confirm subduction zone origin. *Canadian Journal of Earth Sciences*, **28**, 1635–1652.

JOHNSTON, J. D. & PHILLIPS, W. E. A. 1995. Terrane amalgamation in the Clew Bay region, west of Ireland. *Geological Magazine*, **132**, 485–501.

KAMO, S. L., GOWER, C. F. & KROGH, T. E. 1989. Birthdate for the Iapetus Ocean? A precise U–Pb zircon and baddleyite age for the Long Range dikes, southern Labrador. *Geology*, **17**, 602–605.

KARABINOS, P., SAMSON, S. D., HEPBURN, J. C. & STOLL, H. M. 1998. Taconism orogeny in the New England Appalachians: collision between laurentian and the Shelbourne Falls Arc. *Geology*, **26**, 215–218.

KARSON, J., & DEWEY, J. F. 1978, The Coastal Complex, western Newfoundland: an early Ordovician fracture zone. *Geological Society of America Bulletin*, **89**, 1037–1049.

KELLEY, S. P. 1988. The relationship between K–Ar mineral ages, mica grain sizes and movement on the Moine Thrust Zone, NW Highlands, Scotland. *Journal of the Geological Society, London*, **145**, 1–10.

—— & BLUCK, B. J. 1989. Detrital mineral ages from the Southern Uplands using $^{40}$Ar–$^{39}$Ar laser probe. *Journal of the Geological Society, London*, **146**, 401–403.

——, REDDY, S. M. & MADDOCK, R. 1994, Laserprobe $^{40}$Ar/$^{39}$Ar investigation of a pseudotachylite and its host rock from the Outer Isles thrust, Scotland. *Geology*, **22**, 443–446.

KENNEDY, W. Q. 1955. The tectonics of the Morar Anticline and the problem of the North-West Caledonian front. *Journal of the Geological Society, London*, **90**, 359–390.

——1958. The tectonic evolution of the Midland Valley of Scotland. *Geological Society of Glasgow Transactions*, **23**, 106–133.

KINAHAN, G. H., SYMES, R. G., WILKINSON, S. B., NOLAN, J. & LEONARD, H. 1876. *Country around Westport, Erriff Valley, Killary Harbour and western shores of Lough Mask: Explanatory Memoir, Sheets 73, 74, 83 and 84*. Memoir of the Geological Survey, Ireland.

KNELLER, B. C. & AFTALION, M. 1987. The isotopic and structural age of the Aberdeen Granite. *Journal of the Geological Society, London*, **148**, 985–992.

KRABBENDAM, M., LESLIE, A. G., CRANE, A. & GOODMAN, S. 1997. Generation of the Tay Nappe, Scotland by large-scale SE-directed shearing. *Journal of the Geological Society, London*, **154**, 15–24.

KUMAREPELI, P. S., DUNNING, G. R., PINSTON, H. & SHAVER, J. 1989. Geochemistry and U–Pb zircon age of comendite metafelsites of the Tibbit Hill Formation, Quebec Appalachians. *Canadian Journal of Earth Sciences*, **26**, 1374–1383.

LAIRD, J., TRZIENSKI, W. E. & BOTHNER, W. A. 1993. High-pressure Taconian and subsequent polymetamorphism of southern Quebec and northern Vermont. *Department of Geology and Geography, University of Massachusetts Contribution*, **672**, 1–32.

LAMBERT, R. ST. J. 1969. Isotopic studies relating to the Precambrian history of the Moinian of Scotland. *Proceedings of the Geological Society of London*, **1652**, 243–244.

LEAKE, B. E. 1963. A possible fossil in a graphitic marble in the Connemara Schist, Connemara, Co. Galway, Ireland. *Geological Magazine*, **100**, 44–46.

——1986. The geology of SW Connemara, Ireland: a fold and thrust Dalradian and metagabbroic–gneiss complex. *Journal of the Geological Society, London*, **143**, 221–236.

——1989. The metagabbros, orthogneisses and paragneisses of the Connemara complex. *Journal of the Geological Society, London*, **146**, 575–596.

—— & TANNER, P. W. C. 1984. *The geology of the Dalradian and associated rocks of Connemara, Western Ireland*. A report to acompany the 1 : 63 360 geological map and cross sections. Royal Irish Academy, Dublin.

——, —— & SINGH, D. 1983. Major southward thrusting of the Dalradian rocks of Connemara, western Ireland. *Nature*, **305**, 210–213.

——, ——, MCINTYRE, R. M. & ELIAS, E. 1984. Tectonic position of the Dalradian rocks of Connemara and its bearing on the evolution of the Midland Valley of Scotland. *Royal Society of Edinburgh Transactions*, **75**, 165–171.

LEGGETT, J. K. 1978. Eustasy and pelagic regimes in the Iapetus Ocean during the Ordovician and Silurian. *Earth and Planetary Science Letters*, **41**, 163–169.

LEGGO, P. J., TANNER, P. W. G. & LEAKE, B. F. 1969. An isochron study of the Donegal Granite and some Dalradian rocks of Britain. *AAPG Memoirs*, **12**, 354.

LIHOU, J. C. & MANGE-RAJETZKY, M. A. 1996. Provenance of the Sardona Flysch, eastern Swiss Alps: example of high-resolution heavy mineral analysis applied to an ultrastable assemblage. *Sedimentary Geology*, **105**, 141–157.

LIN, S., BARR, S. M., KWOK, Y. Y., CHEN, Y. D., DAVIS, D. W. & VAN STAAL, C. R. 1997. U/Pb geochronological constraints on the geological evolution of the Askey Terrane in Cape Breton and St. Pauls Islands, Nova Scotia. *Geological Association of Canada, Mineralogy, Association of Canada Abstracts*, **22**, A91.

LONERGAN, L. & MANGE-RAJETZKY, M. A. 1994. Evidence for internal zone uroofing from foreland basin sediments, Betic Cordillera, SE Spain. *Journal of the Geological Society, London*, **151**, 515–529.

LONG, C. B., MCCONNELL, B. & ARCHER, J. B. 1995. *The Geology of Connemara and South Mayo*. Geological Survey of Ireland.

MATTINSON, J. M. 1977. U–Pb ages of some crystalline rocks from the Burlington Peninsula, Newfoundland, and implications for the age of the Fleur de Lys metamorphism. *Canadian Journal of Earth Sciences*, **114**, 2316–2324.

MCCAFFREY, R. 1996. Slip partitioning at convergent plate boundaries of SE Asia. *In*: HALL, R. & BLUNDELL, D. J. (eds) *Tectonic Evolution of Southeast Asia*. Geological Society, London, Special Publications, **106**, 3–18.

MCCAIG, A. & REX, D. 1996, Argon-40/Argon-39 ages from the Lewis Hills Massif, Bay of Islands Complex, West Newfoundland. Geological Association of Canada, Special Papers, **41**, 137–145.

MCDOWELL, F. W., MCMAHON, T. P., WARREN, P. Q. & CLOOS, M. 1996. Pliocene Cu–Au-bearing igneous intrusions of the Gunung Bijih (Ertsberg) district, Irian Jaya, Indonesia: K–Ar geochronology. *Journal of Geology*, **104**, 327–340.

MCKENZIE, D. P. 1969. Speculation on the consequences and causes of plate motion. *Geophysical Journal of the Royal Astronomical Society*, **18**, 1–32.

MCKERROW, W. S. & CAMPBELL, C. J. 1960. The stratigraphy and structure of the Lower Palaeozoic rocks of north-west Galway. *Scientiific Proceedings of the Royal Dublin Society*, **1**, 27–52.

MCMANUS, J. 1972. The stratigraphy and structe of the Lower Palaeozoic rocks of eastern Murrisk, Co. Mayo. *Proceedings of the Royal Irish Academy*, **72**, 307–333.

MACKIE, W. 1923. The source of the purple zircons in the sedimentary rocks of Scotland. *Transactions of the Edinburgh Geological Society*, **11**, 200–213.

MAGUIRE, C. K. & GRAHAM, J. R. 1996. Sedimentation and palaeogeographical significance of the Silurian rocks of the Louisburgh–Clare Island succession, western Ireland. *The Royal Society of Edinburgh Transactions*, **86**, 123–136.

MALAIHOLLO, J. H. A. & HALL, R. 1996. The geology and tectonic evolution of the Bacan region, eastern Indonesia. *In*: HALL, R. & BLUNDELL, D. J. (eds) *Tectonic Evolution of Southeast Asia*. Geological Society, London, Special Publications, **106**, 483–497.

MANGE, M. A. & MAURER, H. F. W. 1992. *Heavy Minerals in Colour*. Chapman & Hall.

MANGE-RAJETZKY, M. A. 1995. Subdivision and correlation of monotonous sandstone sequences using high resolution heavy mineral analysis, a case study: the Triassic of the Central Graben. *In*: DUNAY, R. E. & HAILWOOD, E. A. (eds) *Nonbiostratigraphical Methods of Dating and Correlation*. Geological Society, London, Special Publications, **89**, 23–30.

—— & OBERHÄNSLI, R. 1982. Detrital lawsonite and blue sodic amphibole in the Molasse of Savoy, France and their significance in assessing Alpine evolution. *Schweizerische Mineralogische und Petrographische Mitteilungen*, **62**, 415–436.

MENDUM, J. F. & THOMAS, C. W. 1996. *The nature of D2 deformation in the Dalradian rocks of the southern Highlands*. Geological Society, London,

Tectonic Studies Group Annual Meeting, Birmingham. (Abstracts with Programme.)

MENUGE, J. F., WILLIAMS, D. M. & O'CONNOR, P. D. 1995. Silurian turbidites used to reconstruct a volcanic terrain and its Mesoproterozoic basement in the Irish Caledonides. *Journal of the Geological Society, London*, **152**, 269–278.

MILLER, W. M., FALLICK, A. E., LEAKE, B. E., MACINTYRE, R. M. & JENKIN, G. R. T. 1991. Fluid-disturbed hornblende K–Ar ages from the Dalradian rocks of Connemara, western Ireland. *Journal of the Geological Society, London*, **148**, 985–992.

MITCHELL, A. H. G. 1978. The Grampian orogeny in Scotland: arc–continent collision and polarity reversal. *Journal of Geology*, **86**, 643–646.

——1981. The Grampian Orogeny in Scotland and Ireland: almost an ancient Taiwan. *Proceedings of the Geological Society of China*, **24**, 113–119.

——1989, Arc reversal in the Scottish Southern Uplands. *Journal of the Geological Society, London*, **146**, 736–738.

MIYASHIRO, A. 1973. The Troodos ophiolite complex was probably formed in an island arc. *Earth and Planetary Science Letters*, **19**, 218–224.

MOLYNEUX, S. G. 1999. An upper Dalradian microfossil reassessed. *Journal of the Geological Society, London*, in press.

MOORES, E. M. & VINE, F. J. 1971. The Troodos Massif, Cyprus and other ophiolites as oceanic crust: evaluation and implications. *Philosophical Transactions of the Royal Society of London*, **A268**, 433–466.

MORTON, A. C. 1984. Stability of detrital minerals in Tertiary sandstones of the North Sea Basin. *Clay Minerals*, **19**, 287–308.

——1985. Heavy minerals in provenance studies. *In*: ZUFFA, G. G. (ed.) *Provenance of Arenites*. Reidel, 249–277.

——1986. Dissolution of apatite in the North Sea Jurassic sandstones: implications for the generation of secondary porosity. *Clay Minerals*, **21**, 711–733.

—— & JOHNSON, M. J. 1993. Factors influencing the composition of detrital mineral suites in Holocene sands of the Apure River drainage basin, Venezuela. *In*: JOHNSSON, M. J. & BASU, A. (eds) *Processes Controlling the Composition of Clastic Sediments*. Geological Society of America, Special Papers, **284**, 171–185.

MUIR, R. J., FITCHES, W. R., MALTMAN, A. J. & BENTLEY, M. R. 1994. *Precambrian rocks of the Southern Inner Hebrides – Malin Sea region: Colonsay, west Islay, Inishtrahull and Iona*. Geological Society, London, Special Reports, **22**, 54–58.

NOBLE, S. R., HYSLOP, E. K. & HIGHTON, A. J. 1996. High-precision U–Pb monazite geochronolgy of the *c.* 806 Ma Grampian Shear Zone and the implications for the evolution of the Central Highlands of Scotland. *Journal of the Geological Society, London*, **153**, 511–514.

——, HIGHTON, A. J., HYSLOP, E. K. & BARREIRO, B. 1997. A Rodinian connection for the Scottish Highlands? Evidence from U–Pb geochronology of

Grampian Terrane migmatites and pegmatites. *European Union Geosciences, Strasbourg Abstracts*, 727.

OWEN, A. W., HARPER, D. A. T. & ROMANO, M. 1992. The Ordovician biogeography of the Grangegeeth terrane and the Iapetus suture zone in eastern Ireland. *Journal of the Geological Society, London*, **149**, 3–6.

PANKHURST, R. J. 1974. Rb–Sr whole-rock chronology of Caledonian events in northeast Scotland. *Geological Society of America Bulletin*, **85**, 345–350.

——, ANDREWS, J. R., PHILLIPS, W. E. A. & SANDERS, I. S. 1976. Age and structural setting of the Slieve Gamph igneous complex, Co. Mayo, Eire. *Journal of the Geological Society, London*, **132**, 327–336.

PEGLER, G., DAS, S. & WOODHOUSE, J. H. 1995. A seismological study of eastern New Guinea and the western Solomon Sea regions and its tectonic implications. *Geophysical Journal International*, **122**, 961–981.

PETTIJOHN, F. J. 1941. Persistence of heavy minerals and geologic age. *Journal of Geology*, **49**, 610–625.

——, POTTER, P. E. & SIEVER, R. 1973. *Sand and Sandstone*. Springer.

PHILLIPS, W. E. A. 1973. The pre-Silurian rocks of Clare Island, Co. Mayo, Ireland. *Journal of the Geological Society, London*, **129**, 585–606.

——1974. The stratigraphy, sedimentary environments and palaeogeography of the Silurian strata of Clare Island, Co. Mayo, Ireland. *Journal of the Geological Society, London*, **130** 19–41.

——, STILLMAN, C. J. & MURPHY, T. 1976. A Caledonian plate tectonic model. *Journal of the Geological Society, London*, **132**, 579–609.

PIASECKI, M. A. J. & VAN BREEMEN, O. 1979. A Morarian age for the 'younger Moines' of central and western Scotland. *Nature*, **278**, 784–736.

PIDGEON, R. T. & AFTALION, M. 1978. Cogenetic and inherited zircons U–Pb systems in granites. *Journal of Geology, Special Issue*, **10**, 183–220.

—— & JOHNSON, M. R. W. 1974. A comparison of zircon U–Pb and whole rock Rb–Sr systems in three phases of the Carn Chuinneag Granite, Northern Scotland. *Earth and Planetary Science Letters*, **24**, 105–112.

PIGRAM, C. J. & SYMONDS, P. A. 1991. A review of the major tectonic events in the New Guinea Orogen. *Journal of Southeast Asian Earth Sciences*, **6**, 307–318.

POWELL, D., BROOK, M. & BAIRD, A. W. 1983. Structural dating of a Precambrian pegmatite in Moine rocks of northern Scotland and its bearing on the status of the Morarian Orogeny. *Journal of the Geological Society, London*, **140**, 813–823.

POWERS, M. C. 1953. A new roundness scale for sedimentary particles. *Journal of Sediment Petrology*, **23** , 117–119.

——1982. *Comparison chart for estimating roundness and sphericity*. AGI Data Sheet 18, American Geological Institute.

PRYOR, W. A. 1971. Grain shape. *In*: CARVER, R. E. (ed.) *Procedures in Sedimentary Petrology*. Wiley, 131–150.

PUDSEY, C. J. 1984. Fluvial to marine transition in the Ordovician of Ireland – a humid-region fan-delta? *Journal of Geology*, **19**, 143–172.

RANKIN, D. W., DRAKE, A. A., GLOVER, L. *et al.* 1989. Pre-orogenic terranes. *In*: HATCHER, R. D., THOMAS, W. A. & VIELE, G. W. (eds) *The Appalachian–Ouachita Orogen in the United States*. Geological Society of America, 7–100.

RICHARDSON, A. N. & BLUNDELL, D. J. 1996. Continental collision in the Banda arc. Geological Society, London, Special Publications, **106**, 47–60.

ROGERS, G, DEMPSTER, T. J., BLUCK, B. J. & TANNER, P. W. G. 1989. A high-precision U–Pb age for the Ben Vuirich Granite: implications for the evolution of the Scottish Dalradian supergroup. *Journal of the Geological Society, London*, **146**, 1045–1048.

——, PATERSON, B. A., DEMPSTER, T. J. & REDWOOD, S. D. 1995a. U–Pb geochronology of the 'Newer' GABBROS, N. E. Grampians. *In: Programme with Abstracts, Caledonian Terrane Relationships in Britain*. British Geological Survey, Keyworth, Symposium.

——, FRIEND, C. P. L., STRACHAN, R. A., KINNY, P., PATERSON, B. A. & HOLDSWORTH, R. E. 1995b. U–Pb evidence for the presence of Grenvillian events in the Moine of N W Scotland. *EUG 8 Terra Abstracts*, **7**, 353.

ROWLEY, D. B. & KIDD, W. S. F. 1981. Stratigraphic relationships and detrital composition of the Medial Ordovician flysch of western New England: implications for the tectonic evolution of the Taconic Orogen. *Journal of Geology*, **89**, 199–218.

RUSHTON, A. W. A. & PHILLIPS, W. E. A. 1973. A *Protospongia* from the Dalradian of Clare Island, Co. Mayo, Ireland. *Palaeontology*, **16**, 231–237.

RYAN, P. D. & ARCHER, J. B. 1977, The South Mayo Trough: a possible Ordovician Gulf of California-type marginal basin in the west of Ireland. *Canadian Journal of Earth Sciences*, **14**, 2453–2461.

—— & DEWEY, J. F. 1991. A geological and tectonic cross-section of the Caledonides of western Ireland. *Journal of the Geological Society, London*, **148**, 173–180.

——, FLOYD, P. A. & ARCHER, J. B. 1980. The stratigraphy and petrochemistry of the Lough Nafooey Group (Tremadocian), western Ireland. *Journal of the Geological Society, London*, **137**, 443–458.

SANDERS, I. S. 1994. *The northeast Ox Mountains inlier, Ireland*. Geological Society, London, Special Reports, **22**, 63–54.

——, VAN CALSTEREN, P. W. C. & HAWKESWORTH, C. J. 1984. A Grenville Sm–Nd age for the Glenelg eclogite in north-west Scotland. *Nature*, **312**, 439–440.

SHACKLETON, R. M. 1986. Precambrian collision tectonics in Africa. *In*: COWARD, M. P. & RIES, A. C. (eds) *Collision Tectonics*. Geological Society, London, Special Publications, **19**, 329–349.

SILVER, E. A., ABBOTT, L. D., KIRCHOFF-STEIN, K. S., REED, D. L. & BERNSTEIN-TAYLOR, B. 1991. Collision propagation in Papua New Guinea and the Solomon Sea. *Tectonics*, **10**, 863–874.

SMELLIE, J. L. & STONE, P. 1992. Geochemical control on the evolutionary history of the Ballantrae Complex, SW Scotland, from comparison with recent analogues. *In*: PARSON, L. M., MURTON, B. J. & BROWNING, P. (eds) *Ophiolites and their Modern Oceanic Analogues*. Geological Society, London, Special Publications, **60**, 171–178.

SMITHSON, F. 1941. The alteration of detrital minerals in the Mesozoic rocks of Yorkshire. *Geological Magazine*, **78**, 97–112.

SNYDER, D. B., PRASEYTO, H., BLUNDELL, D. J., PIGRAM, C. J., BARBER, A. J., RICHARDSON, A. & TJOKOSAPROETRO, S. 1996, A doubly vergent orogen in the Banda Arc continent–arc collision zone as observed on deep seismic reflection profiles. *Tectonics*, **15**, 34–53.

SOPER, N. J. 1994. Neoproterozoic sedimentation on the northeast margin of Laurentia and the opening of Iapetus. *Geological Magazine*, **131**, 291–299.

—— & ENGLAND, R. W. 1995. Vendian and Riphean rifting in NW Scotland. *Journal of the Geological Society, London*, **152**, 11–14.

—— & HARRIS, A. L. 1997. Proterozoic orogeny questioned: a view from Scottish Highland Field Workshops 1995–96. *Scottish Journal of Geology*, **33**, 187–190.

——, STRACHAN, R. A., HOLDSWORTH, R. E., GAYER, R. A. & GREILING, R. O. 1992. Silurian transpression and the Silurian closure of Iapetus. *Journal of the Geological Society, London*, **149** 871–880.

SPEER, J. A. 1980. Zircon. *In*: RIBBE, P. H. (ed.) *Reviews in Mineralogy, Volume 5, Orthosilicates*. Mineralogical Society of America, 67–112.

SPRAY, J. G. 1997. Thrust related metamorphism beneath the Shetland Islands oceanic fragment, northeast Scotland. *Canadian Journal of Earth Sciences*, **25**, 1760–1776.

—— & DUNNING, G. R. 1991. U/Pb age for the Shetland Islands oceanic fragment; Scottish Caledonides: evidence from anatectic plagiogranites in 'layer 3' shear zones. *Geological Magazine*, **128**, 667–671.

SPRAY, P. 1999. *Geology in South-west Scotland*. British Geological Survey, Keymouth, 214.

STANTON, W. I. 1960. The Lower Palaeozoic rocks of south-west Murrisk, Ireland. *Journal of the Geological Society, London*, **116**, 269–296.

STONE, P. 1996. *Geology in South-west Scotland*. British Geological Survey, Keyworth, 214.

STUKAS, V. & REYNOLDS, P. H. 1974. $^{40}$Ar/$^{39}$Ar dating of the Long Range Dykes, Newfoundland. *Earth and Planetary Science Letters*, **22**, 256–266.

STURT, B. A. 1961. The geological structure of the area south of Loch Tummel. *Journal of the Geological Society, London*, **117**, 131–156.

SURLYK, F. 1991. Tectonostratigraphy of north Greenland. *Greenland Geological Survey Bulletin*, **160**, 25–47.

SUTTON, J., & WATSON, J. 1955. The deposition of the Upper Dalradian rocks of the Banffshire Coast. *Proceedings of the Geologists Association*, **66**, 101–133.

TANNER, P. W. G. 1995. New evidence that the Lower Cambrian Leny Limestone at Callender, Perth-shire, belongs to the Dalradian Supergroup and a reassessment of the exotic status of the Highland Border Complex. *Geological Magazine*, **132**, 473–483.

——1996. Significance of the early fabric in the contact metamorphic aureole of the 590 Ma Ben Vuirich Granite, Perthshire, Scotland. *Geological Magazine*, **133**, 683–695.

—— & LESLIE, A. G. 1994. A pre-D2 age for the 590 Ma Ben Vuirich Granite in the Dalradian of Scotland. *Journal of the Geological Society, London*, **151**, 209–212.

——, DEMPSTER, T. J. & DICKIN, A. P. 1989. Time of docking of the Connemara terrane with the Delany Dome Formation, western Ireland. *Journal of the Geological Society, London*, **146**, 389–392.

——, —— & ROGERS, G. 1997. New constraints upon the structural and isotopic age of the Oughterard Granite, and on the timing of events in the Dalradian rocks of Connemara, western Ireland. *Journal of Geology*, **32**, 247–263.

TECTONIC MAP OF THE WORLD (SHEETS, 11, 12, 17, 18). 1985. Exxon Production and Research Company.

TEMPERLEY, S. & WINDLEY, B. F. 1997. Grenvillian extensional tectonics in northwest Scotland. *Geology*, **25**, 53–56.

THIRLWALL, M. F. & BLUCK, B. J. 1984. Sm–Nd isotope and geological evidence that the Ballantrae 'ophiolite' SW Scotland, is polygenetic. *In*: GASS, I. G., LIPPARD, S. J. & SHELTON, A. W. (eds) *Ophiolites and Oceanic Lithosphere*. Geological Society, London, Special Publications, **13**, 215–230.

THOMPSON, A. B. & ENGLAND, P. C. 1984. Pressure–temperature–time paths of regional metamorphism II: their inferences and interpretations using mineral assemblages in metamorphic rocks. *Journal of Petrology*, **25**, 929–955.

TOMITA, T. 1954. Geologic significance of the colour of granite zircon and the discovery of the Precambrian in Japan. *Memoirs of the Faculty of Science, Kyushu University, Series D, Geology*, **4**, 135–161.

TRENCH, A. & TORSVIK, T. N. 1992. The closure of the Iapetus Ocean and Tornquist Sea: new palaeomagnetic constraints. *Journal of the Geological Society, London*, **149**, 867–876.

TUCKER, R. D. & MCKERROW, W. S. 1995. Early Palaeozoic chronology: a review in light of new U–Pb zircon ages from Newfoundland and Britain. *Canadian Journal of Earth Sciences*, **32**, 368–379.

VAN ANDEL, TJ. H. 1950. *Provenance, Transport and Deposition of Rhine Sediments*. H. Veenman en Zonen.

VAN BREEMEN, O., PIDGEON, R. T. & JOHNSON, M. R. W. 1974. Precambrian and Palaeozoic pegmatites in the Moines of northern Scotland. *Journal of the Geological Society, London*, **130**, 493–507.

——, AFTALION, M., PANKHURST, R. J. & RICHARDSON, S. W. 1979. Age of the Glen Desarry syenite Inverness-shire: diachronous Palaeozoic metamorphism across the Great Glen. *Scottish Journal of Geology*, **15**, 49–62.

VAN STAAL, C. R., DEWEY, J. F., McKERROW, W. S. & MACNIOCAILL, C. 1999. The Cambrian–Silurian tectonic evolution of the northern Appalachians and British Caledonides: history of a complex, southwest Pacific-type segment of Iapetus. *In*: BLUNDELL, D. J. & SCOTT, A. C. (eds) *Lyell: the Present is in the Past*. Geological Society, London, Special Publications, in press.

VANCE, D. & O'NIONS, R. K. 1990. Isotopic geochronology of zoned garnet: growth kinetics and metamorphic histories. *Earth and Planetary Science Letters*, **97**, 227–240.

VEEVERS, J. J. 1984. *Phanerozoic Earth History of Australia*. Oxford University Press.

WELLINGS, S. A. 1998. Timing of deformation associated with the syn-tectonic Dawros–Currywongaun–Doughruagh Complex, NW Connemara, western Ireland. *Journal of the Geological Society, London*, **155**, 25–37.

WILLIAMS, A. 1976. Plate tectonics and biofacies evolution as factors in Ordovician correlation. *In*: BASSETT, M. G. (ed.) *The Ordovician System. Proceedings of a Palaeontological Association Symposium, Birmingham, September 1974*. University of Wales Press of Natural Museum of Wales, 29–66.

WILLIAMS, D. M. & HARPER, D. A. T. 1991. End-Silurian modifications of Ordovician terranes in western Ireland. *Journal of the Geological Society, London*, **148**, 165–171.

—— & RICE, A. H. N. 1989. Low-angle-extensional faulting and the emplacement of the Connemara Dalradian, Ireland. *Tectonics*, **8**, 417–428.

——, HARKIN, J. & HIGGS, K. T. 1996. Implications of new microfloral evidence from the Clew Bay Complex for Silurian relationships in the western Irish Caledonides. *Journal of the Geological Society, London*, **153**, 771–777.

——, ——, ARMSTRONG, H. A. & HIGGS, K. T. 1994. A late Caledonian melange in Ireland: implications for tectonic models. *Journal of the Geological Society, London*, **151**, 307–314.

WILLIAMS, H. & STEVENS, R. K. 1969. Geology of Belle Isle – northern extremity of the deformed Appalachian miogeosynclinal belt. *Canadian Journal of Earth Sciences*, **6**, 1145–1157.

——, DALLMEYER, R. D. & WANLESS, R. K. 1976. Geochronology of the Twillingate Granite and Herring Neck Group, Notre Dame Bay, Newfoundland. *Canadian Journal of Earth Sciences*, **13**, 1591–1601.

——, GILLESPIE, R. T. & VAN BREEMEN, O. 1985. A late Precambrian rift-related igneous suite in western Newfoundland. *Canadian Journal of Earth Sciences*, **22**, 1727–1735.

WINCHESTER, J. A. 1992. Comment on 'Exotic' metamorphic terranes in the Caledonides: tectonic history of the Dalradian block, Scotland. *Geology*, **20**, 764.

WINKLER, W. & BERNOULLI, D. 1986. Detrital high-pressure/low-temperature minerals in late Turonian flysch sequence of the eastern Alps (western Austria): implications for early Alpine tectonics. *Geology*, **14**, 598–601.

WRAFTER, J. P. & GRAHAM, J. R. 1989. Ophiolitic detritus in the Ordovician sediments of South Mayo, Ireland. *Journal of the Geological Society, London*, **146**, 213–215.

YARDLEY, B. W. D. 1976. Deformation and metamorphism of Dalradian rocks and the evolution of the Connemara Cordillera. *Journal of the Geological Society, London*, **132**, 521–542.

—— & SENIOR, A. 1982. Basic magmatism in Connemara: evidence for a volcanic arc? *Journal of the Geological Society, London*, **139**, 67–70.

——, BARBER, J. P. & GRAY, J. R. 1987. The metamorphism of the Dalradian rocks of western Ireland and its relation to tectonic setting. *Philosophical Transactions of the Royal Society of London*, **A231**, 243–268.

# Regional geochemistry, terrane analysis and metallogeny in the British Caledonides

J. A. PLANT,[1] P. STONE[2] & J. R. MENDUM[2]

[1] British Geological Survey, Keyworth, Nottingham NG12 5GG, UK
[2] British Geological Survey, Murchison House, West Mains Road, Edinburgh EH9 3LA, UK

**Abstract:** The Caledonides of northern Britain are commonly divided into an assemblage of terranes based on the interpretation of geology in relation to major fault structures. This paper incorporates metallogenic concepts into the terrane model and uses systematic regional geochemical data to constrain models of terrane configuration and evolution: analyses of one-, two- and three-component geochemical images are used to identify changes in the levels of single elements and their ratios across the orogen. The distribution of sedimentary exhalative (SEDEX) ore deposits and volcanogenic massive sulphide, and mesothermal and epithermal gold mineralization is shown to reflect the presence of long-lived metallogenic provinces which have distinctive geochemistry reflecting their origin as zones of crustal extension and lithospheric thinning.

The geochemistry of the northern Caledonides reflects the predominantly lower crustal nature of the Laurentian continental margin, now exposed in the Lewisian foreland and Caledonized Lewisian inliers in the Moine succession. In contrast, the Avalonian Plate has a more evolved geochemistry, although the ratios of elements associated with basic rocks suggests that they were less fractionated during emplacement into the thinner crust of Avalonia and the Iapetus Suture Zone. The predominantly psammitic Moine Supergroup and the Grampian Group Dalradian have an evolved chemistry consistent with derivation from a higher crustal level than that of the exposed Lewisian complex; basin-scale variations, such as the proportion and composition of arkosic material, are also evident. The regional pattern was intermittently disrupted by the influx of mantle material during deposition of the mineralized Argyll and Southern Highland groups (Dalradian), and later as intrusions were emplaced from early Ordovician (Grampian Orogeny) to early Devonian times.

South of the Southern Upland Fault, the geochemical patterns suggest a complex interplay of sedimentary provenance and depositional environments during the final stages of collision between Laurentia and Avalonia. Provenance changes in early Silurian times probably reflect the initiation of major tectonism in the Scandian Orogen to the northeast and, more locally, a change from a mineralized, extensional back-arc- to a foreland-basin environment as the Laurentian plate overrode Avalonia. Very similar geochemical trends extending from the highest Silurian strata in the Southern Uplands into the Windermere Supergroup of the English Lake District are consistent with sequential deposition in the same foreland basin.

The principal crustal boundaries indicated by the geochemical data are the Moine Thrust Zone, the Grampian–Appin groups' boundary, the Highland Boundary Fault and the Southern Upland Fault. The geochemical evidence provides little support for the Great Glen Fault as a significant terrane boundary, while the Iapetus Suture is shown to be a complex transition zone extending across the Southern Uplands and into the south of the Lake District. Within the former, the Moffat Valley Fault is identified as a major structure of regional significance.

Throughout the orogen there is a direct relationship between geochemistry, tectonic setting and the presence of significant metalliferous mineralization.

The recognition and interpretation of the component terranes of an orogenic belt can be valuable in understanding tectonic evolution and can have a predictive value in terms of metallogeny, while an understanding of the genesis of ore deposit types can, in turn, help to identify tectonic settings (Plant & Tarney 1995).

The definition of terranes across the British sector of the Caledonian Orogen (Fig. 1) has tended to concentrate on major structural lineaments, despite well-established geological links across some of them, and at least seven terranes have been described (e.g. Cope *et al.* 1992 and references cited therein).

*From*: MAC NIOCAILL, C. & RYAN, P. D. (eds) 1999. *Continental Tectonics*. Geological Society, London, Special Publications, **164**, 109–126. 1-86239-051-7/99/$15.00 © The Geological Society of London 1999.

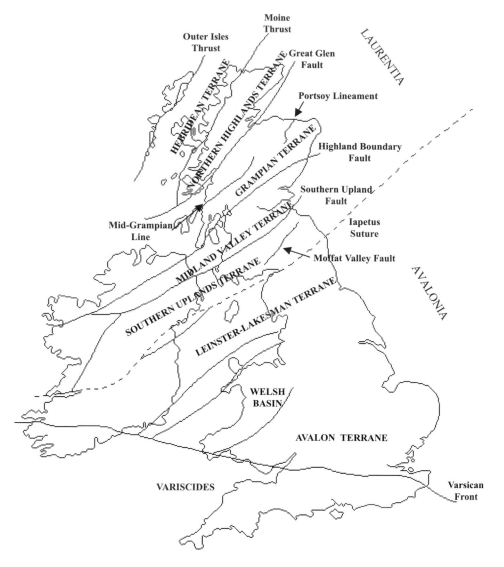

**Fig. 1.** Principal Caledonian terranes and their boundary structures in Britain and Ireland (after Cope *et al.* 1992).

In this paper, the status of the main terranes and their boundary faults are reviewed in relation to the distribution of different types of metalliferous mineralization in the Caledonides. Analysis of regional geochemical data is then used to constrain alternative terrane models and interpret the setting in which sedimentary exhalative (SEDEX) baryte deposits, volcanic massive sulphide (VMS) base-metal mineralization and gold mineralization formed. The regional geochemical data are used in a manner similar to

that used for the interpretation of remotely sensed data, although in addition they provide genetic information on the processes operating during evolution of the orogenic crust and the formation of mineral deposits. It is hoped that this study will encourage others to use systematic geochemical data sets, which are increasingly available, to develop criteria diagnostic of particular tectonic and metallogenic settings which can be combined using artificial intelligence in a GIS environment.

## The Caledonian terrane configuration of northern Britain

The British Caledonides represent a fragment of the orogenic belt formed by the early Palaeozoic collision of Baltica, Laurentia and Avalonia, a fragment of Gondwana. The overall structure of the orogen can be conveniently reviewed as a number of terranes separated by principal terrane-bounding faults (Fig. 1); each has distinctive geological characteristcs (Fig. 2) although not all aspects of any one terrane are unique to that terrane. The Iapetus Suture is generally taken as the boundary between Laurentia and Avalonia, although the two terranes immediately to the north of the suture line could be regarded as suspect.

To the northwest of the Moine Thrust Zone the *Hebridean Terrane* consists of Archaean and early Proterozoic gneissose basement unconformably overlain by undeformed late Proterozoic (Torridonian) rift infill and Cambro-Ordovician shallow-marine-shelf sedimentary assemblages. The *Northern Highlands Terrane*, which has been carried northwest by the Moine Thrust Belt, extends southeast as far as the Great Glen Fault and consists of late Proterozoic clastic shelf and basin sequences of the Moine Supergroup. It shows complex, mid-crustal polyphase deformation and metamorphism ranging from a possible Grenville event at *c.* 1050 Ma to Caledonian folding at *c.* 450 Ma. The status of the Great Glen Fault is particularly controversial. It has been suggested that metamorphic terranes of quite different age and origin are separated (Bluck & Dempster 1991) with significant (*c.* 2000 km) strike-slip displacement (van der Voo & Scotese 1981), although lithological and geochronological similarities have been demonstrated across the fault consistent with displacement of less than *c.* 200 km (Noble *et al.* 1996). The *Grampian Terrane* between the Great Glen and Highland Boundary Faults, includes rocks transitional from Moine-type, sandstone-dominated facies of the Grampian Group Dalradian to turbiditic, deep-water-facies rocks associated with basic volcanic rocks in the higher Southern Highland Group. These reflect the progression from a wide extensional basin to a fragmented and rifted continental margin sequence. The Argyll–Southern Highland Groups' boundary is marked in the southwest Highlands by the Tayvallich basic volcanic rocks, reflecting limited generation of oceanic-like crust at the climax of Dalradian extension (Graham 1976) during initiation of late Precambrian (*c.* 600 Ma) continental rifting which led to the development of the Iapetus Ocean in

Cambrian times. The Dalradian Supergroup was subjected to polyphase metamorphism and deformation culminating in the Grampian Orogeny *c.* 490–440 Ma, probably reflecting an arc–continent collision event (Dewey & Shackleton 1984; Dewey & Ryan 1990; Ryan & Dewey 1991). Obducted arc and ophiolitic material is preserved in western Ireland (Ryan *et al.* 1995) but not in the Grampian Terrane of Scotland. In the Monadhliath Mountains, a basement–cover relationship has been proposed between the basal Dalradian Grampian Group and the underlying gneisses by Piasecki & Temperley (1988), although Lindsay *et al.* (1989) record a continuous stratigraphical succession from the Central Highland migmatites into less-deformed strata of the Grampian Group. Noble *et al.* (1996) consider that the principal tectonic and stratigraphical break occurs higher in the Dalradian.

Geophysical evidence suggests that the crustal basement beneath all of these terranes is composed of gneisses similar to those of the Lewisian (e.g. Barton 1992), although the Portsoy Lineament (Goodman 1994) divides geophysically distinct basement in the Buchan Block, marked by numerous ultramafic and mafic pods, from the rest of the Grampian Highlands. Furthermore, Nd isotope ratios and the characteristics of inherited zircons in granite, suggest that the basement may be younger (or at least reworked) in the southern northwest Highlands and is undoubtedly different southeast of the mid-Grampian Line (Halliday *et al.* 1985). The Colonsay–West Islay basement is composed of syenitic and basic gneisses, dated at *c.* 1800 Ma (Marcantonio *et al.* 1988; Dickin 1992), and may extend southeast beneath the Grampian Terrane. There is little geophysical discontinuity at the Great Glen Fault, although contrasting lamprophyre dyke compositions across it appear to reflect differences in the composition of the lithospheric mantle (Canning *et al.* 1996). Seismic evidence indicates that the principal crustal break occurs further south at, or slightly to the north of, the Highland Boundary Fault, possibly reflecting the limit of the ancient Laurentian craton.

The Highland Boundary Fault is one of the principal tectonic breaks within the British transect of the orogen. To the south, the Caledonian rocks of the *Midland Valley Terrane* are largely obscured by Upper Palaeozoic strata, although dismembered Caledonian ophiolite assemblages are preserved in small outcrops at the northern and southern margins. The Highland Border Complex may represent the remains of a small oceanic basin marginal

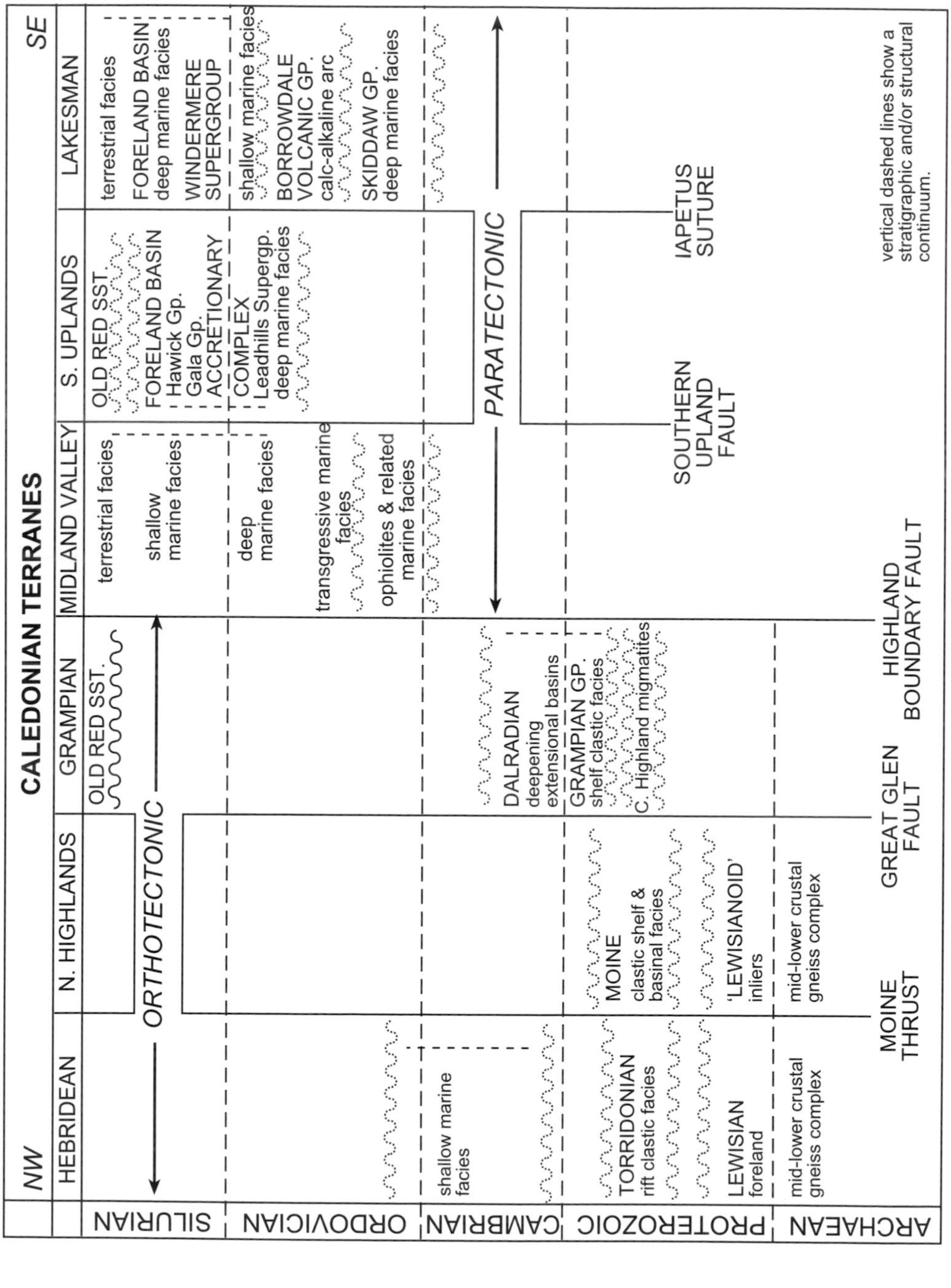

**Fig. 2.** A summary of the geological characteristics of British Caledonian terranes.

to the Laurentian continent (Robertson & Henderson 1984; Curry *et al.* 1984); its relationship to the adjacent Dalradian is poorly understood although some structural properties are common to both (Tanner 1995). The Ballantrae Complex, adjacent to the Southern Upland Fault, has been suggested to represent an arc and marginal basin assemblage (Bluck 1978; Stone & Smellie 1990) with arc–continent collision causing obduction and a reversal of subduction polarity at *c.* 480 Ma, contemporaneous with the Grampian Orogeny further north (Dewey & Shackleton 1984). Other small inliers show a conformable upward progression from marine Silurian to fluviatile Old Red Sandstone facies.

South of the Southern Upland Fault the suspect *Southern Uplands Terrane* formed as an accretionary thrust complex at the Laurentian continental margin during late Ordovician–middle Silurian subduction of the Iapetus Ocean. It has been widely interpreted as a forearc, supra-subduction zone prism (Leggett *et al.* 1979), although the northern (Ordovician) part may well be a subsiding shelf rather than a trench sequence (Armstrong *et al.* 1996). Alternatively, the terrane may have developed from an Ordovician back-arc setting into a mid-Silurian foreland basin following collision of Laurentia with Avalonia (Stone *et al.* 1987). According to both models, southward propagating imbricate thrusts separate lithostratigraphical tracts of steeply inclined strata striking northeast–southwest. Internally, the tracts of turbidite-facies greywacke young towards the north whereas the minimum age of each tract decreases southwards (Rushton *et al.* 1996 and references cited therein). Geophysical evidence indicates the presence of major crustal discontinuities (Kimbell & Stone 1995) to the north of the line of the Iapetus Suture. A particularly significant break which coincides with the Moniaive Shear Zone separates different suites of granitoids and basic dykes (Stone *et al.* 1997*b* and references cited therein). The crustal fragments beneath the Southern Uplands could represent microcontinental and older, arc remnants swept up during closure of the Iapetus Ocean and possibly include a fragment of Avalonia trapped on the 'wrong' side of the suture (Kimbell & Stone 1995; Stone *et al.* 1997*b*).

Following continental convergence in the mid-Silurian, a foreland basin developed on the Avalonian margin as it was thrust beneath Laurentia. Part of the foreland basin sequence is preserved in the south of the Southern Uplands Terrane (Stone *et al.* 1987) and part in the south of the Lake District Lower Palaeozoic Inlier, within the *Lakesman Terrane* (Kneller 1991;

Kneller *et al.* 1993). Previously, on the south side of the Iapetus Ocean, the Lake District succession developed on the leading edge of Avalonia (Cooper *et al.* 1993 and references cited therein). In the north, the Tremadoc–Llanvirn siltstone turbidites of the Skiddaw Group were deposited in an ensialic basin whilst the margin was relatively passive. The initiation (or resumption) of subduction subsequently uplifted the basin fill prior to the eruption of the calc-alkaline, suprasubduction zone Borrowdale Volcanic Group, mainly during the Caradoc. The mid-Silurian Laurentian–Avalonian continental collision was followed by the Acadian Orogeny during the early Devonian.

Plutonic igneous activity associated with the Grampian Orogeny was extensive and east of the Portsoy Lineament in northeast Scotland it included intrusion, at *c.* 470 Ma, of layered tholeiitic mafic and ultramafic intrusions. Unlike apparently similar aged, arc-related mafic–ultramafic intrusive bodies in Connemara (Yardley & Senior 1981; cf. Ryan & Dewey 1991), these Scottish examples are interpreted as products of local extension immediately following arc collision and a flip in subduction polarity. The early muscovite- and garnet-bearing S-type granites of northeast Scotland were intruded immediately following emplacement of the mafic-ultramafic bodies. Extensive calc-alkaline and lesser alkaline magmatism accompanied the final phases of Laurentia–Avalonia collision in the late Silurian–mid-Devonian (425–395 Ma), with late and post-tectonic granitoid plutons emplaced throughout the belt south of the Moine Thrust. On the Avalonian margin of northern England a large compound subvolcanic batholith was intruded during suprasubduction zone volcanicity in the mid-Ordovician (Hughes *et al.* 1996); subsequently, granite plutons were intruded at *c.* 400 Ma as part of the regional, post-collision magmatism. There was also extensive andesitic volcanism in the Grampian and Midland Valley terranes at this time. The youngest plutons are the product of melting after subduction had ceased following continent–continent collision and crustal recovery. At this time, large-scale sinistral movements took place on major northeast trending faults at the culmination of a regime of sinistral transpressive strain (Watson 1984).

## Mineralization and ore deposit types in the British Caledonides

The principal ore deposits and occurrences of mineralization in the Caledonides are documented on the metallogenic map of Britain and

Ireland (British Geological Survey (BGS) 1996) and in the BGS regional geochemical atlas series (BGS 1987, 1990, 1991, 1992, 1993). The locations of the principal deposits and occurrences of base-metal and gold mineralization are shown in Figs 3 and 4.

The earliest significant ore deposits known to have formed in the British Caledonides are the SEDEX deposits, including the world-class Aberfeldy Ba–Zn deposit, hosted by the Ben Eagach Schist in Argyll Group Dalradian rocks (Coats *et al.* 1980). About 0.55 million tons of baryte ore have been worked from the Foss Mine, Aberfeldy, and the Ben Eagach open pit since 1984, and a 7.5 million ton reserve of baryte has been estimated at Duntanlich, Perthshire. The recent discovery of the mineralization, and the recognition of its significance, has emphasized the difference between the more tectonically active setting of the Argyll (and Southern Highland Group) Dalradian compared to the relatively barren rocks of the Moine and Grampian Group Dalradian. The mineralization style evolved into Besshi-type Cu–Zn VMS in younger lithologies of the Argyll Group (Hall 1993), e.g. at Meall Mhor, southwest Highlands.

Subsequent mineralization in the Orthotectonic Caledonides was associated with the extensive igneous activity during the early Ordovician Grampian orogenesis. Concentrations of primary magmatic and remobilized Fe–Cu–Ni mineralization occur locally in the layered syntectonic tholeiitic mafic and ultramafic intrusions of northeast Scotland (Fletcher & Rice 1989). Porphyry style Cu–Mo and Au are associated with small post-tectonic calc-alkaline complexes in the southwest Highlands.

Base-metal veins have been mined historically at Strontian and Tyndrum. The Strontian Ba–Zn–Pb–Ag veins cut the contact of the Strontian Granite with the adjacent Moine psammitic metasediments (BGS 1987). At Tyndrum, vein deposits are associated with fractures at the junction of Appin Group quartzites and Argyll Group schists (Pattrick 1981); the veins predominantly contain Pb (with Zn, Ag and quartz gangue) with minor baryte. Other occurrences of vein mineralization include Pb–F–Cu–Fe–Ba veins at the margins of the Grudie Granite, Sutherland, and Pb–F–Zn veins in Dalradian rocks at the margins of the Ballater Granite, Aberdeenshire.

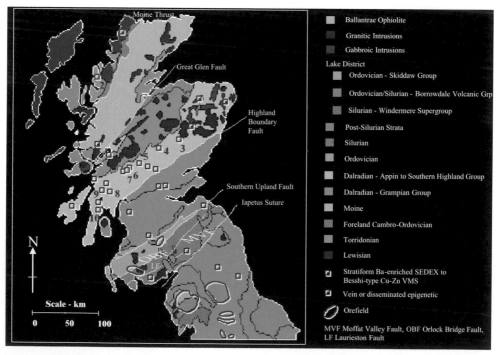

**Fig. 3.** Location of the principal occurrences of base-metal mineralization in northern Britain in relation to simplified solid geology. 1, Scourie; 2, Gairloch; 3, Loch Kander; 4, Aberfeldy; 5, Glen Lyon; 6, Ben Challum; 7, Auchtertyre; 8, McPhun's Cairn; 9, Craignure; 10, Meall Mhor.

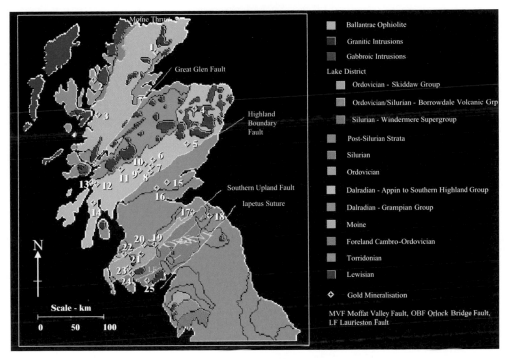

**Fig. 4.** Location of the principal occurrences of Au mineralization in northern Britain in relation to simplified solid geology. 1, Helmsdale; 2, Gairloch; 3, Ratagain; 4, Rhynie; 5, Burn of Fleurs; 6, Calliacher Burn; 7, Glen Almond; 8, Comrie; 9, Lochearnhead; 10, Corrie Buie; 11, Cononish; 12, Lagolochan; 13, Kilmelford; 14, Ardrishaig; 15, Boreland Glen; 16, Airthrey Hill; 17, Stobshiel; 18, Priestlaw; 19, Leadhills–Wanlockhead; 20, Hare Hill;

Synsedimentary mineralization is relatively rare in the Southern Uplands, although stratabound arsenopyrite (enriched in antimony) and pyrite has been described by Gallagher et al. (1989) in Silurian rocks near the old Glendinning stibnite mine, whilst disseminated stratabound Pb–Zn mineralization has been described by Stone & Leake (1984) in Ordovician strata, close to the southern margin of the Loch Doon Pluton. Mineralization directly related to magmatic activity is also less common in the paratectonic Caledonides but porphyry-style mineralization associated with late Caledonian intrusions occurs associated with the Fore Burn Complex (Midland Valley Terrane) (Charley et al. 1989), where it is dominantly of As–Cu–Au type, and at Black Stockarton Moor (Southern Uplands Terrane) west of the Criffell Pluton (Leake & Brown 1979), where it comprises Cu–Mo.

Vein mineralization in the Southern Uplands Terrane occurs mainly in two areas of historical mining (BGS 1993): the Leadhills–Wanlockhead Pb–Zn–Cu Orefield, which is hosted in Ordovician strata, and, to the southwest, the Galloway Orefield which is hosted by late Ordovician and early Silurian strata around the Fleet Pluton.

The extensive vein mineralization of the Lake District Inlier is largely restricted to the older rocks present: the Skiddaw Group hosts mainly Pb–Zn–Ba veins, whereas Cu is dominant in the Borrowdale Volcanic Group (BGS 1992). Wolframite, scheelite and arsenopyrite occur in veins adjacent to the early Devonian Skiddaw Granite Pluton.

In the orthotectonic Caledonides there are many significant occurrences of Au mineralization, including the Cononish Au–Ag deposit, near Tyndrum, Perthshire (Earls et al. 1992), which contains c. 750 000 tonnes of ore grading 10 g/tonne of Au and 43 g/tonne of Ag. The Au mineralization, which generally occurs near to small, late, calc-alkaline complexes and/or major late-Caledonian fault systems, where they cut basin-facies Dalradian rocks, ranges from epithermal acid sulphate (Lagolochan) (Harris et al. 1988; Zhou 1988) and adularia sericite type (Rhynie) (Rice et al. 1995) to the structurally controlled vein mineralization at Cononish.

Significant Au occurrences have been recorded outside the Dalradian outcrop only in the Kildonan area, adjacent to the Strath Halladale Granite in Helmsdale, Sutherland, where alluvial Au occurs (Plant 1970), and associated with the Ratagain Complex in Wester Ross (Alderton 1988). In the Helmsdale area, Read (1931) suggested that the Scaraben Quartzite may be Dalradian and, although this is consistent with some aspects of the geochemistry (Institute of Geological Sciences 1982), Strachan (1988) assigns these rocks to the Moine Supergroup.

In the paratectonic Caledonides a significant alluvial goldfield occurs adjacent to the Leadhills–Wanlockhead base-metal mining field. Mesothermal Au systems occur in several localities (e.g. Glenhead Burn, Moorbrock Hill, Hare Hill), where the Au–As–Sb mineralization is spatially related to late-Caledonian regional strike-slip faults and granitic to dioritic intrusions (Boast *et al.* 1990; Leake *et al.* 1981).

Overall, in the orthotectonic Caledonides, the most important known occurrences of baryte, base metals and gold are associated with basin-facies Dalradian rocks. In the Southern Uplands, mineralization is concentrated mainly in the older strata, the late Ordovician Leadhills and the early Silurian Gala Groups, and in the Lake District in the Ordovician Skiddaw and Borrowdale Volcanic Groups. Rocks of Moine lithofacies in the north, and the late Silurian foreland basin fold-and-thrust belt of the south Southern Uplands and southern Lake District are relatively barren.

## Regional geochemistry in relation to geological evolution

Regional geochemical mapping carried out by the BGS is based on the systematic collection and analysis of stream sediments and waters at a density of approximately one sample per $1.5\,km^2$; rocks and soils are also collected as part of the programme. The data are quality controlled before being entered and stored on the G-BASE database (BGS 1991, 1992, 1993). Studies of the resolution and reproducibility of the data in relation to factors such as the effects of glaciation and other surface processes suggest that the displacement of geochemical anomalies is

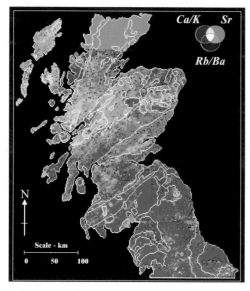

**Fig. 5.** Three-component geochemical map for Ca/K–Rb/Ba–Sr over northern Britain. Three-component images are generated from grids of individual elements (or element ratios) each of which is assigned to one of the primary colours red, green or blue, the highest values being represented by the highest intensity colours and the lowest values by the lowest intensity colours. Composite maps are made according to primary colour interaction: red + green = yellow; red + blue = magenta; blue + green = cyan; red + green + blue = white. Such plots are particularly valuable in identifying cryptic changes in chemistry which would not be apparent from traditional field survey methods.

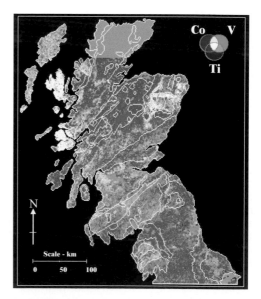

**Fig. 6.** Three-component geochemical map for Co–V–Ti over northern Britain. For explanation see caption to Fig. 4.

generally low (<2 km), with variation in stream-sediment composition reflecting the nature of the underlying bedrock; there is close agreement between the median values of most chemical elements in rocks and stream sediments (Plant *et al.* 1990; Stone *et al.* 1997*a*). Hence, the geochemical data can be used to examine chemical variation across the orogen in a manner comparable to the use of remotely sensed or airborne geophysical data for structural analysis. This is particularly useful over the northern Caledonides which are characterized by complex lithological relationships reflecting polyphase deformation. It is also of value over apparently homogeneous tracts of rocks throughout the orogen where subtle and cryptic compositional changes can be identified. Single-element maps, such as those presented in the BGS's geochemical atlas series, illustrate much of the variation but multi-element comparisons can also be used to highlight differences in lithologies. For example, relative differences in the proportions of elements closely associated with feldspar and/or mica (K–Ba–Rb–Ca–Sr; Fig. 5), those of basic or ultrabasic association (Co–V–Ti or Cr–Ni–Mg; Figs 6 and 7) or of the high field strength (HFS) elements such as Zr–Y–Ti (Fig. 8), which tend to occur in resistate minerals, can be computer enhanced to indicate differences not apparent using single-element plots.

In broad terms, certain plots, especially those based on feldspathic elements such as Ca, K, Sr, Rb and Ba (Fig. 5), distinguish the Laurentian Plate (characterized by high Ca/K ratios and Sr content in the Hebridean Terrane) from the more evolved Avalonian Plate (characterized by higher levels of the more incompatible elements, especially Rb, in the English Lake District). The geochemistry of the intervening cover rocks shows a broadly progressive change from Laurentian to Avalonian characteristics so that the southern part of the Southern Uplands appears more Avalonian than Laurentian. Conversely, the elements of basic association, Co, Ti and V show enrichment of Ti, the most incompatible of the three elements, at the Laurentian rather than the Avalonian margin (Fig. 6). This probably reflects a higher degree of fractionation of basic magmas and hence the thickness of the Laurentian Plate, with lower degrees of fractionation of basic magmas emplaced in the suture zone and in the relatively thin crust of the Avalonian Plate.

## Hebridean Terrane

The Lewisian foreland of the Hebridean Terrane is characterized by low levels of B, Li and Ga, and other incompatible and heat-producing elements such as U, Mo, Be, Rb and K, and high levels of basic elements such as Co, Ti and V and of Sr, Ba and Ca. The exceptionally low levels of the large-ion lithophile elements (LILE), such as U, K and Th, over most of the complex are thought to reflect depletion during a high-pressure granulite event (Weaver & Tarney 1980), while the high level of basic elements reflects the presence of large numbers of basic intrusions in the lower crustal complex.

## Northern Highland Terrane

There is a marked geochemical contrast between the Hebridean Terrane and rocks of Moine lithofacies which have much higher levels of the heat-producing and incompatible elements, and much lower levels of elements associated with basic rocks. Zirconium and Y are also much higher over the Moine than the Lewisian, while Ti levels are exceptionally low. The dominantly psammitic Moine succession lies unconformably on inliers of gneissic rocks that show strong similarities, but also some differences, to the foreland Lewisian. Detritus from these inliers is contained in the basal Moine units but the bulk of the psammites are second-cycle quartzofeldspathic rocks derived from basements of different compositions to those of the presently exposed 'Lewisianoid' inliers. Studies of detrital zircon from the Moine outcrop in Sutherland (Friend *et al.* 1999) show two groups of ages, 1000–1250 and 1500–1750 Ma, consistent with their derivation from a highly evolved source at a higher crustal level than the exposed Archaean foreland basement complex.

The Lewisianoid inliers show varying degrees of Caledonian reworking but the larger ones, e.g. Scardroy and Borgie, contain abundant ultramafic pods and a high proportion of basic gneiss, reflected by high levels of elements such as Cr–Ni–Mg (Fig. 7), and exotic metasedimentary rocks similar to those of the Scourian of the Hebridean Terrane. The Lewisianoid inliers and rocks of the Hebridean Terrane thus appear to have a common heritage, but the geochemical data suggest that they represent different types of gneissose basement. Isotopic dating confirms the Lewisian affinity of the inliers (Moorbath & Taylor 1974) but recent work on the Borgie Inlier (Friend *et al.* 1999) has confirmed the earlier results from Glenelg (Sanders *et al.* 1984), that they were affected by a metamorphic event at *c.* 1050 Ma. Since the inliers are part of the rifted basement on which the Moine was deposited, the latter may also have been affected by the Grenville event.

## Grampian Terrane

The broad geochemical characteristics of the Northern Highlands Terrane extend over the Grampian Group Dalradian to the south of the Great Glen Fault, although levels of Ba and, to a lesser extent, Sr increase; possibly reflecting the juxtaposition of two parts of the same psammite-dominated rift basin sequence as a result of vertical, or only minor transcurrent, movement on the Great Glen Fault. The principal geochemical discontinuity in the ortho-tectonic Caledonides occurs in the Grampian Terrane, between the Grampian Group and the rest of the Dalradian Supergroup. It reflects the transition to the deeper water, basinal lithofacies of the Argyll and Southern Highlands groups. These are clearly distinguished on most single-element plots, including those of B, Cu and As, and on multi-element plots, e.g. by the relative enrichment of Ti on a Zr–Ti–Y map (Fig. 8), by Cu relative to Pb and Zn, and by the generally low levels of incompatible elements such as U–Mo–Be and Sn compared to the Moine and Grampian Group Dalradian.

The marked contrast in the regional geochemistry between the psammite-dominated Moine and Grampian Group Dalradian, and the basin-facies Dalradian, may be summarized as an increase in B, Li, Co, Cr, Cu, Fe, Mg, Mn, Ni, Pb, and Zn, and a decrease in Be, Ba, K La, Rb, Sr, U, Y and Zr. The contrast is attributed to a reduction in the detrital minerals component and an increase in marine clay minerals in the basin facies (Plant *et al.* 1984); there is also a further increase in Li and B over the turbidite divisions of the upper Argyll Group and Southern Highland Group Dalradian. The generally higher levels of basic elements in the Dalradian are interpreted to record an input from volcanic sources. A notable increase in the elements normally associated with ultrabasic–basic rocks, particularly Mg and Co, occurs in a restricted zone near the top of the Argyll Group immediately below and at the level of the Tayvallich Lavas. This geochemical feature extends well beyond the outcrop of the Tayvallich Lavas to include much of the Crinan Subgroup at the base of the Southern Highland Group. The geochemical evidence thus suggests that the Dalradian palaeorift system was more extensive than the mapped outcrop of the Tayvallich Lavas would suggest.

In general, the geochemistry is most consistent with models involving deposition of the Moine and Grampian Group Dalradian on a relatively stable continental mass of normal crustal thickness, whereas the higher parts of the Dalradian

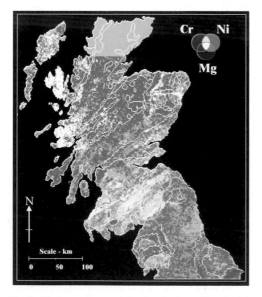

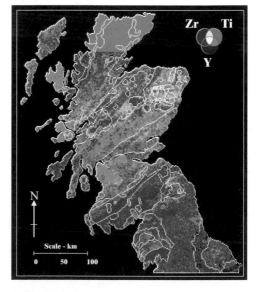

**Fig. 7.** Three-component geochemical map for Cr–Ni–Mg over northern Britain. For explanation see caption to Fig. 4.

**Fig. 8.** Three-component geochemical map for Zr–Ti–Y over northern Britain. For explanation see caption to Fig. 4.

sequences, which also have a distinctive gravity signature, were deposited over thinned continental basement which was progressively stretched and finally ruptured by extensional tectonism. The present Grampian–Appin boundary, which is a marked feature on most geochemical plots, may reflect the boundary between basement comprising the relatively undisrupted Laurentian continental slab and a zone beneath it which is extended, thinned and more variable in composition. The implications for ore genesis are discussed below.

## Southern Uplands Terrane

For many elements the most striking feature of their regional distribution over the Southern Uplands Terrane is a fine northeast–southwest striping trending parallel to the regional bedding strike of the greywacke bedrock and reflecting changes in its composition. Such patterns are particularly well displayed by the Ca/K–Rb/Ba–Sr (Fig. 5), Co–V–Ti (Fig. 6), Cr–Ni–Mg (Fig. 7) and Zr–Ti–Y (Fig. 8) plots. The distribution of B is divided into zones of high values to the north, and lower levels to the south, of the Moffat Valley–Laurieston Fault System. Levels of incompatible elements such as Be and Mo are generally low over the whole terrane.

The exceptionally high levels of elements such as Cr, Ni, Co and Mg are particularly significant in terms of the terrane's origin. The Ballantrae Complex represents oceanic arc material obducted onto the Midland Valley continental block during the early Ordovician (Smellie & Stone 1992 and references cited therein), with subsequent erosion producing high levels of ophiolitic detritus in late Ordovician greywackes in the north of the Southern Uplands Terrane (Floyd 1982). The importance of the ophiolitic provenance declined thereafter but the regional geochemical data provide new evidence of a continuing, intermittent flux of ophiolitic detritus through the Ordovician and early Silurian; the absence of identifiable clasts suggests that much of the material is in the fine-grained matrix of the greywackes. The maximum Cr values, and hence the maximum ophiolitic flux, is associated with the early Llandovery greywacke sequences (Stone et al. 1991, 1997a; BGS 1993), which palaeocurrent patterns (e.g. Greig 1988; Stone 1995) suggest were sourced to the northeast. The Scandian Orogen was beginning to develop at that time (Gee & Roberts 1983; Dallmeyer 1988), so the data for Cr may indicate that erosion of ophiolites was occurring there by early Llandovery times with the detritus car-

ried southwestwards into the Southern Uplands depositional basin. By that time the ophiolites of the Midland Valley Terrane were largely buried beneath thick marine turbidite sequences.

Across the Southern Uplands Terrane levels of the HFS elements, Ti, Zr and Y (Fig. 8), which tend to be concentrated in refractory accessory minerals, closely follow mapped lithostratigraphical boundaries. High Ti levels characteristic of the post-Grampian Group Dalradian rocks extend over the Ordovician quartzo-feldspathic greywackes in the north of the terrane with an abrupt reduction southwards over younger Ordovician greywackes which are rich in andesitic detritus. The Ti levels step down further over the geochemically complex Lower Silurian rocks. The highest Zr levels coincide with the zone of high Cr values over the Lower Silurian rocks, reinforcing the association with the resistate mineral component.

Rubidium, K and Sr plots (Fig. 5) define two greywacke populations in the Southern Uplands (Stone et al. 1993); relatively low levels of Rb and K over the Ordovician sequence north of the Orlock Bridge Fault and higher levels over the Silurian sequence to the south. In contrast, the distribution of Sr shows an abrupt change across the Moffat Valley Fault with higher values to the north. The distribution of Ba reflects the presence of both structures with a zone of relatively high Ba levels between the Orlock Bridge Fault and the Moffat Valley Fault. These data, together with those for B, suggest that the source for the rocks to the south of the Moffat Valley Fault were geochemically more evolved. Complex variations in Rb, Ba, Sr and K levels over the lower parts of the Silurian sequence, where the high levels of Cr occur, are consistent with complex interaction between two or more sedimentary provenances.

The Moffat Valley Fault has hitherto been regarded as one of a large number of strike faults within the Southern Uplands accretionary terrane. The geochemical evidence presented here suggests that it is of much greater importance in regional terms. It is particularly noteworthy that to the north of the Moffat Valley Fault the geochemical signature is most compatible with an extensional tectonic environment, whereas to the south of the fault the geochemistry is markedly different and has more in common with the foreland basin sequence of the Lake District Inlier (Lakesman Terrane; see below). Overall, the geochemistry of the Southern Uplands Terrane seems most consistent with a complex interplay between sedimentary provenances, reflecting large-scale tectonic adjustments along the orogenic belt, superimposed on a change from an extensional back-arc to a load-induced foreland basin formed over the British section of the orogen as Laurentia overrode the Avalonian margin. The Moffat Valley–Lauries-ton Fault System is the most important break between the sequences formed in these two contrasting environments, although it is unlikely that a clear structural divide exists in what was a complex, evolving tectonic framework.

## Lakesman Terrane

The regional geochemistry of the Lake District Inlier is markedly different to that of the Laurentian terranes north of the Moffat Valley Fault. In the north, the predominantly silty turbidites of the Skiddaw Group are characterized by high levels of B, Ga and Li, and of Co, V and Ni; Be levels are also high, although levels of U and Mo are low. Levels of Rb and K, and to a lesser extent Ba, are high. From the detrital heavy mineral component, levels of Y which tends to be concentrated in garnet or monazite, are high, as are those of Ti which occurs in rutile, ilmenite and sphene; conversely, levels of both Zr and Cr are generally low. The geochemistry is compatible with the findings of Cooper *et al.* (1995), that the Skiddaw Group rocks are derived from 'an old inactive continental arc terrane' exposed to the south in the Avalonian hinterland.

The central sector of the Lake District is underlain by the calc-alkaline volcanic assemblage of the Borrowdale Volcanic Group, which formed as a result of south directed Caradocian subduction of Iapetus Ocean crust. The regional geochemistry reflects the andesitic to rhyolitic composition of the volcanics which have higher levels of incompatible elements than igneous rocks of the Laurentian margin.

The geochemistry confirms that neither the Skiddaw or Borrowdale Volcanic Groups, which both formed at the Avalonian margin, had any direct relationship with coeval Laurentian terranes. This situation appears to have changed in the early Silurian, however, as the Iapetus Ocean narrowed and a common depositional system developed over the opposing continental margins. Following continental collision, a foreland basin migrated from the Southern Uplands Terrane across the margin of Avalonia as it was overridden by Laurentia. The Windermere Supergroup, in the south of the Lake District, was largely deposited in this environment and shares geochemical characteristics with the southern part of the Southern Uplands (Stone *et al.* 1993).

## Regional geochemistry and metallogeny

### Orthotectonic Caledonides

Single-element plots, and two- and three-component images of indicative element associations, e.g. of the ore-forming elements Cu–Pb–Zn and petrogenetic elements K–Ba–Sr (Fig. 5), clearly identify the distinctive features of the mineralized zones in the orthotectonic Caledonides which appear to be related to their tectonic setting. The Hebridean and Northern Highlands Terranes, and the Grampian Group of the Grampian Terrane, which are relatively poorly mineralized, are clearly distinguished on almost all individual element plots from basin facies Argyll and Southern Highland Group Dalradian (Grampian Terrane) which host important stratiform Ba–Pb and Cu–Fe–Zn mineralization. Many of the most important occurrences of gold and base-metal vein mineralization also occur in Dalradian rocks of basin facies which are generally enriched in elements of ultrabasic–basic association (Figs 6 and 7), but which have low levels of K, Ba, Sr (Fig. 5) and other incompatible elements. In contrast, rocks of generally barren lithofacies (Moine Supergroup and Grampian Group) in the northwest of the Grampian Terrane have low levels of ultrabasic–basic elements but relatively high levels of K, Ba, Sr and other incompatible elements (Plant *et al.* 1984). Within the Dalradian, the principal occurrences of stratiform and base-metal vein mineralization (Fig. 3) correspond to elevated levels of one or more of the elements of ultrabasic–basic association (Figs 6 and 7). In the case of the stratiform deposits, a zone of high Mg relative to Cr and Ni is particularly distinctive at the regional level, and Sr is enriched relative to K or Ba (Fig. 5).

In the Dalradian, the local controls of the SEDEX deposits are well understood. Evidence such as the concordance of the mineral banding with the adjacent schists (Coats *et al.* 1984) and isotopic results (Hall *et al.* 1991) all indicates that the mineralization was contemporaneous with sedimentation. In the type locality of the Ba–Zn–Pb mineralization near Aberfeldy, the main mineralized zone consists of interbanded barite, quartz–celsian and sulphide-bearing carbonate rock, and the mineralized zone is preceded by *c.* 250 m of strata which contain increased levels of Ba.

The Aberfeldy deposit occurs at, or close to, the top of the Ben Eagach Schist, which was deposited during the climax of the third major basin-deepening episode in the Dalradian (Harris *et al.* 1978). The sequence is overlain by the Ben Lawers Schist at Aberfeldy and the Ardrishaig (= Craignish) Phyllites at Loch Fyne and in Knapdale. These units were deposited during a prolonged period of basin shallowing when sediment influx outpaced basin subsidence. To the northwest of Loch Fyne the stratabound Ni-pyrrhotite deposits of Coillebraghad and Craignure in the Ardrishaig Phyllites are thought to form part of a continuous zone containing nickeliferous pyrrhotite and chalcopyrite, with subordinate pyrite and pentlandite (Wilson & Flett 1921). The formation of the Pyrite Zone at the top of the Ardrishaig Phyllites coincided with the climax of shallowing of the Easdale Basin. As discussed above, the Easdale Subgroup, which hosts the SEDEX to Besshi-type stratabound mineralization, is clearly identified by single- and three-component geochemical plots.

A major basic magmatic contribution to the Aberfeldy mineralization is unlikely because of the paucity of Cu or other elements of basic association in the deposits, and the rising magma is envisaged principally as a heat source driving the hydrothermal system. Elsewhere, e.g. during deposition of the Pyrite Zone and associated mineralization in South Knapdale, hydrothermal fluids had more direct access to underlying and nearby basic volcanic rocks, as reflected by an increase in the Cu content of the mineralization. The distinctive regional Mg and Co anomaly mentioned above, which occurs over volcanic units and other rock types stratigraphically above the deposits, may indicate exhalation of fluids which had interacted with juvenile magmas, or the incorporation of basic detritus in the sedimentary rocks. The stratabound Ba–Zn–Pb mineral deposits, tholeiitic volcanic rocks and the chemistry of the Dalradian succession generally, particularly in the Argyll–Southern Highland Group sequence, are thus related to rifting in Laurentia as a result of major extension leading to localized crustal rupture associated with the early development of the Iapetus Ocean. It is likely that this was preceded by thinning of the lithosphere and partial melting of the upper mantle, associated with a linear zone of anomalous heat flow and tectonism, giving rise to hydrothermal activity and mineralization. The style of stratabound mineralization thus reflects the interplay of magmatism and the temperature of intermittent sub-seafloor–seafloor geothermal systems, with tectonic factors controlling basin shallowing and deepening. The local evidence on the processes involved in metallogenesis fits well into such a regional model. The results suggest that geochemistry provides powerful exploration criteria which could be used to iden-tify settings prospective for stratabound SEDEX and VMS ore deposits in similar orogenic basins worldwide.

The principal known occurrences of Au in the orthotectonic Caledonides cover a range of types but are mainly of lower Devonian age. They all lie within the basin facies Dalradian rocks which, as noted above, have distinctive geochemical characteristics. The association between Au mineralization and the distribution of the pathfinder elements As, Sb and Bi is striking at both the regional and local scale (Plant et al. 1989, 1991). Arsenic is enriched at the regional scale and is further enriched, in some cases together with Sb and Bi, around mineralized centres. All of the known Au occurrences in the Dalradian are hosted by, or occur near to, rocks enriched in Sr relative to Ba and K (although Sr is generally low in the Dalradian) (Fig. 5), and enriched in Cu relative to Pb or Zn but also Zn relative to Ba and Pb. Overall, there is no relationship between the chemistry of the igneous intrusions (including the late or post-tectonic Caledonian granites), which vary from diorites to evolved granites, and the presence of stratiform or base-metal or Au vein mineralization in the Dalradian.

*Paratectonic Caledonides*

The trace-element enrichment characteristics of the northern sector of the Southern Uplands Terrane (the Leadhills Group and most of the Gala Group), including Au and its pathfinder elements, are most consistent with a zone of suprasubduction extension. This association suggests that a greater potential exists there for the discovery of ore deposits than in the putative foreland basin sequence to the south (mostly Hawick and Riccarton Groups) which is relatively barren. As in the northern Caledonides, the regional geochemistry suggests that mineralization has occurred as a result of the enrichment of elements in hydrothermal systems associated with the emplacement of granite complexes into sedimentary rocks which acted as crustal reservoirs for ore-forming elements. Hence, the Loch Doon Pluton, which intrudes the Leadhills Group, and the Cairnsmore of Fleet Pluton, which intrudes the Gala Group, have haloes of Pb–Zn–Cu enrichment whereas the Criffell–Bengairn complex, which intrudes the Hawick Group (foreland basin) turbidites, has no such halo (Stone et al. 1995).

Mineralization at the Avalonian margin is concentrated in the Skiddaw and Borrow-dale Volcanic Groups, mainly as Pb–Zn–Ba veins in the former and Cu veins in the latter, reflecting the regional trace-element geochemistry. Pb–Zn–Cu levels are fairly high across the Skiddaw Group outcrop and are locally concentrated around mineralized zones. High background levels of As and the other Au pathfinders, Sb and Bi, are also characteristic of the Skiddaw Group, with a particularly important zone coinciding with high B levels along the Causey Pike Fault, a major eastnortheast–westsouthwest structure (Cooper *et al.* 1988). The levels and associations of base metal and Au pathfinder elements of the Skiddaw Group are closely comparable with those of the basin facies Dalradian rocks to the north of the suture (Plant *et al.* 1991). Across the Borrowdale Volcanic Group the Au pathfinders, As–Sb–Bi, and the base metals, Cu–Pb–Zn, are all elevated with particularly high levels of Cu in the vicinity of the mineralized sites near Coniston. The Windermere Supergroup, which it is suggested is the tectonostratigraphic extension of the Southern Uplands foreland basin sequence, is barren.

## Conclusions

The application of regional geochemical data to terrane analysis in the British Caledonides provides new insights into the evolution of this sector of the orogen and its ore deposits. Several crustal units with distinct geochemical characteristics are identified, and the importance of the Moine Thrust Belt, the Grampian–Appin groups' boundary, the Highland Boundary Fault and the Southern Upland Fault as major crustal boundaries is confirmed. Analysis of the geochemical data provides little support for the Great Glen Fault as a major discontinuity, calling into question its status as a terrane boundary, as well as palaeogeographical reconstructions which involve large amounts of strike-slip movement along it. In the case of the Iapetus Suture Zone, many different element associations suggest that at the present erosion level it is a complex zone extending from the Southern Upland Fault to the southern Lake District rather than a simple lineament. The Southern Uplands appear, on the basis of the regional geochemistry, to be divided into a northern and a southern sector, although the boundary does not coincide uniquely with any particular structure. However, a major linear discontinuity (identified by the geochemistry) corresponds with the Moffat Valley Fault, and this structure seems most likely to mark the boundary between an active/extensional margin and a foreland basin. The geochemical evidence suggests that the latter subsequently extended southwards to

form the depositional basin for much of the Windermere Supergroup in the Lake District.

Regional geochemical patterns clearly characterize the major crustal blocks and reflect their tectonic history and potential for metalliferous mineralization. Mineralization is generally associated with zones of crustal extension and lithospheric thinning which have distinctive geochemical characteristics. The most significant known ore deposits are associated with the Argyll Group Dalradian and are thought to reflect increased heat flow associated with basic volcanism and rifting of the continental crust during the initiation of the Iapetus Ocean. Gold and porphyry-style mineralization, which formed during the Grampian orogenic event, is associated mainly with post-tectonic granite magmatism and major transcurrent faulting, although the geochemical evidence suggests that Au and its pathfinder elements were remobilized from the Argyll and Southern Highland Group Dalradian which acted as crustal reservoirs. Only minor mineralization is known to be associated with the earlier Moine and Grampian Group intracratonic basins, or with the Lewisian or younger gneissic basement of Laurentia. The relatively high levels of ore-forming elements in the Leadhills and Gala Groups of the Southern Uplands, and the Skiddaw Group of the Lake District also reflect extensional depositional environments, comparable to those of the mineralized Dalradian sequences of the orthotectonic Caledonides. The foreland basin sequences of the southern Southern Uplands and the southern Lake District are relatively barren of mineralization.

The results presented here suggest that systematic high-resolution geochemical data of the quality prepared by the BGS could have wide application in understanding the tectonic and metallogenic evolution of orogenic belts worldwide, and might be of particular value in further understanding the Caledonides of Ireland which are along-strike to the southwest of the region studied here. Similar studies over other geologically well-constrained sections of orogenic belts could lead to the identification of geochemical signatures diagnostic of particular tectonic and metallogenic settings worldwide.

This paper is published by permission of the Director, British Geological Survey (NERC). We thank P. Green, D. Flight and J. Freeman for assistance with image processing and the referees, C. Stillman and C. van Staal, for thorough and helpful reviews. Many of the concepts discussed in this paper relating large scale geochemical patterns to crustal processes and plate tectonics derive substantially from the influential ideas and research of Professor J. Dewey FRS.

# References

ALDERTON, D. H. M. 1988. Ag–Au–Te mineralisation in the Ratagain complex, north-west Scotland. *Transactions of the Institution of Mining and Metallurgy (Section B: Applied earth science)*, **97**, B171–B180

ARMSTRONG, H. A., OWEN, A. W., SCRUTTON, C. T., CLARKSON, E. N. K. & TAYLOR, M. 1996. Evolution of the Northern Belt, Southern Uplands: implications for the Southern Uplands controversy. *Journal of the Geological Society, London*, **153**, 197–205.

BARTON, P. J. 1992. LISPB revisited: a new look under the Caledonides of northern Britain. *Geophysical Journal International*, **110**, 371–391

BLUCK, B. J. 1978. Geology of a continental margin 1: the Ballantrae Complex. *In*: BOWES, D. R. & LEAKE, B. E. (eds) *Crustal Evolution in Northwestern Britain and Adjacent Regions*. Geological Journal, Special Issue, **10**, 151–162

—— & DEMPSTER, T. J. 1991. Exotic metamorphic terranes in the Caledonides: tectonic history of the Dalradian block, Scotland. *Geology*, **19**, 407–436.

BOAST, A. M., HARRIS, M. & STEFFE, D. 1990. Intrusive-hosted gold mineralization at Hare Hill, Southern Uplands, Scotland. *Transactions of the Institution of Mining and Metallurgy (Section B: Applied earth science)*, **99**, B106–B112.

BRITISH GEOLOGICAL SURVEY 1987. *Regional Geochemical Atlas: Great Glen*. BGS.

——1990. *Regional Geochemical Atlas: Argyll*. BGS.

——1991. *Regional Geochemistry of the East Grampians Area*. BGS.

——1992. *Regional Geochemistry of the Lake District and Adjacent Areas*. BGS.

——1993. *Regional Geochemistry of Southern Scotland and Part of Northern England*. BGS.

— —, COLMAN, T. B., SCRIVENER, R. C., MORRIS, J. H., LONG, C. B., O'CONNOR, P. J., STANLEY, G. & LEGG, I. C. (compilers). 1996. *Metallogenic Map of Britain and Ireland, 1:1 500 000*. BGS.

CANNING, J. C., HENNEY, P. J., MORRISON, M. A. & GASKARTH, J. W. 1996. Geochemistry of late Caledonian minettes from Northern Britain: implications for the Caledonian sub-continental lithospheric mantle. *Mineralogical Magazine*, **60**, 221–236.

CHARLEY, M. J., HAZLETON, R. E. & TEAR, S. J. 1989. Precious metal mineralisation associated with the Fore Burn igneous complex, Ayrshire, south-west Scotland. *Transactions of the Institution of Mining and Metallurgy (Section B: Applied earth science)*, **98**, B48.

COATS, J. S., PEASE, S. F., GALLAGHER, M. J. & GROUT, A. 1984. Stratiform barium enrichment in the Dalradian of Scotland. *Economic Geology*, **79**, 1585–1595.

——, SMITH, C. G., FORTEY, N. J., GALLAGHER, M. J., MAY, F. & McCOURT, W. J. 1980. Stratabound barium–zinc mineralisation in Dalradian Schist near Aberfeldy. *Transactions of the Institution of Mining and Metallurgy (Section B: Applied earth science)*, **89**, B110–B122.

COOPER, A. H., MILLWARD, D., JOHNSON, E. W. & SOPER, N. J. 1993. The early Palaeozoic evolution of north-west England. *Geological Magazine*, **130**, 711–724.

——, RUSHTON, A. W. A., MOLYNEUX, S. G., HUGHES, R. A., MOORE, R. M. & WEBB, B. C. 1995. The stratigraphy, correlation, provenance and palaeogeography of the Skiddaw Group (Ordovician) in the English Lake District. *Geological Magazine*, **132**, 185–211.

COOPER, D. C., LEE, M. K., FORTEY, N. J., COOPER, A. H., RUNDLE, C. C., WEBB, B. C. & ALLEN, P. M. 1988. The Crummock Water aureole: a zone of metasomatism and source of ore metals in the English Lake District. *Journal of the Geological Society, London*, **145**, 523–540.

COPE, J. C. W., INGHAM, J. K. & RAWSON, P. F. 1992. *Atlas of Palaeogeography and Lithofacies*. Geological Society of London, Memoir, **13**.

CURRY, G. B., BLUCK, B. J., BURTON, C. J., INGHAM, J. K., SIVETER, D. J. & WILLIAMS, A. 1984. Age, evolution and tectonic history of the Highland Border Complex, Scotland. *Transactions of the Royal Society of Edinburgh: Earth Sciences*, **75**, 113–133.

DALLMEYER, R. D. 1988. Polyphase tectonothermal evolution of the Scandinavian Caledonides. *In*: HARRIS, A. L. & FETTES, D. J. (eds) *The Caledonian – Appalachian Orogen*. Geological Society, London, Special Publication, **38**, 365–379

DEWEY, J. F. & RYAN, P. D. 1990. The Ordovician evolution of the South Mayo Trough, Western Ireland. *Tectonics*, **9**, 887–903.

—— & SHACKLETON, R. M. 1984. A model for the evolution of the Grampian tract in the early Caledonides and Appalachians. *Nature*, **312**, 115–121.

DICKIN, A. P. 1992. Evidence for an Early Proterozoic crustal province in the North Atlantic region. *Journal of the Geological Society, London*, **149**, 483–486.

EARLS, G., PARKER, R. T. G., CLIFFORD, J. A. & MELDRUM, A. H. 1992. The geology of the Cononish gold–silver deposit, Grampian Highlands of Scotland. *In*: BOWDEN, A. A., EARLS, G., O'CONNOR, P. G. & PYNE, J. F. (eds) *The Irish Mining Industry 1980–1990 – A Review of the Decade*. Irish Association for Economic Geology, 89–103.

FLETCHER, T. A. & RICE, C. M. 1989. Geology, mineralization (Ni–Cu) and precious metal geochemistry of Caledonian mafic and ultramafic intrusions near Huntly, northeast Scotland. *Transactions of the Institution of Mining and Metallurgy (Section B: Applied earth science)*, **98**, B185–B200.

FLOYD, J. D. 1982. Stratigraphy of a flysch succession: the Ordovician of W. Nithsdale, SW Scotland. *Transactions of the Royal Society of Edinburgh: Earth Sciences*, **73**, 1–9.

FRIEND, C. R. L., KINNY, P. D. & STRACHAN, R. A. 1999. U–Pb SHRIMP geochronology of tectonothermal events within the basement and cover of the Caledonian thrust nappes, north Sutherland. *Journal of the Geological Society, London*, in press.

GALLAGHER, M. J., STONE, P. & DULLER, P. R. 1989. Gold-bearing arsenic–antimony concentration in Silurian greywackes, south Scotland. *Transactions of the Institution of Mining and Metallurgy (Section B: Applied earth science)*, **98**, B58–B60.

GEE, D. G. & ROBERTS, D. 1983. Timing of deformation in the Scandinavian Caledonides. *In*: SCHENK, P. E. (ed.) *Regional Trends in the Geology of the Appalachian–Caledonian–Hercynian–Mauritanide Orogen*. NATO ASI Series C, **116**, Reidel, 279–292.

GOODMAN, S. 1994. The Portsoy–Duchray Hill Lineament: a review of the evidence. *Geological Magazine*, **131**, 407–415.

GRAHAM, C. M. 1976. Petrochemistry and tectonic significance of Dalradian metabasaltic rocks of the S.W. Scottish Highlands. *Journal of the Geological Society, London*, **132**, 61–84.

GREIG, D. C. 1988. Geology of the Eyemouth district. *Memoir of the British Geological Survey*, Sheet 34 (Scotland).

HALL, A. J. 1993. Stratiform mineralization in the Dalradian of Scotland. *In*: PATTRICK, R. A. D. & POLYA, D. A. (eds) *Mineralization of the British Isles*. Chapman & Hall.

——, BOYCE, A. J., FALLICK, A. E. & HAMILTON, P. J. 1991. Isotopic evidence of the depositional environment of Late Proterozoic stratiform baryte mineralisation, Aberfeldy, Scotland. *Chemical Geology (Isotopic Geoscience Section)*, **87**, 99–114.

HALLIDAY, A. N., STEPHEN, W. E., HUNTER, R. H., MENZIES, M. A., DICKIN, A. P. & HAMILTON, P. J. 1985. Isotopic and chemical constraints on the building of the deep Scottish lithosphere. *Scottish Journal of Geology*, **21**, 456–491.

HARRIS, A. L., BALDWIN, C. T., BRADBURY, H. J., JOHNSON, H. D. & SMITH, R. 1978. Ensialic basin sedimentation – the Dalradian Supergroup. *Crustal Evolution in Northwestern Britain and Adjacent Regions*. Geological Journal, Special Issue, **10**, 115–138.

HARRIS, M., KAY, E. A., WIDNALL, M. A., JONES, E. M. & STELLE, G. B. 1988. Geology and mineralisation of the Lagolochan intrusive complex, western Argyll, Scotland. *Transactions of the Institution of Mining and Metallurgy (Section B: Applied earth science)*, **97**, B15–B21.

HUGHES, R. A., EVANS, J. A., NOBLE, S. R. & RUNDLE, C. C. 1996. U–Pb chronology of the Ennerdale and Eskdale intrusions supports sub-volcanic relationships with the Borrowdale Volcanic Group (Ordovician, English Lake District). *Journal of the Geological Society, London*. **153**, 33–38.

INSTITUTE OF GEOLOGICAL SCIENCES 1982. *Regional Geochemical Atlas: Sutherland*. IGS.

KIMBELL, G. S. & STONE, P. 1995. Crustal magnetization variations across the Iapetus Suture Zone. *Geological Magazine*, **132**, 599–609.

KNELLER, B. C. 1991. A foreland basin on the southern margin of Iapetus. *Journal of the Geological Society, London*, **148**, 204–210.

——, KING, L. M. & BELL, A. M. 1993. Foreland basin development and tectonics on the northwest margin of eastern Avalonia. *Geological Magazine*, **130**, 691–697.

LEAKE, R. C. & BROWN, M. J. 1979. Porphyry-style copper mineralisation at Black Stockarton Moor, south-west Scotland. *Transactions of the Institution of Mining and Metallurgy (Section B: Applied earth science)*, **88**, B177–B181.

——, AULD, H. A., STONE, P. & JOHNSON, C. E. 1981. Gold mineralisation at the southern margin of the Loch Doon granitoid complex, south-west Scotland. *Mineral Reconnaissance Programme Report, Institute of Geological Sciences*, **46**.

LEGGETT, J. K., MCKERROW, W. S. & EALES, M. H. 1979. The Southern Uplands of Scotland: a Lower Palaeozoic accretionary prism. *Journal of the Geological Society, London*, **136**, 755–776.

LINDSAY, N. G., HASELOCK, P. J. & HARRIS, A. L. 1989. The extent of Grampian orogenic activity in the Scottish Highlands. *Journal of the Geological Society, London*, **146**, 733–735.

MARCANTONIO, F., DICKIN, A. P., MCNUTT, R. H. & HEAMAN, L. M. 1988. A 1800-million-year-old Proterozoic gneiss terrane in Islay with implications for the crustal structure evolution of Britain. *Nature*, **335**, 62–64.

MOORBATH, S. & TAYLOR, P. N. 1974. Lewisian age for the Scardroy mass. *Nature*, **250**, 41–43.

NOBLE, S. R., HYSLOP, E. K. & HIGHTON, A. J. 1996. High precision U–Pb monazite geochronology of the *c.* 806 Ma Grampian Shear Zone and the implications for the evolution of the Central Highlands of Scotland. *Journal of the Geological Society, London*, **153**, 511–514.

PATTRICK, R. A. D. 1981. *The vein mineralisation at Tyndrum, Scotland and a study of substitution in tetrahedrites*. PhD Thesis, University of Strathclyde.

PIASECKI, M. A. J. & TEMPERLEY, S. 1988. The Central Highland Division. *In*: WINCHESTER, J. A. (ed.) *Later Proterozoic Stratigraphy of the Northern Atlantic Region*. Blackie, Glasgow and London. 46–53.

PLANT, J. A. 1970. *Distribution of gold in alluvium of Strath Halladale, Sutherland*. Sutherland County Council.

—— & TARNEY, J. 1995. Mineral deposit models and primary rock geochemical characteristics. *In*: HALE, M. & PLANT, J. A. (eds) *Drainage Geochemistry, Handbook of Exploration Geochemistry*, **6**. G. S. Govett.

——, WATSON, J. V. & GREEN, P. M. 1984. Moine–Dalradian relationships and their palaeotectonic significance. *Proceedings of the Royal Society of London*, **A395**, 185–202.

——, BREWARD, N., FORREST, M. D. & SMITH, R. T. 1989. The gold pathfinder elements As, Sb and Bi – their distribution and significance in the southwest Highlands of Scotland. *Transactions of the Institution of Mining and Metallurgy (Section B: Applied earth science)*, **98**, B91–B101.

——, ——, SIMPSON, P. R. & SLATER, D. 1990. Regional geochemistry and the identification of metallogenic provinces: examples from lead–zinc–barium, tin–uranium and gold deposits. *Journal of Geochemical Exploration*, **39**, 195–224.

——, COOPER, D. C., GREEN, P. M., REEDMAN, A. J. & SIMPSON, P. R. 1991. Regional distribution of As, Sb and Bi in the Grampian Highlands of Scotland and English Lake District: implications for gold metallogeny. *Transactions of the Institution of Mining and Metallurgy (Section B: Applied earth science)*, **100**, B135–B147.

READ, H. H. 1931. The geology of central Sutherland (east-central Sutherland and south-west Caithness). Explanation of Sheets 108 and 109. *Memoir of the Geological Survey of Great Britain (Scotland)*.

RICE, C. M., ASHCROFT, W. A., BATTEN, D. J. *et al.* 1995. Devonian auriferous hot spring system, Rhynie, Scotland. *Journal of the Geological Society, London*, **52**, 229–250.

ROBERTSON, A. H. & HENDERSON, W. G. 1984. Geochemical evidence for the origins of igneous and sedimentary rocks of the Highland Border, Scotland. *Transactions of the Royal Society of Edinburgh: Earth Sciences*, **75**, 135–150.

RUSHTON, A. W. A., STONE, P. & HUGHES, R. A. 1996. Biostratigraphical control of thrust models for the Southern Uplands of Scotland. *Transactions of the Royal Society of Edinburgh: Earth Sciences*, **86**, 137–152.

RYAN, P. D. & DEWEY, J. F. 1991. A geological and tectonic cross-section of the Caledonides of western Ireland. *Journal of the Geological Society, London*, **148**, 173–181.

——, SOPER, N. J., SNYDER, D. B., ENGLAND, R. W. & HUTTON, D. H. W. 1995. The Antrim – Galway line; a resolution of the Highland Border Fault enigma of the Caledonides of Britain and Ireland. *Geological Magazine*, **132**, 171–184.

SANDERS, I. S., VAN CALSTEREN, P. W. C. & HAWKESWORTH, C. J. 1984. A Grenville Sm–Nd age for the Glenelg eclogite in north-west Scotland. *Nature*, **312**, 439–440.

SMELLIE, J. L. & STONE, P. 1992. Geochemical control on the evolutionary history of the Ballantrae Complex, S.W. Scotland, from comparisons with recent analogues. *In*: PARSON, L. M., MURTON, B. J. & BROWNING, P. (eds). *Ophiolites and Their Modern Oceanic Analogues*. Geological Society, London, Special Publications, **60**, 171–178.

STRACHAN, R. A. 1988. The metamorphic rocks of the Scaraben area, East Sutherland and Caithness. *Scottish Journal of Geology*, **24**, 1–13.

STONE, P. 1995. Geology of the Rhins of Galloway district. *Memoir of the British Geological Survey*, Sheets 1 and 3 (Scotland).

—— & LEAKE, R. C. 1984. Disseminated and epigenetic Pb–Zn mineralisation in Ordovician mudstone, Galloway. *Scottish Journal of Geology*, **20**, 181–190.

—— & SMELLIE, J. L. 1990. The Ballantrae ophiolite, Scotland: an Ordovician island arc–marginal basin assemblage. *In*: MALPAS, J. A., MOORES, E. M., PANAYIOTOU, A. & XENOPHONTOS, C. (eds) *Ophiolites – Oceanic Crustal Analogues*. Proceed-

ings of the Troodos 87 Symposium, Geological Survey Dept, Nicosia, Cyprus, 533–546.

——, GREEN, P. M. & WILLIAMS, T. M. 1997a. Relationship of source and drainage geochemistry in the British paratectonic Caledonides: an exploratory regional assessment. *Transactions of the Institution of Mining and Metallurgy (Section B: Applied earth science)*, **106**, B78–B84.

——, KIMBELL, G. S. & HENNEY, P. J. 1997b. Basement control on the location of strike-slip shear in the Southern Uplands of Scotland. *Journal of the Geological Society, London*, **154**, 141–144.

——, FLOYD, J. D., BARNES, R. P. & LINTERN, B. C. 1987. A sequential back-arc and foreland basin thrust duplex model for the Southern Uplands of Scotland. *Journal of the Geological Society, London*, **144**, 753–764.

——, COOK, J. M., MCDERMOTT, C., ROBINSON, J. J. & SIMPSON, P. R. 1995. Lithostratigraphic and structural controls on the distribution of As and Au in the southwest Southern Uplands, Scotland. *Transactions of the Institution of Mining and Metallurgy (Section B: Applied earth science)*, **104**, B111–B119.

——, GREEN, P. M., LINTERN, B. C., SIMPSON, P. R. & PLANT, J. A. 1993. Regional geochemical variation across the Iapetus Suture Zone: tectonic implications. *Scottish Journal of Geology*, **29**, 113–121.

——, ——, ——, PLANT, J. A., SIMPSON, P. R. & BREWARD, N. 1991. Geochemistry characterizes provenance in southern Scotland. *Geology Today*, **7**, 177–181.

TANNER, P. W. G. 1995. New evidence that the Lower Cambrian Leny Limestone at Callander, Perthshire, belongs to the Dalradian Supergroup, and a reassessment of the 'exotic' status of the Highland Border Complex. *Geological Magazine*, **132**, 473–483.

VAN DER VOO, R. & SCOTESE, C. 1981. Palaeomagnetic evidence for a large (~2000 km) sinistral offset along the Great Glen Fault during Carboniferous times. *Geology*, **9**, 583–589.

WATSON, J. V. 1984. The ending of the Caledonian orogeny in Scotland. *Journal of the Geological Society, London*, **141**, 193–214.

WEAVER, B. L. & TARNEY, J. 1980. Rare-earth geochemistry of Lewisian granulite-facies gneisses, northwest Scotland: implications for the petrogenesis of the Archaean lower continental crust. *Earth and Planetary Science Letters*, **51**, 279–296.

WILSON, G. V. & FLETT, J. S. 1921. The lead, zinc, copper and nickel ores of Scotland. *Memoir of the Geological Survey, Scotland: Special Reports on the Mineral Resources of Great Britain*, **17**.

YARDLEY, B. W. D. & SENIOR, A. 1981. Basic magmatism in Connemara, Ireland: evidence for a volcanic arc? *Journal of the Geological Society, London*, **139**, 67–70.

ZHOU, J. 1988. A gold and silver-bearing subvolcanic centre in the Scottish Caledonides near Lagolochan, Argyllshire. *Journal of the Geological Society, London*, **145**, 225–234.

# Taconian seismogenic deformation in the Appalachian Orogen and the North American Craton

NICHOLAS RAST, FRANK R. ETTENSOHN & DIANA E. RAST

*Department of Geological Sciences, 101 Slone Building, University of Kentucky, Lexington, KY 40506-0053, USA*

**Abstract:** Originally the paradigm of plate tectonics was largely based on seismological interpretation of earthquakes. It therefore behoves that geological literature should pay more attention to the detection of seismic activity in ancient rocks. The effects of earthquakes on relatively young Tertiary and Holocene deposits have been recognized, commonly in the context of engineering works. In this paper, observational evidence for seismogenic structures in the Lower Palaeozoic carbonate belt of E. North America is presented. The structures range from minor faults and liquefaction effects to breccias and melanges. It is sometimes possible to suggest the dynamics of faults on which earthquakes occurred. Although most of the deformation happened at the margins of the craton and the growing orogenic belt, intracratonic faulting, earthquakes and synsedimentary deformation also took place.

Plate tectonics as a paradigm emerged as a result of advances in the understanding of the distribution of present-day seismic events and their effects on rocks. It is therefore strange that in the literature relatively little attention has been paid to the identification and placing of such events in the geological past. The emergence of palaeoseismology as a new discipline is thus very recent, yet it has already acquired specialized terminology, although the text by McCalpin (1996) is claimed to be the first volume in English that is concerned with palaeoseismology. Yet even this book deals essentially with very young geological phenomena, barely touching rocks older than the Pleistocene or even Holocene.

In this paper, it is attempted to examine evidence for old earthquakes in the Palaeozoic rocks of the edge of the Interior Platform of the United States that underwent an orogenic episode, some 450+ Ma old, and which inferentially was affected by earthquake activity. The earthquakes that affected submarine sediments and produced liquefaction were particularly significant. In Holocene palaeoseismology this effect has been recognized by engineers concerned with lakes, dams, rivers, etc., but, although widely reported (Wallace 1987; Obermeier *et al.* 1990; Sims & Garvin 1995) in engineering publications, remained less noticed in specifically geological literature. Much of the development of palaeoseismology was pursued in Russia and has recently been briefly reviewed by McCalpin & Nelson (1996). The current palaeoseismological

studies (cf. Hancock & Michetti 1997) are commonly carried out within the context of another recently emerged science known as *active tectonics* (Keller & Pinter 1996), which overlaps seismology, tectonics and geomorphology.

Existing faults commonly, prior to and after earthquake-generating movements, undergo continuous creep which is *aseismic*. However, they are in places entirely locked, and abrupt motion on these faults is intermittent and *coseismic*, being reflected in one or several earthquakes. The coseismic motion may be distributed areally, producing a widespread change in the elevation of the land so affected. This can be detected by satellite radar interferometry. An earthquake break may also be accompanied by *preseismic* or *postseismic* creep. In relation to an earthquake one may have the preseismic creep, then a coseismic pulse, followed by creep due to postseismic afterslip (Fig. 1). There also may be an interseismic interval between each of two earthquakes where no creep takes place. Therefore, seismicity and creep deformation are commonly cyclic. While preseismic and postseismic deformation is recognized in the nature of distortion, and does not produce earthquakes or shocks, it can be detected by a continuous strain of the stratified sedimentary rock: the development of pull-apart structures in the case of tension and overriding repetition of the same horizon in the case of compression. Transcurrent effects are not easily recognized in cross-section but may be detected in map view.

*From:* MAC NIOCAILL, C. & RYAN, P. D. (eds) 1999. *Continental Tectonics.* Geological Society, London, Special Publications, **164**, 127–137. 1-86239-051-7/99/$15.00 © The Geological Society of London 1999.

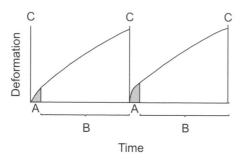

**Fig. 1.** Modern view of an earthquake cycle illustrating cumulative deformation in time: interval C, coseismic slip; interval A, postseismic creep; interval B, interseismic deformation and build-up of stress.

Palaeoseismic evidence may be either primary or secondary, the former includes fold scarps, fissures and folds, and the evidence can be geomorphic or stratigraphic, while the secondary evidence, also called seismogravitational, consists of sand blows, landslides, sand dykes, soft sediment deformed features and turbidites, the classic example of which was produced by the 1929 Grand Banks Earthquake off Newfoundland.

In shallow-marine sediments, features that are particularly significant as indicators of palaeoseismicity are known as *seismites*, a term introduced by Seilacher (1969, 1984). It is pointed out by Vittori *et al.* (1991) that surface faulting is manifested by fracture at the surface, where the hypocentral depth is <20 km and reflects an earthquake of magnitude $M = 6.5$. Nevertheless, even smaller earthquakes may produce surface phenomena if the affected sediment is capable of liquefaction, hydroplastic deformation or producing large landslides.

The evidence of fault traces in older rocks is questionable, since: (1) it is difficult to determine the age of faulting in older sedimentary rocks and to relate the fault to other possible effects; (2) a direct evidence for the intensity of shaking is missing. The phenomena, generated by shaking include geomorphic changes such as rockfalls and landslides on land, as well as liquefaction features such as sand dykes, load structures, anomalous silt deposits and turbidites. Some of the secondary effects are not easily explained by seismic activity and may have been produced by other processes; therefore, some of the secondary evidence is, on its own, ambiguous. The generation of evidence partly depends on the magnitude, and therefore depth, of the earthquake foci (Wells & Coppersmith 1994). At epicentral sites, earthquakes require a threshold magnitude of $M = 5$ to cause liquidization or liquefaction (Joyner & Boone 1988), although smaller earthquakes (Keefer 1984) may cause landsliding. The structures more widely observed in marine sediments are slumps, related turbidites, ball-and-pillow structures, convolutions and load casts, all of which may have been initiated by seismic shaking. These structures, although in places enigmatic, form the core of evidence for palaeoseismicity in the Ordovician of Kentucky, where in time it is related to the Taconian orogeny.

## Taconian Orogeny in the Appalachians

The Middle–Upper Ordovician tectonic episode, known as the Taconian orogeny (Bird & Dewey 1970; Rodgers 1971), has left direct and indirect evidence for deformation from Newfoundland to the southern Appalachians of the USA. It appears to be the first significant mountain building period of the Appalachian cycle and lasted from Late Cambrian to Late Ordovician times. Throughout this period there is evidence of vigorous tectonic activity resulting in the formation of either direct tectonic manifestations, e.g. faults or the generation of slumps, and indirect palaeoseismic effects such as olistostromes, breccias and melanges (Rast & Horton 1989), margined to the west by a belt of seismites (Rast & Ettensohn 1995; Pope *et al.* 1997). The evidence of a seismic origin for these features is especially striking in and around Lexington (Kentucky), although geometrically similar features occur as far away as Virginia. Most of these seismites occur in carbonate rocks and have been mentioned from time to time as local, regionally unimportant structures.

In the northern Appalachians, Taconian overthrusts are, in places, overprinted by the Acadian (Devonian) or the slightly earlier Salinic (Silurian) structures, and are generally unaffected by the later Alleghanian orogenic events. In western Newfoundland, as pointed out by Cawood *et al.* (1995), the Taconian orogeny is Early–Middle Ordovician, involving ophiolite abduction, and is considered as an early phase (Taconian I). Taconian I is followed by the subduction of an Ordovician arc under Laurentia producing local magmatism and terminating in a general deformation (Taconian II) and subduction that ended before the earliest of the Caradoc. In Salinian times, extensional collapse took place.

Further west, in Quebec, an earlier Late Cambrian–Early Ordovician orogenic event, accompanied by seismic activity (Bailey *et al.* 1928), has been recognized. All of these more or less local compressional episodes have recently been related (MacNiocaill *et al.* 1997) to interactions with Iapetan arcs which, in the process of the closure of the Iapetus Ocean, collided one after the other with Laurentia, each causing a laterally fairly continuous zone of deformation, earthquake activity and mountain building. This suggests that the Taconian and Acadian were not separate episodes, but tended to merge one into another (cf. van Staal & de Roo 1995). In the Ordovician, seismicity was recorded in the numerous so-called olistostromic melanges. Bailey *et al.* (1928) examined the conglomeratic breccias exposed near Quebec City which contained numerous blocks of carbonate interbedded with shales. The conglomeratic blocks are associated with convoluted calcareous strata, which are probably seismites because the convolutions lack regular vergence. Bailey *et al.* (1928) proposed that deformation was caused by submarine slip triggered by earthquake activity and advanced a detailed refutation of the, then current, hypothesis of glacial or shoreline deposition.

In the southern Appalachians, at the border with Laurentia, a series of rocks referred to as melanges have been reported. In particular, Lash (1987), described both bedding-parallel and cross-cutting lithologies involving blocks of carbonate in shaley matrix, some of which he interpreted as dewatering channels associated with the formation of mud volcanoes. It is proposed here that these structures arose as a consequence of earthquakes and are, therefore, seismites, and indicate approximate positions of earthquake epicentres. Lash (1987) positionally attributed the melanges to the formation of accretionary complexes. The sites of these earthquakes are related here to the position and occurrence of structures interpreted as seismites. The affected area, though large, cannot be specified since the younger sedimentary cover hides much of the earlier country rock.

Inland, within the former Laurentian border to the orogen, Pope *et al.* (1997) have proposed four areas of concentration of diagnostic seismites, three in Kentucky and the fourth in Virginia (Fig. 2), encompassing: Martinsburg, Garrard, Lexington and Fairview seismites. These seismites are particularly prominent in the Middle and Upper Ordovician carbonates where they are interbedded with shales. In this paper the Lexington seismites that are well exposed in

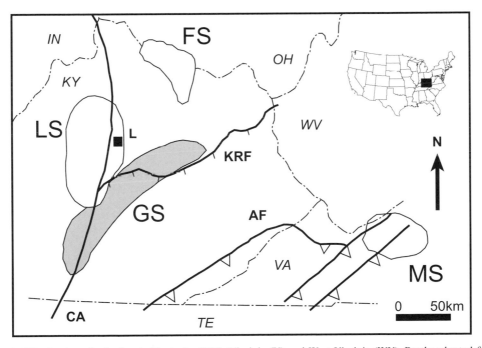

**Fig. 2.** Distribution of seismites in Kentucky (KY), Virginia (V) and West Virginia (WV). Partly adapted from Pope *et al.* (1997). Areas of noted seismites (S) are: FS, Fairview; GS, Garrard; LS, Lexington; MS, Martinsburg. Note that much of the area between the Garrard and Martinsburg seismites is covered by Carboniferous strata. CA, Cincinnati Arch; KRF, Kentucky River Fault Zone; AF, Appalachian Front. Normal faults are ticked and thrusts are marked with triangles.

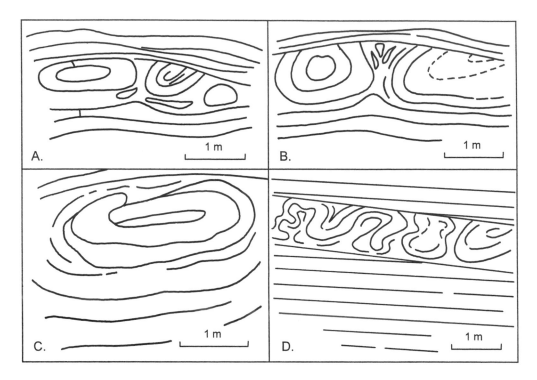

**Fig. 3.** Morphology of seismites: (**a**) ball and pillow; (**b**) tulip; (**c**) flow-rolled pillow; (**d**) convoluted bed, cross-cutting normal stratification.

the city of Lexington and in the surrounding countryside of central Kentucky are especially emphasized. Seilacher (1969, 1984) suggested that, in many instances, graded bedding, typical of turbidites, is originally seismogenic. His term seismite, encompassed soft-sedimentary deformation as a result of fluidization caused by seismic shocks as well as graded bedding. In the Ordovician of Kentucky, several types of soft-sediment structures have been recognized as seismites (Rast & Ettensohn 1995) and similar structures have been observed elsewhere (Seilacher 1984; Davenport & Ringrose, 1987; Hibsch *et al.* 1997). Some of the structures have been previously reported in literature and given specific names, such as 'ball-and-pillow', 'convolution', 'molar tooth', etc. The geometry of ball and pillow structures in Kentucky limestones is emphasized by flat and elongated chert nodules that clearly follow flow laminations. The flow laminations of calcisiltite and micrite are normally wrapped around the pillows (Fig. 3), and both the pillows and laminations are cut off by fine, almost homogeneous, calcarenite. Similar, almost homogeneous, calcarenites also cut across convolutions. The hydrodynamic interpretation attributes such structures to the phenomena of fluidization or liquefaction.

Fluidization occurs when the pore fluid, forced upwards, carries grains of sediment against the force of gravity, leading ultimately to the sepa-ration of the grains and water, and dragging the sediment towards the upper surface of the layer (Lowe 1976), internally generating the laminations and giving rise to planar or vertical channels of dewatering. The higher the upward velocity produced, the greater the number of suspended grains carried to the surface, giving rise to small sand volcanoes. It is suggested that shaking generated by earthquakes will produce fluidization, expelling some sediment. The sediment grains ultimately precipitate on the bed as a fine homogeneous sediment covering the exits of dewatering channels. The chert nodules are then produced by the upward-moving fluid which, presumably, is rich in silica.

The Ordovician succession in central Kentucky ranges from the Lower Carddocian to the Ashgillian. This succession consists generally of carbonates, but there are shaley horizons toward its bottom and also in the middle (Fig. 4). At the very bottom of the succession, the carbonates, although occasionally shaley, do not show evidence of seismic activity. The bedding is generally uniform and undisturbed. In Kentucky, this part of the succession is referred to as the High Bridge Group and the constituent carbonates have been interpreted as peritidal. Seismites occur throughout the Lexington Limestone, but are especially well exposed *c.* 1 km south of Camp Nelson in a part of the Lexington Limestone known as the Curdsville Member, which is a fairly well-bedded limestone with some strongly disturbed, laterally discontinuous horizons. The Lexington Limestone shows rapid facies changes,

varying from micritic to coarse fragmental rocks which are, from place to place, interrupted by thin 'shaley' greenish-grey layers of K-metabentonite that also occur in the underlying carbonate of the High Bridge Group, having been explained as beds of volcanic ash (Huff *et al.* 1992). The structures interpreted as seismites in the Lexington Limestone are mainly beds broken into ball and pillows, which are usually abruptly truncated by the overlying beds (Fig. 3a). They do not occur in the High Bridge Group, but are very distinctive in the Lexington Limestone, especially in the Brannon, the Curdsville and Clays Ferry units.

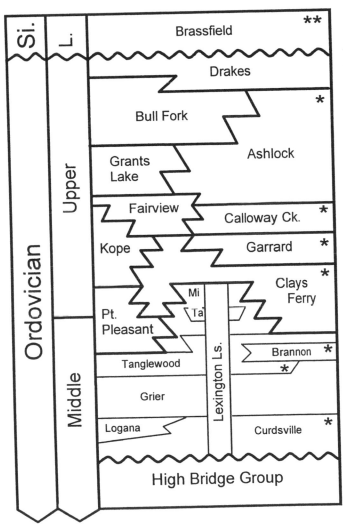

**Fig. 4.** The Ordovician succession of Kentucky. The positions of seismites are indicated by *. The positions of possible bolide impact structures are marked by **.

## Interpretations of Seismites in East-Central USA

Modern sedimentation, especially in shallow-marine and lacustrine environments, has been interpreted not only in terms of the environment but also in terms of deformation affecting it. In particular, there is now much evidence that earthquake activity plays a substantial role (Davenport & Ringrose 1987). The principal effect, induced by earthquakes, is the extensive liquefaction of the newly deposited, uncon-solidated sediment. The two most important submechanisms involved are liquefaction and fluidization. The former describes a momentary suspension of solid particles in the fluid phase (water), while the latter is the change in the fluid phase such that it produces support of the solid phase by liquid drag. Both may be followed by the development of hydroplasticity. In the late 1970–1980s, the concept of liquefaction became widely employed by geologists (e.g. Einsele *et al.* 1991), who were searching for an explanation of catastrophic and episodic sedimentation invol-ving storms or earthquake-related events such as tsunamis.

Concepts of storm deposition were used to explain the origins of ancient deposits on the North American Craton, with special emphasis on Lower Palaeozoic clastic and carbonate sedi-ments (Ettensohn *et al.* 1986; Ettensohn 1992). However, the influence of earthquakes in cra-tonic areas was not discussed. Meanwhile, Seila-cher (1969, 1984) inferred that soft-sediment disturbances observed in siliciclastic carbonate deposits were produced through the trigger effect of seismic shocks or associated marine seismic waves (tsunamis).

Davenport & Ringrose (1987), in an excava-tion of an outwash terrace in Perthshire, Scot-land, used the following features as evidence for effects of fluidization induced by earthquakes: (1) destruction and modification of depositional fabric; (2) partial or complete loss of primary lamination; (3) reorientation of some laminae and fragments as evidence of liquefied or fluid-ized flow; (4) formation of subvertical flow channels indicating the presence of small-scale sand or mud volcanoes (sand boils). The tops of deformed layers of such disturbed sediments are often cut across by a homogeneous overlying thin bed of generally undisturbed rock. In places, such truncation surfaces developed small-scale thrusts. Thus, the heavily disturbed layers formed ball-and-pillow structures which have overlying, structureless layers that reflect the accumulation of muddy–silty sediment thrown into suspension by the liquid escape channels.

Usually, seismites are found in areas affected by, or at the borders of, orogenic deformation. It has, however, long been recognized that in the USA significant post-orogenic seismicity has also occurred in association with mid-continental structures (e.g. the New Madrid–Reelfoot region of Kentucky and Tennessee). Moreover, it is possible that the relatively recent earthquake activity is the result of reactivation of Precam-brian–Cambrian faults (Hamilton & Zoback 1982). In central Kentucky, faults that are prone to reactivation are widely reported (Black 1986; Shumaker 1986). The reactivation appears to influence the sedimentary cover on the craton, both syn- and postdepositionally. Evi-dence for syndepositional effects is in the form of facies changes, disjunctive unit distribution, abrupt changes in facies and the existence of local unconformities. However, in much recent literature the abundant ball-and-pillow struc-ture, and other liquefaction structures, in central Kentucky have been attributed to slope instabil-ity or pressure changes induced by storms. A seismic origin has rarely been considered for Ordovician ball-and-pillow structures [see, how-ever, Pope *et al.* (1997)]. Preliminary investiga-tions of the Ordovician stratigraphy of the Lexington Limestone (Kulp 1995) have yielded an informative map of the distribution of seismites of this period (Fig. 2), suggesting an original location of earthquake epicentres in a shallow-marine environment.

Leeder (1987) has pointed out that a com-plex of dewatering pipes and surface sand volcanoes commonly develop in the neighbour-hood of active faults, affecting unconsolidated sediments. In Kentucky, dewatering produced saucer-shaped bodies (Fig. 4a) suspended by fluidized material. At the same time, the coarser material underwent hydroplastic deformation to produce tulip (Fig. 4b) and pillow structures (Fig. 4c). Sims (1975), who had advanced criteria for the recognition of seismogenically disturbed sediments, suggested that: (1) seismites are pro-duced in seismically active regions and occupy individual, relatively thin horizons, under- and overlain by undeformed strata of locally wide extent; (2) there is little evidence for slope failure and the section containing the deformed horizon is generally flat lying, (3) lacustrine deposits of seismogenic areas commonly have ball-and-pillow structures.

In addition, it has been observed that, occasionally, layers with disturbed structures gently cross-cut the sediments containing them (Fig. 4d). Moreover, in Kentucky, the most common rock successions developing seismites are either sandstones or carbonates involving mixtures of water-saturated calcisiltites, and finer calcarenites, respectively, while sequences of fine micrites and shales, because of their poor permeability, were so cohesive that they did not respond to seismic shock and are free from seismites. Seismites do not develop in calcirudite successions because of their efficient intragranular drainage.

Seilacher (1991) has indicated that there are sedimentological criteria to distinguish between seismites, current-induced inundites (floods) and storm-generated tempestites. He suggested that the historical underestimation of seismogenic sediments may be due to the fact that they were often deleted by subsequent events such as storms. If caused by tsunamis such sediments developed a pronounced homogeneity and, if extensive [currently known as *homogenites* (Kostens & Cita 1981)], are not amenable to easy, unambiguous interpretation.

## Examples of Seismites in the Lexington Area

In the Lexington area there are a large number of exposures of ball and pillow structures (Fig. 1), interpreted here as seismites, that are covered by thin beds of mixed calcilutites and calcisiltites. The latter strata are, if massive, interpreted as exposures of tsunami deposits and sometimes contain large fragments. The seismites, in many localities exposed in road cuts, commonly contain numerous siliceous, cherty inclusions, some of which dimensionally follow flow laminations in the calcilutites and calcisiltites acquiring laminae-parallel orientation. In some cases, the calcarenitic ball-and-pillow structures show faint internal laminae, suggesting water-escape channels. The cross-cutting, overlying homogeneous carbonates, coarser than the calcilutites but usually finer than calcarenites, are in some instances, probably tsunamigenic homogenites or products of seiches. Two general outcrops will be examined in detail (Fig. 2): (1) the structures in the Curdsville Member of the Lexington Limestone south of Camp Nelson; (2) the seismites on Route 4, Lexington, associated with the Brannon Member. A brief and partial description of seismites on Leestown Pike (Fig. 2) will follow.

### Camp Nelson outcrops

A description of the general geology of this area, situated some 20 miles south-southwest of Lex-

ington, is available in Kuhnhenn & Haney (1986): for the general succession see Fig. 4. Seismites are particularly well developed in the Curdsville Member, where they occur, on the east side of Highway 27, in lensoid sheets with internal convolutions and ball-and-pillow structures, in places appearing to occur in successive cross-cutting horizons, possibly indicating a strong shock being succeeded by an aftershock. At some points, the inception of dewatering pipes is nucleated by ripples and mounds on the underlying beds.

Exposures of seismites on the west side of the road have open cross-cutting joints which indicate that convolutions have no sense of preferred overturning, thus implying that they were produced by strong vibrations associated with earthquakes rather than by slumping. Seismically disturbed layers are 1–2 m thick and extend laterlly up to 100 m.

### Route 4 outcrops

These outcrops, affecting the Brannon Member of the Lexington Limestone, are situated under an overpass along circular Route 4, one of the main roads in Lexington, at the intersection with Route 421. A very well-displayed series of pillow ball-and-pillow structures resemble those described by Davenport & Ringrose (1987). It is possible to detect a succession of at least three layers of seismites separated by thin homogeneous interlayers of silty micrite, each possibly representing deposits of a minor tsunami. The hydroplastic ball and pillow structures of this outcrop vary from saucer shaped to pillows (Fig. 4e).

### Leestown Pike outcrops

These outcrops also occur at the intersection of Leestown Pike with Route 4, and in the underpath and some ramps along Route 421. These exposures are situated at a dangerous traffic intersection, so future observers must be careful. Both seismites and homogenites with suspended blocks, produced possibly by tsunamis, can be observed on either side of the pike, producing local unconformities (Fig. 5).

## Manifestations of Seismicity

Early observations on Californian earthquakes gave rise to a model known as the seismic deformation cycle, involving a gradual accumulation of elastic strain culminating in its release

A paleoseismic history recorded in lacustrine sediments. *Journal of Geodynamics*, **24**, 259–280.

HUFF, W. D., BERGSRTOM, S. M. & KOLATA, D. R. 1992. Gigantic Ordovician volcanic ash fall in North America and Europe; biological, tectono-magmatic and event-stratigraphic significance. *Geology*, **20**, 875–878.

JOYNER, W. B. & BOORE, D. M. 1988. Measurement, characterization, and prediction of strong ground motion. *In*: VON THUN, J. L. (ed.) *Earthquake Engineering and Soil Dynamics, 11 Recent Advances in Ground-Motion Evaluation*. American Society of Civil Engineers and Geotechnicians, Special Publication, **20**, 43–103.

KEEFER, D. K. 1984. Landslides caused by earthquakes. *Geological Society of America, Bulletin*, **95**, 405–421.

KELLER, E. A. & PINTER, N. 1996. *Active Tectonics*. Prentice Hall.

KOSTENS, K. A. & CITA, M. B. 1981. Tsunami induced sediment transport in the abyssal Mediterranean sea. *Geological Society of America, Bulletin*, **89**, 591–604.

KUHNHENN, G. L. & HANEY, D. C. 1986. Middle Ordovician High Bridge Group and Kentucky River fault system in central Kentucky. *In*: NEATHERLY, T. L. & DRAWER, O. (eds) *Centennial Field Guide, 6*. Geological Society of America, 25–30.

KULP, M. A. 1995. *Paleoenvironmental interpretations of the Brannon Member, Middle–Upper Ordovician Lexington Limestone, central Bluegrass region of Kentucky*. M.S. Thesis, University of Kentucky.

LASH, G. G. 1987. Diverse mélanges of an ancient subduction complex. *Geology*, **15**, 652–655.

LEEDER, M. R. 1987. Sediment deformation structures and the paleotectonic analysis of sedimentary basins, with a case-study from the Carboniferous of northern England. *In*: JONES, M. E. & PRESTON, R. M. F. (eds) *Deformation of Sediments and Sedimentary Rocks*. Geological Society, London, Special Publications, **29**, 137–146.

LOWE, D. R. 1976. Subaqueous liquefied and fluidized sediment flows and their deposits. *Sedimentology*, **23**, 785–808.

MCALPIN, J. P. (ed.) 1996. *Paleoseismicity*. Academic Press.

—— & NELSON, A. R. 1996. Introduction to Paleoseismology. *In*: MCALPIN, J. P. (ed.) *Paleoseisrnicity*. Academic Press, San Diego, 1–32.

MACNIOCAILL, C., VAN DER PLUIJM, S. A. & VAN DER VOO, R. 1997. Ordovician paleogeography and the evolution of the Iapetus Ocean. *Geology*, **25**, 159–162.

MALO, M. & KIRKWOOD, D. 1995. Faulting and progressive strain history of the Gaspe' Peninsula in post-Taconian time: A review. *In*: HIBBARD, J. P., VAN STAAL, C. R. & CAWOOD, P. A. (eds) *Current Perspectives in the Appalachian–Caledonian Orogen*. Geological Association of Canada, Special Paper, **41**, 267–282.

MASSONNET, D., ROSSI, M., CARMONA, C., ARDAGNA, F., PELTZER, G., FEIGI, K. & RABAUTE, T. 1993.

The displacement field of the Landers earthquake mapped by radar interferometry. *Nature*, **364**, 138–142.

OBERMEIER, S. F., JACOBSON, R. B., SMOOT, J. P., WEEMS, R. E., GOHN, G. S., MONROE, J. E. & POWERS, D. S. 1990. Earthquake-induced liquefaction features in the coastal setting of South Carolina and in the fluvial setting of the New Madrid seismic zone. *United States Geological Survey Professional Paper*, **1504**, 1–44.

PLAFKER, G. 1972. Great earthquakes, tsunamis, and tectonic deformation in some circum-Pacific areas. *Geological Society of America, Programs and Abstracts*, **4**, 3218–3219.

POPE, M. C., READ, F. J., SAMBACH, R. & HOFMANN, H. J. 1997. Late Middle to Late Ordovician seismites of Kentucky, southwest Ohio and Virginia: Sedimentary recorders of earthquakes in the Appalachian basin. *Geological Society of America, Bulletin*, **109**, 489–503.

RAST, N. & ETTENSOHN, F. R. 1995. Effects of seismic disturbance on epicontinental depositional systems in the Ordovician and Devonian rocks of central Kentucky: *Geological Society of America, Abstracts with Programs*, **27**, A381.

—— & HORTON, J. W., JR 1989. Mélanges and olistostromes in the Appalachians of the United States and mainland Canada: An assessment. *In*: HORTON, J. W., JR. & RAST, N. (eds) *Mélanges and Olistostromes of the US Appalachians*. Geological Society of America, Special Paper, **228**, 1–15.

RODGERS, J. 1971. *The Taconic Orogeny. Geological Society of America, Bulletin*, **82**, 1141–1178.

SEILACHER, A. 1969. Fault-graded beds interpreted as seismites. *Seismology*, **13**, 155–159.

——1984. Sedimentary structures tentatively attributed to seismic events. *Marine Geology*, **55**, 1–12.

——1991. Events and their signatures; an overview. *In*: EINSELE, G., RICKEB, W. & SEILACHER, A. (eds) *Cycles and Events in Stratigraphy*. Springer, 222–226.

SHUMAKER, R. C. 1986. Structural development of Paleozoic continental basins of eastern North America. *International Basement Tectonics Association, Salt Lake City, UT, United States*, **6**, 82–95.

SIMS, J. D. 1975. Determining earthquake recurrence intervals from deformational structures in young lacustrine sediments. *Tectonophysics*, **29**, 161–163.

—— & GARVIN, C. D. 1995. Recurrent liquefaction induced by the 1989 Loma Prieta earthquake and 1990 and 1991 aftershocks: Implications for paleoseismicity studies. *Bulletin of the Seismological Society of America*, **85**, 51–65.

SYLVESTER, A. G. 1986. Near-field tectonic geodesy. *In*: Active Tectonics: Studies in Geophysics (R. E. Wallace, chairman). National Academy Press, WASHINGTON, DC, 164–180.

VAN STAAL, C. R. & DE ROO, J. A. 1995. Mid-Paleozoic tectonic evolution of the Appalachian Central Mobile Belt in Northern New Brunswick, Canada: Collision, extensional collapse and dextral transpression. *In*: HIBBARD, J. P., VAN STAAL, C. R. & CRAWFORD, P. A. (eds) *Current Perspectives in the Appalachian–Caledonian Orogen*. Geological Association of Canada, Special Paper, **41**, 367–389.

VITTORI, E., LABINI, S. S. & SERVA, L. 1991. Paleoseismology; review of the state-of-the-art. *Tectonophysics*, **193**, 9–32.

WALLACE, R. E. 1987. *A perspective of paleoseismology*. United States Geological Survey, Open File Report, **69–315**, 7–16.

WELLS, D. L. & COPPERSMITH, K. J. 1994. Empirical relationships among magnitude, rupture, length, rupture area, and surface displacement. *Bulletin of the Seismological Society of America*, **84**, 974–1002.

# Thermal and mechanical models for the structural and metamorphic evolution of the Zanskar High Himalaya

M. P. SEARLE,[1] D. J. WATERS,[1] M. W. DRANSFIELD,[1]
B. J. STEPHENSON,[1] C. B. WALKER,[1] J. D. WALKER[1] & D. C. REX[2]

[1] *Department of Earth Sciences, Oxford University, Parks Road, Oxford OX1 3PR, UK*
[2] *Department of Earth Sciences, University of Leeds, Leeds LS2 9JT, UK*

**Abstract:** The regional Barrovian facies metamorphic rocks of the High Himalayan Slab in Zanskar are bounded along the base by the southwest vergent Main Central Thrust (MCT) with its characteristic zone of inverted isograds, and along the top of the slab by the northeast dipping Zanskar Shear Zone (ZSZ), part of the South Tibetan Detachment (STD) System of normal faults. Summarized here are the results of systematic mapping combined with detailed $P$–$T$–$t$ data from the Zanskar Himalaya, and models for the thermal and mechanical evolution of the middle and deep crustal rocks of the Himalaya are discussed. Temperatures and pressures increase dramatically up-structural section across the MCT Zone from biotite through garnet, staurolite and kyanite grade to sillimanite + muscovite, and decrease along the top of the stab beneath the ZSZ normal faults. In Zanskar, peak $P$–$T$ conditions of the M1 kyanite-grade rocks are 550–680°C and 9.5–10.5 kbar, and M2 sillimanite-grade gneisses were formed at 650–770°C and 4.5–7 kbar. The core of the High Himalayan Zone is a 30 km wide zone of sillimanite + K-feldspar-grade gneisses, migmatites and anatectic leucogranites with a right way-up isograd sequence above and an inverted sequence below.

Thermal models of thrusting a hot slab over a cold slab and frictional heating along the MCT are not supported by the structural and thermobarometric data. Mechanical models of post-metamorphic structural disruption by folding and thrusting of a pre-existing, right way-up metamorphic sequence are compatible with structural and $P$–$T$ data. Ductile shearing along the MCT Zone has structurally condensed the isograds, and final motion along the ZSZ postdated leucogranite crystallization (21.5–19.5 Ma; U–Pb monazite ages) and emplacement in the footwall. Crustal shortening and thickening, resulting in prograde metamorphism, lasted from the time of India–Asia collision at 54–50 Ma until at least 30–25 Ma. Crustal melting is constrained at 20.8–19.5 Ma along the core of the stab. Late stage out-of-sequence thrusting within the High Himalayan Slab effectively maintained the crustal thickening process enabling $P$–$T$ conditions to remain high for 10–5 Ma. Rapid exhumation between 21–18.5 Ma was accompanied by removal of 18–25 km of overburden by erosion at exhumation rates of 6–10 mm yr$^{-1}$. This early Miocene period of high exhumation rates, rapid erosion and exhumation of rocks buried at >25 km depth is interpreted to indicate the uplift of high mountains and erosion of deep valleys, probably with high precipitation during that time.

The Himalaya are probably the largest, youngest and best orogenic belt to study the 3D structure of overthickened continental crust resulting from collisional processes (Fig. 1). Following the India–Asia collision, c. 54–50 Ma ago (Garzanti & van Haver 1988; Searle *et al.* 1988, 1997*a*), folding and thrusting of the Indian Plate resulted in crustal shortening and thickening. The deeper levels of the crust are now exposed along the High Himalayan Range, which is bounded along the base by the Main Central Thrust (MCT) Zone, a crustal-scale, southwest vergent reverse fault associated with condensed and inverted

metamorphic isograds, and along the top by the Zanskar Shear Zone (ZSZ), a crustal-scale, shallow-angle, northeast dipping normal fault zone. Earlier mapping of isograds and $P$–$T$ data in the Zanskar region, largely by Honegger (1983), Searle (unpublished maps 1981–1996), Herren (1987), Staubli (1989), and Kundig (1989) led to the proposed thermal model published by Searle & Rex (1989; Fig. 2). This model, based largely on the field mapping of metamorphic index minerals and isograds, proposed a large-scale, recumbent southwest verging fold affecting all the isograds, coupled with ductile shearing

*From*: MAC NIOCAILL, C. & RYAN, P. D. (eds) 1999. *Continental Tectonics*. Geological Society, London, Special Publications, **164**, 139–156. 1-86239-051-7/99/$15.00 © The Geological Society of London 1999.

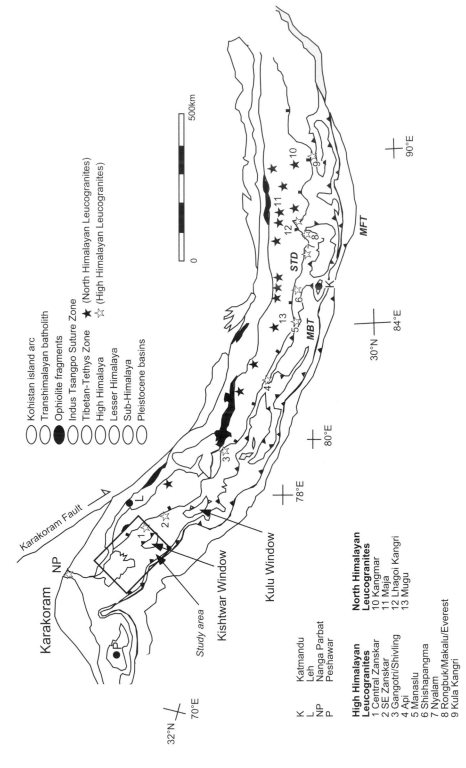

**Fig. 1.** Sketch map of the western Himalaya showing the major structures and the location of the Zanskar Himalaya.

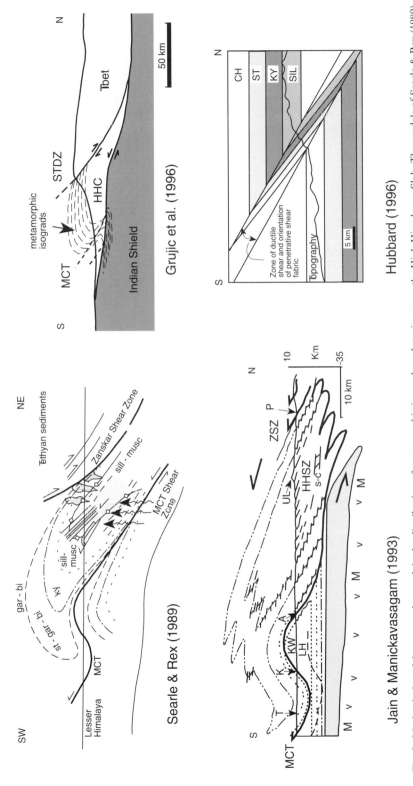

**Fig. 2.** Mechanical models suggested to explain the distribution of metamorphic isograds and strain across the High Himalayan Slab. The models of Searle & Rex (1989) and Jain & Manickavasagam (1993) are based on the Zanskar Himalaya, the model of Grujic *et al.* (1996) is based on the Bhutan Himalaya and the model of Hubbard (1989, 1996) is based on the Khumbu Himalaya, central Nepal.

along the MCT at the base of the slab and normal faulting along the ZSZ at the top of the slab, both of which had the affect of condensing the isograd spacing by shearing. The net result was the southwestward expulsion of the middle and deep crustal metamorphic rocks, migmatites and leucogranites bounded by these two major shear zones.

The problem of the inverted metamorphism along the MCT Zone in the Himalaya has long been debated. Models to explain the inversion of isograds can broadly be grouped into three types: (1) thermal models whereby a hot slab is thrust directly onto cold footwall rocks (e.g. LeFort 1975, 1981); (2) thermal consequenses of shear heating by friction along the MCT (e.g. Arita 1983; Molnar & England 1990; England *et al.* 1992); or (3) mechanical models whereby earlier, right way-up metamorphic isograds have been structurally inverted by later, post-metamorphic folding and/or thrusting of isograds. Several of the latter category of models are shown in Fig. 2. The model of Jain & Manickavasagam (1993), from the Zanskar Himalaya, is remarkably similar to the section of Searle & Rex (1989), except that the former mark the ZSZ as a thrust rather than a normal fault and isograds are marked as cutting across folds in the footwall to the MCT, neither of which are correct. Hubbard (1996) presented a model for the mechanical inversion of metamorphic isograds by ductile shearing along the MCT Zone in the central Nepal Himalaya (Fig. 2). She considered only the base of the High Himalayan Slab in her example, where isograds from sillimanite through kyanite, staurolite, garnet and biotite have been structurally inverted by ductile thrusting, with isograds being parallel to shear fabrics. The three observations of Hubbard (1989, 1996) – a zone of distributed deformation along the MCT Zone, metamorphic textures being overprinted by deformation fabrics related to the final MCT movement and isograds subparallel to MCT-related shear fabrics – are all identical to the model of Searle & Rex (1989) and Searle *et al.* (1992). A model was proposed by Grujic *et al.* (1996) to explain the structure and metamorphism of the Bhutan Himalaya (Fig. 2); however, the geometry of the isograds, the MCT and the South Tibetan Detachment (STD), and the scale of the fold structure, are all very similar to Searle & Rex's (1989) model.

The objective of this paper is firstly to summarize the field structural data from the Zanskar Himalaya, secondly to discuss the *P–T* profiles across the High Himalayan Slab in conjunction with the *T–t* cooling histories, and finally to propose a comprehensive model for the structural and thermal evolution which satisfies all the structural and *P–T–t* data from the area.

## Isograd pattern in Zanskar

A summary isograd map of the High Himalayan Slab in Zanskar is shown in Fig. 3 and a structural cross-section is shown in Fig. 4 along which detailed thermobarometry and dating studies have been carried out (Searle *et al.* 1992; Dransfield 1994; Stephenson 1997). The structure, metamorphism and geochronology of the southeast Zanskar and Lahoul area is presented in Walker *et al.* (1999). The overall structure of the metamorphic isograds in the High Himalayan Slab in central and western Zanskar is a very large-scale, southwest verging recumbent anticline with higher grade rocks exposed along the axis, which corresponds roughly to the highest topography. Metamorphic grade decreases northeastwards away from the axis, with isograds right way-up and condensed along the Zanskar normal fault at the top of the slab. Metamorphism also decreases southwestwards, away from the axis where the isograds are inverted, with higher grade rocks on top of lower grade rocks along the MCT Zone at the base. The axis of the structure plunges northwest and the trace of the isograds has been mapped around the fold closure in western Zanskar and eastern Kashmir (Searle *et al.* 1988, 1992; Searle & Rex 1989). Several post-metamorphic faults have disrupted the pattern, notably the Warwan Backthrust in the northwest, where late-stage, northeast directed backthrusting has placed the Lesser Himalayan low-grade sediments of the Kashmir Basin over high-grade gneisses of the High Himalaya.

As a result of the northeast dipping isograd fold axial plane along the High Himalaya and the northwest plunge of the fold axis in northwest Zanskar, higher structural levels can be seen in western Zanskar overlying the deepest structural levels along the axis of the fold in central Zanskar. Approximately 25–30 km of structural profile through the exhumed middle and deep crust is exposed in this area. Western Zanskar shows three large-scale, southwest vergent recumbent nappes – the Sankoo, Tapshah and Donara Folds – the axial planes of which have been folded around the underlying Suru Dome (Searle *et al.* 1988, 1992; Searle & Rex 1989). The highest nappe, the Sankoo Fold, is in garnet + biotite grade; the structurally deeper

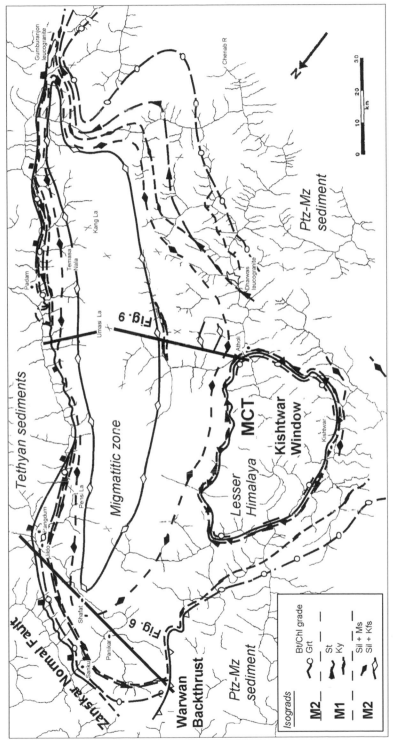

**Fig. 3.** Metamorphic isograd map of the Zanskar–Kishtwar Himalaya, based on the mapping of Honegger (1983), Herren (1987), Searle *et al.* (1988, 1992), Staubli (1989), Kundig (1989), Guntli (1993), Dransfield (1994), Stephenson (1999) and Walker *et al.* (1999). The locations of the two *P–T–t* profiles in Fig. 6 (Suru Valley) and Fig. 9 (Umasi-la) are shown.

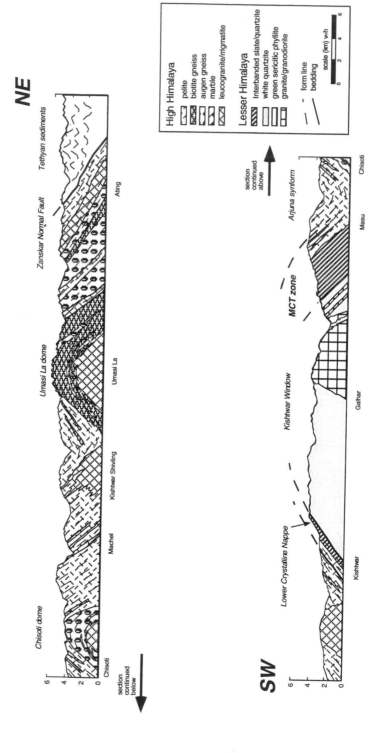

**Fig. 4.** Cross-sections along the Kishtwar Window-Umasi-la profile *P–T–t* profile discussed in detail.

Tapshah and Donara Folds, and the deepest level Suru Dome are all in kyanite grade. To the southeast of this area, in central Zanskar, sillimanite is the dominant aluminium silicate across most of the width of the slab. The deepest levels of the slab correspond to the sillimanite + K-feldspar zone where migmatization is widespread (Searle & Fryer 1986). This is the source for melting of the Himalayan leucogranites, some of which have coalesced into dykes, sheet intrusives and a few larger plutons which have migrated upwards during simple shear from the source into overlying rocks. Both migmatitic leucosomes from the middle of the slab (Umasi-la) and higher level leucogranites from the Suru Valley have been dated, using U–Pb monazite, as 20.8–19.5 Ma (Noble & Searle 1995).

Although mapping of metamorphic isograds in Zanskar clearly shows the fold pattern identified, this does not imply one giant fold structure in the rocks. The large scale fold affects the isograds, which have been superimposed on all earlier structures including earlier recumbent folding. Metamorphism affected overthickened crust which was deformed by similar tight and isoclinal folding and thrusting as presently seen in the upper crustal Tethyan shelf sediments north of the ZSZ (Searle et al. 1997a). Syn- and post-metamorphic deformation has folded the isograds within the slab into a series of domes and basins. Domes, such as the Kishtwar Window culmination, are related to thrust ramping above the MCT, and structural depressions, such as the Chamba basin, southeast of the Kishtwar Window, show high-level, low-grade metamorphism (Fig. 3). Several sillimanite grade granite gneiss domes have also been mapped in the central part of the slab, northeast of the Kishtwar Window (Kundig 1989; Stephenson 1997). Foliation wraps around these domes, symmetrically creating high-strain triple junctions in between the domes. The contacts between granitic cores and gneissic margins are usually concordant but mylonitized, indicating late-stage extension around the domes similar to metamorphic core complexes.

## Time constraints on metamorphism and melting

Regional metamorphism following the India–Asia collision resulted from the increase in pressure and temperature as a consequence of crustal thickening. Field, metamorphic and textural evidence from along the Himalayan chain shows that two periods of metamorphism can

be distinguished, an earlier (sometimes called Eo-Himalayan) kyanite-grade event and a later (sometimes called Neo-Himalayan) sillimanite-grade event (Hodges et al. 1988a, b). It is likely that these two metamorphic events were both part of a continuing metamorphism, which progressively went from a higher pressure to a higher temperature with time. The later sillimanite grade metamorphism coincided with mid- or lower crustal melting along the High Himalayan chain, and the formation of migmatites and leucogranite magmas. The timing of crystallization of these leucogranites span c. 24–12 Ma, but the majority along the High Himalaya are between c. 21–17 Ma [see review in Searle (1996)].

The timing of peak metamorphism can be constrained by U–Pb dating of metamorphic monazites which have a closing temperature of c. 700°C, high enough to record the time of peak temperature in kyanite- or sillimanite-grade rocks. Walker et al. (1999) dated metamorphic monazites from the top, middle and near the base of the Himalayan Slab in southeast Zanskar, which were all similar, c. 32–30 Ma. In western Zanskar, Vance & Harris (1999) obtained Sm–Nd ages of 33–31 and 31–28 Ma for cores and rims of two garnets, which are taken to indicate the time of prograde garnet growth of M1 kyanite-grade assemblages. The timing of crustal melting is known from U-Pb dating of magmatic monazites, xenotimes and uraninites, from both migmatitic leucosomes and leucogranite bodies. Noble & Searle (1995) obtained U–Pb ages of 20.6–19.5 Ma from leucogranitic melt pods in migmatites from the Umasi-la and 20.8 ± 0.3 Ma from higher level leucogranitic melt pods at Shafat (Fig. 3). These early Miocene ages probably record the timing of peak temperature during the M2 sillimanite-grade metamorphism. Metamorphic textures and geochronology suggest that temperatures remained high for 10–5 Ma between peak M1 at c. 33–25 Ma and peak M2 at c. 21–20 Ma.

Cooling histories from Zanskar are recorded by K–Ar and $^{40}$Ar/$^{39}$Ar geochronology (Searle et al. 1992; Vance & Kelley 1994), and also by fission-track dating of zircons and apatites (Kumar et al. 1995). These geochronological data have been compiled from the Shafat and Umasi-la localities to obtain the cooling histories from each area. The thermochronology data have been combined with the geobarometry data to constrain the depths and times of metamorphism and melting, thereby giving rates of exhumation and erosion.

Thermobarometry and K–Ar and $^{40}$Ar/$^{39}$Ar geochronology was carried out, mainly along

**Fig. 5.** Panorama of the upper Suru Valley in northwest Zanskar showing the middle part of the profile around Shafat and the upper part of the profile around Rangdum and Jildo. The ZSZ runs through Jildo at the far right of the photograph.

two profiles across the entire High Himalayan Slab, in order to test the various thermal and mechanical models proposed. The first crosses from eastern Kashmir (Bobang Gali pass) to the Suru Valley at Panikar, and extends to Jildo and Rangdum at the top of the slab (Fig. 5). The second profile extends from the Kishtwar Window at Atholi across the Umasi-la to the Zanskar Valley and crosses the sillimanite-grade anatectic core of the High Himalaya. Detailed petrology and thermobarometric details can be found in Searle *et al.* (1992); Dransfield (1994); and Stephenson (1997).

## Kashmir–Suru Valley profile

### Pressure–Temperature (P–T) array

P–T–t profiles across the Kashmir–Suru valley section are shown in Fig. 6. The base of the slab, the true MCT, is not exposed in this section because late-stage, northeast directed backthrusting has been superimposed. Consequently, all the lower grade section along the base has been faulted out. From Bobang Peak, eastward to the Shafat region, kyanite is the stable aluminium silicate phase across some 35 km of the profile.

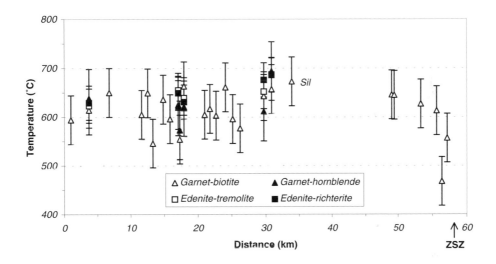

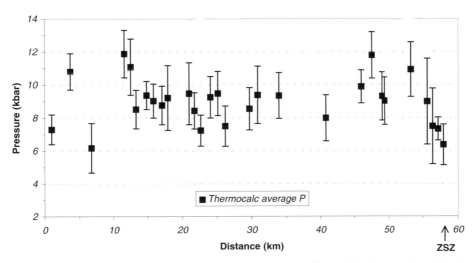

**Fig. 6.** Geothermobarometric data for the east Kashmir–Suru Valley profile, see Fig. 3 for location. Point 0 at the left-hand end corresponds to the Warwan Backthrust, the southern limit of High Himalayan metamorphic rocks in this area. The single sillimanite-bearing sample is indicated on the *T* profile.

The major southwest vergent nappe structures – the Tapshah and Donara recumbent
anticlines, which overlie the Suru Dome – are
all in kyanite grade (Searle *et al.* 1992). Pelites,
marbles and amphibolites that make up these
structures are thought to be metamorphosed
equivalents of the Tethyan shelf units north of
the ZSZ. The amphibolites are probably metamorphosed equivalents of the Permian Panjal
volcanics, which are present both in the Kashmir
Lesser Himalaya to the southwest, and in the
Zanskar Shelf to the northeast of the High
Himalayan range (Honegger 1983).

Individual geothermometers used are the
garnet–biotite Fe–Mg exchange thermometer
of Bhattacharya *et al.* (1992), the garnet-horn-
blende Fe–Mg exchange thermometer of
Graham & Powell (1984) and the two hornblende–plagioclase thermometers of Holland &
Blundy (1994). Temperatures and pressures have
also been determined by the 'average *T* and *P*'
method of Powell & Holland (1988) and Holland & Powell (1990) using the computer programme THERMOCALC. This incorporates
self-consistent calibrations of several more conventional geothermometers (e.g. garnet–biotite,
garnet–hornblende) and geobarometer equilibria (e.g. garnet–plagioclase–kyanite–quartz,
Newton & Haselton 1981; garnet–muscovite–
biotite–plagioclase, Ghent & Stout 1981).

Temperatures across this profile remain consistently high, between 550 and 700°C, and
pressures mainly lie in the 7.5–12 kbar range.
Temperatures are >550°C for a distance of
55 km along this profile to the ZSZ at the top
of the slab, where they decrease rapidly to the
normal fault zone. The highest temperatures
(650–700°C) are recorded near Shafat in the
Suru Valley, where kyanite occurs as a product
of garnet breakdown during decompression at
high temperatures (Searle *et al.* 1992). Sillimanite appears together with migmatites in the core
of the High Himalaya to the southeast of Shafat.
This anatectic core zone widens to the southeast,
where it continues for at least 150 km along the
axis of the High Himalaya. Most of the larger
granites around Shafat are probably pre-Hima-
layan, K-feldspar megacrystic biotite granites,
although many also have garnet ± tourmaline.
Muscovite + biotite + garnet ± tourmaline leucogranites also occur as migmatitic leucosomes and
higher level intrusive sheets (Searle & Rex 1989).
Their occurrence is restricted to the sillimanite
zone for *in situ* melting, although sheet intrusives
do intrude up into the kyanite grade rocks.

### Temperature–Time (T–t) Cooling history

Leucogranites from Shafat have been dated by
U–Pb on monazites at $20.8 \pm 0.3$ Ma (Noble &

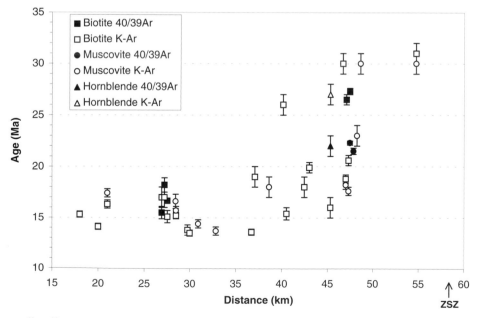

**Fig. 7.** $^{40}$Ar/$^{39}$Ar ages of hornblende, muscovite and biotite for the central part of the Kashmir–Suru Valley profile. Cooling ages are older toward the top of the slab beneath the ZSZ and young progressively toward the southwest, and the deeper structural levels in the Suru dome.

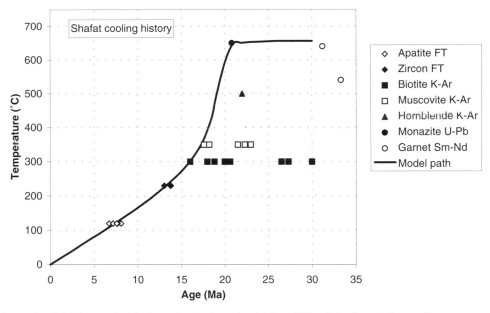

**Fig. 8.** Cooling history of rocks from the Shafat region in the middle of the Suru Valley profile.

Searle 1995). This probably coincides with the timing of peak sillimanite grade M2 metamorphism. K–Ar and $^{40}Ar/^{39}Ar$ ages of hornblendes, muscovites and biotites [data in Searle *et al.* (1992)] from the east side of the Suru Dome in the west, to Jildo in the east, along this profile, have been plotted on Fig. 7 with respect to location along the profile, and in Fig. 8 the cooling path of Shafat rocks has been plotted on a *T–t* plot. Most biotites contain some excess Ar and hence give old, geologically meaningless, ages. Muscovites are less disturbed and indicate a diachronous cooling across the High Himalayan Slab. Cooling through 300°C, the closing temperature for muscovite in the Ar system, occurred at ≤29 Ma at the top of the slab beneath the ZSZ, ≤21 Ma for the Shafat area in the middle and ≤14 Ma for the core of the Suru Dome, the lowest structural level in this profile. The cooling curve for the Shafat area (Fig. 8) shows a rapid cooling between 21 and 18.5 Ma, which was accompanied by removal of 18–25 km of overburden by erosion, at exhumation rates of 6–10 mm yr$^{-1}$. It is speculated that this timing was coincidental to the timing of rapid movement along the MCT, an idea that is supported by the $^{40}Ar/^{39}Ar$ ages of hornblendes growing along ductile MCT-related fabrics in central Nepal (Hubbard & Harrison 1989). Slower cooling and more steady-state erosion from 16–0 Ma is recorded by the younger fission-track zircon and apatite ages (Kumar *et al.* 1995).

## Kishtwar Window–Umasi-la profile

### P–T array

*P–T* profiles across the Kishtwar Window–Umasi-la section are shown in Fig. 9. This profile crosses the central part of Zanskar Range and the deepest exposed parts of the High Himalayan crust. The MCT has been domed around the Kishtwar Window by southwest vergent thrust ramping. Foliation along the hanging wall dips northeast, parallel to the MCT along the base and parallel to the ZSZ along the top of the slab. The central part of the slab shows a number of granitic gneiss domes within the sillimanite-grade rocks, where the foliation is folded around the domes. A prominent zone of post-metamorphic faulting along the Dharlang Valley shows an anomalous zone of kyanite + garnet + muscovite gneisses separating zones of high-grade sillimanite + K-feldspar gneisses on either side (Stephenson 1997). The central part of the profile along the crest of the High Himalaya shows a wide (15–28 km) zone of sillimanite–K-feldspar gneisses and migmatites with abundant leucogranitic melt pods, dykes and pegmatite veins (Searle & Rex 1989; Noble & Searle 1995). Pressures and temperatures were calculated using the same conventional thermobarometers as the previous profile, as well as the 'average-*T* and *P*' method using *Thermocalc*. Around the Kishtwar

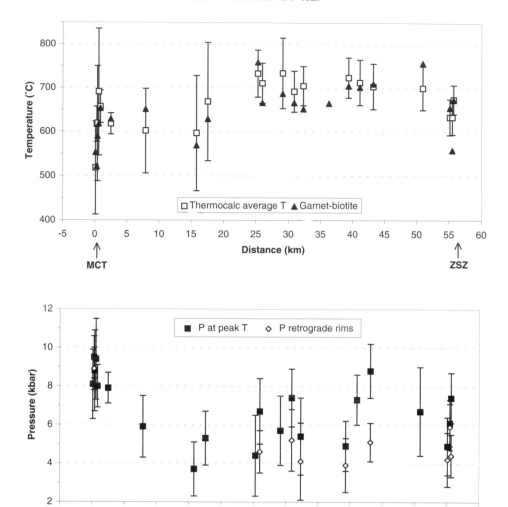

**Fig. 9.** Thermobarometric data for the Kishtwar Window–Umasi-la profile, see Fig. 3 for location. The MCT Zone is between the points −5 and 0 on the scale, corresponding to the MCT Zone at the margin of the Kishtwar Window.

Window, mineral data from Staubli (1989) were also recalculated using these thermobarometers.

Metamorphic grade increases towards the centre of the slab from both the base (MCT) and the top (ZSZ). At the MCT Zone (located from −5 to 0 on Fig. 9), temperatures show a sharp increase from 525 to 650°C across a narrow zone only 0.5 km wide, corresponding to a high-strain mylonitic shear zone (Stephenson 1997). The condensed temperature array across the MCT Zone confirms the field data, that ductile shearing has affected the metamorphic isograds after peak metamorphism (Searle &

Rex 1989; Stephenson 1997). Temperatures increase towards the middle of the slab, where peak temperatures are 700–750°C, corresponding to the sillimanite + K-feldspar migmatite zone of *in situ* partial melting. Temperatures remain high across 55 km of the slab before dropping off along the ZSZ at the top of the slab, where isograds have also been strongly sheared and condensed (Searle 1986; Herren 1987; Dransfield 1994).

The highest pressures occur in the MCT Zone, corresponding to the narrow zone of kyanite grade gneisses formed during the early M1 metamorphism. Pressures increase rapidly across the MCT Zone from 6 to 8.5 kbar from the garnet through the staurolite to the kyanite zone within a section only 0.5 km wide, then decrease gently towards the middle of the slab where they spread between 4 and 7 kbar. The lower pressures in the middle of the slab correspond to the sillimanite-grade M2 metamorphism which shows higher temperatures (700–750°C) but lower pressures than M1 kyanite-grade metamorphism. Garnets from rocks in the hot interior of the slab are commonly strongly resorbed and partly replaced by biotite ± sillimanite. The texture and P–T pattern are consistent with decompression at relatively high temperatures.

## T–t Cooling history

The precise timing of peak metamorphism of the kyanite-grade rocks above the MCT and below the ZSZ cannot yet be accurately constrained, but it is presumed that the timing is similar to both the Sm–Nd garnet ages reported by Vance & Harris (1999) from the Suru Valley to the northwest and the U–Pb metamorphic monazite ages reported by Walker et al. (1999) from the southeast Zanskar area, which are c. 35–30 Ma. The timing of peak temperature during sillimanite-grade M2 metamorphism is most likely coincidental with the timing of anatectic melting in the sillimanite + K-feldspar migmatites. The age of crustal melting is accurately constrained by the U–Pb monazite ages from migmatitic leucosomes from the Umasi-la locality which are 20.6–19.5 Ma (Noble & Searle 1995). The cooling history of rocks from the Umasi-la locality are plotted on Fig. 10: K–Ar and $^{40}Ar/^{39}Ar$ dating of muscovites and biotites [data in Searle et al. (1992)] reveal a pattern very similar to the Shafat locality. Rapid cooling from 700 to 300°C between 20 and 16 Ma is followed by a shallower cooling curve from 16 to 0 Ma. The $^{40}Ar/^{39}Ar$ muscovite age of 15.8 Ma and the biotite ages of 18.3–16.2 Ma reflect cooling through the 350 and 300°C isotherms, although it is uncertain to what degree the biotites are affected by excess Ar. The fission-track data (two zircons, five apatites; Kumar et al. 1995) are consistent with uniform cooling for the past 12 Ma and extrapolate back to 300°C at c. 16 Ma, consistent with the biotite cooling ages.

## Structural evolution

The structural and metamorphic evolution of the High Himalayan zone in western and central Zanskar can be summarized by four main stages, discussed below.

### Stage 1 (50–30 Ma)

This stage, following the collision of the Indian Plate with Asia along the Indus Suture Zone (ISZ), is dominated by crustal thickening and

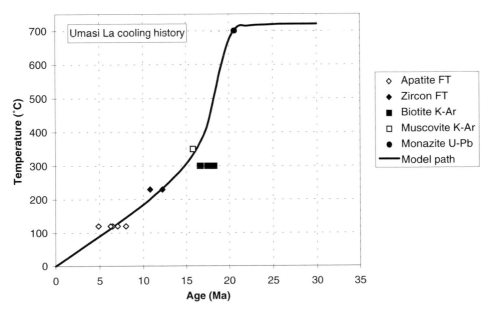

**Fig. 10.** Cooling history of rocks from the Umasi-la region in the middle of the High Himalayan Slab.

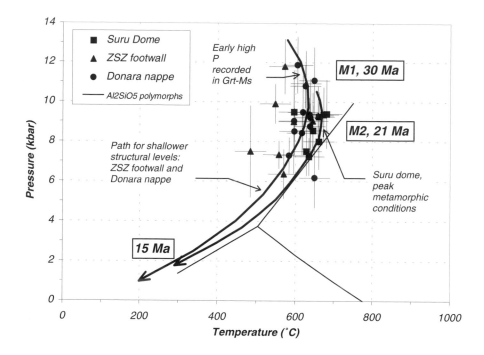

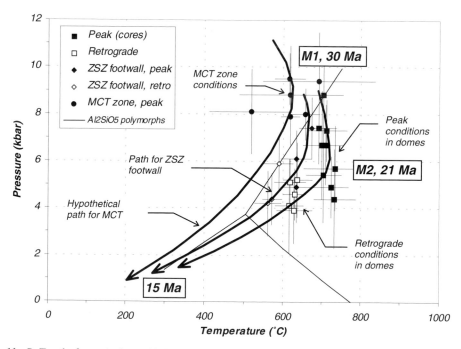

**Figure 11.** *P–T* paths for rocks from: (**a**) the Suru Valley profile; (**b**) the Umasi-la profile. Data is compiled from Searle *et al.* (1992) and Stephenson (1997). The aluminium silicate triple point is taken from Holdaway & Mukhopadhyay (1993). See text for discussion.

shortening processes along the north Indian continental margin. Folding and thrusting of the Mesozoic and early Tertiary passive margin sediments is magnificently exposed along the Zanskar Range and structural profiles from the ISZ to the ZSZ indicate a minimum shortening of 56% (or 126 km) in the upper crust (Searle 1986; Searle et al. 1997a). Similar shortening and thickening in the lower crust resulted in prograde metamorphism (M1) reaching kyanite grade across the High Himalayan Zone. Kyanite gneisses in the Donara Nappe contain early porphyroblasts and fabrics, not affected by later recrystallization, which record prograde peak P–T conditions of c. 620–650°C and 11 kbar (Fig. 11). This metamorphism relates to early crustal thickening, probably before movement on the MCT and ZSZ. Stacking of the Donara fold nappe caused these early assemblages to be transported piggyback to higher structural levels above the Suru Dome.

## Stage 2 (30–21 Ma)

Initiation of normal fault motion along the ZSZ during this period could have resulted in the ending of the Stage 1 burial phase and the beginnings of lower P–higher T M2 metamorphism characteristic of the early Miocene. The initiation of the proto-MCT may also have also occured around this time. Rocks in the MCT zone followed a clockwise P–T trajectory during the early M1 metamorphism and were quenched by the underthrusting of cold footwall rocks along the MCT.

In the Suru Dome, a younger dominant fabric [S3 of Searle et al. (1992)] formed during exhumation through 650°C and 7 kbar. Some of the overlying rocks of the Donara Nappe re-equilibrated at 600°C and 7–8 kbar at this stage (Fig.11a). Peak (M1) metamorphism is thought to be coincident with Sm–Nd ages from garnets in kyanite-grade metapelites from the Suru Valley which are c. 32–28 Ma (Vance & Harris 1999), and also with U–Pb ages of metamorphic monazites from southeast Zanskar which are 32–30 Ma. (Walker et al. 1999).

Rocks in the interior of the slab around the Umasi-la continued on the prograde path with steepening of the thermal gradient due to rise of the domes, culminating in a peak M2 metamorphism at 700–750°C and 4–7 kbar (Fig. 11b). The climax of M2 metamorphism was widespread, melting in the highest grade sillimanite + K-feldspar gneisses and migmatites along the axial core of the High Himalayan Zone. Hodges et al. (1988a, b) suggested that the high temperatures of sillimanite-grade meta-

morphism in Nepal were buffered by widespread anatexis. Timing of crustal melting in Zanskar is well constrained from U–Pb dating of magmatic monazites, both from migmatitic leucosomes at the Umasi-la, which are 20.6–19.5 Ma, and from higher level leucogranite sheets near Shafat in the Suru Valley, which are $20.8 \pm 0.3$ Ma (Noble & Searle 1995). During this period, temperatures increased and pressures decreased from the peak kyanite-grade M1 assemblages to the M2 sillimanite-grade assemblages. Rocks from the top of the slab, near the ZSZ, appear to follow a steeper retrograde path from 650°C at 6 kbar (Fig. 11b), consistent with the rapid onset of cooling from the top as the footwall of the ZSZ and juxtaposition of hot footwall rocks with cold hanging-wall rocks across the ZSZ. The timing of the cooling stage is constrained by the $^{40}Ar/^{39}Ar$ mica ages, some of which have been disturbed by excess Ar.

## Stage 3 (21–18.5 Ma)

Following the time of crustal melting and the crystallization of leucogranites, a period of very rapid exhumation resulted from both mechanical erosion and tectonic unroofing along the footwall of the ZSZ normal fault. Leucogranites were generated at pressures of up to 5 kbar at depths of 3–15 km beneath the ZSZ, and a few of the larger plutons were intruded up to higher levels, but were ponded along the footwall of the ZSZ. In Zanskar, like almost everywhere else along the High Himalayan chain, no leucogranites intrude across the normal fault into the overlying Tethyan sediments, indicating that final motion along the ZSZ postdated leucogranite melting and emplacement. Melting of the Gumburanjun leucogranite in southeast Zanskar has been constrained at $21.7 \pm 0.47$ Ma from U–Pb dating of magmatic monazites, xenotimes and uraninites (Walker et al. 1999) indicating that melting did occur simultaneously along the whole axis of the Zanskar High Himalaya at this time. $^{40}Ar/^{39}Ar$ dating of muscovites from western Zanskar (Searle et al. 1992) show a rapid cooling from 21 to 18.5 Ma, during which time some 18–25 km of material was removed in <3 Ma from this region at exhumation rates of c. 9 mm yr$^{-1}$. $^{40}Ar/^{39}Ar$ dating of muscovites from southeast Zanskar (Walker et al. 1999) shows that the ZSZ in this sector was active at 25–19.5 Ma at slip rates of 0.5 mm yr$^{-1}$.

## Stage 4 (18.5–0 Ma)

Following this early Miocene period of high erosion, rapid exhumation and rapid normal-fault motion along the ZSZ, there followed a more relaxed steady erosion rate from 18.5 Ma to the present time. Fission-track dating of zircons and apatites (Kumar *et al.* 1995) reveal a shallower cooling curve and exhumation rates of *c.* 0.4 mm yr$^{-1}$ during this time. It is somewhat surprising that, given the present-day high topography, deep erosion and extremely high monsoonal rainfall along the southern slopes of the Himalaya, the average erosion–exhumation rates appear to be lower during the last 18 Ma than they were during the early Miocene. If much of the present-day extreme topography along the High Himalayan chain (7000–8000 m high peaks and deeply incised valleys) was formed during the Quaternary, it may indicate a very recent change in the boundary conditions, either tectonic or climatic, or both. The most obvious of these is the increase in glaciation during the Pleistocene. The present-day High Himalaya marks the divide between monsoon-dominated climate and high erosion south of the range from high, arid, low erosion rates north of the range.

## Conclusions

*P–T* information derived from thermobarometry and *T–t* histories derived from thermochronology, have been used to constrain the tectonic evolution of the Zanskar Himalaya. The timing of prograde growth of metamorphic garnet (Vance & Harris 1997) and metamorphic monazites (Walker *et al.* 1999) record the minimum age of peak metamorphic conditions. The U–Pb ages of magmatic monazites, xenotimes and uraninites from both migmatitic leucosomes and from leucogranites record the timing of melting (Noble & Searle 1995; Walker *et al.* 1999). $^{40}$Ar/$^{39}$Ar dating of hornblendes and muscovites (and some biotites, those without excess Ar), and fission-track dating of zircon and apatite record the lower temperature cooling ages. These geochronological data have been plotted on *T–t* graphs to show the cooling histories of each area.

Crustal thickening along the north Indian Plate margin during the middle and late Eocene (50–35 Ma) led to regional kyanite-grade metamorphism (M1) which reached peak temperatures probably *c.* 30 Ma ago. Temperatures remained high, but pressures decreased as M2 sillimanite-grade metamorphism reached peak temperatures *c.* 21–20 Ma ago. Crustal melting in the sillimanite + K-feldspar zone along the axis of the High Himalaya led to migmatization and the separation of leucogranitic melt pods, sheets and dykes, and eventually to a few larger intrusive plutons. Melting occurred at 20.8–19.5 Ma during the early Miocene and was

followed by a phase of rapid exhumation from 20 to 18.5 Ma. This may coincide with initiation of rapid motion along the ZSZ normal fault at the top of the slab and possibly also the MCT along the base of the slab. These two major shear zones are undoubtedly linked mechanically and are almost certainly related temporally, although, as yet, the precise timing of slip along the MCT cannot be constraint.

Syn- and post-peak metamorphic shortening deformed the metamorphic isograds by large-scale, southwest verging folding of the previously 'frozen-in' isograds. Isograds are parallel to the shearing fabrics along both the top and base of the slab, and the spacing between isograds has been dramatically compressed by high ductile strain along the MCT and ZSZ. The axis of this fold plunges northwest along the axis of the High Himalaya in northwest Zanskar, allowing the 3D mapping of this structure. The distribution of *P–T* data across the Zanskar Himalaya does not support the concept of frictional heating along the MCT, nor of 'instantaneous' thrusting of a hot slab over a cold slab. The distribution of *P–T* data across the Zanskar, combined with thermochronology, does support the concept of post-metamorphic deformation of isograds with inverted *P–T* conditions and condensed isograds along the MCT at the base, and upward decreasing *P–T* conditions and condensed isograds along the ZSZ at the top of the slab. Consistent high temperatures across a 55 km width at the centre of the High Himalayan structure are also consistent with the thermal structure of the Searle & Rex (1989) large-scale folded isograd model.

This work was carried out using NERC grant GT5/96/13/E to M.P.S., NERC D.Phil. studentship grants to B.J.S., C.B.W. and J.D.W., and a Shell grant to M.W.D. We are particularly grateful to Norman Charnley for help with the microprobe, Steve Noble for U–Pb dating and Phil Guise for help with argon dating on our samples. Finally, many thanks to Paul Ryan for organizing a stimulating conference and to John Dewey for many years of discussions, cricket matches, good humour and much more.

## References

ARITA, K. 1983. Origin of the inverted metamorphism of the Lower Himalayas, central Nepal. *Tectonophysics*, **95**, 43–60.

BHATTACHARYA, A., MOHANTY, L., MAJI, A., SEN, S. K. & RAITH, M. 1992. Non-ideal mixing in the phlogopite–annite binary: constraints from experimental data on Mg–Fe partitioning and a reformulation of the biotite–garnet geothermometer. *Contributions to Mineralogy and Petrology*, **111**, 87–93.

DRANSFIELD, M. W. 1994. Extensional exhumation of high-grade metamorphic rocks in western Norway and the Zanskar Himalaya. D.Phil. Thesis, Oxford University.

ENGLAND, P. C., LEFORT, P., MOLNAR, P. & PECHER, A. 1992. Heat sources for Tertiary metamorphism and anatexis in the Annapurna–Manaslu region, central Nepal. *Journal of Geophysical Research*, **97**, B2, 2107–2128.

FERRY, J. M. & SPEAR, F. S. 1978. Experimental calibration of the partitioning of Fe and Mg between biotite and garnet. *Contributions to Mineralogy and Petrology*, **66**, 113–117.

GARZANTI, E. & VAN-HAVER, T. 1988. The Indus clastics: forearc basin sedimentation in the Ladakh Himalaya (India). *Sedimentary Geology*, **59**, 237–249.

GHENT, E. D. & STOUT, M. Z. 1981. Geobarometry and geothermometry of plagioclase–biotite–garnet–muscovite assemblages. *Contributions to Mineralogy and Petrology*, **76**, 92–97.

GRAHAM, C. M. & POWELL, R. 1984. A garnet–hornblende geothermometer: calibration, testing and application to the Pelona Schist, southern California. *Journal of Metamorphic Geology*, **2**, 13–31.

GRUJIC, D., CASEY, M., DAVIDSON, C., HOLLISTER, L., KUNDIG, R., PAVLIS, T. & SCHMID, S. 1996. Ductile extension of the Higher Himalayan crystalline in Bhutan: evidence from quartz microfabrics. *Tectonophysics*, **260**, 21–43.

GUNTLI, P. 1993. Geologie und Tectonik des Higher und Lesser Himalaya im Gebiet von Kishtwar, SE Kashmir (NW Indien). PhD Thesis, ETH Zurich.

HERREN, E. 1987. Zanskar Shear Zone: Northeast–southwest extension within the Higher Himalayas (Ladakh, India). *Geology*, **15**, 409–413.

HODGES, K. V., HUBBARD, M. & SILVERBERG, D. S. 1988a. Metamorphic constraints on the thermal evolution of the central Himalayan orogeny. *Philosophical Transactions of the Royal Society, London*, **A326**, 257–280.

—, LEFORT, P. & PECHER, A. 1988b. Possible thermal buffering by crustal anatexis in collisional orogens: thermobarometric evidence from the Nepalese Himalaya. *Geology*, **16**, 707–710.

HOLDAWAY, M. J. & MUKHOPADHYAY, B. 1993. A reevaluation of the stability relations of andalusite: thermochemical data and phase diagram for the aluminium silicates. *American Mineralogist*, **718**, 298–315.

HOLLAND, T. J. B. & BLUNDY, J. 1994. Non-ideal interactions in calcic amphiboles and their bearing on amphibole–plagioclase thermometry. *Contributions to Mineralogy and Petrology*, **116**, 433–447.

—— & POWELL, R. 1990. An enlarged and updated internally consistent thermodynamic dataset with uncertainties and correlations: the system $K_2O$–$Na_2O$–$CaO$–$MgO$–$MnO$–$FeO$–$Fe_2O_3$–$Al_2O_3$–$TiO_2$–$SiO_2$–$C$–$H_2$–$O_2$. *Journal of Metamorphic Geology*, **8**, 89–124.

HONEGGER, K 1983. Strukturen und Metamorphose im Zanskar Kristallin (Ladakh – Kashmir, Indien). PhD Thesis, ETH Zurich.

HUBBARD, M. S. 1989. Thermobarometric constraints on the thermal history of the Main Central Thrust zone and Tibetan Slab, eastern Nepal Himalaya. *Journal of Metamorphic Geology*, **7**, 19–30.

——1996. Ductile shear as a cause of inverted metamorphism: Example from the Nepal Himalaya. *Journal of Geology*, **104**, 493–499.

—— & HARRISON, T. M. 1989. $^{40}Ar/^{39}Ar$ age constraints on deformation and metamorphism in the MCT zone and Tibetan Slab, eastern Nepal Himalaya. *Tectonics*, **8**, 865–880.

JAIN, A. K. & MANICKAVASAGAM, R. M. 1993. Inverted metamorphism in the intracontinental ductile shear zone during Himalayan collision tectonics. *Geology*, **21**, 407–410.

KUMAR, A., LAL, N., JAIN, A. K. & SORKHABI, R. S. 1995. Late Cenozoic–Quaternary thermo-tectonic history of Higher Himalayan Crystalline (HHC) in Kishtwar Padar–Zanskar region, NW Himalaya: evidence from fission track ages. *Journal of the Geological Society of India*, **45**, 375–391.

KUNDIG, R. 1989. Domal structures and high-grade metamorphism in the Higher Himalayan Crystalline, Zanskar region, northwest Himalaya, India. *Journal of Metamorphic Geology*, **7**, 43–55.

LEFORT, P. 1975. Himalayas: the collided range. Present knowledge of the continental arc. *American Journal of Science*, **275a**, 1–44.

——1981. Manaslu leucogranite: a collisional signature of the Himalaya, a model for its genesis and emplacement. *Journal of Geophysical Research*, **86**, 10 545–10 568.

MOLNAR, P. & ENGLAND, P. 1990. Temperatures, heat flux, and frictional stress near major thrust faults. *Journal of Geophysical Research*, **95**, 4833–4856.

NEWTON, R. C. & HASELTON, H. T. 1981. Thermodynamics of the garnet–plagioclase–$Al_2SiO_5$–quartz geobarometer. *In*: NEWTON, R. C., NAVROTSKY, A. & WOOD, B. J. (eds) *Thermodynamics of Minerals and Melts*. Springer, 131–147.

NOBLE, S. R. & SEARLE, M. P. 1995. Age of crustal melting and leucogranite formation from U–Pb zircon and monazite dating in the western Himalaya, Zanskar, India. *Geology*, **12**, 1135–1138.

POWELL, R. & HOLLAND, T. J. B. 1988. An internally consistent thermodynamic dataset with uncertainties and correlations: 3. Applications to geobarometry, worked examples, and a computer program. *Journal of Metamorphic Geology*, **6**, 173–204.

SEARLE, M. P. 1986. Structural evolution and sequence of thrusting in the High Himalayan, Tibetan-Tethys and Indus suture zones of Zanskar and Ladakh, Western Himalaya. *Journal of the Geological Society, London*, **8**, 923–936.

——1996. Cooling history, erosion, exhumation and kinematics of the Himalaya–Karakoram–Tibet orogenic belt. *In*: YIN, A. N. & HARRISON, T. M. (eds) *The Tectonics of Asia*. Cambridge University Press, 110–137.

—— & FRYER, B. J. 1986. Garnet, tourmaline and muscovite-bearing leucogranites, gneisses and migmatites of the Higher Himalayas from Zanskar, Kulu, Lahoul and Kashmir. *In*: COWARD, M. P.

& RIES, A. C. (eds) *Collision Tectonics.* Geological Society, London, Special Publications, **19**, 83–94.

—— & REX, A. J. 1989. Thermal model for the Zanskar Himalaya. *Journal of Metamorphic Geology*, **7**, 127–134.

——, COOPER, D. J. W. & REX, A. J. 1988. Collision tectonics of the Ladakh–Zanskar Himalaya. *Philosophical Transactions of the Royal Society, London*, **A326**, 117–150.

——, CORFIELD, R. I, STEPHENSON, B. J. & MCCARRON, J. 1997a. Structure of the north Indian continental margin in the Ladakh–Zanskar Himalayas: implications for the timing of obduction of the Spontang ophiolite, India–Asia collision and deformation events in the Himalaya. *Geological Magazine*, **134**, 297–316.

——, WATERS, D. J., REX, D. C. & WILSON, R. N. 1992. Pressure, temperature and time constraints on Himalayan metamorphism from eastern Kashmir and western Zanskar. *Journal of the Geological Society, London*, **149**, 753–773.

——, PARRISH, R. R., HODGES, K. V., HURFORD, A., AYRES, M. W. & WHITEHOUSE, M. J. 1997b. Shisha Pangma leucogranite, South Tibetan Himalaya: Field relations, geochemistry, age, origin and emplacement. *Journal of Geology*, **195**, 295–317.

STÄUBLI, A. 1989. Polyphase metamorphism and the development of the Main Central Thrust. 7. *Journal of Metamorphic Geology*, **7**, 73–93.

STEPHENSON, B. J. 1997. The tectonic and metamorphic evolution of the Main Central Thrust zone and High Himalaya around the Kishtwar and Kulu windows, northwest India. D.Phil. Thesis, Oxford University.

VANCE, D. & HARRIS, N. 1999. Timing of prograde metamorphism in the Zanskar Himalaya. *Geology*, **27**, 295–398.

—— & KELLEY, S. 1994. Ar–Ar constraints on erosional versus extensional unroofing in orogenic belts: the Zanskar Himalaya, NW India. *Mineralogical Magazine*, **58A**, 930–932.

WALKER, J. D., MARTIN, M. W., BOWRING, S. A., SEARLE, M. P., WATERS, D. J., & HODGES, K. V. 1999. Metamorphism, melting and extension: age constraints from the High Himalayan slab of SE Zanskar and NW Lahaul. *Journal of Geology*, in press

# Separate episodes of eclogite and blueschist facies metamorphism in the Aegean metamorphic core complex of Ios, Cyclades, Greece

M. A. FORSTER & G. S. LISTER

*Australian Crustal Research Centre, Monash University, Melbourne 3168, Australia*

**Abstract:** Blueschists and eclogites outcrop together throughout the Cyclades, Aegean Sea, Greece. Their coexistence has been explained by compositional variation between different lithologies and a single metamorphic episode under transitional eclogite–blueschist facies conditions has been proposed. However, a detailed study of the fabrics and microstructures in an exceptionally well-preserved eclogite boudin reveals a distinct sequence of several different mineral growth events, all under conditions of high-pressure metamorphism. The earliest metamorphic mineral paragenesis recognized is a blueschist facies, glaucophane-bearing assemblage. Eclogite facies metamorphism followed, which is suggested here as being pervasive throughout the rock mass. This period of metamorphic mineral growth was overprinted by widespread blueschist facies retrogression, with hydration $\pm$ metasomatism of earlier formed (eclogite facies) assemblages. These hydration reactions were responsible for a period of porphyroblastic mineral growth in the blueschist facies (prior to $c.\,40\,\mathrm{Ma}$) which largely obliterated the earlier formed eclogite facies assemblages. Omphacite + garnet assemblages relict from the eclogite facies event are preserved only within the cores of the mafic boudins because fluid was not able to penetrate them. Evidence for the early blueschist assemblage is now preserved only as inclusions within the eclogitic garnets in the core of the boudin. This pattern of metamorphic mineral growth may have significant tectonic implications, as it is consistent with the pattern of metamorphic evolution across the high-pressure metamorphic belt of the Aegean region.

There are numerous exposures (Fig. 1) of eclogite and blueschist facies mineral assemblages throughout the Cyclades, Aegean Sea, Greece (Dürr *et al.* 1978; Blake *et al.* 1981; Papanikolaou 1984, 1987; Rodgers 1984; Avigad & Garfunkel 1991). Previous research has ascribed these to the effects of a single metamorphic 'event' (M1) (Henjes-Kunst 1980; Van der Maar 1980*a, b*; Van der Maar & Jansen 1981, 1983; Matthews & Schliestedt 1984; Schliestedt 1986; Schliestedt & Matthews 1987; Bröcker *et al.* 1993; Grütter 1993), with the consensus view that compositional variation between different lithologies explains the coexistence of eclogite and blueschist facies minerals in adjacent rock units. However, when individual occurrences of eclogites are examined in terms of the evolution of their fabrics and microstructures, a different picture emerges. A number of different metamorphic assemblages can be recognized and these appear to be the results of successive episodes of metamorphic mineral growth.

Ridley (1984) and Ridley & Dixon (1984) observed that, in northern Syros, a blueschist assemblage had been overprinted by omphacite and garnet. They interpreted this as an example of progressive metamorphism, as conditions changed from blueschist to eclogite facies. However, something different is seen on other islands (e.g. in northern Ios or northern Sifnos), where eclogitic mineral parageneses appear to be overprinted by porphyroblasts grown under blueschist facies conditions (Forster & Lister 1996*c*; Lister & Raouzaios 1996). Irrespective of whether or not eclogite and blueschist facies mineral assemblages can theoretically coexist in adjacent rock units (Evans & Brown 1986, 1987; Bröcker *et al.* 1990; Carswell 1990; Evans 1990), these observations suggest that a sequence of different metamorphic events has been responsible for the superimposition of blueschist facies assemblages over the products of eclogite facies metamorphism, and vice versa. This has significant implications in respect to the tectonic evolution of the Cyclades.

The blueschist facies overprint varies considerably in its degree of pervasiveness throughout the Cyclades. On Ios, in the central Cyclades, there is only rare preservation of omphacite–garnet assemblages in the high-pressure rocks (Fig. 2). These occur in mafic boudins that are $<10^{-6}$–$10^{-8}$ of the total volume of the rock mass

*From*: MAC NIOCALL, C. & RYAN, P. D. (eds) 1999. *Continental Tectonics*. Geological Society, London, Special Publications, **164**, 157–177. 1-86239-051-7/99/$15.00 © The Geological Society of London 1999.

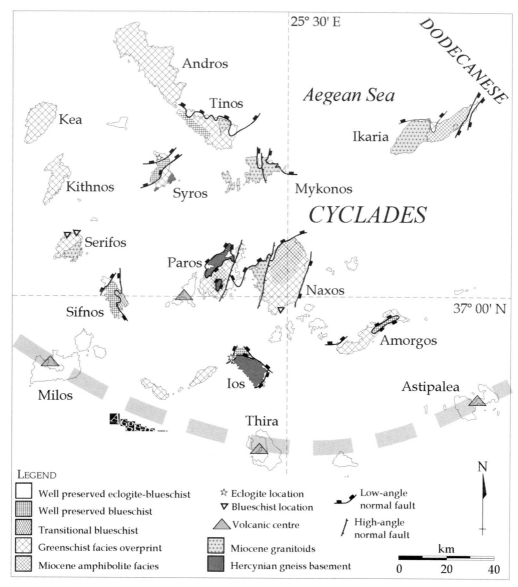

**Fig. 1.** Exposures of eclogite and blueschist facies mineral assemblages throughout the Cyclades, Aegean Sea, Greece.

(Fig. 3). Yet there is no reason to believe that the entire rock mass was not once subject to the same metamorphic conditions. On other islands, widespread development of eclogite facies mineral assemblages can still be observed (e.g. on Syros and northern Sifnos). However, over much of the Cyclades the early formed eclogitic assemblages have been totally overprinted, with blueschist and (later) greenschist facies parag005eneses obliterating evidence for the earliest part of the metamorphic history.

Although there are well-preserved tracts of eclogites overprinted by blueschist facies assemblages on islands such as Syros and Sifnos, the most complete record of the effects of the early metamorphic evolution of the rocks of the Cycladic blueschist belt has been found within the Varvara Boudin on Ios (Fig. 4). This paper is therefore devoted to the description of the sequence of events that has been recognized in the Varvara Boudin and in the enclosing schists. The goal is to demonstrate the episodic nature of

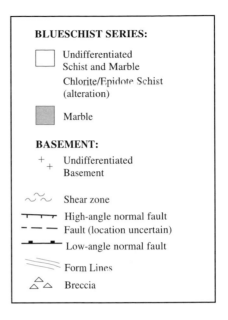

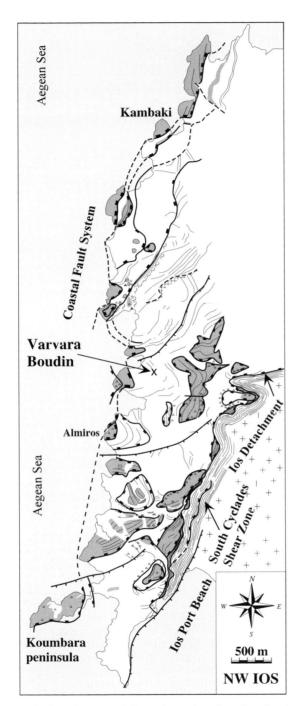

**Fig 2.** Overview map of the geology of northern Ios, showing the location of the eclogite boudins that were examined in this study.

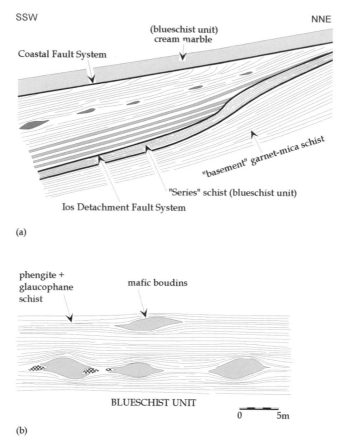

(a)

(b)

**Fig. 3.** The Varvara Boudin is one of a number of boudins formed by the disruption of a mafic layer within the schists as the result of intense later deformation. Schematic cross-section through northwest Ios, showing the occurrence of the eclogite boudins in respect to: (**a**) the two detachment systems reported by Forster & Lister (1999); and (**b**) the geometry of the eclogite boudins and their mantle of glaucophane and phengite schist.

the mineral growth events and to show how the record of events preserved inside the Varvara Boudin can be linked to record of events elsewhere. This is difficult to accomplish because the surrounding lithologies have been pervasively metamorphosed under blueschist and/or greenschist facies conditions. There are no relict mineral assemblages left to suggest that the country rock has been metamorphosed to higher pressures or temperatures.

## Geology of Ios

The schists, carbonates and metavolcanics of the Cyclades eclogite–blueschist belt occur beneath low-angle normal (detachment?) faults, where they have been drawn out from beneath a carbonate platform (Papanikolaou 1987) and structurally higher non-metamorphic rocks which

include remnants of the Cyclades ophiolite nappe (Dürr *et al.* 1978; Avigad & Garfunkel 1989, 1991; Lister & Forster 1996). The eclogite–blueschist belt has, in turn, been dissected by several different generations of detachment faults (Avigad & Garfunkel 1989; Avigad 1993; Gautier *et al.* 1993; Gautier & Brun 1994; Forster & Lister 1996*a, b*, 1999). Each tectonic slice has different states of preservation of the high-pressure minerals and individual slices yield different patterns of $^{40}Ar/^{39}Ar$ geochronology (Kreuzer *et al.* 1978; Wijbrans & McDougall 1986, 1988; Maluski *et al.* 1987; Wijbrans *et al.* 1990; Bröcker *et al.* 1993; Matthews & Okrusch 1993; Baldwin 1996; Baldwin & Lister 1999).

On a regional scale, focusing solely on the eclogite–blueschist belt and comparing different islands, it appears that higher level slices have best preserved the oldest metamorphic assemblages (Forster & Lister 1999). Conversely, the

**Fig. 4.** The Varvara Boudin outcrops as massive omphacite + garnet rock with a skin of coarse glaucophane prisms, mantled by glaucophane-bearing schists.

effects of younger metamorphic events become increasingly dominant in successively lower tectonic slices. Yet, elements of the entire metamorphic (and deformational) history appear to be present in each slice. In respect to the youngest metamorphism to overprint these rocks (i.e. the Miocene amphibolite–greenschist facies overprint), the highest temperatures are evident in rocks from the deepest structural levels. $^{40}Ar/^{39}Ar$ apparent ages show a pattern of variation that involves the appearance of progressively younger ages in successively lower slices (Baldwin 1996; Baldwin & Lister 1999). On Ios, in particular, there are several low-angle faults and several distinct tectonic slices. Forster (1996) and Forster & Lister (1996a, b, 1999) have described a 'telescoped' or attenuated crustal section, but it is difficult to correlate these slices with those observed on other islands. A more complete crustal section may have been preserved elsewhere, but there are no reports of a downwards increase in metamorphic grade and apparent ductility, as observed on Ios (Forster 1996).

The lowest structural levels of Ios consist of a recumbently folded 'basement' gneiss complex. This is derived from a Hercynian granite complex that originally intruded into a garnet–mica schist (amphibolite facies). This lithology now mantles the 'basement' gneiss complex. The

highest structural levels of the 'basement' core zone are bounded by a domed Miocene ductile shear zone (the South Cyclades Shear Zone; Lister et al. 1984). This shear zone is itself cut by a domed low-angle normal fault (the Ios Detachment; Forster & Lister 1999). This fault juxtaposes high-pressure rocks of the 'Scries' (part of the Cycladic blueschist belt) against the mylonitized 'basement' gneisses in the South Cyclades Shear Zone. A schematic map and cross-section of northwest Ios are shown in Figs 2 and 3.

Ios has been described as an Aegean metamorphic core complex because it has many of the structural components of equivalent(?) structures in the Colorado River extensional corridor (Lister et al. 1984; Vandenberg & Lister 1996). There is also some similarity with the active core complexes recognized in the Solomon Sea (Hill et al. 1992, 1995; Baldwin et al. 1993), as evident in Fig. 5 which is a schematic illustration of the 3D geology of Ios in a juvenile stage of its geomorphological evolution. The most prominent feature is the 'basement' dome defined by the bowed-up c. 0.5–1.0 km thick South Cyclades Shear Zone. This shear zone acts as a carapace to the 'basement' schists and gneisses. The Ios Detachment Fault caps the South Cyclades Shear Zone and juxtaposes the

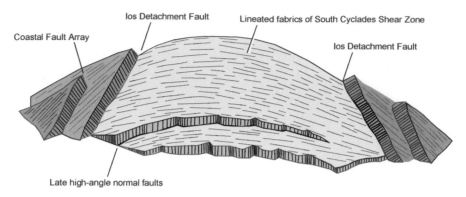

**Fig. 5.** The geomorphology of Ios, at a juvenile stage in its development. A 'basement' gneiss dome is mantled by the bowed and domed South Cyclades Shear Zone, capped by the Ios Detachment Fault. Multiple generations of low-angle normal faults affect the overlying metasedimentary 'series'. The core complex has been transected by several generations of high-angle normal faults at a later stage of its development.

'basement' gneisses against the rocks of the overlying (metasedimentary) 'Series' (Forster & Lister 1999). The island has been subsequently dissected by several generations of high-angle normal faults.

*Metamorphic evolution of Ios*

Previous workers on Ios have described the metamorphic petrography of the rocks of the 'basement' and of the 'series' on the island (Henjes-Kunst 1980; Van der Maar 1980a, b; 1981; Van der Maar & Jansen 1981, 1983; Grütter 1993; Vandenberg & Lister 1996). These authors have recognized three distinct episodes of metamorphism.

M0 is described as a Hercynian amphibolite facies metamorphism that affected the rocks of the 'basement'. M1 is described as an Eocene high-pressure–low temperature (HP–LT) blueschist facies metamorphism that affected both 'basement' and 'series', although in the 'basement' much of the evidence for this event has been obliterated by later metamorphism. M2 is an Oligo-Miocene greenschist facies metamorphism that overprints the M1 mineral assemblages to a variable degree in the upper plate but dominates the petrology of the lower plate.

M0 (amphibolite facies metamorphism) does not affect the metasediments above the Ios Detachment. The deposition and/or extrusion of the sediments and volcanics in the 'series' is thought to postdate amphibolite facies metamorphism of the basement (Van der Maar 1980a, b; Van der Maar & Jansen 1981, 1983).

M1 (eclogite–blueschist facies metamorphism) has had a major influence on the rocks immediately above the Ios Detachment Fault. This work will show that several different episodes of meta-

morphic mineral growth have been involved and that preserved eclogite boudins previously associated with blueschist facies metamorphism (Van der Maar 1980a, b; Van der Maar & Jansen 1981, 1983) may, in fact, represent the effects of an earlier period of eclogite facies metamorphism.

M2 (greenschist facies metamorphism) has also had a major effect on the rocks above the Ios Detachment Fault but its influence is non-pervasive. This is in sharp contrast to the rocks of the 'basement', where the effects of M2 are difficult to avoid. In general, M2 (greenschist facies) assemblages are mid- to low greenschist facies (albite ± chlorite ± actinolite), and were strongly controlled in their spatial distribution by fluid access along shear zones, faults and depositional layering.

M1 took place prior to $c.\,40$ Ma, based on available geochronology (Kreuzer *et al.* 1978; Henjes-Kunst & Kreuzer 1982; Andriessen *et al.* 1987; Baldwin 1996; Baldwin & Lister 1999). The greenschist facies overprint (M2) may have occurred at $c.\,20$–14 Ma (Van der Maar & Jansen 1983; Baldwin & Lister 1998).

The composition of Na-amphiboles, Na-pyroxenes, garnets and white micas have been determined using Energy Dispersive Spectroscopy (EDS) analysis on the Monash ARL SEMQ Microprobe. Analysis of these results shows that these minerals have the same, or similar, compositions to those published by Van der Maar (1980a, b).

*Deformation*

The dominant fabric on Ios (in both the basement and cover) appears to be a differentiated crenulation cleavage. This is, at least, a second-generation fabric, referred to as S2 by previous workers (Vandenberg & Lister 1996); however, this fabric may in fact be a relict S2 (i.e. a fabric that has been substantially modified by subsequent deformations). These later deformations (if they exist) *sensu stricto* should be distinguished as distinct events, but only if they can be shown to consistently overprint D2 fabrics, as is the case with the D3 deformation on Sifnos (Lister & Raouzaios 1996). Sifnos D3 substantially modifies the locally identified S2 fabric, but it also produces mesoscopic folding in large-scale recumbent folds. In these zones, new axial plane cleavages develop, and overprinting relationships show that the dominant S2 fabric has been stretched and accentuated, and locally mylonitized, during D3 deformation. The same may be true on many other islands, such as Ios, Syros, Andros, Tinos, Amorgos and Folegandros.

Another issue, for Ios, is that there is no certainty that only one major deformation event preceded the formation of the S2 fabric. The deformed and metamorphosed Hercynian basement complex has been subject to a protracted pre-Alpine history. However, for the rocks of the 'series', such considerations may not apply. Nevertheless, the deformation history has the potential for more complexity than has presently been convincingly documented. At some future date the different deformation events may need relabelling.

The formation and/or the modification of S2 appears to have taken place under high-pressure conditions, in the epidote–blueschist facies (Evans 1990), based on the stability of the relevant metamorphic mineral paragenesis during this deformation. Phengitic white mica and Na-amphiboles define the S2 fabric. The L2 lineation is defined by the preferred orientation of Na-amphibole crystals and it is often intensely developed at the margins of the boudins examined in this study (Fig. 6). L2 has northwest–southeast trends, except where the S2 fabric is overprinted by younger deformations. It has

previously been argued that D2 is the result of Eocene collision (Vandenberg & Lister 1996).

Several folding events appear to have taken place subsequent to this deformation, but only one pervasively overprinting fold generation has been reported in the 'basement' domain (Vandenberg & Lister 1996). In the 'series' there is evidence for several fold generations that overprint the dominant (S2) fabric (Forster 1996).

Immediately adjacent to the Varvara Boudin there are several, gently north plunging, reclined F3 folds. Glaucophane needles are abruptly reoriented in their axial zones, into north–south trending orientations, whereas on the fold limbs these porphyroblasts retain northwest–southeast trend characteristic of L2 (Fig. 7). As a result, the L2 glaucophane lineation alternates from northwest–southeast to north–south trends (on a scale of $c.0.5$ m). Under the microscope, it can be seen that reorientation in the hinge region occurs by a process of rigid-body rotation and that the individual glaucophane prisms appear to have forced their way though the adjacent matrix. Glaucophane, white mica and epidote within the hinge zones display evidence for brittle deformation and some recrystallization. White micas are kinked and bent. The clinozoisite porphyroblasts are folded, whereas the growth of albite porphyroblasts overprints the axial fabrics. The D3 deformation event thus occurred after the clinozoisite growth event but prior to the onset of M2 greenschist facies metamorphism.

D4 in the 'basement' domain was associated with north–south stretching of the Aegean continental crust (Lister et al. 1984; Vandenberg & Lister 1996) and the South Cyclades Shear Zone (Lister et al. 1984; Vandenberg & Lister

Mafic Boudin

**Fig. 6.** Later deformation of the eclogite boudins results in the progressive reorientation of M1A glaucophane prisms into an intensely defined northwest–southeast trending lineation.

**Fig. 7.** The northwest–southeast trending L2 glaucophane lineation is refolded by D3 folds and, in their axial zones, it is reoriented into north–south trends.

1996) formed at this time. D4 deformation in the 'series' is typified by S-C fabrics and the formation of localized shear zones. These have intensely developed north–south trending lineations, and pre-existing porphyroblasts are pulled apart and the fractures infilled by albite, suggesting that this deformation occurred during the M2 event. L2 is reoriented into north–south trends wherever it is affected by the D4 deformation.

Island-scale doming took place during D5 as the result of east–west shortening superimposed on previously developed geometries. The 'compressional' characteristics of D5 are confirmed by small-scale folds formed by buckling, reported on Ios and on the adjacent island of Naxos (Lister & Forster 1996; Vandenberg & Lister 1996).

Subsequent events in the 'series' are the result of the operation of the Coastal Fault System. These low-angle normal faults appear to have formed during and after the formation of the Ios Dome, since they cut (with gentle dips) through already moderately dipping fabrics on the dome flanks (Forster 1996; Forster & Lister 1999). This geometry places bedding and cleavage into the shortening field of the shear zones associated with the formation of the Coastal Fault System and, in consequence, a variety of complex (dominantly kink- and chevron-style) folding geometries develop. The low-angle faults of the Coastal Fault System are also associated with the forma-tion of narrow ductile shear zones containing spectacular phyllonitic fabrics, local folding, breccias and extensive epidote–chlorite alteration.

## Evidence for multiple high-pressure metamorphic growth events

This study has identified several episodes of metamorphic mineral growth in what has been previously grouped as one event (Van der Maar & Jansen 1981, 1983). Therefore, the description of metamorphic events given here will depart from the traditional M0, M1, M2 nomenclature of previously published work. The blueschist facies 'event' (traditionally M1) has in this study been divided into an early blueschist facies event, M1a, an eclogite facies event M1b and a blueschist facies event M1c. There is even some evidence for a later albite–epidote blueschist event, M1d, but this may have been transitional into the greenschist facies (Evans 1990). M1a and M1b represent the prograde pressure–temperature (P–T) path, with M1a being a relict blueschist facies event and M1b representing peak metamorphism at eclogite facies conditions.

## The Varvara Boudin

Small boudins or lenses of eclogite facies assemblages occur within glaucophanitic or retrogressed glaucophanitic schists in the lower levels of the 'Series' above the Ios Detachment Fault (Fig. 3). The boudins consist of assemblages containing omphacite and garnet, overprinted by later formed Na-amphibole bearing assemblages and an even later greenschist facies overprint. The eclogitic lenses and boudins have been found within 1–2 km of one another, but only in the lower structural levels of the 'series'. They may represent the disrupted remnants of mafic intrusions in the schist packages, since the boudins are found at or about the same structural level. The boudins usually occur in layers that are structurally above the Ios Detachment Fault, although they do not occur in the immediate vicinity of this fault. There is no apparent structural feature that determines the location of these boudins. Instead, they appear to be randomly distributed within the schist packages (Fig. 3b).

The eclogitic assemblages were studied at four localities, three of which are adjacent to later faults and are consequently retrogressed. However, one boudin is exceptionally well preserved. It occurs on a hillside, to the southwest of Agia Varvara, so it will henceforth be referred to as the Varvara Boudin (Fig. 4). This boudin is exceptional in the degree of preservation of the early eclogite facies assemblages and therefore defines an important location for the geology of Ios (Forster & Lister 1996c). Similar boudins and lenses, c. 1 m wide by tens of metres laterally, with the same mineral assemblage as the core (see below) of the Varvara Boudin, are found in the northern regions of Syros. However, on Syros no zonation from core to rim occurs as in the boudins on Ios.

On Ios, the preserved mafic boudins display a broad zonation from core to rim, but the transition from eclogite facies mineralogy of the core zone (omphacite + garnet) to the Na-amphibole-dominated rim can be quite abrupt (Fig. 8a). This zonation is not consistent from the centre of the boudin to the rim, as the boudins have undergone later fracturing and retrogression, deformation and fluid access appears to have controlled the alteration pattern. Generally, the core of the boudin consists of a matrix of clinopyroxene (omphacite/aegirine–augite) and garnet porphyroblasts, with lesser amounts of later(?) sphene, and even later actinolite. The clinopyroxene and garnet assemblage is overprinted by stilpnomelane- and riebeckite-bearing assemblages, and these

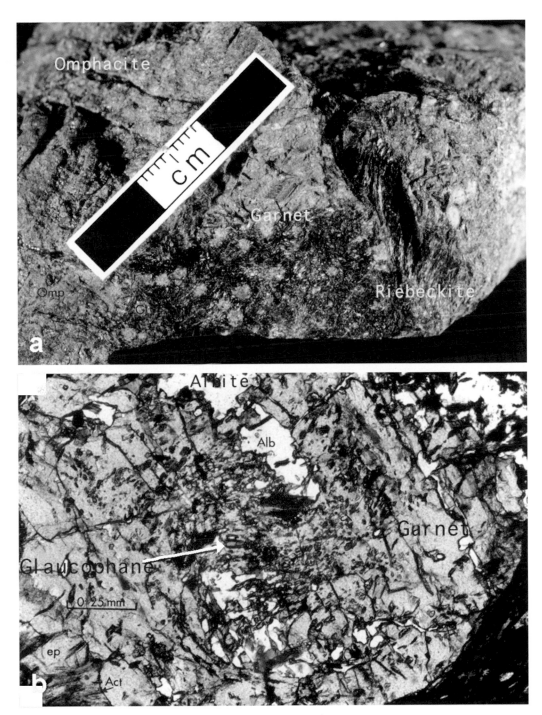

**Fig. 8.** (a) Metabasic sample of eclogite (omphacite + garnet) and blueschist facies mineral assemblages defined by a skin of Na-amphibole + garnet. (b) Micrograph of garnet growth in the eclogite boudin core. The core of the garnet displays curved inclusion trails (S1), defined by Na-amphibole. The garnet rim has few and randomly oriented inclusions.

also increase in abundance away from the core (Fig. 9a and b).

The boudins are mantled by a rim comprising coarse-grained Na-amphibole and garnet. This blueschist assemblage rim becomes progressively more deformed outwards until it forms a mantling skin defined by highly lineated blueschist. This is the S2 foliation and L2 lineation. The intensity of D2 deformation increases markedly towards the outside of the mantling rim and can be seen to progressively align randomly oriented Na-amphibole prisms, from core to rim (Fig. 6). The trend of the resultant lineations are in a northwest–southeast direction, which is identical to the orientation of L2 lineations recognized in surrounding schists. The rim of the boudin shows a distinct change in mineral assemblage, with the dominant minerals being glaucophane/crossite and, to a lesser degree, epidote.

Garnets occur throughout the boudin. In the core, garnets are well preserved with pressure shadows of albite and actinolite. However, the garnets closer to the outside of the rim are partially, or completely, replaced by albite ± chlorite. Amphiboles are commonly up to 5–6 mm in length and garnets commonly <2 mm in diameter. Monomineralic veins (up to 20 mm thick) of Na-amphiboles and/or epidote occur with random orientations throughout the boudin.

## The Boudin core

The core of the boudin contains a coarse omphacite groundmass with (c. 2–3 mm) garnet porphyroblasts (M1b). This paragenesis appears to have grown in a distinct period of metamorphic mineral growth. The core has a massive appearance but there is a weakly developed grain-shaped fabric evident in the coarse omphacite matrix, defined by the shape orientation of (125 × 50 μm) elongate omphacite crystals. This fabric appears to have been imposed during the D2 deformation.

Fabric intensity in the omphacite groundmass increases away from the core, deformation progressively increasing the intensity of the alignment of minerals that overprint the omphacite–garnet matrix, such as Na-amphibole (M1c) and actinolite (M2) (Fig. 6). The alignment of glaucophane prisms appears to be the result of bulk rotation during deformation. The actinolite fabric is the result of later deformation. The grain boundaries separating omphacite matrix crystals have been subsequently altered and have been replaced by fine actinolitic amphibole. Thus, it is difficult to determine whether

the omphacite crystals were metastable during the early part of later deformation. However, since some degree of recrystallization must have taken place to allow development of the shape fabric in the omphacite matrix, it must be concluded that the omphacite crystals were (meta)stable during this later deformation, which hence must have taken place under transitional eclogite–blueschist facies conditions.

The garnets of the inner core of the boudin are generally well preserved, idioblastic and wrapped by the omphacite-defined fabric (Fig. 8b). The garnets have two distinct types of pressure shadows. These formed during deformation events that took place under different conditions because they have different mineral parageneses growing within them. Pressure shadows that contain albite, glaucophane and sphene may have developed during deformation under transitional blueschist–greenschist facies conditions. Pressure shadows that contain epidote with small actinolite needles developed during later deformation, but now under greenschist facies conditions. The matrix foliation has thus been imprinted by deformation events under different metamorphic conditions, one in which omphacite remained (meta)stable and the other in which limited retrogression around omphacite grain boundaries took place, as described above. Both the development of the fabric and the growth of the pressure shadow minerals post-date the growth of the (M1b) garnet–omphacite paragenesis.

## Relics of early blueschist metamorphism

The garnet porphyroblasts of the boudin core preserve evidence of a more complex history than does the omphacite matrix and two generations of growth can be identified within the garnet porphyroblasts. These porphyroblasts have a core zone (diameter c. 1 mm) mantled by a rim zone (thickness c. 1 mm). The core of the garnet has abundant inclusions but the mineralogy of the inclusions is difficult to decipher owing to their small size. The more distinct crystals have been identified optically as glaucophane, epidote and quartz. The rim of the garnets contain some inclusions (epidote, rutile and opaques) but the orientation of the inclusions in the rim zone appears to be random. The boundary between the rim and the core is occasionally defined by elongate glaucophane needles that appear to wrap the rim.

Unlike the garnet rim, the inclusions in the garnet core are sometimes aligned in a distinct 'S' shape (Fig. 8b), implying that the garnet core grew during, or subsequent to, a deformation event (D1). More detail on this aspect might reveal more of the complexity in the history of the evolution of the metamorphism, since this observation seems to imply a glaucophane fabric (M1a) that predates the development of the (M1b) omphacite–garnet paragenesis. When the garnets are not zoned they have the composition (almandine) and microstructural characteristics of the rim zone garnets.

## Transition from eclogite to blueschist facies

Progressive overprinting of the omphacite–garnet paragenesis by more hydrous minerals takes place with increasing distance from the preserved cores of the Varvara Boudin. The first stage of overprinting is defined by the large blue amphibole prisms. These clearly overprint the early omphacite–garnet paragenesis. However, as later deformation increased in its intensity, these prisms were progressively rotated towards the plane of the grain-shaped fabric that is now evident in the omphacite matrix. It is difficult to find samples without a grain-shaped fabric evident in the omphacite, and in most cases the blue amphibole prisms are inclined $c. <10°$ steeper than this grain-shaped fabric. These observations suggests a period of porphyroblastic growth of blue amphibole after the omphacite–garnet paragenesis (M1b) but prior to the later deformation event that develops the omphacite fabric, and partially aligns the blue amphibole prisms. Subsequent retrogression during deformation leads to the development of green amphibole.

The next stage of overprinting is defined by alteration zones that can most readily be explained as the result of fluid ingress while high-pressure conditions still prevailed. Several centimetres from the boundary of the outer rim, large ($<2$–3 mm) (and spectacular) stilpnomelane crystals radiate from the garnet porphyroblasts (Fig. 9a). The garnets at these locations are partially replaced (sometimes almost completely) by clots of albite $\pm$ epidote $\pm$ blue amphibole. Veins with these mineral parageneses also exist.

Towards the outer rim, an increased number of large Na-amphibole prisms overprint the omphacite–garnet paragenesis (Fig. 9b). These are poorly aligned with the omphacite foliation, with up to $\pm40°$ variation in orientation. These may have developed coevally with the stilpnomelane-bearing assemblages, (but at the very outer rim there are no clear textural relationships), with the large Na-amphiboles being the dominant feature and the stilpnomelane only minor.

Monomineralic veins of large Na-amphibole crystals ($<1$ cm length) cut through the boudins. Pockets or lenses of finer grained Na-amphibole with euhedral garnets (1–2 mm) occur at the outer rim of the boudins. The large Na-amphiboles are Mg-rich riebeckite rather than glaucophane. These regions have simpler mineral assemblages than the glaucophanites or glaucophanitic schists, and consist of either Na-amphibole and garnet, or just Na-amphibole. They display zoning with Mg-rich cores and Fe-rich rims, which is typical of the effect of the M1c overprint on the eclogitic boudins. The Na-amphibole crystals in these veins (riebeckite) are black in hand-specimen, unlike the blue-grey Na-amphiboles (glaucophane–crossite) found elsewhere throughout the area. The Na-amphiboles at the outer transition zone of the eclogite boudins are the same colour, size and composition as the veins of black Na-amphiboles. Euhedral terminations can be found, attesting to the presence of fluid-filled cavities during growth (cf. Barnicoat 1988). At times, the riebeckite crystals form a 'skin' over the omphacite–garnet assemblage, the skin essentially being a continuation (or deformed equivalent) of veins cutting the boudin. However, the veins are monomineralic, whereas the skin appears to have abundant garnets.

## Transition from blueschist to greenschist facies

The major change from the core to the rim of the Varvara Boudin is the immediate loss of the omphacite garnet paragenesis. The percentage of Na-amphibole increases markedly in the outer zone and at times the rim consists dominantly of Na-amphiboles similar to those in the core. Generally, however, the Na-amphiboles in the rim are much smaller, elongate and aligned in an non-spaced intense foliation. These glaucophanes are similar in composition to those found in the glaucophanitic schists. These Na-amphiboles are paler in colour with little or no zonation. They have retrogressed along fractures, and at times at their terminations, to bright green amphibole. Their composition is Fe-rich glaucophane to ferroglaucophane.

The first stage of overprinting of the omphacite–garnet assemblage is defined by what are now small elongate green amphiboles ($<1000$ m $\times 250\,\mu$m) and larger zoned amphibole crystals with a blue-lavender core. Hypotheses for the appearance of these zoned amphiboles are that: (1) there were two stages of overprinting amphibole growth; (2) there was a progressive

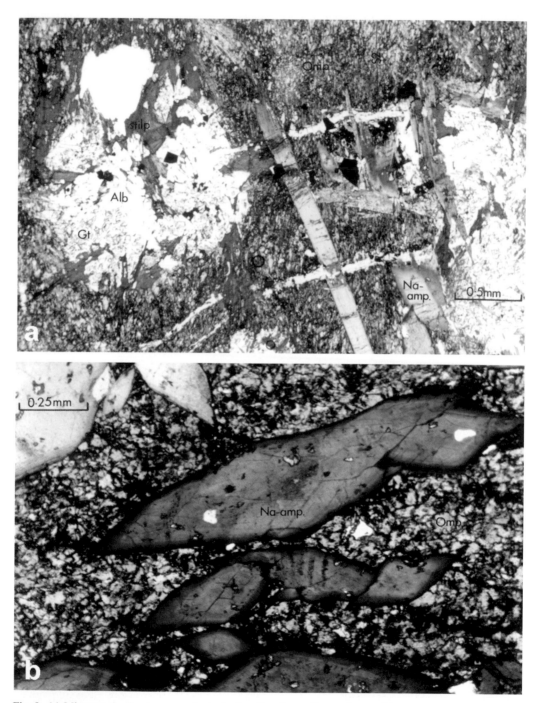

**Fig. 9.** (**a**) Micrograph of outer zone of eclogite boudin where stilpnomelane and Na-amphibole overprint the omphacite–garnet assemblage. (**b**) Micrograph of the outer transition zone of the eclogite boudin, where zoned Na-amphibole overprint the omphacite groundmass. Omp, omphacite; Gt, garnet; Alb, albite; Na-amp, Na-amphibole.

change of conditions during growth; or (3) there was subsequent replacement and progressive alteration of the mineral chemistry during retrogression. Of these possibilities, (3) appears the most likely. The amphibole prisms appear to have been altered to blue–green amphibole and only the larger crystals preserve their original Na-amphibole core. The same green amphibole anastomoses amongst the boundaries of the omphacite grains, adding to the definition of the fabric. In addition, pressure shadows are developed at the ends of the blue amphibole prisms and these contain the same green amphibole.

The transition from blueschist to greenschist facies conditions was intimately associated with the migration of fluids, as evidenced by veining and more subtle expressions of replacement, as described above (cf. Schliestedt & Matthews 1987; Bröcker 1990; Rubie 1990; Matthews & Okrusch 1993). Veins form either as the result of dilation, in which case it is possible to find coarse-grained fibre textures, with the grain shape parallel to the direction of opening, or as replacement zones spreading from the small crack which allowed fluid ingress. The eclogite boudins are cross-cut by veins of both epidote and Na-amphibole (glaucophane–riebeckite). Monomineralic veins of epidote cut both the boudin and the Na-amphibole veins. This cross-cutting relationship defines the relative timing between the M1b eclogite assemblage, M1c blueschist assemblages and late (M1d?) epidote growth.

## Metamorphism of the country rock

The following timing relations have been established in the schist surrounding the Varvara Boudin. As already indicated, M1c involved initially a period of porphyroblastic glaucophane growth, but most amphibole was aligned into northwest–southeast trends during later deformation, i.e. D2 (Fig. 6). Porphyroblastic growth of glaucophane, white mica, clinozoisite and garnet has been observed. Conditions during subsequent deformation (D2?) were probably blueschist facies, as the high-pressure assemblages remained stable through this event and some glaucophane grew in pressure shadows in the eclogite Varvara Boudin. The modified S2 fabric is observed in glaucophane schists and in the rim zone of the earlier eclogitic boudins.

### Garnets in glaucophane schists and gneisses

Glaucophane schists, in general, consist of differing modal proportions of glaucophane, epidote (Fe-rich), white mica, sphene, garnet and accessory minerals of rutile, magnetite, calcite and quartz with minor chlorite. Amounts of either glaucophane, epidote or white mica can differ considerably from one rock to another. Some assemblages contain significant proportions of quartz (>20%), which occurs as layers with serrated grain boundaries.

Garnet porphyroblasts occur in most of the assemblages, although not all the assemblages show the same generation of garnet growth, size or alteration. The garnets in the glaucophanitic schist are much smaller (c. 700 μm) compared to the eclogitic garnets (commonly c. >2 mm) and large garnets (c. 6 mm) of the massive glaucophane-rich assemblages. Garnets in glaucophanitic schists are typically small euhedral grains with only minor inclusions, displaying one generation of growth. These garnets are similar to that of the rim (second-generation) growth in the eclogitic garnets in the core of the Varvara Boudin, and/or to the cores of the massive glaucophane-rich assemblages. There is no albite replacement in these porphyroblasts, however, as is the case with the eclogitic garnets.

Layers of coarse-grained garnet–glaucophanites in the country rock display a history not seen in other glaucophane-rich layers. The most interesting feature is the garnet porphyroblasts. These are unusual in that they often occur as layers or veins only one garnet thick (<3 cm long layer and generally <3 mm diameter), as well as in minor isolated occurrences. This layering could be due to garnet nucleation and growth on a band of more favourable composition, e.g. Mn rich. Alternatively, the garnet layer may be associated with fluid infiltration along a layering. These garnets are idioblastic, mostly with abundant inclusions which display distinctive inclusion trails. These trails are defined by elongate glaucophane and white mica, plus non-elongate epidote, clinozoisite and sphene. At the rim of the porphyroblast the inclusion trails curve into parallelism with the orientation of the (S2) fabric outside the garnets.

This observation suggests that S2 formed prior to the growth of these garnets and that subsequent deformation has strongly modified the S2 fabric (as observed elsewhere in the Cyclades, e.g. on Sifnos; Lister & Raouzaios (1996); see also previous discussion). In some instances the garnets are zoned with a core that is almost inclusion free, with abundant inclusions in the rims, similar to the pattern observed in the Varvara Boudin. This zonation pattern suggests that the cores of the garnet porphyroblasts are relict from the earlier episode of eclogite facies metamorphism. The (modified) S2 fabric wraps the garnets in the regions of most intense fabric development and is overprinted in the pressure shadows. These pressure shadows display a greenschist facies metamorphic assemblage of albite and chlorite.

### Silky glaucophane layers

Another occurrence of glaucophane is in layers with a silky appearance due to their smaller grain size and intense fabric. These narrow layers are dominantly glaucophane ($500 \times c. 25 \mu m$) with minor amounts of white mica, epidote and albite. They generally are more highly deformed than the more competent adjacent layers (such as the coarse-grained glaucophanite layers), and thus display an intense foliation and lineation defined by the elongate glaucophane crystals. Shear zones with north–south trending mineral L4 lineations commonly occur in these layers.

### Epidote growth events

Several epidote growth events overprint the glaucophane-rich assemblages. The earliest event (M1d) occurs as abundant clinozoisite crystals and is seen to overprint the glaucophane defined (S2) fabric (Fig. 10a). These clinozoisite crystals are large ($<4$ mm) and occur non-pervasively throughout the area. When clinozoisite crystals do occur they are abundant ($c. 20\%$ modal proportion). The late epidote veins in the eclogitic boudins which overprint the glaucophane veins could be the same generation as this period of epidote growth.

The second epidote growth event is a separate and quite distinct growth event. It produces much smaller crystals ($c. <250 \mu m$) and these are more Fe rich than the clinozoisite crystals. This epidote growth is a distinctly younger event and more pervasive than the previous clinozoisite growth event.

### The greenschist facies overprint

The greenschist facies overprint is mineralogically defined by the growth of albite, epidote, actinolite and chlorite plus accessories of magnetite or haematite. Glaucophane schists are typically observed with greenschist facies overprinting but the glaucophane-bearing assemblages remain dominant until they are overprinted by D4 extensional fabrics. The change from blueschist metamorphic assemblages to greenschist metamorphic assemblages can be observed at varying stages of overprinting (Fig. 10b). With increasing intensity of D4 the glaucophane assemblage (relict blueschist) is completely lost.

The earliest grown albite (M2) overprints D3 folds and microscale (D4) shear zones in the glaucophane schist assemblage. Where the albite overprints the glaucophane micro-shears, it infills crystals which have been pulled apart during D4. The timing of albite growth can thus be constrained to have occurred post-D3 and early D4. By the time an S-C fabric has developed, albite is typically the dominant mineral, defining porphyroblasts ($c. 1.5 \times 0.75$ mm) around which the S-C fabric anastomoses. The albites generally have pressure shadows which help define the S-component of this fabric. The C-plane is defined by white micas, chlorite and glaucophane in differing amounts. Extensional deformational features differ from site to site. Fabrics can range from classic S-C bands to an anastomosing fabric, or a more planar fabric reflecting the effect of intense shearing (Fig. 10c).

Albite is typically the dominant mineral in retrograded glaucophanite, usually $c. <1.5$ mm and often poikioblastic with a relict glaucophane-rich S2 fabric. It is associated with actinolite needles ($c. <250 \times 25 \mu m$) which are pale green under plane polarized light, unlike actinolite associated with glaucophane or omphacite in the M1b and M1c assemblages. These actinolites resemble actinolite in the pressure shadows of garnets in the eclogitic assemblages. Garnets resemble those described in the M1b assemblage in the boudins and are generally ($<1.5$ mm) euhedral crystals. They contain abundant inclusions.

### M2 garnets in phengite schists

Some variation in respect to the above is seen in the phengite schists, which preserve a distinct layering of quartz and white micas, often with garnet porphyroblasts. Relict (S1) white mica occurs in the microlithons of a decussately recrystallized, differentiated crenulation cleavage. At times, intense overprinting during the D4 shearing event is evident. Garnets that grew in these pelitic lithologies resemble the garnets in the garnet–mica schist below the Ios Detachment Fault, at the contact between the upper and the lower plate (Vandenberg & Lister 1996). They grew as (locally abundant) small euhedral crystals ($c. 100 \mu m$) pre- or early in the later M2 greenschist facies metamorphism. The euhedral form is best preserved where these garnet porphyroblasts have been overgrown by even later developed albite porphyroblasts. Elsewhere

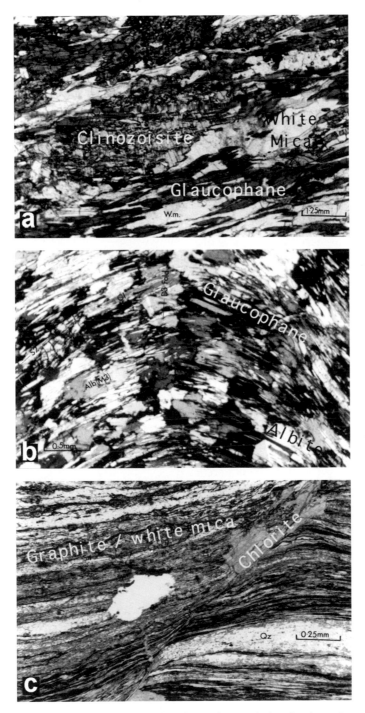

**Fig. 10.** (a) Micrograph of late-M2 clinozoisite mineral growth overprinting the glaucophane-defined S2 fabric. (b) Albite porphyroblasts growing over a lineation defined by altered glaucophane prisms. (c) Chloritization along a discrete shear band through the modified S2 fabric.

in the fabric the euhedral garnet porphyroblasts are retrogressed and corroded during ongoing M2 metamorphism and D4 deformation.

## Later alteration

Many rocks of the greenschist facies assemblages are no longer solid but crumble to the touch, as the result of late chloritization and subsequent alteration to clay-forming minerals. This is most evident with an increase in chlorite along shear planes or layers. Greenschist assemblages are often overprinted by strong chloritization. This occurs in the footwall of major faults, in particular in association with large scale low-angle faults along the coastal section (Forster & Lister 1996b, c). Local retrogression adjacent to smaller faults, most likely due to fluid influx, is observed. Chloritization also occurs at the microscale as fluid infiltrates along discrete shear bands, shear zones or layering.

## Discussion

This study of the Varvara Boudin presents results that appear to be of wider significance than just for Ios. The same metamorphic assemblages occur across the Cyclades blueschist belt; however, an understanding of regional structures is needed to understand their spatial occurrence. For example, lenses of eclogite facies assemblages on northern Syros are the same as those in the Varvara Boudin on Ios, while large areas of eastern Syros have eclogite facies mineralogy which appears to be the same as on northern Sifnos, but similar tracts do not occur on Ios.

However, there are significant variations in mineralogy from place to place. For example, the zonation of boudins and lenses including minerals such as stilpnomelane in the growth sequence has not been recognized in other localities. However, the basic elements of the sequence of metamorphic mineral parageneses appear to be the same across a wide region and therefore typical of the metamorphic evolution of this high pressure terrane as a whole. Eclogite–blueschist terranes in other regions have similar characteristics (Carswell 1990) and similar sequences may be evident elsewhere.

## Implications for other eclogite–blueschist terranes

The coexistence of blueschist and eclogite mineral assemblages need not imply transitional blueschist–eclogite facies conditions. If micro-structural examination indicates a succession of metamorphic mineral parageneses, conditions during one event have to be estimated independently of those during other events. Van der Maar (1980a, b) suggests that the maximum blueschist facies conditions are between 300 and 400°C with a pressure range of 9–11 kbar. The petrogenetic grid, set out by Carswell (1990), shows this to be at the lowest pressure boundary for these mineral assemblages, suggesting that pressures may have been $\gg$11 kbar during the period of eclogite facies metamorphism.

Although eclogites may be derived from mafic material and the protolith of the blueschist may be metasediments (Carswell 1990), this does not mean that the transition from eclogite to blueschist facies parageneses is compositionally dependent. The entire rock mass may once have been converted to eclogite facies mineral parageneses. Yet there is no obvious evidence of relict eclogite facies mineralogy in the schists surrounding the Varvara Boudin. This may be because schist packages have been completely retrogressed.

Mafic boudins are, in general more competent, and thus offer little or no encouragement in respect to fluid ingress. Mafic eclogites consisting primarily of omphacite and garnet have to hydrate to equilibrate in the blueschist facies. The preservation of the eclogite boudins thus depends on the absence of fluid, or the inability of fluid to penetrate the boudin cores. Carswell (1990) suggests that mineral reactions may not necessarily be dependent on $P$–$T$ conditions but on the timing of when fluid becomes available at reaction sites. Relict boudins are preserved due to their inaccessibility to fluid interaction.

This is not true in the surrounding schist packages and thus they typically reveal the younger part of the orogenic history. However, it is known that the eclogite and the blueschist assemblages are not coeval because, at the margins of the mafic boudins, eclogite facies mineralogy is overprinted by blueschist facies assemblages. The blueschist minerals form fabrics that can be traced into the surrounding schists. Therefore, the different metamorphic events can be (relatively) timed.

Carswell (1990) suggests that the metabasic rocks undergo hydration reactions while the metapelites are undergoing dehydration reactions, thus acting as a source of fluid for the metabasic reactions. This conjecture cannot be substantiated on Ios, since there is no micro-structural evidence of the prior existence of eclogite facies mineral assemblages in the metapelites. However, elsewhere in the Cyclades (e.g. on northern Sifnos), there is abundant evidence for Fe-metasomatism during blueschist retrogression of the earlier formed eclogite facies assemblages, both in the metapelites and in the metabasic units. Circumstantial evidence suggests fluid movement (and metasomatism) during high-pressure metamorphism was an important factor in these rocks and there is no reason to assume that fluid was not externally introduced. Note that, although the minor occurrence and apparent random distribution of eclogite boudins appears to characterize blueschist–eclogite terranes (Carswell 1990), this may reflect more the relative abundance of mafic boudins and country rock.

## Prograde v. retrograde metamorphism

The accepted view is that the dominant eclogite–blueschist assemblages represent peak metamorphic conditions during the prograde part of the $P-T$–time (t) path (Ernst 1977, 1988). However, this is contradicted by microstructural observations on Ios. Microstructures indicate that the eclogite assemblage (M1b) represents peak metamorphic conditions. This implies that the dominant blueschist assemblage (M1c) seen on Ios is a retrograde assemblage. Some blueschist assemblages in the Cyclades may have formed prior to the eclogite facies assemblages, during prograde metamorphism (e.g. on Syros; Ridley, 1984; Ridley & Dixon 1984). However, on Ios, the only evidence for prograde blueschist facies metamorphism (M1a) is found in inclusion trails in the cores of eclogitic garnets.

The concept of prograde v. retrograde metamorphism has the implication that there are essentially two portions to the $P-T-t$ path. It is assumed that during prograde metamorphism the mineral assemblages continually adjust until conditions of maximum entropy are attained. It might be expected that some relicts of early metamorphic mineral assemblages can be preserved, e.g. in the cores of garnet porphyroblasts as observed for the M1a assemblage in the Varvara Boudin. In general, however, the conditions of 'peak' metamorphism are assumed to have resulted in the conversion of the bulk of the rock mass into an appropriate mineral paragenesis.

Prograde reactions typically involve dehydration of earlier formed assemblages, and the pervasive release of fluid may ultimately be linked to the penetrative conversion of the rock mass that takes place during prograde metamorphism. Retrograde reactions typically involve hydration and the introduction of fluid along ductile shear zones is the important element governing whether or not conversion of the 'peak metamorphic assemblage' takes place. Numerous

studies have linked the operation of shear zones with the formation of microscopic dilatancy (e.g. Rutter & Brodie 1985; Urai et al. 1986; Zhang et al. 1995). This dynamically maintained permeability is essential for the fluid ingress that has been shown to be the significant factor determining whether or not conversion of metamorphic mineral assemblages from blueschist to greenschist facies takes place (Matthews & Schliestedt 1984; Schliestedt & Matthews 1987; Bröcker et al. 1993). Fluid ingress also appears to have controlled retrogression of blueschist assemblages on Ios, and the later greenschist facies overprints are localized along and adjacent to the pathways that allowed fluid movement (i.e. in particular lithologies, ductile shear zones and/or faults).

## Metamorphic episodes and their implications

In the case of the Varvara Boudin, the $P-T-t$ path followed can be described as a classic Alpine loop, involving isothermal decompression (Fig. 11a). However, this path is not compatible with available $^{40}Ar/^{39}Ar$ geochronology (Wijbrans & McDougall 1986, 1988; Wijbrans et al. 1990; Lister & Raouzaios 1996). Cooling has to have taken place after the period of high-pressure metamorphism. Changes in temperature might be the result of several phenomena: short episodes of heat, tectonic juxtaposition of thinner slabs of differing thermal history and/or the influx of large volumes of pervasive fluid could all bring in heat, or at least trigger metamorphic reactions. However, the sequence of mineral growth events recognized may still mark individual points along a smoothly varying $P-T-t$ path (Fig. 11b).

The $P-T-t$ path can be even more complex. Each episode of metamorphic mineral growth may take place under conditions of temperature and/or pressure that are transiently applied to the rock mass (Fig. 11c). Baldwin & Lister (1999) examined rocks from the Ios basement and argue that a succession of transient thermal pulses occurred. These pulses most likely affected rocks examined in the overlying blueschist unit (i.e. the Varvara Boudin).

The significant issue is to constrain the nature of the $P-T-t$ path between each episode of metamorphic mineral growth, since, as shown in Fig. 11b and c, radically different circumstances might apply. Regional metamorphism need not always involve long time periods and the conditions of metamorphism (pressure, temperature and fluid activity) may only be transiently maintained (Baldwin & Lister 1999).

**(a)**

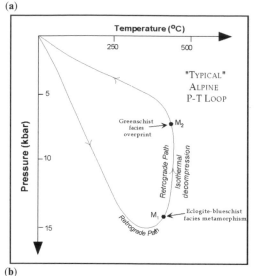

**(b)**

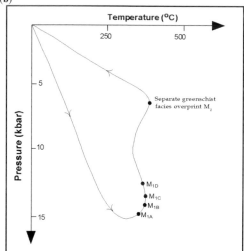

**(c)**

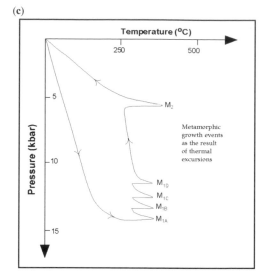

**Fig. 11.** Suggested *P–T* paths for the rocks above the Ios Detachment Fault. (**a**) A classic Alpine loop involves isothermal decompression, with the eclogite–blueschist facies rocks passing through the greenschist facies during the exhumation cycle. The survival of old $^{40}Ar/^{39}Ar$ apparent ages means that this *P–T* path is not feasible. (**b**) Cooling must take place between the period of high pressure metamorphism and the later greenschist facies overprint. The sequence of metamorphic mineral growth events take place under high pressure on the exhumation path, but *P–T* conditions might vary smoothly between each growth episode. (**c**) Schematic representation of a *P–T* path where each period of metamorphic mineral growth takes place under transiently maintained conditions (i.e. as the result of a thermal pulse). There may be a link between large-scale tectonic process and these periods of metamorphism.

## The origin of eclogite facies metamorphism

Coherent high pressure metamorphic terranes, such as the Cycladic blueschist belt, may have their genesis in the roots of mountain belts produced by significant overthickening of the continental crust during collisional orogeny. Dewey *et al.* (1993) show that there is no difficulty in attaining such a great thickness of continental crust. The metamorphic reactions that produce eclogite facies parageneses significantly increase crustal density. In addition, the total gravitational potential energy is lowered by significant thickening of the mantle lithosphere and by consequent reduction of the lithospheric geotherm.

The subsequent collapse and destruction of the mountain belt may take place in episodes, with each episode initially involving a pulse of metamorphic mineral growth. If peak metamorphism involved pervasive eclogite facies metamorphism, the first major episode of blueschist facies retrogression may be the first signal that orogenic collapse has been initiated. In this case, the timing of orogenesis in the Aegean continental crust during the Alpine–Himalaya collision needs to be re-evaluated, since the widespread appearance of Eocene ages may, in fact, not date the onset of orogenesis, but the onset of orogenic collapse.

Extensional tectonism in the Aegean may have thus commenced much earlier than has been documented. The last epoch of continental extension may be of relatively minor importance in comparison to the impact of these earlier events.

## Conclusion

The results of the detailed study of the Varvara Boudin imply that the eclogite facies and blueschist facies events are not cogenetic, but represent separate metamorphic episodes. There is no reason to ascribe high-pressure metamorphism in the Cyclades to a single transitional eclogite–blueschist facies growth event.

The M1a metamorphic event(?) involved the growth of glaucophane, white mica and epidote. It is only seen as inclusion trails in the cores of garnets in the cores of eclogitic boudins. M1a most likely represents the blueschist facies assemblage which formed as the rocks were being overthrust, before they reached peak metamorphic-grade eclogite facies conditions.

M1b is the peak (eclogite facies) metamorphic event attained by the rocks on Ios. The metamorphic event is recorded in the omphacite–garnet core of boudins in glaucophanitic schists. This period of eclogite facies metamorphism (M1b) is distinct from the period of regional blueschist facies metamorphism (M1c).

The eclogitic mineral assemblage is overprinted by a distinct and separate blueschist facies event (M1c), during which porphyroblastic growth of glaucophane, garnet, white mica and clinozoisite took place. The blueschist facies mineral assemblages grew before and during the early stages of a major deformation event.

M1b assemblages may once have been pervasive prior to the formation of the now dominant fabric. However, M1b assemblages have now been obliterated, except within the cores of relatively few massive eclogitic boudins.

The effects of the M1c blueschist facies overprint may have been pervasive and the pattern of reaction implies that fluid ingress was critical in allowing these reactions to proceed. Hydration ± metasomatism of earlier formed eclogite facies assemblages appears to have caused blueschist facies retrogression.

The findings from the Varvara Boudin on Ios can be extrapolated across the high-pressure belt in the Agean, particularly to the west (Sifnos) and northwest (Syros). Syros has distinct lenses of eclogitic mineral assemblages, with a metamorphic history the same as that from the Varvara Boudin on Ios. The larger scale implications of this work are that the period of blueschist facies retrogression (M1c)

may, in fact, mark the onset of orogenic collapse rather than the timing of collision.

## References

ANDRIESSEN, P. A. M., BANGA, G. & HEBEDA, E. H. 1987. Isotopic age study of pre-Alpine rocks in the basal units on Naxos, Sikinos and Ios, Greek Cyclades. *Geologie en Mijnbouw*, **66**, 3–14.

AVIGAD, D. 1993. Tectonic juxtaposition of blueschists and greenschists in Sifnos Island (Aegean Sea) – implications for the structure of the Cycladic blueschist belt. *Journal of Structural Geology*, **15**, 1459–1469.

—— & GARFUNKEL, Z. 1989. Low-angle faults above and below a blueschist belt – Tinos Island, Cyclades, Greece. *Terra Nova*, **1**, 182–187.

—— & ——1991. Uplift and exhumation of high-pressure metamorphic terrains: the example of the Cycladic blueschist belt (Aegean Sea). *Tectonophysics*, **188**, 357–372.

BALDWIN, S. L. 1996. Contrasting *P–T–t* histories for blueschists from the western Baja terrane and the Aegean: effects of synsubduction exhumation and backarc extension. *In*: BEBOUT, G. E., SCHOLL, D. W., KIRBY, S. H. & PLATT, J. P. (eds) *Subduction Top to Bottom*. Geophysical Monograph, **96**, 135–141.

—— & LISTER, G. S. 1999. Thermochronology of the South Cyclades Shear Zone, Ios, Greece: the effects of ductile shear in the argon partial retention zone (PRZ). *Journal of Geophysical Research*, in press.

——, ——, HILL, E. J., FOSTER, D. A. & McDOUGALL, I. 1993. Thermochronologic constraints on the tectonic evolution of active metamorphic core complexes, D'Entrecasteaux Islands, Papua New Guinea. *Tectonics*, **12**, 611–628.

BARNICOAT, A. C. 1988. The mechanism of veining and retrograde alteration of Alpine eclogites. *Journal of Metamorphic Geology*, **6**, 545–558.

BLAKE, M. C., BONNEAU, M., GEYSSANT, J., KIENAST, J. R., LEPVEIER, C., MALUSKI, H. & PAPANIKOLAOU, D. 1981. A geological reconnaissance of the Cycladic blueschist belt, Greece. *Geological Society of America Bulletin*, **92**, 247–254.

BRÖCKER, M. 1990. Blueschist-to-greenschist transition in metabasites from Tinos Island, Cyclades, Greece: compositional control or fluid infiltration? *Lithos*, **25**, 25–39.

——, KREUZER, H. & CARSWELL, D. A. 1990. Eclogite and the eclogite facies: definitions and classifications. *In*: CARSWELL, D. A. (ed.) *Eclogite Facies Rocks*. Chapman & Hall, 1–13.

——, ——, MATTHEWS, A. & OKRUSCH, M. 1993. $^{40}Ar/^{39}Ar$ and oxygen isotope studies of polymetamorphism from Tinos Island, Cycladic blueschist belt, Greece. *Journal of Metamorphic Geology*, **11**, 223–240.

BURCHFIEL, B. C., CHEN, Z., HODGES, K. V., LIU, Y., ROYDEN, L. H., DENG, C. & XU, J. 1992. The South Tibetan detachment system, Himalayan orogen; extension contemporaneous with and parallel to shortening in a collisional mountain belt. *Geological Society of America, Special Paper*, **269**.

CARSWELL, D. A. 1990. Eclogite and the eclogite facies: definitions and classifications. *In*: CARSWELL, D. A. (ed.) *Eclogite Facies Rocks*. Chapman & Hall, 1–13.

DEWEY, J. F., RYAN, P. D. & ANDERSEN, T. B. 1993. Orogenic uplift and collapse, crustal thickness, fabrics and metamorphic phase changes: the role of eclogites. *In*: PRICHARD, H. M., ALABASTER, T., HARRIS, N. B. W., NEARY, C. R. (eds) *Magmatic Processes and Plate Tectonics*. Geological Society, London, Special Publications, **76**, 325–343.

DÜRR, S., ALTHERR, R., KELLER, J., OKRUSCH, M. & SEIDEL, E. 1978. The median Aegean crystalline belt: stratigraphy, structure, metamorphism, magmatism. *In*: CLOOS, H., ROEDER, D. & SCHMIDT, K. (eds) *Alps, Appenines, Hellenides*. IUGS Report, **38**, 455–477.

ERNST, W. G. 1977. Tectonics and prograde versus retrograde *P–T* trajectories of high pressure metamorphic belts. *Rendiconti della Societa Italiana de Mineralogia e Petrologia*, **33**, 191–220.

——1988. Tectonic history of subduction zones inferred from retrograde blueschist *P–T* paths. *Geology*, **16**(12), 1081–1084.

EVANS, B. W. 1990. Phase relations of epidote–blueschist. *Lithos*, **25**, 3–23.

—— & BROWN, E. H. 1986. Blueschists and eclogites. *Geological Society of America Memoir*, **164**.

—— & ——1987. Reply to blueschists and eclogites. *Geology*, **15**, 773–775.

FORSTER, M. A. 1996. *Deformation and metamorphism of the upper plate, Ios, Cyclades, Greece*. MSc Thesis, Monash University.

—— & LISTER, G. S. 1996*a*. Detachment fault systems on Ios: Traverse 3 – Tower Hill. *In*: LISTER, G. S. & FORSTER, M. A. (eds) *Inside the Aegean Metamorphic Core Complexes: A Field Trip Guide Illustrating the Geology of the Aegean Metamorphic Core Complexes*. Australian Crustal Research Centre, Technical Publication, **45**, 41–46.

—— & ——1996*b*. Detachment fault systems on Ios: Traverse 4 – Port Beach shear zone to Koumbara Peninsula. *In*: LISTER, G. S. & FORSTER, M. A. (eds) *Inside the Aegean Metamorphic Core Complexes: A Field Trip Guide Illustrating the Geology of the Aegean Metamorphic Core Complexes*. Australian Crustal Research Centre, Technical Publication, **45**, 47–52.

—— & ——1996*c*. Detachment fault systems on Ios: Traverse 5 – Varvara Boudin to the 'Goat Beach'. *In*: LISTER, G. S. & FORSTER, M. A. (eds) *Inside the Aegean Metamorphic Core Complexes: A Field Trip Guide Illustrating the Geology of the Aegean Metamorphic Core Complexes*. Australian Crustal Research Centre, Technical Publication, **45**, 53–60.

—— & ——1999. Detachment faults in the Aegean metamorphic core complex of Ios, Cyclades, Greece. *In*: RING, U., BRANDON, M. T., LISTER, G. S. & WILLET, S. D. (eds) *Exhumation Processes:*

*Normal Faulting, Ductile Flow and Erosion*. Geological Society, London, Special Publications, **154**, 305–324.

GAUTIER, P. & BRUN, J. P. 1994. Ductile crust exhumation and extensional detachments in the central Aegean (Cyclades and Evvia Islands). *Geodynamica Acta (Paris)*, **7/2**, 57–85.

——, —— & JOLIVET, L. 1993. Structure and kinematics of Upper Cenozoic extensional detachment on Naxos and Paros (Cyclades islands, Greece). *Tectonics*, **12**, 1180–1194.

GRUJIC, D., CASEY, M., CAMERON, D., HOLLISTER, L. S., RAINER, K., PAVLIS, T. & SCHMID, S. 1996. Ductile extrusion of the Higher Himalayan Crystalline in Bhutan; evidence from quartz microfabrics. *Tectonophysics*, **260**, 21–43.

GRÜTTER, H. S. 1993. *Structural and metamorphic studies on Ios, Cyclades, Greece*. PhD Thesis, University of Cambridge.

HENJES-KUNST, F. 1980. *Alpidische einforming des preAlpidischen kristallins und seiner Mesozoischen hulle auf Ios [Kykladen, Greichenland]*. PhD Thesis, University of Braunschweig.

—— & KREUZER, H. 1982. Isotopic dating of pre-Alpidic rocks from the island of Ios (Cyclades, Greece). *Contributions to Mineralogy and Petrology*, **80**, 245–253.

HILL, E. J., BALDWIN, S. & LISTER, G. S. 1992. Unroofing of active metamorphic core complexes in the D'Entrecasteaux Islands, Papua New Guinea. *Geology*, **20**, 907–910.

——, —— & ——1995. Magmatism as an essential driving force for formation of active metamorphic core complexes in eastern Papua New Guinea. *Journal of Geophysical Research, B Solid Earth and Planets*, **100 (B6)**, 10 441–10 451.

KREUZER, H., HARRE, W., LENZ, H., WENDT, I. & HENJES-KUNST, F. 1978. K/Ar and Rb/Sr Daten von Mineralen aus dem polymetamorphen Kristallin der Kykladen, Insel Ios (Griechenland). *Fortschritte der Mineralogie*, **56**, 69–70.

LAVECCHIA, G. 1988. The Tyrrhenian–Apennines system; structural setting and seismotectonogenesis. *Tectonophysics*, **147**, 263–296.

——, STOPPA, F. & KARNER, G. 1991. Apennine compression engendered by Tyrrhenian extension; large-scale inversion tectonics of the Italian Peninsula. *Terra Abstracts*, **3**, 233.

——, BROZZETTI, F., BARCHI, M., MENICHETTI, M. & KELLER, J. V. A. 1994. Seismo-tectonic zoning in east-central Italy deduced from an analysis of the Neogene to present deformations and related stress fields. *Geological Society of America Bulletin*, **106**, 1107–1120.

LISTER, G. S. & FORSTER, M. A. (eds) 1996. *Inside the Aegean Metamorphic Core Complexes: A Field Trip Guide Illustrating the Geology of the Aegean Metamorphic Core Complexes*. Australian Crustal Research Centre, Technical Publication, **45**.

—— & RAOUZAIOS, A. 1996. The tectonic significance of a porphyroblastic blueschist overprint during Alpine orogenesis: Sifnos, Aegean Sea, Greece. *Journal of Structural Geology*, **18**, 1417–1435.

——, BANGA, G. & FEENSTRA, A. 1984. Metamorphic core complexes of Cordilleran type in the Cyclades, Aegean Sea, Greece. *Geology*, **12**, 221–225.

MALUSKI, H., BONNEAU, M. & KIENAST, J. R. 1987. Dating the metamorphic events in the Cycladic area: $^{40}$Ar/$^{39}$Ar data from metamorphic rocks of the island of Syros (Greece). *Bulletin de la Societe Géologique de France*, **8**, 833–842.

MANN, P. & GORDON, M. B. 1996. Tectonic uplift and exhumation of blueschist belts along transpressional strike-slip fault zones. *In*: BEBOUT, G. E. SCHOLL, D. W., KIRBY, S. H. & PLATT, J. P. (eds) *Subduction top to bottom*. Geophysical Monograph, **96**, 141–154.

MANCKTELOW, N. S. 1992. Neogene lateral extension during convergence in the Central Alps: evidence from interrelated faulting and backfolding around the Simplonpass (Switzerland). *Tectonophysics*, **215**, 295–317.

MATTHEWS, A. & OKRUSCH, M. 1993. $^{40}$Ar/$^{39}$Ar and oxygen isotope studies of polymetamorphism from Tinos Island, Cycladic blueschist belt, Greece. *Journal of Metamorphic Geology*, **11**, 223–240.

—— & SCHLIESTEDT, M. 1984. Evolution of the blueschist and greenschist facies rocks of Sifnos, Cyclades, Greece. *Contributions to Mineralogy and Petrology*, **88**, 150–163.

PAPANIKOLAOU, D. 1984. The three metamorphic belts of the Hellenides: a review and a kinematic interpretation. *In*: DIXON, J. E. & ROBERTSON, A. H. F. (eds) *The Geological Evolution of the Eastern Mediterranean*. Geological Society, London, Special Publications, **17**, 551–561.

——1987. Tectonic evolution of the Cycladic blueschist belt (Aegean Sea, Greece). *In*: HELGESON, H. C. (ed.) *Chemical Transport in Metasomatic Processes*. Reidel, 429–450.

PLATT, J. P. 1987. The uplift of high-pressure low-temperature metamorphic rocks. *Philosophical Transactions of the Royal Society of London*, **A321**, 87–103.

——1993. Exhumation of high-pressure rocks: a review of concepts and processes. *Terra Nova*, **5**, 119–133.

—— & ENGLAND, P. C. 1993. Convective removal of lithosphere beneath mountain belts: thermal and mechanical consequences. *American Journal of Science*, **293**, 307–336.

RIDLEY, J. 1984. Evidence of a temperature-dependent 'blueschist' to 'eclogite' transformation in high pressure metamorphism of metabasic rocks. *Journal of Petrology*, **25**, 852–870.

—— & DIXON, J. E. 1984. Reaction pathways during the progressive evolution of a blueschist metabasite: the role of chemical disequilibrium and restricted range phenomenon. *Journal of Metamorphic Geology*, **2**, 115–128.

RODGERS, J. 1984. Discussion of 'A geological reconnaissance of the Cycladic blueschist belt, Greece'. *Geological Society of America Bulletin*, **95**, 117–121.

RUBIE, D. C. 1990. Role of kinetics in the formation and preservation of eclogites. *In*: CARSWELL, D. A. (ed.) *Eclogite Facies Rocks*. Chapman & Hall, 111–140.

RUTTER, E. H. & BRODIE, K. H. 1985. The permeation of water in hydrating shear zones. *In*: THOMPSON, A. B. & RUBIE, D. C. (eds) *Metamorphic Reactions, Kinetics, Textures and Deformation*. Springer.

SCHLIESTEDT, M. 1986. Eclogite–blueschist relationships as evidenced by mineral equilibria in the high-pressure metabasic rocks of Sifnos (Cycladic Islands), Greece. *Journal of Petrology* **27**, 1437–1459.

—— & MATTHEWS, A. 1987. Transformation of blueschist to greenschist facies rocks as a consequence of fluid infiltration Sifnos (Cyclades) Greece. *Contributions to Mineralogy and Petrology*, **97**, 237–250.

SEDLOCK, R. L. 1996. Syn-subduction forearc extension and blueschist exhumation in Baja, California, México. *In*: BEBOUT, G. E., SCHOLL, D. W., KIRBY, S. H. & PLATT, J. P. (eds) *Subduction top to bottom*. Geophysical Monograph, **96**, 155–162.

URAI, J. L., SPIERS, C. J., ZWART, H. J. & LISTER, G. S. 1986. Weakening of rock salt by water during long-term creep. *Nature*, **324**, 554–557.

VAN DER MAAR, P. A. 1980a. Metamorphism on Ios and the geological history of the southern Cyclades, Greece. *Geologica Ultrajectina*, **28**.

——1980b. The geology and petrology of Ios, Cyclades Greece. *Annales Geologiques pays Helleniques*, **30**, 206–224.

—— & JANSEN, J. B. H. 1981. *Geological map of Greece, Island of Ios, 1 : 50 000 scale*. Institute for Geology and Mineral Exploration (IGME), Athens.

—— & —— 1983. The geology of the polymetamorphic complex on Ios, Cyclades, Greece, and its significance for the Cycladic massif. *Geologische Rundschan*, **72**, 283–299.

VANDENBERG, L. C. & LISTER, G. S. 1996. Structural analysis of basement tectonites from the Aegean metamorphic core complex of Ios, Cyclades, Greece. *Journal of Structural Geology*, **18**, 1437–1454.

WIJBRANS, J. R. & McDOUGALL, I. 1986. $^{40}$Ar/$^{39}$Ar dating of white micas from an Alpine high pressure metamorphic belt on Naxos (Greece): the resetting of the argon isotopic system. *Contributions to Mineralogy and Petrology*, **93**, 187–194.

—— & ——1988. Metamorphic evolution of the Attic Cycladic metamorphic Belt on Naxos (Cyclades, Greece) utilising the $^{40}$Ar/$^{39}$Ar age spectrum measurements. *Journal of Metamorphic Geology*, **6**, 571–594.

——, SCHLIESTEDT, M. & YORK, D. 1990. Single grain argon laser probe dating of phengites from the blueschist to greenschist transition on Sifnos (Cyclades, Greece). *Contributions to Mineralogy and Petrology*, **104**, 582–594.

ZHANG, S., COX, S. F. & PATERSON, M. S. 1995. The influence of room temperature deformation on porosity and permeability in calcite aggregates. *Journal of Geophysical Research*, **99**, 15 761–15 775.

# The Northern Sacramento Mountains, southwest United States. Part I: Structural profile through a crustal extensional detachment system

V. PEASE[1] & J. ARGENT[2]

Department of Earth Sciences, University of Oxford, Parks Road, Oxford OX1 3PR, UK
[1] Present address: Department of Earth Sciences, Uppsala University, Villavägen 16, S-752 36 Uppsala, Sweden
[2] Present address: Amerada Hess Ltd, 33 Grosvenor Place, London SW1X 7HY, UK

**Abstract:** Miocene exhumation of the Sacramento Mountains metamorphic core complex in the Colorado River extensional corridor, southwest United States, produced a range of deformational fabrics: pervasive, synthetic ductile deformation (with associated syntectonic intrusion) occurred at depth, discrete antithetic ductile through to brittle extensional shear zones developed at intermediate crustal levels and brittle cataclasis occurred at upper crustal levels. The region divides into three structural domains on the basis of these fabrics: (1) the eastern domain is dominated by a penetrative, ductile fabric with a consistent mineral alignment and stretching lineation, and sense of shear synthethic with detachment fault motion (top-to-northeast); (2) the central domain is characterized by a continuum of overprinting ductile through to brittle extensional shear zones with motion antithetic (top-to-southwest) to detachment faulting; and (3) the western domain, where deformation occurred entirely by brittle cataclasis. Structural and petrofabric data from the Sacramento Mountains suggest that increasing structural depth/temperature is exposed in the direction of tectonic transport (N60E), consistent with detachment faulting. This region is interpreted as a profile through a crustal-scale shear zone in which the central domain of discrete antithetic shear zones formed in response to a combination of non-coaxial shear and vertical shortening, and represents an exhumed brittle–ductile transition. Rapid uplift and cooling of the northern Sacramento Mountains in the Miocene facilitated the preservation of this relatively transient regime.

Deformation associated with the exhumation of metamorphic core complexes has received considerable attention since their recognition as a principle feature of extreme continental extension in the late 1970s (e.g. Crittenden *et al.* 1980). However, with the exception of a few studies (e.g. Wernicke & Axen 1988; Bartley *et al.* 1990; Daniel *et al.* 1994; Manning & Bartley 1994; Fletcher *et al.* 1995), most projects have been concerned with the geometry of the detachment faults or hanging-wall structure and stratigraphy. The structures associated with the internal deformation of the footwall have, until recently, received little detailed attention. Features that have commonly been observed within the footwalls to major detachment faults include: the formation of a penetrative, co-linear mylonitic fabric (Lister & Davis 1989), subsequent folding of this fabric (Reynolds & Lister 1990), brittle layer-parallel contraction with the formation of subvertical brittle shear zones (Spencer 1984; Wernicke & Axen 1988; Axen & Wernicke 1991; Manning & Bartley 1994; Fletcher *et al.* 1995) and antithetic shear zones

(Reynolds & Lister 1990). Late-stage, post-mylonitic structures have been attributed to either isostatic rebound of the footwall during regional-scale unroofing (Lister & Davis 1989; Reynolds & Lister 1990; Spencer 1984) or footwall flexure as it passes through a 'rolling hinge' (Bartley *et al.* 1990; Manning & Bartley 1994; Fletcher *et al.* 1995; Axen & Bartley 1997).

The Sacramento Mountains metamorphic core complex in the Mojave Desert of southeast California provides an opportunity to view extensional deformation in the continental lithosphere with increasing structural depth/temperature. Traversing northeastward through the footwall, in the direction of tectonic transport, deeper structural levels of the extended continental crust are exposed. This paper documents a continuum of penetrative, top-to-the-northeast, ductile deformation with associated syntectonic intrusion; discrete, top-to-the-southwest, ductile through to brittle extensional shear zones; and brittle cataclasis. All fabrics developed during the footwall exhumation of the Sacramento Mountains core complex. A remarkable profile

*From*: MAC NIOCAILL, C. & RYAN, P. D. (eds) 1999. *Continental Tectonics*. Geological Society, London, Special Publications, **164**, 179–198. 1-86239-051-7/99/$15.00 © The Geological Society of London 1999.

through a crustal-scale detachment fault system
is preserved and suggests that discrete ductile
through brittle extensional shear zones, developed directly below the detachment fault in the
northern Sacramento Mountains, result from
deformation within the brittle–ductile transition
zone at the onset of extension.

## Regional setting

Cenozoic extension, effectively doubling the
width of the Basin-and-Range Province in the
United States, has been accommodated by
numerous 'tilt domains' (Stewart 1980). The
Sacramento Mountains are located within
the Colorado River extensional corridor (Howard & John 1987), one such tilt domain in the
southern Basin-and-Range (Fig. 1). In contrast
to northern regions, where active extensional
faulting continues to this day, the southern Basin
and Range is generally seismically quiescent. The
Colorado River extensional corridor, stretching

from southern Nevada to western Arizona, encompasses a belt of extreme mid-Tertiary extension accommodated on a series of low-angle
(<30°), northeast dipping, regional-scale, normal
faults (Howard & John 1987). These *detachment
faults* are exposed along the flanks of metamorphic core complexes of the Dead, Homer,
Sacramento, Chemehuevi, Whipple, Buckskin
and Harcuvar Mountains (Fig. 1). The core
complexes are terranes of exhumed Precambrian
basement intruded by numerous Cretaceous and
Tertiary plutonic suites, juxtaposed against a
hanging wall of tilted fault blocks, mostly
comprised of Tertiary clastic sediments and
volcanic rocks. The tectonic transport direction of the hanging wall to these detachment
faults is consistently northeastward throughout the extensional corridor (Howard & John
1987). The onset and cessation of extension
across the corridor is reasonably well constrained
by the Miocene sedimentary record (Nielson
& Beratan 1995) and thermochronologic cooling
studies of the exhumed crystalline footwall

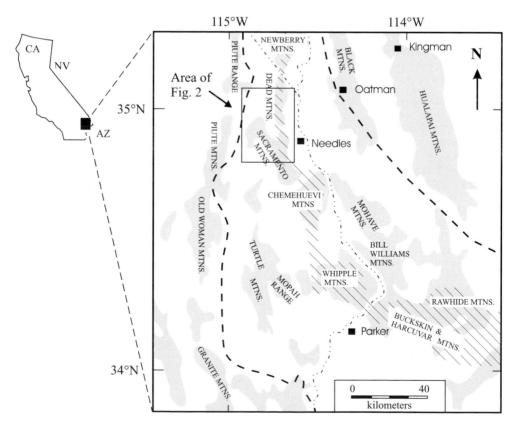

**Fig. 1.** Location of the Sacramento Mountains within the Colorado River extensional corridor (bold dashed lines) of Howard & John (1987). Nearby ranges and axial zone of metamorphic core complexes (light hatchure) also shown.

terranes (Foster *et al.* 1990; John & Foster 1993; Pease *et al.* 1999), commencing at *c.* 24 Ma and ceasing by *c.* 12 Ma.

## Geology of the Sacramento Mountains

The Sacramento Mountains exhibit characteristics typical of a 'Cordilleran-style' metamorphic core complex (cf. Crittenden *et al.* 1980). A basal detachment fault skirts around the flanks of the range, separating a hanging wall of mid-Miocene, synextensional sedimentary and volcanic strata, from a crystalline footwall comprising Precambrian and Cretaceous schists and gneisses intruded by a Miocene plutonic suite and associated dykes (Fig. 2). A northeasterly (N60E) tectonic transport direction of the hanging wall is clearly defined by striations on the detachment surface, stratal and fault dips in the hanging wall, and transport parallel corrugations of the basal detachment fault surface (McClelland 1984; Spencer 1985).

The Sacramento Mountains detachment fault (SMDF) system is composed of a number of imbricate low-angle detachment faults (Spencer & Turner 1983; McClelland 1984). The development of multiple generations of detachment faults has also been recognized in the Chemehuevi Mountains (John 1987) and in the Whipple Mountains (Davis & Lister 1988). In the northern Sacramento Mountains, these faults are poorly exposed and their cross-cutting relationships have been inferred from outcrop pattern and structural position. The basal detachment fault of the SMDF system separates autochthonous footwall rocks (exposed in the northwest and northeast part of the range) from an intervening low-lying region of mid-Miocene sedimentary and volcanic strata, and their depositional basement (Fig. 2). The hanging wall is segmented into three allochthonous plates by a number of low-angle faults, all of which are truncated by the basal detachment (Spencer & Turner 1983; McClelland 1984). The structurally lowest of these hanging wall plates comprises poorly exposed and deeply weathered depositional basement of Precambrian porphyritic granite, which intrudes highly deformed pelitic schists and gneisses. The middle hanging wall plate, with a total stratigraphic thickness of *c.* 300 m (Spencer & Turner 1983), consists dominantly of Tertiary pyroclastic flows, volcanoclastic sandstone and, locally, thin bands of lacustrine carbonate. A welded ash flow within this middle plate contains phenocrysts of adularescent sanidine and accessory sphene, characteristic of the regionally correlative Peach

Springs Tuff of Young & Brennan (1974) which has been dated at $18.5 \pm 0.2$ Ma ($^{40}Ar/^{39}Ar$ sanidine; Nielson *et al.* 1990). The third, and structurally highest, allochthonous plate of the hanging wall comprises a section of tilted Tertiary boulder conglomerate, breccia, sandstone and mudstone. This uppermost plate is a highly faulted, discontinuous section of coarse, red, clastic sediments with an estimated stratigraphic thickness of 1 km (Fedo & Miller 1992). Fedo & Miller (1992) propose that these sediments were deposited in an active asymmetric graben developed synchronous with detachment faulting. This sequence of sediments is overlain by a flat-lying, $14.6 \pm 0.2$ Ma (K–Ar whole rock; Spencer & Turner 1983), andesitic flow which forms the prominent plateau of Flattop Mountain (Fig. 2). This untilted andesite and the extrusion of the Eagle Peak rhyodacite onto the exhumed fault surface of the SMDF at *c.* 14.3 Ma (K–Ar biotite and sanidine; Simpson *et al.* 1991) define the cessation of low-angle faulting in the western region of the Sacramento Mountains to be *c.* 14 Ma.

The autochthonous footwall to the basal SMDF includes the regionally correlative Fenner Gneiss, a coarse-grained, porphyritic biotite granite (Hazzard & Dosch 1937; Bender *et al.* 1990). In the northwest Sacramento Mountains, the Fenner Gneiss contains minor screens of older Precambrian gneisses and schists, and comprises most of the exposure north of Interstate 40 (Fig. 2). South of Interstate 40, Late Cretaceous granodioritic gneiss and schists ($72 \pm 0.4$ Ma, U–Pb zircon; Pease *et al.* 1999) are the dominant lithologies, and are interpreted to represent the easternmost limit of Laramide orogenesis. Leucogranite sheets and granitic plutons intruding these gneisses and schists were also variably deformed during the Late Cretaceous. Undeformed, two-mica garnet granites ($73 \pm 27$ Ma, U–Pb zircon and monazite; Pease *et al.* 1999) also intrude these gneisses and schists, but postdate Laramide tectonism.

In the northeast Sacramento Mountains, basement lithologies below the basal SMDF include Precambrian and Cretaceous rocks which are intruded by a voluminous Miocene plutonic suite. The calc-alkaline, metaluminous Eagle Wash Intrusive Complex (EWIC after Schweitzer 1991) is predominantly granodiorite, dated at *c.* 19 Ma (SHRIMP U–Pb zircon and U–Pb sphene; Pease *et al.* 1999), with lesser amounts of leucogranite, diorite and granite (Pease 1997). In regions of low strain, magma mingling textures between all compositional phases of the EWIC are preserved and suggest that all compositional phases are relatively contemporaneous

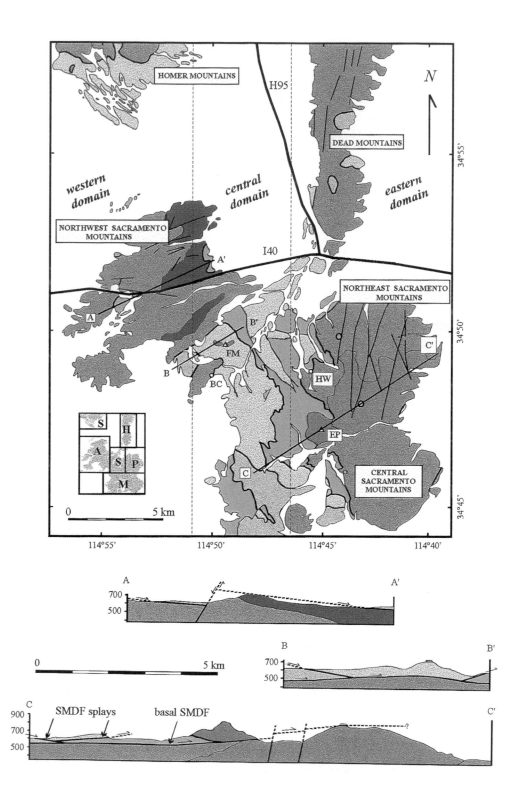

(Pease 1997). The EWIC generally has a well-developed protomylonitic to mylonitic fabric with abundant xenoliths and rafts of Cretaceous(?) and Precambrian gneisses and schists aligned with this mylonitic foliation.

In both footwall regions, basement lithologies have been intruded by Miocene dyke swarms. In the northwest Sacramento Mountains, east–west striking, subvertical dykes cross-cut the footwall. These dykes are best observed south of Interstate 40 where resistant, silicic compositions form prominent ridges. The swarm is generally porphyritic and ranges in composition from basalt to rhyolite (Argent 1996). Typical of calc-alkaline magmatic suites, the dyke swarm is principally metaluminous, becoming marginally peraluminous with increasing silica content, and has a consistently high (>2.0 wt%) potassium oxide content (Argent 1996). Comparable east–west striking, subvertical dykes in the Piute Range and Homer Mountains yield dates of 17–19 Ma (K–Ar biotite; Spencer 1985). This suggests that the dykes were emplaced during the window of active extension, although not orthogonal to the regional tectonic transport direction of N60E.

In the northeast Sacramento Mountains, undeformed, subvertical dykes of basaltic to rhyolitic composition strike north–south, following the trend of local, near-vertical faults. A similarly oriented dyke in the central Newberry Mountains yielded a date of c. 18.5 Ma (K–Ar biotite; Spencer 1985). A dyke compositionally correlative with the Eagle Peak rhyodacite, and therefore c. 14.3 Ma (Simpson et al. 1991), cuts across the entire footwall in a roughly north–south orientation. These data suggest that dyke intrusion in the northeast region occurred between 18–14 Ma, also contemporaneous with detachment faulting.

Miocene deformation of the northern Sacramento Mountains footwall can be divided into three structural domains (Fig. 2). These three domains show a spatial and temporal progression from ductile through to brittle deformation. In the northeast Sacramento Mountains, the eastern domain, footwall deformation is dominated by a penetrative, top-to-the-northeast mylonitic fabric. This is overprinted by brittle structures associated with Miocene footwall exhumation. In the westernmost part of the range, the western domain, ductile deformation associated with Miocene exhumation is entirely absent. Instead, Miocene extensional deformation is solely accommodated by brittle fracturing and dyke emplacement. The central domain separates these two contrasting structural regions, where discrete, top-to-the-southwest shear zones change from ductile to brittle in character. The basal SMDF truncates all of these footwall fabrics.

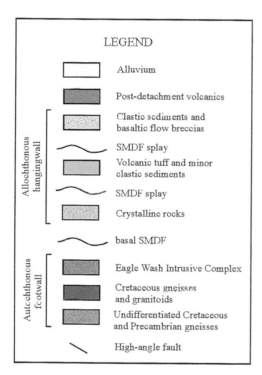

Fig. 2. Simplified geologic map of, and cross-sections through, the northern Sacramento Mountains. Compiled from: S, Spencer & Turner (1983); M, McClelland (1984); A, Argent (1996); P, Pease (1997); H, K. Howard (pers. comm.). Triangles represent local peaks: FM, Flattop Mountain; EP, Eagle Peak. Site locations: BC, Browns Camp; HW, Hacienda Wash. Interstate 40 (I40) and Highway 95 (H95) indicated. Western structural domain, central structural domain and eastern structural domain are shown. Lines of simplified cross-sections: A–A′, B–B′ and C–C′. Note expanded scale on cross-sections to show geologic detail.

## Top-to-the-northeast, non-coaxial, ductile deformation

The northeast Sacramento Mountains is dominated by a pervasive mylonitic deformation of footwall rocks. This protomylonitic to ultramylonitic (Sibson 1977; Wise et al. 1984) fabric is well developed in the c. 19 Ma EWIC, and variably overprints earlier fabrics associated with Cretaceous(?) and Precambrian lithologies. Strain in this region is notably heterogeneous,

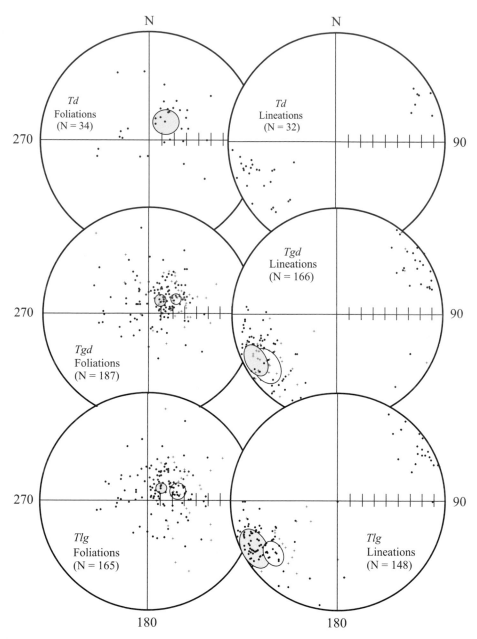

**Fig. 3.** Eastern domain equal-area stereonets (lower hemisphere projection). Foliations associated with EWIC lithologies (diorite, *Td*; granodiorite, *Tgd*; leucogranite, *Tlg*) plotted as poles and lineations as trend and plunge. Data from the west side of the eastern domain is represented by crosses and the open $\alpha_{95}$, the east side by dots and the shaded $\alpha_{95}$.

increasing eastwards, with the highest relative strain observed occurring at lithologic contacts between units and along generally subparallel ultramylonitic shear zones.

The mylonitic fabric is characterized by a well-developed, gently southwest dipping foliation (145°, 20°SW) and a subhorizontal, southwest plunging, mineral stretching lineation (240°, 15°). All compositional phases of the EWIC display this fabric (Fig. 3). The foliation is defined by a millimetre-scale segregation of tabular plagioclase and amphibole. The consistently oriented lineation, defined by a shape-preferred orientation of amphibole and stretching of biotite along the basal (0 0 1) cleavage plane, becomes more pronounced to the east. Tight to isoclinal, small scale folds are present within the EWIC only on the east side of the range (Pease 1997).

Mylonitic deformation in the northeast Sacramento Mountains followed a number of earlier events. Pre-Tertiary deformation associated with Precambrian and Cretaceous lithologies is manifest as a transposed schistosity or gneissic layering (Pease 1997). Precambrian rocks on the west side of the northeast Sacramento Mountains dip moderately to the west (170°, 30°W), whereas Precambrian rocks in the eastern region have the same orientation as EWIC lithologies, i.e. 150°, 10°SW. This subparallelism with the mylonic fabric is manifest by the gradual transposition from west to east of the more steeply, west dipping mylonitic foliation, to the shallower, southwest dipping orientation. This is interpreted to reflect the rotation of an older fabric(s) associated with the pre-Tertiary lithologies into alignment with the Miocene ductile fabric. Pre-Tertiary, amphibolite facies fabrics are variably preserved in the northest Sacramento Mountains. The original orientation of this older fabric(s) is not known, but the mineralogy associated with its higher metamorphic grade can be recognized: Precambrian schist preserves a relict upper amphibolite facies assemblage of sillimanite + biotite + garnet + quartz + plagioclase (Pease 1997). Recrystallization and polygonization of amphibole, plagioclase and quartz are associated with Cretaceous amphibolite and felsic layers of Precambrian schist. Whether the higher metamorphic grade represents original Proterozoic $P-T$ conditions or a later (Mesozoic) thermotectonic event is unknown (e.g. Hoish et al. 1988; Foster et al. 1992). Pre-Tertiary lithologies also preserve, in regions of low Tertiary strain, structures which record a more complex, multi-stage deformation history, i.e. small-scale (<5 m) sheath and refolded folds, irregular and ptygmatic folds, and recrystallized quartz veins.

Shear sense indicators in the northeast Sacramento Mountains include $\sigma$- and $\delta$-porphyroclasts, C-S fabrics, deformed veinlets, and parasitic folds (Simpson & Schmid 1983; Passchier & Simpson 1986; Hanmer & Passchier 1991). The majority (71%) of shear sense indicators show top-to-the-northeast motion, regardless of lithology or geographic distribution (Pease 1997). This suggests that deformation occurred under conditions of dominantly non-coaxial shear directed N60E, consistent with the regional tectonic transport direction. Less than 1% (five

of 415) of the samples evaluated in the northeast Sacramento Mountains show top-to-the-southwest shear sense. The shear direction could not be determined for a significant proportion (22%) of the samples due to: (1) grain-to-grain interference resulting in poorly defined fabrics; or (2) the presence of opposing shear directions.

## Metamorphic conditions associated with top-to-the-northeast deformation

The EWIC has a well-developed mylonitic fabric. The temperature and pressure associated with the emplacement of the granodioritic phase of the EWIC is $680 \pm 19°C$ at $3.3 \pm 0.75$ kbar (Pease et al. 1999). Pease et al. (1999) have demonstrated that the EWIC was intruded at c. 19 Ma and cooled to temperatures c. <105°C by c. 14 Ma, well below the temperatures necessary for pervasive mylonite fabric development. Consequently, mylonitic deformation in the EWIC must have occurred between c. 19 Ma and c. 14 Ma.

The conditions of mylonitization can be inferred by comparing phase mineralogy and fabrics from the deformed rocks to their undeformed equivalents preserved in domains of low strain (Anderson 1988). In undeformed Tgd, the stable mineral assemblage is hornblende + plagioclase + biotite ± potassium feldspar + accessory sphene, magnetite, apatite and zircon. Thus, undeformed Tgd reflects crystallization consistent with amphibolite facies conditions. Deformation of the EWIC apparently occurred entirely in the subsolidus field, as magma mixing and mingling textures are preserved in domains of low strain, but prior to the exhumation of the footwall. Amphibole may be fractured parallel to the (0 1 1) cleavage plane (Fig. 4), be bent or kinked, or display grain–boundary sliding. Cataclastic deformation of hornblende is widespread up to 600°C (Brodie & Rutter 1985). Undulose extinction, deformation bands and subgrain recrystallization along deformation bands in plagioclase (Fig. 5), suggest temperatures >450°C (Brodie & Rutter 1985; Simpson 1985). Quartz generally displays well-developed dynamic recrystallization fabrics (Fig. 6). Fine grained ribbons, undulose extinction, parallel and elongate subgrains, etc., generally indicate temperatures >350°C, although may also occur under lower greenschist conditions (Tullis 1983; Simpson 1985). Thus, deformation associated with mylonite development appears to record upper- to middle-greenschist facies conditions, i.e. temperatures

**Fig. 4.** Eastern domain microfabrics: stretched amphibole in *Tgd*. Cataclastic deformation in amphibole is widespread up to 600°C (Brodie & Rutter 1985). Field of view, 1.35 mm.

**Fig. 5.** Eastern domain microfabrics: low-temperature plasticity of plagioclase in *Tgd*. Undulose extinction, deformation bands and subgrain recrystallization along deformation bands suggest temperatures in excess of >450°C (Simpson 1985; Brodie & Rutter 1985). Field of view, 3.5 mm.

**Fig. 6.** Eastern domain microfabrics: dynamic recrystallization of quartz in *Tgd*. Fine-grained ribbons, undulose extinction, parallel and clongate subgrains generally indicate temperatures >350°C (Tullis 1983; Simpson 1985). Field of view, 3.5 mm.

>450°C (low-temperature plasticity of plagioclase) but <600°C (brittle deformation of hornblende). These higher temperature features are overprinted by variably developed brittle fracturing of feldspar, sphene and apatite. The long crystallographic axes in these minerals are aligned parallel to the stretching lineation, with extensional microfractures developed perpendicular to the long axes. These brittle microfractures are filled with quartz (locally epidote) rather than feldspar. Although the complex relationship between temperature, pressure, fluid pressure, composition, grain size and strain rate controlling fabric development allows a rock, at a given temperature, to deform in a ductile manner at a low strain rate and in a brittle manner at a high strain rate, Pease *et al.* (1999) have shown that the EWIC cooled rapidly. It is therefore suggested that this brittle microdeformation developed under lower temperature conditions, i.e. <450°C, and hence below temperatures allowing low-temperature plasticity in feldspar.

It is worth noting that the strongest textural evidence for the syntectonic emplacement of plutons is the preservation of a continuous transition from submagmatic to high-temperature, solid-state deformation (Patterson *et al.* 1989). Many plutons emplaced during regional deformation do not preserve such evidence, however, because of the transitory nature of the submagmatic state and the obscuring effects of post-emplacement deformation (Miller & Paterson 1994). Miller & Paterson (1994) suggest that syntectonic fabrics are most likely preserved in plutons cooled at slow to moderate rates, deformed at high strain rates during emplacement, or intruded during the waning stages of regional deformation. The EWIC was intruded relatively early during the regional deformation and then cooled rapidly (Pease *et al.* 1999), circumstances not conducive to the preservation of syntectonic fabrics.

## Transitional zone of top-to-the-southwest, non-coaxial deformation

As previously mentioned, pervasive ductile deformation is not present in the footwall of the northwest Sacramento Mountains. Ductile deformation in the central domain is predominantly associated with mid-Miocene dykes in a narrow zone immediately below the basal SMDF, as a series of ductile through to brittle shear zones. These discrete ductile and brittle structures are overprinted by brittle cataclasis and pervasive fracturing directly associated with ongoing detachment faulting.

Near the basal SMDF in the central domain, the east–west trending, subvertical Tertiary dyke swarm dramatically changes orientation to a moderate southwestward dip and develops an internal, wall-parallel fabric. These southwest dipping dykes exhibit a range of foliation types from tectonically induced 'mylonitic' fabrics, to those produced primarily by magmatic flow. Mylonitic fabrics within these dykes and shear zones consistently indicate top-to-the-southwest shear on a southwest dipping mylonitic foliation. Kinematic indicators are, without exception, opposite to the northeast directed shear associated with detachment faulting. Ductile through to brittle extensional shear zones are also developed within the country rock, paralleling these sheared dykes and preserving the same kinematics. The ductile shear zones are consistently overprinted by brittle shear zones of a similar orientation and shear sense. This contrasts with previous regional syntheses that have considered the presently exposed levels of

footwall in the northwest Sacramento Mountains to be pervasively mylonitized during Tertiary extension (Davis *et al.* 1980; Davis & Lister 1988; Lister & Davis 1989; John & Mukasa 1990).

### Ductile shear zones

The first extensional structures to develop within the central domain are localized ductile extensional shear zones (Fig. 7). These are discrete, continuous shear zones <0.15 m wide. The displacement across each of these shear zones is estimated to be 0.50–1.0 m using offset granitic veins and compositional banding within the Cretaceous granodioritic gneiss; they dip moderately to the west (173°, 30°W; Fig. 8). Mineral stretching lineations generally plunge south-southwest (207°, 26°; Fig. 8), oblique to the shear plane foliation. Kinematic indicators (including the deflection of the foliation into the shear zone, asymmetric σ-porphyroclasts of feldspar, mica fish, asymmetric extensional shear bands and inclined quartz mosaics) are consistently top-to-the-southwest, opposite to the motion of detachment faulting. Within these shear zones, quartz is strung out into sinuous

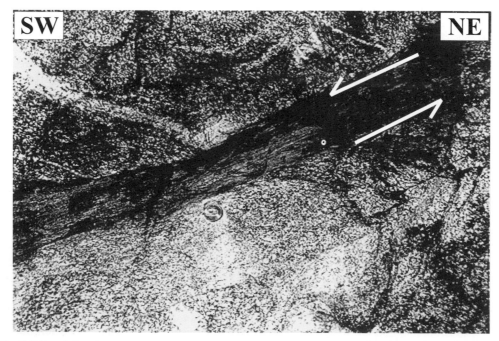

**Fig. 7.** Central domain ductile shear zone. In Cretaceous granodioritic gneiss north of Interstate 40, the sweep of the foliation into the discrete, continuous shear zone indicates a top-to-the-southwest sense of shear (opposite to the overriding detachment fault). Coin for scale (diameter 20 mm).

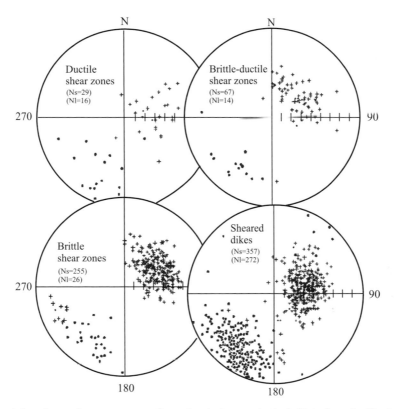

**Fig. 8.** Central domain equal-area stereonets (lower hemisphere projection). Data from ductile shear zones, brittle–ductile shear zones, brittle fractures and sheared dykes shown. Poles to shear zones represented by crosses and associated mineral stretching lineation (trend and plunge) by dots.

ribbons of very fine-grained mosaics or extremely fine individual elongated grains, inclined to the S-planar fabric with irregular consertal grain boundaries and undulose extinction patterns. Ribbons are drawn out parallel to the C-plane. Feldspar, usually completely sericitized, forms asymmetric $\sigma$-porphyroclasts. Porphyroclast tails comprise fragments from the original clast, sinuous quartz ribbons, and 'matted' intergrowths of biotite and chlorite ribbons. Biotite and chlorite can also occur as individual laths aligned to form the intervening S-planes. Both have a strong preferred optical orientation. Epidote and calcite occur, infilling microfractures within feldspar and as fine, anhedral grains intergrown with biotite ribbons.

### Brittle–ductile shear zones

Brittle–ductile shear zones differ from the ductile shear zones in that they form discrete fracture surfaces with bounding wall rocks showing duc-

tile, normal drag deformation (Fig. 9). The displacement across individual shear zones ranges from 0.15 to 0.50 m. Brittle–ductile shear zones dip moderately to the southwest (142°, 31°SW; Fig. 8). Mineral stretching lineations on the fracture plane generally plunge to the southwest (222°, 33°; Fig. 8). Kinematic indicators, including normal drag of the foliation towards the fracture and displaced aplite veins, consistently indicate a top-to-the-southwest sense of shear.

### Brittle fractures

Brittle fractures with minor displacements (0.01–0.1 m) are by far the most common manifestation of top-to-the-southwest deformation. The orientation and kinematics associated with brittle shear zones are identical to the ductile and brittle–ductile shear zones (Fig. 8). These fractures dip moderately to the southwest (150°, 32°SW). Striations on the fracture plane generally plunge to the south-southwest (212°, 28°).

**Fig. 9.** Central domain brittle–ductile shear zone. In Cretaceous granodiortic gneiss north of Interstate 40, the sweep of the foliation into the discrete shear zone indicates a top-to-the-southwest sense of shear, opposite to the overriding detachment fault.

The offset of compositional layers within the granodioritic gneiss along these fractures indicate a consistent top-to-the-southwest sense of shear.

## Sheared dykes

The occurrence of sheared dykes is the most dramatic feature of the central structural domain and has not been previously documented within the Colorado River extensional corridor. These sheared dykes, basaltic to rhyolitic porphyries, are compositionally indistinguishable from their undeformed equivalents (Argent 1996). The absence of cross-cutting relationships between the deformed and undeformed dykes suggests that they are contemporaneous. The most striking feature associated with these dykes is the distinct strain partitioning between the dykes and their host rocks. In most exposures, strain is taken up entirely by the dyke, which develops a penetrative, wall-parallel magmatic or proto-mylonitic to ultramylonitic foliation (Fig. 10). The sheared dykes dip consistently southwestward. In most cases, the internal foliation is subparallel with the dyke walls, but is occasionally observed oblique to the dyke walls. Lineations relating to this fabric plunge consistently to the south-southwest (214°, 26°; Fig. 8). Where the foliation is oblique to the dyke walls, the dykes tend to be relatively steeply dipping (>40°). In these more steeply dipping dykes, the foliation is orientated at a shallower angle and sweeps sigmoidally into a steeper orientation towards the walls of the dyke. Foliation planes are defined by the preferred orientation of biotite and chlorite. The lineation can be either a mineral alignment lineation defined by the orientation of hornblende or feldspar long axes, or a mineral stretching lineation defined by feldspar, quartz aggregate trails, and elongate clots of microcrystalline biotite and chlorite. In addition to these mineral stretching and alignment lineations, highly strained dykes with a more schistose fabric possess a crenulation lineation and cleavage. The crenulation lineations are consistently oriented perpendicular to the mineral stretching and alignment lineations. Kinematic indicators (including $\sigma$- and $\delta$-porphyroclasts, sweep of the foliation, and asymmetric extensional shear bands) define a consistent top-to-the-southwest sense of shear.

The foliation developed in the sheared dykes is heterogeneous and varies in intensity. Dykes with a weakly developed foliation are occasionally less intensely deformed at the dyke margin

**Fig. 10.** Central domain sheared dyke. Strain is completely partitioned onto this dykelet, which intrudes a brittle fracture. Note the development of a wall-parallel foliation within the dyke. Top-to-the-southwest shear from porphyroclasts is obvious in thin-section. Coin for scale (diameter 20 mm).

than in the dyke interior. A parallel, penetrative mylonitic fabric is sometimes developed in the country rock immediately adjacent to mylonitized dykes (<0.25 m from the dyke wall). Discrete ductile shear zones or fractures may also develop proximal to dyke margins. Xenoliths of country rock assimilated by dykes are either completely mylonitized and show well-developed C-S fabrics and asymmetric extensional shear bands (Fig. 11), or have remained completely undeformed, with the magmatic to protomylonitic fabric of the dyke appearing to 'flow' around the xenolith. The southwest dipping fractures that have dykes intruding along them also show greater displacements (up to 1.5 m). These observations suggest that the intrusion of the dykes was synchronous with the development of both the ductile and brittle shear zones, and that these discontinuities were used as conduits for dyke propagation. Thus, the dykes were emplaced during active extensional deformation within a zone of top-to-the-southwest, non-coaxial shear.

Both magmatic flow and solid state processes appear to be responsible for the fabrics preserved in the dykes. Magmatic flow is defined as deformation by the displacement of melt, with consequent rigid-body rotation of phenocrysts, without sufficient interference between them to cause plastic deformation (Paterson *et al.* 1989). Fabrics formed by magmatic flow can be recognized by a number of microstructural features: the main indication is the presence of a preferred orientation of primary igneous phases which show no evidence of plastic deformation or recrystallization of the aligned phenocrysts or interstitial groundmass (Paterson *et al.* 1989). For this to occur, sufficient melt must be present during deformation to allow the phenocrysts to rotate freely into alignment. In the case of these porphyritic dykes, euhedral plagioclase and orthoclase feldspar phenocrysts showing oscillatory zoning and lamellar twinning are undeformed, aligned with the foliation, and form a weak mineral alignment fabric. In addition, the consistent orientation of acicular hornblende phenocrysts defines a common mineral alignment lineation in these dykes. This foliation is also deflected around undeformed xenoliths, a further indication of a magmatic fabric. For the most part, wall rock appears undeformed with respect to the Miocene deformation, with strain partitioned completely onto the dykes. Such comprehensive strain partitioning suggests that,

**Fig. 11.** Central domain sheared xenolith. Mylonitized Cretaceous granodioritic gneiss within a sheared dyke. A top-to-the-southwest sense of shear is indicated by $\sigma$-porphyroclasts and asymmetric extensional shear bands within the xenolith. Scale bar, 5 cm.

during deformation, the dykes were much less competent than their host rocks, i.e. at a higher temperature.

Solid-state deformation is also recorded in these dykes, as indicated by $\sigma$- and $\delta$-porphyroclasts of feldspar, plastic deformation, and recrystallization of quartz and biotite, C-S fabrics in xenoliths (see above), etc. All of these define a top-to-the-southwest sense of shear, consistent with the observed offsets of wall-rock markers across the dykes and shear sense associated with the ductile through to brittle shear zones. Thus, these dykes preserve fabrics generated by magmatic flow, as well as by solid-state deformation which completly overprints the original magmatic textures. Both of these processes were operating during the emplacement and deformation of the dykes within the central domain.

### Metamorphic conditions associated with top-to-the-southwest deformation

Metamorphic conditions during the formation of these top-to-the-southwest shear zones is inferred from microstructural textures (the solid-state deformation of quartz and feldspar), as well

as the growth of new metamorphic phases within the shear zones. Quartz was dynamically recrystallized into microcrystalline ribbons, whereas feldspar deformed principally by brittle fracturing and minor recrystallization. This was accompanied by the growth of small laths of biotite, chlorite and epidote. These observations indicate that lower greenschist facies conditions ($350-400°C$ at $2-3$ kbar) prevailed during deformation (Simpson 1985).

Pease *et al.* (1999) show that, near the onset of extension, the maximum temperature of the footwall in the central domain was between $c.\,315$ and $510°C$. This is consistent with the lack of pervasively developed mylonites within the footwall of the northwest Sacramento Mountains and with the development of discrete ductile to brittle–ductile shear zones and fractures in this domain. The brittle fractures developed at lower temperatures than the ductile and brittle–ductile structures. Although deformation in a ductile regime may be transformed to a brittle regime if the strain rate is increased, it is clear that the region cooled rapidly (Pease *et al.* 1999). Consequently, an increase in strain rate, although permissible, is not required and given the cooling history of the region seems unlikely. Metamorphic conditions during the development of the central structural domain are, therefore, considered to be lower greenschist, certainly within the temperature range for semi-ductile to brittle deformation.

## Brittle deformation

In the westernmost part of the northern Sacramento Mountains, Miocene extensional deformation is accommodated along low-angle faults, high-angle faults, fractures, joints, tension gashes and by dyke emplacement. Ductile deformation is entirely absent. These brittle fabrics overprint the pervasive mylonitic fabric in the eastern domain, as well as the brittle–ductile fabrics in the central domain.

The basal SMDF (Fig. 2), best exposed west of Browns Camp (114°49.37 W. Lat., 34°48.75 N. Long.) and along Hacienda Wash (114°46.86 W. Lat., 34°50 N. Long.), currently dips ≤25°. Near Browns Camp, the gouge zone is c. 1 m thick with principal and conjugate Riedel fractures and synthetic shears consistent with top-to-the-northeast transport of the hanging wall relative to the footwall. Striations on the fault surface plunge gently northeast and southwest (061°, 17° and 241°, 9°), consistent with striae on detachment faults throughout the Colorado River extensional corridor (McClelland 1984; John 1987; Davis & Lister 1988), which indicate a regional tectonic transport direction of N60E. Rocks below the basal SMDF form a breccia zone of varying thickness from 1 to 20 m, constituting fault gouge, microbreccia, breccia and cataclasites. Using the classification of Sibson (1977), these fault rocks represent a depth-dependent series of shallow (0–5 km, gouge and breccia) to intermediate (5–10 km, breccia and cataclasite) crustal levels. Gouge-and-breccia fault rocks preserved in the central domain consist of recrystallized quartz and feldspar in a matrix of haematite + chlorite ± epidote ± albite ± sericite ± copper oxides. In the eastern domain, a wide range of gouge, microbreccia, breccia and cataclasites are observed. Pseudotachylites (Sibson 1977) have not been observed in the northern Sacramento Mountains.

The youngest brittle structural features in the footwall of the northeast Sacramento Mountains are the regionally unique, north–south to northeast–southwest trending, near-vertical faults. These faults define the unusual north–south physiographic trend of the northeast Sacramento Mountains, continue into the Dead Mountains (Spencer 1985) and the central Sacramento Mountains (Schweitzer 1991), and are absent elsewhere in the extensional corridor (Fig. 2). Some of these faults cut the basal SMDF, clearly indicating some post-detachment deformation. Dykes inferred to be Miocene in age (see above) intrude along these structures, however, suggesting that these faults are also, in part, contemporaneous with detachment faulting. Fault plane slickenlines indicate that these faults have a component of strike-slip and dip-slip motion. When relative motion along these faults can be determined (using offset markers, smooth direction on striae or small drag folds), it is dominantly dextral with small displacements (<20 m). Sinistral strike-slip motion is also developed, although to a much lesser extent. The relative motion on dip-slip faults is typically down-to-the-east, east of the range axial high, and down-to-the-west, west of the range axial high, suggesting that domal uplift was occurring at the time of high-angle faulting. The maximum offset observed on normal dip-slip faults is c. 100 m. This figure is consistent with the depth of Quaternary alluvium on the east side of the range front, where a homesteader drilled a water-well to 110 m without encountering basement lithologies (Landowner, pers. comm.). East–west faults have not been observed in the northeast Sacramento Mountains.

In the northwest Sacramento Mountains, north–south trending faults are absent and east–west trending, subvertical brittle faults cross the range, truncating the basal SMDF. Again, last motions from fault plane slickenlines indicate that these faults have a component of strike-slip or oblique-slip to the normal dip-slip motion. The relative motion is dominantly sinistral, with displacements of 100–200 m. Fault gouge, breccia and/or protocataclasite (up to 1.0 m thick) are common.

## Discussion

### Antithetic shear zones

The dykes in which the discrete shear zones and fractures of the central domain formed are thought to be 17–19 Ma. These fabrics therefore developed during detachment faulting but prior to truncation by the basal SMDF. They possess a shear sense opposite to the overall kinematic regime associated with the SMDF system. Consequently, these structures are regarded as antithetic. Antithetic shear zones have been documented in the up-dip region of the footwall in the South Mountains (Reynolds & Lister 1990). In the Whipple Mountains there are numerous antithetic shear zones below the mylonitic front (Davis & Lister 1988) and only 65% of the kinematic indicators within the mylonitic front display synthetic (top-to-the-northeast) shear sense (Davis & Lister 1986). Although these antithetic structures could indicate 'true' top-to-the-southwest transport, this is unlikely for the following reasons. Firstly, throughout the Colorado River extensional corridor, rotated hanging-wall blocks dip consistently to the southwest, indicating top-to-the-northeast transport across the region. Secondly, these antithetic structures are restricted to a narrow zone proximal to the basal SMDF in the central domain and their orientation is parallel to the mylonitic foliation of the eastern domain. This precludes their being a manifestation of, for example, more or less symmetrical doming with both top-to-the-southwest and top-to-the-northeast transport.

## Mechanism of formation

All of the antithethic structures in the central domain dip to the southwest. Relative motion between the intervening blocks (described above) is consistently top-to-the-southwest. Shear zones bounding these blocks are approximately parallel to one another. Geometric restrictions therefore require that these shear zones rotate similarly to one another during progressive deformation. A geometric model for the structural evolution of the central domain is one of continuous shear

zone initiation and progressive rotation within a cooling crust (Fig. 12). During extension, ductile shear zones form at depth. Concomitant with extension and exhumation, the crust is cooled and brittle fractures form approximately parallel to the ductile shear zones. This is consistent with observations that some of the ductile shear zones have the same dip as the brittle shear zones, that brittle shear zones cross-cut the earlier ductile shear zones at small angles, and that displacement on brittle shear zones is less than that on ductile shear zones. Subvertical shortening during extension is also indicated by the development of extensional crenulation shear bands within the dykes and ductile shear zones. It is suggested that the formation of southwest dipping shear zones requires a combination of top-to-the-northeast directed non-coaxial shear and a component of subvertical shortening.

The sequence of strain recorded by these shear zones and fractures may result from: (1) the migration of a rolling hinge; or (2) deformation within the uppermost levels of a ductile shear zone. In a rolling hinge, flexural rotation and isostatic uplift of the footwall results from the tectonic denudation of the hanging wall during normal faulting (Spencer 1984; Buck 1988; Hamilton 1988; Wernicke & Axen 1988). Moderate- to high-angle faults are rotated to lower angles in response to isostatic forces in the footwall. New, higher angle faults propagate when the rotated faults can no longer overcome frictional resistance. The 'hinge' which separates

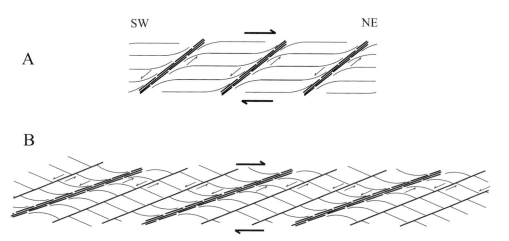

**Fig. 12.** Development of antithetic shear zones. (**a**) Ductile shear zones form. Continuing deformation rotates blocks bounded by shear zones similarly. The crust cools as a result of faulting and exhumation, and the central domain moves into the brittle regime. (**b**) The development of brittle–ductile shear zones and brittle fractures with the same orientation and kinematics as the ductile shear zones suggests that all structures formed in a regime of consistently oriented non-coaxial shear and decreasing temperature.

the shallow, rotated fault segments from the steeper, unrotated fault segments, 'rolls' through the footwall, migrating towards the tectonically denuding hanging wall. The deformational structures predicted by this model are well documented (e.g. Axen & Wernicke 1991; Manning & Bartley 1994; Axen & Bartley 1997) and principally include: (1) a detachment fault consisting of a series of smaller fault segments, which progressively young from shallower to higher angles and towards the tectonic transport direction; (2) a footwall in which the upper portion displays post-mylonitic, extensional brittle deformation, overprinted by contractional brittle deformation.

Footwall deformation in the northern Sacramento Mountains does not support the migration of a rolling hinge through the region. Though extension in the northern Sacramento Mountains appears to have progressed through time from southwest to northeast, the youngest fault in the region (the basal SMDF) is the most shallowly dipping structure, truncating earlier and more steeply dipping fault splays. Late, contractional deformational structures are completely absent from the footwall of the northern Sacramento Mountains. Although antithetic structures may develop in rolling hinge models if far-field stresses within the footwall override local stresses generated within the rolling hinge (Manning & Bartley 1994), footwall strain developed in response to the migration of a rolling hinge should also postdate the formation of the detachment fault. However, the basal SMDF truncates all ductile and brittle–ductile structures, and almost all brittle structures, and is, therefore, a young structural feature.

It is considered here that the observed strain within the northern Sacramento Mountains results from deformation within a crustal-scale shear zone (Fig. 13). In the uppermost levels of an extending crust, deformation would be transitional from ductile to brittle. Such a transition, observed in the central domain of the northwest Sacramento Mountains, is apparent in the overprinting relationships of successive, discrete ductile through to brittle shear zones. The similar orientations of these extensional structures demonstrate that they have formed in a zone of consistently oriented, non-coaxial shear. Their top-to-the-southwest shear sense and the presence of extensional crenulation shear bands suggest that they also formed with a component of subvertical shortening. This interpretation agrees with the thermal profile of the footwall in the central domain at the onset of extension (315–510°C; Pease et al. 1999), the limited extent of this transitional domain between strictly

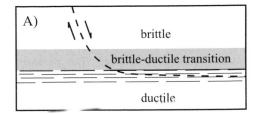

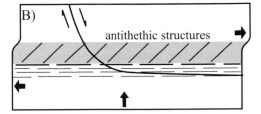

**Fig. 13.** Structural evolution of the northern Sacramento Mountains. (**a**) Ambient, predetachment faulting conditions. (**b**) During top-to-the-northeast, non-coaxial shear and subvertical shortening, deformation in the brittle–ductile transition results in the formation of discrete antithetic shear zones whose overprinting relationships suggest decreasing temperatures. (**c**) Continued detachment faulting exposed the three structural domains seen today in the northern Sacramento Mountains.

brittle deformation of the footwall to the west and pervasive ductile deformation of the footwall to the east, and the relative timing of exhumation of footwall rocks in the Sacramento Mountains (Pease et al. 1999). Furthermore, it is consistent with experimental results which indicate that the brittle–ductile transition in quartzo-feldspathic rocks takes place gradually between 300 and 500°C (Tullis & Yund 1977). The parallelism and identical kinematic shear sense between the sequence of ductile through to brittle structures developed in the central domain suggests that the kinematic framework has also remained constant throughout the evolution of this crustal section. Consequently, the orientation of the footwall was not significantly changed during the formation of these structures.

Structural depth in the northern Sacramento Mountains increases northeastwards in the direction of tectonic transport (Pease et al. 1999).

—— & MUKASA, S. 1990. Footwall rocks of the mid-Tertiary Chemehuevi detachment fault: A window into the middle crust in the southern Cordillera. *Journal of Geophysical Research*, **95**, 463–475.

LISTER, G. & BALDWIN, S. 1993. Plutonism and the origin of metamorphic core complexes. *Geology*, **21**, 607–610.

—— & DAVIS, G. 1989. The origin of metamorphic core complexes and detachment faults formed during Tertiary continental extension in the northern Colorado River region, U.S.A. *Journal of Structural Geology*, **11**, 65–94.

McCLELLAND, W. 1984. *Low-angle Cenozoic faulting in the central Sacramento Mountains, San Bernardino County, California*. MSc Thesis, University of Southern California.

MANNING, A. & BARTLEY, J. 1994. Postmylonitic deformation in the Raft River metamorphic core complex, northwestern Utah: Evidence of a rolling hinge. *Tectonics*, **13**, 596–612.

MILLER, R. & PATERSON, S. 1994. The transition from magmatic to high-temperature solid-state deformation: Implications from the Mount Stuart batholith, Washington. *Journal of Structural Geology*, **16**, 853–865.

NIELSON, J. & BERATAN, K. 1995. Stratigraphic and structural synthesis of a Miocene extensional terrane, southeast California and west-central Arizona. *Bulletin of the Geological Society of America*, **107**, 241–252.

——, LUX, D., DALRYMPLE, G. & GLAZNER, A. 1990. The age of the Peach Springs Tuff, southeastern California and western Arizona. *Journal of Geophysical Research*, **95**, 571–580.

PARSONS, T. & THOMPSON, G. 1993. Does magmatism influence low-angle normal faulting? *Geology*, **21**, 247–250.

PASSCHIER, C. & SIMPSON, C. 1986. Porphyroclast systems as kinematic indicators. *Journal of Structural Geology*, **8**, 831–843.

PATTERSON, S., VERNON, R. & TOBISCH, O. 1989. A review of criteria for the identification of magmatic and tectonic foliations in granitoids. *Journal of Structural Geology*, **11**, 349–363.

PEASE, V. 1997. *Geology of the northeast Sacramento Mountains, California*. DPhil. Thesis, University of Oxford.

——, FOSTER, D., WOODEN, J., O'SULLIVAN, P., ARGENT, J. & FANNING, C. 1999. The northern Sacramento Mountains, southwest United States. Part II: Exhumation history and detachment faulting. *This volume.*

REYNOLDS, S. & LISTER, G. S. 1990. Folding of mylonitic zones in Cordilleran metamorphic core complexes: Evidence from near the mylonitic front. *Geology*, **18**, 216–219.

SCHWEITZER, J. 1991. *Structural evolution of crystalline lower plate rocks, central Sacramento Mountains, southeastern California*. PhD Thesis, Virginia Polytechnic Institute and State University.

SIBSON, R. 1977. Fault rocks and fault mechanisms. *Journal of the Geological Society, London*, **133**, 191–213.

SIMPSON, C. 1985. Deformation of granitic rocks across the brittle–ductile transition. *Journal of Structural Geology*, **71**, 503–511.

—— & SCHMID, S. 1983. An evaluation of criteria to deduce the sense of movement in sheared rocks. *Bulletin of the Geological Society of America*, **94**, 1281–1288.

——, SCHWEITZER, J. & HOWARD, K. 1991. A reinterpretation of the timing, position, and significance of part of the Sacramento Mountains detachment fault, southeastern California. *Geological Society of America Bulletin*, **103**, 751–761.

SPENCER, J. 1984. The role of tectonic denudation in the warping and uplift of low-angle normal faults. *Geology*, **12**, 95–98.

——1985. Miocene low-angle normal faulting and dike emplacement, Homer Mountain and surrounding areas, southeastern California and southernmost Nevada. *Geological Society of America Bulletin*, **96**, 1140–1155.

—— & TURNER, R. 1983. *Geologic map of part of the northwestern Sacramento Mountains, southeastern California*. US Geological Survey Open-File Report, **83–614**.

STEWART, J. 1980. Regional tilt patterns of late Cenozoic basin-range fault blocks, western United States. *Bulletin of the Geological Society of America*, **91**, 460–464.

TULLIS, J. 1983. Deformation of feldspars. *In*: RIBBE, P. (ed.) *Feldspar Mineralogy*. Reviews of Mineralogy, **2**, 297–323.

—— & YUND, R. 1977. Experimental deformation of dry Westerly Granite. *Journal of Geophysical Research*, **82**, 5705–5718.

WERNICKE, B. & AXEN, G. 1988. On the role of isostasy in the evolution of normal fault systems. *Geology*, **16**, 848–851.

WISE, D., DUNN, D., ENGELDER, J. *et al.* 1984. Fault-related rocks: Suggestions for terminology. *Geology*, **12**, 391–394.

YOUNG, R. & BRENNAN, W. 1974. Peach Springs Tuff: Its bearing on structural evolution of the Colorado Plateau and development of Cenozoic drainage in Mohave County, Arizona. *Bulletin of the Geological Society of America*, **85**, 83–90.

# The Northern Sacramento Mountains, southwest United States. Part II: Exhumation history and detachment faulting

V. PEASE,[1,2] D. FOSTER,[3] J. WOODEN,[4] P. O'SULLIVAN,[3]
J. ARGENT[1,5] & C. FANNING[6]

[1] *Department of Earth Sciences, University of Oxford, Parks Road, Oxford OX1 3PR, UK*
[2] *Present address: Department of Earth Sciences, Uppsala University, Villavägen 16,
S-752 36 Uppsala, Sweden*
[3] *Department of Earth Sciences, La Trobe University, Bundoora,Victoria 3083,
Australia and Department of Geology, University of Florida, Gainesville,
FL 32611, USA*
[4] *United States Geological Survey, 345 Middlefield Road, Menlo Park, CA 94025, USA*
[5] *Present address: Amerada Hess Limited, 33 Grosvenor Place, London SW1X 7HY, UK*
[6] *Research School of Earth Sciences, Australian National University, Canberra,
ACT 0200, Australia*

**Abstract:** Thermochronologic and thermobarometric data reveal the timing, distribution and intensity of thermal events associated with detachment faulting in the Sacramento Mountains metamorphic core complex. In the northwest Sacramento Mountains, cooling rates of $c.\,100°C\,Ma^{-1}$ are associated with Late Cretaceous plutonism followed by cooling of the crust by thermal conduction. Post-Late Cretaceous cooling slowed to $c.\,1–6°C\,Ma^{-1}$. Finally, the region records average cooling rates of $38–53°C\,Ma^{-1}$ between $c.\,20$ and $15\,Ma$. In contrast, the thermal profile of the northeast Sacramento Mountains is dominated by syntectonic Tertiary plutonism followed by very rapid cooling. A granodioritic suite intruded at $c.\,680°C$ and $c.\,3\,kbar$ at $c.\,20\,Ma$, records cooling to $<100°C$ by $c.\,15\,Ma$. Such rapid cooling and exhumation suggests that unroofing by tectonic denudation was the driving mechanism for the final cooling. The similarity of the miocene cooling profiles between these two areas clearly suggests that the Sacramento Mountains experienced a regional cooling event associated with tectonic unroofing driven by regional Miocene crustal extension. Estimates of the initial angle of the Sacramento Mountains detachment fault using palaeothermal gradients suggest that it was active at a dip of $25°$.

Shallow-dipping ($<30°$) detachment faults, traceable for tens of kilometres, have been widely recognized in the continental crust for some time (Crittenden *et al.* 1980), although it remains controversial whether these regional scale, low-angle normal faults initiate or slip at shallow dips (Wernicke 1995). Thermochronology from the footwall of such faults suggests that some of them initiate at low angles (e.g. Foster *et al.* 1990*a,b*; Dokka 1993; John & Foster 1993). Thermochronological techniques relate temperature to time and, when combined with thermobarometric data (relating temperature to pressure), also constrain the exhumation history of the footwall of the northern Sacramento Mountains. Furthermore, the thermochronologic data can be used to define the original dip of the Sacramento Mountains detachment fault (SMDF).

As normal faulting proceeds and the footwall is progressively unroofed, minerals from deeper crustal levels in the footwall and along the fault surface should pass through their respective closure temperatures later in time. This results in apparent mineral ages that decrease down-section/down-dip. Isotopic systems used as thermochronometers include uranium–lead (U–Pb), argon ($^{40}Ar/^{39}Ar$) and fission-track (FT) dating. These radioactive systems, applied to a variety of minerals, define a set of closure temperatures ranging from magmatic to subsolidus conditions, from $>800°C$ to below $60°C$. Thermobarometric analysis of the Eagle Wash Intrusive Complex (EWIC after Schweitzer 1991) utilized the

*From*: MAC NIOCAILL, C. & RYAN, P. D. (eds) 1999. *Continental Tectonics*. Geological Society, London, Special Publications, **164**, 199–237. 1-86239-051-7/99/$15.00 © The Geological Society of London 1999.

Al-in-hornblende barometer and hornblende–plagioclase thermometer.

Unravelling the exhumation and cooling history of the footwall in the northern Sacramento Mountains metamorphic core complex is important for understanding the structural and tectonic evolution of the region, especially that associated with Miocene detachment faulting. To determine the denudation and cooling history of the northern Sacramento Mountains metamorphic core complex and to calculate the angle of faulting associated with the SMDF, samples were collected along transects oriented parallel to the inferred tectonic transport direction in order to sample increasing structural depths in the footwall. Our results indicate that tectonic denudation (unroofing) via active low-angle (<30) normal faulting during the Miocene was responsible for the rapid exhumation and cooling of metamorphic core complexes in the region.

## Regional framework

The Sacramento Mountains are located in the Colorado River extensional corridor (Fig. 1), a region of extreme (*c.* 100%) crustal extension (Howard & John 1987) within the Basin-and-Range Province of the southwest United States. Detachment faults exposed in this region are typically subhorizontal or domed, dip at low angles (10–30°), and separate upper-plate Tertiary rocks and their depositional basement from footwalls of originally deeper seated crystalline rocks (e.g. Coney 1980; Davis *et al.* 1980; Armstrong 1982). The SMDF forms part of a rooted, imbricate, asymmetric, top-to-the-northeast mid-Tertiary shear system (Davis *et al.* 1980; Howard *et al.* 1982; John 1982; Spencer 1985; Howard & John 1987; Davis & Lister 1988). Cumulative slip on the fault system increases to the northeast across the region and totals 40–77 km (Howard & John 1987; Davis

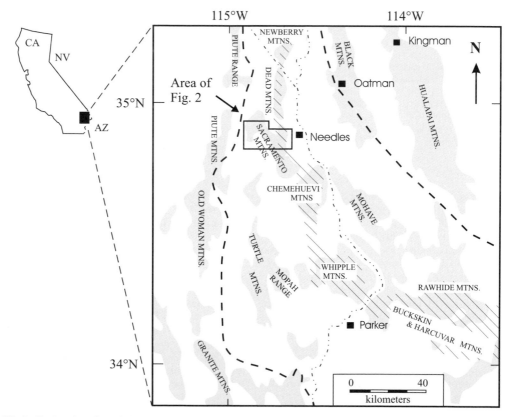

**Fig 1.** Regional setting of the Sacramento Mountains. Limits of the Colorado River extensional corridor of Howard & John (1987) shown by dashed lines, metamorphic core complexes by hatchure. Extension migrated from south to north, beginning at *c.* 26 Ma in the Buckskin–Harcuvar Mountains and at *c.* 18 Ma in the Newberry Mountains.

1988; Spencer & Reynolds 1991). Extension in the corridor progressed from south to north, based on K–Ar ages of the oldest volcanic units in the synextensional Tertiary basins (Brooks & Martin 1985; Howard & John 1987; Spencer & Reynolds 1991), crystallization ages of the oldest synextensional plutons in the lower plates of core complexes (Anderson *et al.* 1988; Foster *et al.* 1990*a*; Howard *et al.* 1996), and the oldest $^{40}Ar/^{39}Ar$ cooling ages directly related to denudation of footwall rocks (Foster *et al.* 1990*a*). Extension began at *c.* 26 Ma in the Buckskin and Harcuvar Mountains (Spencer & Reynolds 1991; Lucchitta & Suneson 1993), and at *c.* 18 Ma in the Newberry Mountains (Howard *et al.* 1994). The ages of untilted volcanic rocks, thermochronology and flat-lying sediments in Miocene basin deposits suggest that rapid extension throughout the corridor had ceased by *c.* 12 Ma (Howard & John 1987; Davis & Lister

1988; Richard *et al.* 1990; Simpson *et al.* 1991; Foster *et al.* 1993; John & Foster 1993; Nielson & Beratan 1995).

## The Sacramento Mountains

In the northern Sacramento Mountains (Fig. 2), the SMDF separates upper-plate, hanging wall rocks (Tertiary volcanic and sedimentary rocks, and their depositional crystalline basement) from a footwall of Proterozoic gneisses, Cretaceous granites and Miocene plutonic rocks (McClelland 1982, 1984; Spencer & Turner 1983; Spencer 1985; Argent 1996; Pease 1997). The extrusion of the *c.* 14.3 Ma rhyodacite of Eagle Peak onto the exhumed fault surface of the SMDF (K–Ar biotite and sanidine; Simpson *et al.* 1991), and the untilted 14.6 ± 0.2 Ma basalt of Flattop Mountain (K–Ar whole rock; Spencer 1985)

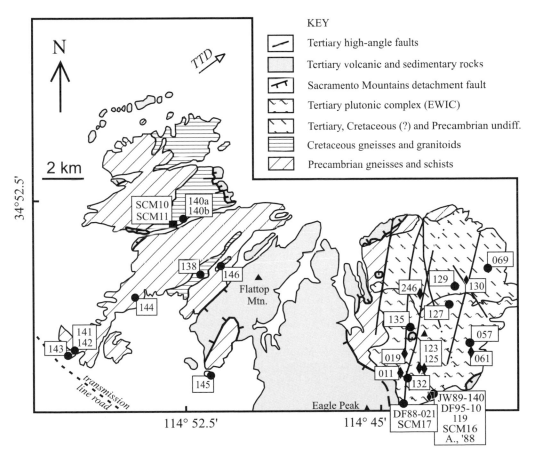

**Fig. 2.** Simplified geology, sample locations and sample numbers. The footwall of the northwest and northeast Sacramento Mountains are separated by an intervening block of hanging wall (shown in grey). Locations of thermochronologic samples are represented by circles, thermobarometric samples by diamonds and published data of others by squares (refer to text). Tectonic transport direction (TTD) shown.

capping tilted, syntectonic volcanogenic sediments in the hanging wall, suggests that by *c.* 14 Ma, low-angle normal faulting in the western part of the northern Sacramento Mountains had ceased.

Lower-plate crystalline rocks in the northeast Sacramento Mountains comprise Precambrian and Cretaceous(?) gneisses to the north and a syntectonic intrusive complex to the south (Fig. 2). The gneisses are thought to be similar in age and composition to those exposed in the northwest Sacramento Mountains (Pease 1997). The EWIC is dominated by granodiorite, with lesser amounts of leucogranite, diorite and granite. Magma mingling and mixing textures between these compositional phases are preserved in low-strain domains. Xenoliths (<0.3 m in size) and rafts (>10 m in size) of Cretaceous(?) and Precambrian material are abundant in the EWIC and show variable degrees of resorption.

A pervasive mylonitic fabric is developed in the EWIC which records subsolidus, upper greenschist through to lower greenschist/subgreenschist facies conditions (Pease & Argent 1999). The mylonitic foliation, striking *c.* 145°SE and dipping 12°SW, is defined by a millimetre-scale segregation of tabular plagioclase and amphibole and is subparallel to the detachment fault (Pease & Argent 1999). The shape-preferred orientation of amphibole and a biotite stretching lineation define a well-developed and consistent lineation trending 241°, 14°SW (Pease & Argent 1999). This lineation direction is co-linear with the regional tectonic transport direction of N60E, which is defined by striations on the detachment surface, stratal and fault dips in the hanging wall, and transport-parallel corrugations of the basal detachment fault surface (Spencer 1985). The majority of shear sense indicators show top-to-the-northeast motion, also consistent with the regional tectonic transport direction of N60E. Taken together, the radiometric ages presented here and the petrofabrics from the EWIC presented by Pease & Argent (1999), indicate that the EWIC was intruded shortly before the onset of detachment faulting in the northeast Sacramento Mountains. The EWIC is, therefore, ideal for providing information related to conditions at the time of detachment faulting.

## Exhumation and cooling history

### U–Pb analyses

*Analytical methods.* The data presented in this investigation are from conventional thermal ioni-

zation mass spectrometric analyses of zircon, monazite and sphene, with the exception of samples VP94-129 and DF95-10 (which were analysed by the ion-microprobe technique). For conventional data, weighed aliquots of size fractions of pure, hand-picked mineral separate were washed, dissolved, spiked, and the U and Pb separated by anion exchange chromatography. All U–Pb chromatographic procedures use pre-cleaned Biorad AG1 × 8, 200–400 mesh, resin (chloride form). Mass spectrometry for the conventional data was performed using a Finnigan MAT262 multi-collector mass spectrometer at the United States Geological Survey, Menlo Park. Lead (zircon, monazite, sphene and common Pb in feldspar) was loaded onto single Re filaments with silica-gel activator. U was loaded onto the evaporation filament of a double Re filament assembly. Repeated analysis of NBS981 and NBS982 standards indicate a reproducible mass fractionation of $-0.12\%$ per atomic mass unit (amu). Overall, $2\sigma$ analytical errors for the feldspar analyses are $0.1\%$ on all ratios to $^{204}$Pb. Measured uncertainties in the $^{206}$Pb/$^{204}$Pb ratios of the zircon fractions range from 0.6 to 5.2% ($2\sigma$); uncertainties in $^{238}$U/$^{235}$U are 0.1% ($2\sigma$). The uncertainties associated with monazite and sphene are equal to, or better than, these values. Age calculations use the decay constants of Steiger & Jäger (1977). Ion-probe analytical methods for the Australian National University SHRIMP (Sensitive High-Resolution Ion MicroProbe) I and II are fully described in Compston *et al.* (1984, 1992) and Williams & Claesson (1987). Closure temperatures associated with U–Pb systematics in zircon, monazite and sphene are $>900°$C (Lee *et al.* 1997), *c.* 700°C (Copeland *et al.* 1988) and $>600°$C (Cherniak 1993), respectively.

*U–Pb results.* Analytical data are presented in Tables 1 and 2. In the northeast Sacramento Mountains (Fig. 2), the biotite schist (VP94-129), collected from country rock near the north-central intrusive contact of the EWIC, is associated with amphibolite and minor migmatite. This sample has a homogeneous population of uniformly small (<63 μm), uncoloured and euhedral zircon grains. The uniformly small grain size of these zircons precluded sorting into size populations for conventional U–Pb analysis, consequently SHRIMP analysis was used. Cathodoluminesence images reveal well-defined, euhedral cores truncated by thin, rounded overgrowths. Only cores were analysed and several grains are concordant at 1.6–1.7 Ga, while the remainder show ≤5% discordance (Fig. 3). The discordance may be associated with

**Table 1a.** *U–Pb analytical data, conventional*

| Sample (fraction/grain spot) (μm) | U (ppm) | Pb (ppm) | $^{207}Pb/^{235}Pb$ | ±2σ (%) | $^{206}Pb/^{238}U$ | ±2σ (%) | $^{207}Pb/^{206}Pb$ | ±2σ (%) | Age estimates (Ma) | | | $^{206}Pb/^{204}Pb$ |
|---|---|---|---|---|---|---|---|---|---|---|---|---|
| | | | | | | | | | $^{207}Pb–^{235}U$ | $^{206}Pb–^{238}U$ | $^{207}Pb–^{206}Pb$ | |
| **VP94-069b** | | | | | | | | | | | | |
| Zircon | | | | | | | | | | | | |
| 80–100 | 433.1 | 56.75 | 1.7370 | 0.05 | 0.13084 | 0.03 | 0.09629 | 0.04 | 1022 | 793 | 1515 | 14498 |
| 100–130 | 386.51 | 51.26 | 1.7572 | 0.04 | 0.13214 | 0.03 | 0.09645 | 0.02 | 1030 | 800 | 1557 | 12658 |
| 130–163 | 370.3 | 51.1 | 1.8353 | 0.07 | 0.13690 | 0.06 | 0.09723 | 0.03 | 1058 | 827 | 1572 | 8000 |
| >163 | 349.9 | 50.27 | 1.9139 | 0.04 | 0.14183 | 0.03 | 0.09787 | 0.03 | 855 | 1086 | 1584 | 12048 |
| Monazite | | | | | | | | | | | | |
| 100–130 | 1087 | 84.11 | 0.06271 | 0.72 | 0.00706 | 0.07 | 0.06440 | 0.66 | 62 | 45 | 755 | 123.2 |
| 130–163 | 1154 | 82.11 | 0.05793 | 0.53 | 0.00678 | 0.07 | 0.06199 | 0.49 | 57 | 48 | 674 | 122.1 |
| >163 | 811 | 172.3 | 0.29247 | 0.14 | 0.02446 | 0.05 | 0.08674 | 0.12 | 261 | 156 | 1355 | 461.7 |
| **VP94-127** | | | | | | | | | | | | |
| Zircon | | | | | | | | | | | | |
| <100 | 340.4 | 40.43 | 1.57076 | 0.07 | 0.11640 | 0.05 | 0.09787 | 0.04 | 959 | 710 | 1584 | 9050 |
| >100 | 273.4 | 37.23 | 1.78662 | 0.04 | 0.13211 | 0.03 | 0.09808 | 0.02 | 1041 | 800 | 1588 | 8330 |
| sphene | 2.38 | 53.86 | 0.0193 | 5.9 | 0.002926 | 0.30 | 0.0484 | 5.6 | 19.4 | 18.8 | 91.3 | 23.1 |
| **VP94-138** | | | | | | | | | | | | |
| Zircon | | | | | | | | | | | | |
| 63–80 | 353.4 | 55.86 | 2.0616 | 0.03 | 0.15735 | 0.02 | 0.09502 | 0.02 | 1136 | 942 | 1528 | 6579 |
| 80–100 | 318.8 | 50.6 | 2.0590 | 0.07 | 0.15701 | 0.03 | 0.09511 | 0.05 | 1135 | 940 | 1530 | 9000 |
| 100–130 | 279.9 | 44.62 | 2.0579 | 0.06 | 0.15678 | 0.05 | 0.09520 | 0.01 | 1135 | 939 | 1532 | 6040 |
| >130 | 209.8 | 29.42 | 1.7901 | 0.10 | 0.13606 | 0.03 | 0.09542 | 0.09 | 1042 | 822 | 1536 | 4975 |
| **monazite** | | | | | | | | | | | | |
| fraction size? | 594.7 | 105.0 | 0.1618 | 0.13 | 0.01782 | 0.13 | 0.0659 | 1.2 | 152 | 114 | 801 | 110.4 |
| **VP94-140a** | | | | | | | | | | | | |
| Zircon | | | | | | | | | | | | |
| <63 | 245.7 | 7.9 | 0.33128 | 0.44 | 0.02970 | 0.06 | 0.08088 | 0.40 | 291 | 189 | 122 | 1324 |
| 63–80 | 233.85 | 12.38 | 0.62035 | 0.14 | 0.05027 | 0.11 | 0.08951 | 0.08 | 490 | 316 | 141 | 3871 |
| 100–130 | 196.1 | 21.19 | 1.3626 | 0.15 | 0.10293 | 0.04 | 0.09602 | 0.13 | 873 | 632 | 1548 | 5747 |
| 130–163 | 195.4 | 22.2 | 1.4419 | 0.07 | 0.10875 | 0.04 | 0.09616 | 0.05 | 907 | 666 | 1551 | 6426 |
| **sphene** | 52.59 | 1.95 | 0.07270 | 1.1 | 0.01097 | 0.08 | 0.04808 | 1.1 | 71.3 | 70.3 | 103.1 | 42.4 |
| **JW89-140** | | | | | | | | | | | | |
| Sphene | 436.77 | 4.11 | 0.0194 | 2.3 | 0.002947 | 0.12 | 0.0478 | 2.2 | 19.5 | 19.0 | 90.1 | 57.5 |

Analyses reported at 2σ precision; data was obtained following techniques described by Whitehouse *et al.* (1992); mineral size fractions given in μm and only non-magnetic fractions were used. All ages calculated using the decay constants of Steiger & Jäger (1977).

**Table 1b.** *U–Pb analytical data, ion probe*

| Sample (fraction/grain spot) (μm) | U (ppm) | Th (ppm) | Pb (ppm) | $^{207}Pb/^{235}U$ | ±1σ (%) | $^{206}Pb/^{238}U$ | ±1σ (%) | $^{207}Pb/^{206}Pb$ | ±1σ (%) | $^{207}Pb-^{235}U$ | $^{206}Pb-^{238}U$ | $^{207}Pb-^{206}Pb$ | $(^{204}Pb/^{206}Pb) \times 10^{-5}$ |
|---|---|---|---|---|---|---|---|---|---|---|---|---|---|
| **VP94-129** Zircon | | | | | | | | | | | | | |
| 1.1 | 1071 | 15 | 261 | 3.503 | 0.105 | 0.2575 | 0.0066 | 0.0987 | 0.0013 | 1528 ± 24 | 1477 ± 34 | 1599 ± 25 | 1.0 |
| 2.1 | 533 | 93 | 135 | 3.495 | 0.090 | 0.2562 | 0.0062 | 0.0989 | 0.0006 | 1526 ± 20 | 1470 ± 32 | 1604 ± 10 | 2.4 |
| 3.1 | 220 | 35 | 72 | 5.368 | 0.160 | 0.3281 | 0.0085 | 0.1187 | 0.0014 | 1880 ± 26 | 1829 ± 41 | 1936 ± 21 | 8.8 |
| 4.1 | 1469 | 30 | 358 | 3.481 | 0.083 | 0.2573 | 0.0060 | 0.0981 | 0.0004 | 1523 ± 19 | 1476 ± 31 | 1589 ± 7 | 3.2 |
| 5.1 | 1677 | 29 | 400 | 3.290 | 0.078 | 0.2522 | 0.0058 | 0.0946 | 0.0003 | 1479 ± 18 | 1450 ± 30 | 1521 ± 6 | 0.4 |
| 6.1 | 326 | 198 | 104 | 4.477 | 0.175 | 0.2892 | 0.0088 | 0.1123 | 0.0024 | 1727 ± 32 | 1637 ± 44 | 1837 ± 39 | 2.8 |
| 7.1 | 5559 | 607 | 1554 | 3.917 | 0.090 | 0.2864 | 0.0065 | 0.0992 | 0.0002 | 1617 ± 19 | 1624 ± 33 | 1609 ± 3 | – |
| 8.1 | 2054 | 64 | 496 | 3.378 | 0.080 | 0.2538 | 0.0059 | 0.0965 | 0.0003 | 1499 ± 19 | 1458 ± 30 | 1558 ± 6 | 1.1 |
| 9.1 | 630 | 190 | 163 | 3.065 | 0.088 | 0.2575 | 0.0069 | 0.0863 | 0.0007 | 1424 ± 22 | 1477 ± 35 | 1346 ± 15 | 3.4 |
| 10.1 | 2923 | 56 | 962 | 5.003 | 0.353 | 0.3448 | 0.0163 | 0.1052 | 0.0049 | 1820 ± 60 | 1910 ± 78 | 1719 ± 89 | – |
| 11.1 | 749 | 111 | 224 | 4.321 | 0.471 | 0.3038 | 0.0190 | 0.1032 | 0.0084 | 1697 ± 90 | 1710 ± 94 | 1682 ± 159 | 5.6 |
| **DF95-10** Zircon | | | | | | | | | | | | | |
| 1.1 | 162 | 108 | 1 | – | – | 0.0029 | 0.0001 | – | – | – | 18.99 ± 26 | – | 222.2 |
| 2.1 | 174 | 190 | 1 | – | – | 0.0029 | 0.0001 | – | – | – | 18.52 ± 26 | – | 255.6 |
| 3.1 | 67 | 88 | 19 | 2.7301 | 0.096 | 0.2263 | 0.0069 | 0.0875 | 0.0012 | 1337 ± 26 | 1315 ± 26 | 1372 ± 26 | 7.1 |
| 4.1 | 1117 | 3462 | 6 | – | – | 0.0029 | 0.0001 | – | – | – | 18.63 ± 26 | – | 71.5 |
| 5.1 | 644 | 1402 | 3 | – | – | 0.0029 | 0.0001 | – | – | – | 18.93 ± 26 | – | 17.5 |
| 6.1 | 150 | 233 | 43 | 2.6013 | 0.083 | 0.2149 | 0.0064 | 0.0878 | 0.0007 | 1301 ± 24 | 1255 ± 26 | 1378 ± 15 | 3.8 |
| 7.1 | 228 | 527 | 1 | – | – | 0.0028 | 0.0001 | – | – | – | 18.14 ± 26 | – | 219.3 |
| 8.1 | 610 | 1251 | 3 | – | – | 0.0029 | 0.0001 | – | – | – | 18.81 ± 26 | – | 26.0 |
| 9.1 | 495 | 631 | 2 | – | – | 0.0029 | 0.0001 | – | – | – | 18.16 ± 26 | – | 0.1 |
| 10.1 | 165 | 336 | 1 | – | – | 0.0026 | 0.0001 | – | – | – | 16.73 ± 26 | – | 184.3 |
| 11.1 | 218 | 464 | 1 | – | – | 0.0028 | 0.0001 | – | – | – | 18.16 ± 26 | – | 119.5 |
| 9.2 | 257 | 585 | 1 | – | – | 0.0027 | 0.0001 | – | – | – | 17.76 ± 26 | – | 0.1 |
| 13.1 | 411 | 1121 | 2 | – | – | 0.0030 | 0.0001 | – | – | – | 19.54 ± 26 | – | 0.1 |
| 14.1 | 112 | 113 | 31 | 2.8009 | 0.0917 | 0.2320 | 0.0071 | 0.0876 | 0.0007 | 1356 ± 25 | 1345 ± 26 | 1373 ± 16 | 0.1 |
| 15.1 | 170 | 195 | 1 | – | – | 0.0030 | 0.0001 | – | – | – | 19.45 ± 26 | – | 190.4 |
| 16.1 | 196 | 112 | 1 | – | – | 0.0030 | 0.0001 | – | – | – | 19.63 ± 26 | – | 6.0 |
| 17.1 | 100 | 130 | 28 | 2.6404 | 0.0864 | 0.2172 | 0.0067 | 0.0882 | 0.0007 | 1312 ± 25 | 1267 ± 26 | 1386 ± 16 | 2.1 |
| 18.1 | 89 | 86 | 24 | 2.7629 | 0.0920 | 0.2303 | 0.0070 | 0.0870 | 0.0009 | 1346 ± 25 | 1336 ± 26 | 1361 ± 20 | 9.6 |
| 19.1 | 162 | 90 | 42 | 3.0667 | 0.0959 | 0.2427 | 0.0072 | 0.0916 | 0.0006 | 1424 ± 24 | 1401 ± 26 | 1460 ± 13 | 4.5 |

Ion-probe data reported at 1σ precision. Analyses were performed on the SHRIMP II (sensitive high-resolution ion micro-probe) instrument at the Australian National University, Canberra, using previously described analytical techniques (Compton *et al.* 1984; Williams & Claesson 1987). Analyses are denoted by grain-spot number. All ages calculated using the decay constants of Steiger & Jäger (1977). Common Pb correction was made using the $^{208}Pb/^{206}Pb$ ratio or by extrapolation along a common Pb mixing line using the measured $^{207}Pb/^{206}Pb$ ratio [after Tera & Wasserburg (1972)]. (–), $^{204}Pb$ was not detected or ratios could not be calculated for Tera–Wasserburg-corrected analyses.

**Table 2.** *Feldspar Pb isotopic data*

| Sample | $^{206}Pb/^{207}Pb$ | $^{207}Pb/^{204}Pb$ | $^{208}Pb/^{204}Pb$ |
|---|---|---|---|
| VP94-069b | 17.715 | 15.538 | 38.681 |
| VP94-127 | 18.532 | 15.599 | 39.069 |
| VP94-138 | 17.665 | 15.534 | 38.279 |
| VP94-140a | 17.510 | 15.505 | 38.404 |
| VP94-140b | 18.304 | 15.585 | 39.054 |
| VP94-141 | 17.910 | 15.560 | 38.896 |
| VP94-142 | 17.675 | 15.548 | 38.762 |
| VP94-159 | 18.268 | 15.605 | 39.053 |
| VP94-160 | 18.013 | 15.589 | 39.021 |

Initial Pb ratios obtained following techniques described by Whitehouse *et al.* (1992).

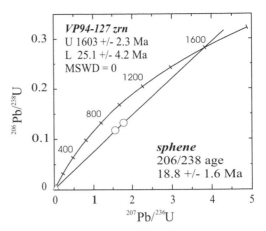

**Fig. 4.** U–Pb concordia diagram of VP94–127 (zircon and sphene). Zircon represented by circles and sphene by triangles. Only zircon data used in fitting discordia.

high temperatures from a post-1.7 Ga high-grade metamorphic event (Burchfiel & Davis 1981; John & Mukasa 1990; Foster *et al.* 1992) or from contact metamorphism associated with the emplacement of the EWIC. The variation in concordant ages from 1.6 to 1.7 Ga likely reflects heterogeneity in the source rock, preserved in spite of amphibolite-grade metamorphism.

VP94-127 is from the fine- to medium grained, weakly deformed (protomylonitic), granodioritic phase of the EWIC adjacent to the biotite schist sample (VP94-129). The zircon population from VP94-127 is varied in size, colour (clear, smoky and pink), and shape (euhedral to subhedral, variable aspect ratios), suggesting that inheritance has occurred. The zircon discordia is defined by two points which have a limited spread in U–Pb composition, resulting in imprecise age information (Fig. 4); nevertheless, a Middle Proterozoic upper intercept (*c.* 1.6 Ga) and a lower intercept of *c.* 25 Ma can be defined.

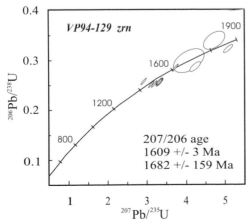

**Fig. 3.** U–Pb concordia diagram of VP94–129 (zircon ion-probe data).

Given the range of zircon grains present and visible xenoliths of biotite schist present in the field, these data are interpreted to reflect simple binary mixing between an older inherited Pb component and a younger, non-inherited Pb component from the EWIC granodioritic magma. Thus, the upper intercept is in agreement with the known age of resorbed material (see VP94-129 above) and the lower intercept suggests an Oligocene–Miocene age for the EWIC. Sphene data from VP94-127 is concordant (Fig. 4) and yields a $^{206}Pb/^{238}U$ age of $18.8 \pm 1.6$ Ma.

DF95-10 (zircon) and JW89-40 (sphene) are both from a fine to medium-grained, weakly deformed (protomylonitic), granodioritic phase of the EWIC at the quarry in the south of the northeast part of the range (Fig. 2). SHRIMP analysis was utilized for the zircon population from this location, as the results from VP94-127 suggested that inheritance was likely. SHRIMP results show a bimodal distribution of Middle Proterozoic and Miocene ages. The older population shows concordant ages of 1.35–1.45 Ga (Fig. 5a). The younger population is concordant at $18.7 \pm 0.4$ Ma (Fig. 5b). These data are interpreted to reflect simple binary mixing between an inherited Pb component (resorbed from Proterozoic country rock) and a younger, non-inherited Pb component from the EWIC granodioritic magma. Sphene data at this location (JW89-40) are concordant (Fig. 5a) and yield a $^{206}Pb$–$^{238}U$ age of $19.2 \pm 0.6$ Ma. The sphene and zircon ages are within error and represent a crystallization age of *c.* 19 Ma for the EWIC at the quarry location.

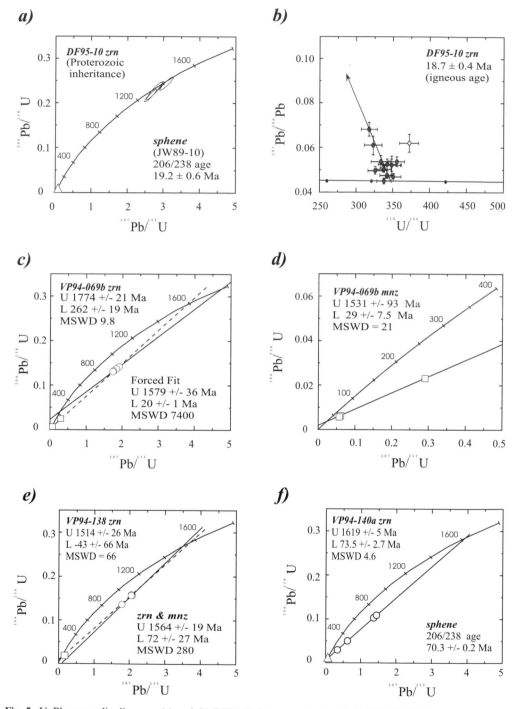

**Fig. 5.** U–Pb concordia diagram. (**a**) and (**b**) DF95-10 (zircon and sphene); (**c**) VP94-069b (zircon); (**d**) VP94-069b (monazite); (**e**) VP94-138 (zircon and monazite); (**f**) VP94-140a (zircon and sphene). Zircon represented by circles, monazite by squares and sphene by triangles. Refer to text for details on line fits. Symbols larger than $2\sigma$ error unless otherwise indicated.

Sample VP94-069b is from a coarse-grained, mylonitic leucogranite sheet on the eastern side of the northeast part of the range (Fig. 2). This sample is thought to represent the leucocratic phase of the EWIC. Both the zircon and monazite data from this sample are discordant (Figs 5c, d). The zircon upper intercept is c. 1.8 Ga and the lower intercept is c. 260 Ma. The lower intercept, having no known geologic significance in this region, appears unreasonable and is likely a function of the very narrow range of U–Pb composition exhibited by these zircon populations. Consequently, the discordia was forced through a 20 Ma lower intercept and recalculated. The 20 Ma intercept is used because it represents a known thermal event, roughly the time of EWIC intrusion (from VP94-127 and DF95-10 above). The upper intercept resulting from the forced fit agrees well with the known age of surrounding country rocks (1.6–1.7 Ga), though the mean square of the weighted deviation (MSWD) is poor. The poor MSWD may reflect heterogeneity in the source or a more complex, multi-stage evolutionary history for this sample. The former is favoured for reasons outlined below.

Monazite results from this sample (Fig. 5d) have an upper intercept of c. 1.5–1.6 Ga and a lower intercept of $29 \pm 7.5$ Ma. The monazite data are poorly constrained because the discordia is essentially a two-point line. These data can be interpreted as: (1) a mixing line between an inherited Pb component (from Proterozic country rock) and a younger, non-inherited Pb component (the c. 20 Ma EWIC); or (2) reworking of Proterozic crustal material during a later high-grade (amphibolite) event(s). There are no known high-temperature thermal events (igneous intrusion/metamorphism) of Oligocene age in the region, but there is the intrusion of the EWIC and other Miocene plutonic rocks in the immediate vicinity (Davis et al. 1980, 1982; Foster et al. 1996a; Howard et al. 1996). Consequently, the former interpretation is favoured. Recent studies have shown that monazite, similar to zircon, can be inherited or xenocrystic (e.g. Copeland et al. 1988; Parrish 1990; Harrison et al. 1995). The general agreement (upper intercept at c. 1.6 Ga and a lower intercept as young as c. 22 Ma) between the zircon data (forced fit) and monazite data for this sample suggests that the leucogranite is Miocene in age with a strong, inherited crustal component.

Samples VP94-138 and VP94-140a are from the northwest Sacramento Mountains. VP94-138 is from a medium-grained, undeformed two-mica granite. The zircon population varied in size, shape and colour, suggesting that inheritance was likely. The zircon data are discordant, defining essentially a two-point discordia with an upper intercept of $1514 \pm 26$ Ma (Fig. 5e). The lower intercept is $-43 \pm 66$ Ma. These data define a limited range in compositional space, thus the poor line fit. Only a single monazite size population was analysed and, if combined with the zircon data, an upper intercept of $1564 \pm 19$ Ma and a lower intercept of $72 \pm 27$ Ma are obtained. The data suggest: (1) the presence of a large component of inherited Pb (xenocrystic mid-Proterozoic zircon); or (2) significant, high-temperature reworking of a mid-Proterozoic terrane. The lack of deformation in this granite favours the former interpretation. The lower intercept of the combined data suggests a Late Cretaceous age for this pluton. This is consistent with the presence of Cretaceous plutonism throughout the region which was derived, in part, by the partial melting/assimilation of 'continental' country rocks (Foster et al. 1989; Anderson & Cullers 1990; John & Wooden 1990; Miller et al. 1990; Foster et al. 1992). A Late Cretaceous age of c. 72 Ma is also supported by $^{40}Ar/^{39}Ar$ data (see following section).

The results from VP94-140a are also discordant (Fig. 5f). This is a medium- to coarse-grained, sphene-bearing granodioritic gneiss, in sharp contact with Precambrian augen gneiss and intruded by an undeformed Late Cretaceous two-mica granite (VP94-138 above). The zircon population varied in size, shape and colour, suggesting that inheritance was likely. The zircon data define a discordia with an upper intercept of $1619 \pm 5$ Ma and a lower intercept of $73.5 \pm 2.7$ Ma; sphene is concordant at $70.3 + 2.7$ Ma. Thus, this sample appears to represent: (1) syntectonic Late Cretaceous magmatism with a large component of inherited Pb (mid-Proterozoic zircons); or (2) Proterozoic crust which experienced significant high-temperature reworking at c. 70 Ma. Argent (1996) identified three episodes of deformation in the northwest Sacramento Mountains based on structural arguments and deformational fabrics: (1) an older amphibolite facies deformation; (2) a Cretaceous greenschist facies deformation; (3) a Miocene lower greenschist deformation. Both of the older fabrics are present in this unit. Cretaceous greenschist-to amphibolite-grade metamorphism, resulting from Laramide compression (thrusting and magmatism), is well documented in the region (Burchfiel & Davis 1981; John & Mukasa 1990; Foster et al. 1992). This information is consistent with the interpretation that this gneiss represents older crust reworked under amphibolite-grade conditions in the Cretaceous and intruded by post-kinematic Late Cretaceous plutons.

*Discussion of U–Pb results.* There is good within-site and between-site agreement of ages for the EWIC, especially considering the rather limited range of U–Pb composition from both zircon and monazite populations. At *c.* 600°C (the closure temperature of sphene), the ages associated with sphene from EWIC lithologies are within analytical uncertainty. Combining the sphene data, the best age for the granodiorite at the closure temperature for sphene is calculated to be $19.0 \pm 0.6$ Ma ($2\sigma$). A similar age of $18.7 \pm 0.4$ Ma ($2\sigma$) is obtained from zircon at the quarry site, and the combined zircon and sphene data from the quarry site suggest a crystallization age for the granodiorite of *c.* 19 Ma. Zircon and monazite ages obtained from samples further to the northeast, while poorly constrained, appear slightly older but also suggest Miocene crystallization ages for all compositional phases of the EWIC.

## $^{40}Ar/^{39}Ar$ analyses

*Analytical methods.* $^{40}Ar/^{39}Ar$ analyses were performed at La Trobe University, Melbourne, Australia. The neutron flux ($J$) associated with each irradiation was calculated using biotite GA1550 (McDougall & Roksandic 1974) as the monitor mineral. Samples and flux monitors were wrapped in Al foil and placed in a quartz vial with the flux monitors spaced at 0.5–1.0 cm intervals over the length of the tube. The vials were sealed and irradiated at the Soreq Nuclear Research Centre, IRR-1 reactor, Israel. For the biotite samples, single grains or small numbers of grains were loaded into milled pits in a copper plate and placed into a gas extraction line fitted with a glass viewport. Gas was extracted from these samples using a defocused 6 watt (W) Ar ion laser with stepwise heating accomplished by varying the power output of the laser. Gas extraction for the hornblende samples was performed in a computer controlled double-vacuum resistance furnace. Extracted gas was expanded into a stainless steel clean-up line and purified for *c.* 10 min with Zr–Ti getters. Ar isotopes were measured using a VG3600 mass spectrometer operating at a sensitivity of $1 \times 10^{-3}$ A torr$^{-1}$, using a Daly photomultiplier set at a gain of 100 over the Faraday cups. The data were corrected for machine background, line blank and mass discrimination. Extraction line blanks of atmospheric composition were typically $9 \times 10^{-17}$ to $1 \times 10^{-16}$ mol $^{40}Ar$ for the laser and $2 \times 10^{-16}$ mol Ar for the furnace. The value for mass discrimination, determined by analysing atmospheric Ar, was 1.0105 for 1 amu. Biotite and hornblende samples were analyzed

using the laser-fusion technique and only hornblende samples were also analysed using the incremental-heating technique. Errors associated with age calculations were propagated using the method of Dalrymple & Lanphere (1971).

Information about the thermal history of these samples can be inferred from the form of age spectra [see McDougall & Harrison (1988) for a review]. In principle, the $^{40}Ar/^{39}Ar$ technique is capable of distinguishing between disturbed and undisturbed samples, identifying the presence of excess Ar, and of recovering the crystallization or reheating age of a sample that has experienced a thermal or chemical perturbation if Ar loss is <15–20% (Harrison 1983). Age gradients revealed by step-heating analysis of hornblende may result from slow cooling, partial Ar loss due to reheating or from the presence of excess $^{40}Ar$. Samples that yield flat-age spectra, referred to as 'plateaus', have traditionally been regarded as undisturbed since cooling (e.g. see Fleck *et al.* 1977). In some cases, however, these plateau may result from experimental artefacts such as homogenization of gas in minerals that are unstable in the vacuum furnace (Gaber *et al.* 1988; Harrison *et al.* 1985). Non-ideal forms, such as saddle-shaped spectra (Lanphere & Dalrymple 1976) and large-domain, trapped-Ar effect spectra (Foster *et al.* 1990*a*), result from incorporation of trapped Ar with a non-atmospheric composition. Isochron analysis is commonly an effective method to reveal both the trapped $^{40}Ar/^{36}Ar$ and the ages of samples so affected (Roddick 1978; Heizler & Harrison 1988), providing significant radiogenic spread between steps.

Isochron analyses have been performed on samples analysed with the incremental-heating technique, but were not feasible for samples analysed with the laser (too few data points). In the case of laser analyses, concordant steps from replicate analyses were used to generate weighted mean ages. In the case of disturbed spectra, minimum (or maximum) ages have been assigned on the basis of the youngest (or oldest) age release step recorded. Generally, greater confidence can be placed in plateau and isochron ages than in weighted mean and minimum–maximum ages. The closure temperature of the K–Ar isotopic system in hornblende ranges from 530 to 490°C for cooling rates $\leq 200°C$ Ma$^{-1}$ (Harrison 1981) and from 350 to 280°C for typical igneous biotite at cooling rates of $\leq 200°C$ Ma$^{-1}$ (Harrison *et al.* 1985; Grove & Harrison 1996).

*$^{40}Ar/^{39}Ar$ results.* A summary of $^{40}Ar/^{39}Ar$ results is presented in Table 3. Analytical data

**Table 3.** *Summary of $^{40}Ar/^{39}Ar$ analytical results*

| Sample no. | Mineral separated | Age (Ma) $\pm 2\sigma$ | Distance (km) | Type of age* | UTM† Northerly | Easterly |
|---|---|---|---|---|---|---|
| **Northeast Sacramento Mountains** | | | | | | |
| VP94-057b | hornblende | 20.1 ± 0.5 | 16.5 | i(4) | 3 853 884 | 710 016 |
| VP94-069b | biotite | na | 20.3 | na | 3 857 370 | 711 526 |
| VP94-127 | hornblende | 20.6 ± 0.5 | 16.75 | i(4) | 3 855 745 | 708 756 |
| | biotite | 14.0 ± 0.6 | 16.75 | w(5) | | |
| VP94-129‡ | biotite | 13.0 ± 0.2 | 17.0 | w(4) | 3 856 006 | 708 917 |
| VP94-132 | hornblende | 21.3 ± 0.4 | 12.5 | p(5) | 3 851 955 | 706 696 |
| | biotite | 14.2 ± 0.4 | 12.5 | w(3) | | |
| VP94-135 | hornblende | 20.8 ± 0.8 | 14.25 | i(5) | 3 855 544 | 706 893 |
| | biotite | 13.4 ± 1.4 | 14.25 | m(1) | | |
| **Northwest Sacramento Mountains** | | | | | | |
| VP94-138 | biotite | 16.4 ± 0.3 | 8.7 | p(8) | 3 857 082 | 696 608 |
| VP94-140a | hornblende | ≤69.0 | 7.6 | m(2) | 3 859 258 | 694 215 |
| | biotite | c. 15? | 7.6 | tf(2) | | |
| VP94-140b | hornblende | ≤22.6 | 7.6 | m(1) | 3 859 258 | 694 215 |
| VP94-141 | biotite | 70.2 ± 1.0 | 0.5 | w(3) | 3 853 405 | 689 919 |
| VP94-142 | biotite | 70.6 ± 0.6 | 0.5 | w(3) | 3 853 405 | 689 919 |
| VP94-143 | biotite | 68.2 ± 0.7 | 0.4 | w(3) | 3 853 352 | 689 761 |
| VP94-144 | hornblende | ≥77 | 4.3 | m(1) | 3 855 653 | 693 110 |
| | biotite | c. 70? | 4.3 | na | | |
| VP94-146 | hornblende | ≤62.8 | 10.3 | m(1) | 3 857 722 | 698 120 |
| | biotite | 28.8 ± 0.6 | 10.3 | w(3) | | |

Distance down-dip from an arbitrary reference line (the steel tower transmission line road), perpendicular to the transport direction (Fig. 2).
* p, plateau; i, isochron; w, weighted mean; m, maximum or minimum; tf, total fusion; all with the number of steps used in parentheses; na, not applicable.
† Universal Transmercator coordinate system (metres), zone 11. Sample DF88–021, UTM N 3850549 and E 706780.
‡ Sample from Precambrian biotite schist at the intrusive contact.

are presented in Table 4. Release spectra and isochron plots can be found in Pease (1997). Samples in the northeast Sacramento Mountains are predominantly from the weakly deformed (protomylonitic), medium grained, granodioritic phase of the EWIC (Fig. 2). The two exceptions are VP94-129, a sample of Precambrian biotite schist, and VP94-069b, a coarse-grained, mylonitic leucogranite correlative with the leucocratic phase of the main EWIC body to the south. Samples from the northwest Sacramento Mountains are predominantly from medium- to coarse-grained Precambrian and Cretaceous gneisses, and medium grained, undeformed, granitic intrusions (Fig. 2). One sample (VP94–140b) is from a later cross-cutting granodioritic dyke, part of the Miocene dyke swarm in the region (Argent 1996, Spencer & Turner 1983), and possibly correlative with the EWIC to the northeast. Results from other workers are reported for comparison, e.g. hornblende and biotite $^{40}Ar/^{39}Ar$ data from Foster *et al.* (1990*b*), sample numbers SCM10, SCM11, SCM16 and SCM17.

Hornblende samples from the northeast Sacramento Mountains were analysed by the incremental-heating technique. These samples yield isochron or plateau ages within $2\sigma$ error of each other of 20–21 Ma (Table 4 and Fig. 6), and are concordant with U–Pb data or are slightly (c. 1%) older. All samples display saddle-shaped (concave) spectra with concordant steps, comprising variable percentages of the gas released, for the intermediate temperature steps. Isochron analysis suggests that VP94-057b contains excess radiogenic Ar ($^{40}Ar/^{36}Ar$ intercept c. 330), which would explain the shape of the release spectra (Lanphere & Dalrymple 1976). The surrounding Precambrian country rock provides a likely source for this older, non-atmospheric component of $^{40}Ar$.

Biotite samples from this region (all analysed by the laser step-heating and fusion techniques) display more irregular degassing patterns, with some flat spectra (VP94-127 and VP94-132) as well as irregular spectra (VP94-069b, VP94-129 and VP94-135). In the case of irregular spectra,

**Table 4.** $^{40}Ar/^{39}Ar$ *analytical data*

| T (°C) [or watts (W)] | $^{36}Ar$ (mol) | $^{37}Ar$ (mol) | $^{39}Ar$ (mol) | $^{40}Ar$ (mol) | $^{40}Ar^*$ (%) | $^{40}Ar^*/^{39}Ar(K)$ | Cumulative $^{39}Ar$ (%) | Calculated age (Ma $\pm 1\sigma$) | Ca/K |
|---|---|---|---|---|---|---|---|---|---|
| **VP94-057b, hornblende, J = 1.0572 × 10$^{-2}$** | | | | | | | | | |
| 0.5 W | $4.717 \times 10^{-18}$ | $2.799 \times 10^{-15}$ | $6.070 \times 10^{-16}$ | $1.725 \times 10^{-15}$ | 33.1 | 0.943 | 29.25 | $17.90 \pm 4.35$ | 8.79 |
| 1.0 W | $6.739 \times 10^{-18}$ | $7.695 \times 10^{-15}$ | $1.469 \times 10^{-15}$ | $2.222 \times 10^{-15}$ | 40.2 | 0.610 | 100.00 | $11.61 \pm 0.64$ | 9.99 |
| Total | $1.146 \times 10^{-17}$ | $1.049 \times 10^{-14}$ | $2.076 \times 10^{-15}$ | $3.947 \times 10^{-15}$ | | 0.708 | | $13.45 \pm 1.72$ | |
| **VP94-057b, hornblende, J = 1.0572 × 10$^{-2}$** | | | | | | | | | |
| 4.0 W | $2.085 \times 10^{-17}$ | $2.791 \times 10^{-14}$ | $5.253 \times 10^{-15}$ | $1.040 \times 10^{-14}$ | 63.8 | 1.268 | 100.00 | $24.03 \pm 0.47$ | 1.01E + 1 |
| Total | $2.085 \times 10^{-17}$ | $2.791 \times 10^{-14}$ | $5.253 \times 10^{-15}$ | $1.040 \times 10^{-14}$ | | 1.268 | | $24.03 \pm 0.47$ | |
| **VP94-057b, hornblende, J = 1.0572 × 10$^{-2}$** | | | | | | | | | |
| 800 | $6.453 \times 10^{-17}$ | $4.418 \times 10^{-15}$ | $4.565 \times 10^{-15}$ | $3.323 \times 10^{-14}$ | 3.179 | 3.32 | 43.6 | $59.6 \pm 40.79$ | 1.84 |
| 900 | $4.289 \times 10^{-17}$ | $8.851 \times 10^{-15}$ | $3.310 \times 10^{-14}$ | $1.584 \times 10^{-14}$ | 24.7 | 1.183 | 5.73 | $22.42 \pm 0.82$ | 5.09 |
| 980 | $9.810 \times 10^{-17}$ | $1.034 \times 10^{-13}$ | $1.830 \times 10^{-14}$ | $4.406 \times 10^{-14}$ | 54.4 | 1.316 | 19.01 | $24.93 \pm 0.58$ | 1.08E + 1 |
| 1010 | $1.153 \times 10^{-16}$ | $1.892 \times 10^{-13}$ | $3.378 \times 10^{-14}$ | $5.802 \times 10^{-14}$ | 69.4 | 1.196 | 43.52 | $22.67 \pm 0.35$ | 1.07E + 1 |
| 1040* | $7.640 \times 10^{-17}$ | $1.164 \times 10^{-13}$ | $2.180 \times 10^{-14}$ | $3.712 \times 10^{-14}$ | 66.2 | 1.131 | 59.34 | $21.45 \pm 0.23$ | 1.02E + 1 |
| 1070* | $5.209 \times 10^{-17}$ | $5.144 \times 10^{-14}$ | $9.328 \times 10^{-15}$ | $2.218 \times 10^{-14}$ | 50.6 | 1.208 | 66.11 | $22.89 \pm 0.41$ | 1.05E + 1 |
| 1100* | $6.081 \times 10^{-17}$ | $4.561 \times 10^{-14}$ | $7.704 \times 10^{-15}$ | $2.383 \times 10^{-14}$ | 41.1 | 1.277 | 71.70 | $24.18 \pm 1.17$ | 1.13E + 1 |
| 1130* | $4.596 \times 10^{-17}$ | $3.382 \times 10^{-14}$ | $5.672 \times 10^{-15}$ | $1.795 \times 10^{-14}$ | 40.6 | 1.289 | 75.81 | $24.43 \pm 1.32$ | 1.14E + 1 |
| 1170* | $5.016 \times 10^{-17}$ | $3.189 \times 10^{-14}$ | $5.483 \times 10^{-15}$ | $1.938 \times 10^{-14}$ | 37.7 | 1.338 | 79.79 | $25.34 \pm 1.10$ | 1.11E + 1 |
| 1450 | $2.904 \times 10^{-16}$ | $1.534 \times 10^{-13}$ | $2.785 \times 10^{-14}$ | $1.308 \times 10^{-13}$ | 44.5 | 2.100 | 100.00 | $39.62 \pm 0.50$ | 1.05E + 1 |
| Total | $8.967 \times 10^{-16}$ | $7.384 \times 10^{-13}$ | $1.378 \times 10^{-13}$ | $4.024 \times 10^{-13}$ | | 1.465 | | $27.72 \pm 0.54$ | |
| **VP96b-057b, hornblende, J = 1.0647 × 10$^{-2}$** | | | | | | | | | |
| 0.3 | $6.783 \times 10^{-21}$ | 6.928E-20 | $2.105 \times 10^{-15}$ | $2.013 \times 10^{-14}$ | 99.9 | 9.549 | 29.04 | $174.88 \pm 2.44$ | 6.25E − 5 |
| 0.5 | $4.833 \times 10^{-18}$ | $4.359 \times 10^{-17}$ | $2.068 \times 10^{-15}$ | $2.524 \times 10^{-14}$ | 94.3 | 11.508 | 57.55 | $208.75 \pm 3.71$ | 4.01E − 2 |
| 0.7 | $4.959 \times 10^{-18}$ | $1.366 \times 10^{-17}$ | $1.335 \times 10^{-15}$ | $2.750 \times 10^{-14}$ | 94.6 | 19.486 | 75.97 | $340.47 \pm 5.97$ | 1.94E − 2 |
| 1.0 | $3.626 \times 10^{-18}$ | $4.519 \times 10^{-17}$ | $1.028 \times 10^{-15}$ | $2.660 \times 10^{-15}$ | 59.5 | 1.538 | 90.15 | $29.34 \pm 1.46$ | 8.35E − 2 |
| 4.0 | $1.363 \times 10^{-18}$ | $6.186 \times 10^{-18}$ | $7.140 \times 10^{-16}$ | $4.112 \times 10^{-15}$ | 90.2 | 5.192 | 100.00 | $97.17 \pm 3.53$ | 1.65E − 1 |
| Total* | $1.479 \times 10^{-17}$ | $1.644 \times 10^{-16}$ | $7.251 \times 10^{-15}$ | $7.964 \times 10^{-14}$ | | 10.373 | | $189.20 \pm 3.42$ | |
| **VP94-069b, biotite, J = 1.0647 × 10$^{-2}$** | | | | | | | | | |
| 0.3 | $1.194 \times 10^{-17}$ | $7.016 \times 10^{-18}$ | $3.013 \times 10^{-15}$ | $9.829 \times 10^{-15}$ | 63.8 | 2.081 | 23.54 | $39.53 \pm 0.90$ | 4.42E − 3 |
| 0.5 | $3.438 \times 10^{-19}$ | $4.919 \times 10^{-17}$ | $3.059 \times 10^{-15}$ | $8.983 \times 10^{-15}$ | 98.6 | 2.895 | 47.44 | $54.76 \pm 1.03$ | 3.06E − 2 |
| 0.8 | $6.783 \times 10^{-21}$ | $6.665 \times 10^{-18}$ | $2.147 \times 10^{-15}$ | $1.070 \times 10^{-14}$ | 99.8 | 4.973 | 66.22 | $93.07 \pm 1.11$ | 5.90E − 3 |
| 4.0 | $4.827 \times 10^{-18}$ | $6.141 \times 10^{-17}$ | $4.580 \times 10^{-15}$ | $1.115 \times 10^{-14}$ | 86.8 | 2.114 | 100.00 | $40.15 \pm 1.04$ | 255E − 2 |
| Total* | $1.712 \times 10^{-17}$ | $1.243 \times 10^{-16}$ | $1.280 \times 10^{-14}$ | $4.067 \times 10^{-14}$ | | 2.772 | | $52.48 \pm 1.01$ | |

**VP94-127, biotite, $J = 1.0659 \times 10^{-2}$**

| Step | | | | | | | | | |
|---|---|---|---|---|---|---|---|---|---|
| 0.3 W | $4.608 \times 10^{-17}$ | $4.657 \times 10^{-16}$ | $9.469 \times 10^{-15}$ | $2.078 \times 10^{-14}$ | 34.2 | 0.751 | 23.61 | $14.36 \pm 0.24$ | $9.34\text{E}-2$ |
| 0.5 W | $6.497 \times 10^{-18}$ | $1.050 \times 10^{-16}$ | $8.294 \times 10^{-15}$ | $7.682 \times 10^{-15}$ | 74.0 | 0.685 | 44.29 | $13.12 \pm 0.64$ | $2.40\text{E}-2$ |
| 0.8 W | $4.568 \times 10^{-18}$ | $4.351 \times 10^{-16}$ | $1.166 \times 10^{-14}$ | $1.036 \times 10^{-14}$ | 86.2 | 0.766 | 73.36 | $14.65 \pm 0.15$ | $7.09\text{E}-2$ |
| 4.0 W | $6.298 \times 10^{-18}$ | $7.182 \times 10^{-16}$ | $1.068 \times 10^{-14}$ | $9.799 \times 10^{-15}$ | 80.5 | 0.738 | 100.00 | $14.13 \pm 0.17$ | $1.28\text{E}-1$ |
| Total | $6.344 \times 10^{-17}$ | $1.724 \times 10^{-15}$ | $4.010 \times 10^{-14}$ | $4.862 \times 10^{-14}$ | | 0.738 | | $14.13 \pm 0.28$ | |

**VP94-127, biotite, $J = 1.0659 \times 10^{-2}$**

| Step | | | | | | | | | |
|---|---|---|---|---|---|---|---|---|---|
| 0.25 W | $6.783 \times 10^{-21}$ | $8.450 \times 10^{-18}$ | $1.130 \times 10^{-17}$ | $7.442 \times 10^{-17}$ | 98.1 | 6.469 | 0.09 | $120.30 \pm 12.66$ | 1.42 |
| 0.5 W* | $4.071 \times 10^{-18}$ | $2.200 \times 10^{-16}$ | $3.054 \times 10^{-15}$ | $3.225 \times 10^{-15}$ | 62.3 | 0.658 | 24.68 | $12.60 \pm 0.67$ | $1.37\text{E}-1$ |
| 0.7 W* | $5.42 \times 10^{-18}$ | $1.079 \times 10^{-15}$ | $5.485 \times 10^{-15}$ | $5.498 \times 10^{-15}$ | 71.5 | 0.717 | 68.84 | $13.74 \pm 0.50$ | $3.74\text{E}-1$ |
| 4.0 W* | $3.899 \times 10^{-18}$ | $6.795 \times 10^{-16}$ | $3.871 \times 10^{-15}$ | $3.889 \times 10^{-15}$ | 70.9 | 0.712 | 100.00 | $13.64 \pm 0.35$ | $3.34\text{E}-1$ |
| Total | $1.340 \times 10^{-17}$ | $1.987 \times 10^{-15}$ | $1.242 \times 10^{-14}$ | $1.269 \times 10^{-14}$ | | 0.706 | | $13.53 \pm 0.51$ | |

**VP94-127, hornblende, $J = 1.0584 \times 10^{-2}$**

| Step | | | | | | | | | |
|---|---|---|---|---|---|---|---|---|---|
| 0.25 W | $1.029 \times 10^{-17}$ | $1.817 \times 10^{-16}$ | $9.041 \times 10^{-17}$ | $3.556 \times 10^{-15}$ | 14.9 | 5.882 | 2.42 | $108.96 \pm 19.96$ | 3.82 |
| 0.5 W | $5.695 \times 10^{-18}$ | $6.865 \times 10^{-16}$ | $2.241 \times 10^{-16}$ | $1.028 \times 10^{-15}$ | $-58.0$ | 0.001 | 8.42 | $0.019 \pm 5.846$ | 5.83 |
| 0.8 W | $7.910 \times 10^{-18}$ | $9.014 \times 10^{-16}$ | $1.968 \times 10^{-15}$ | $3.197 \times 10^{-15}$ | 51.1 | 0.832 | 61.06 | $15.82 \pm 0.85$ | 8.73 |
| 4.0 W | $5.056 \times 10^{-18}$ | $6.892 \times 10^{-16}$ | $1.456 \times 10^{-15}$ | $2.223 \times 10^{-15}$ | 59.4 | 0.910 | 100.00 | $17.29 \pm 1.16$ | 9.02 |
| Total | $2.895 \times 10^{-17}$ | $1.677 \times 10^{-14}$ | $3.739 \times 10^{-15}$ | $1.000 \times 10^{-14}$ | | 0.935 | | $17.77 \pm 1.73$ | |

**VP94-127, hornblende, $J = 1.0584 \times 10^{-2}$**

| Step | | | | | | | | | |
|---|---|---|---|---|---|---|---|---|---|
| 800 | $2.153 \times 10^{-16}$ | $7.031 \times 10^{-15}$ | $7.314 \times 10^{-15}$ | $7.968 \times 10^{-14}$ | 20.8 | 2.271 | 4.55 | $42.85 \pm 0.88$ | 1.83 |
| 900 | $2.728 \times 10^{-17}$ | $4.617 \times 10^{-15}$ | $3.217 \times 10^{-15}$ | $1.032 \times 10^{-14}$ | 25.5 | 0.819 | 6.55 | $15.57 \pm 1.01$ | 2.73 |
| 980 | $5.672 \times 10^{-17}$ | $4.693 \times 10^{-14}$ | $1.324 \times 10^{-14}$ | $2.593 \times 10^{-14}$ | 50.8 | 0.997 | 14.77 | $18.93 \pm 0.25$ | 6.75 |
| 1020 | $9.169 \times 10^{-17}$ | $1.362 \times 10^{-13}$ | $3.092 \times 10^{-14}$ | $4.796 \times 10^{-14}$ | 67.8 | 1.056 | 33.95 | $20.04 \pm 0.32$ | 8.39 |
| 1040 | $6.614 \times 10^{-17}$ | $1.180 \times 10^{-13}$ | $2.582 \times 10^{-14}$ | $3.499 \times 10^{-14}$ | 73.0 | 0.993 | 49.97 | $18.87 \pm 0.27$ | 8.71 |
| 1060 | lost | | | | | | | | |
| 1090* | $6.194 \times 10^{-17}$ | $7.182 \times 10^{-14}$ | $1.482 \times 10^{-14}$ | $2.829 \times 10^{-14}$ | 57.1 | 1.094 | 59.16 | $20.77 \pm 0.55$ | 9.24 |
| 1110* | $5.006 \times 10^{-17}$ | $5.736 \times 10^{-14}$ | $1.133 \times 10^{-14}$ | $2.205 \times 10^{-14}$ | 55.3 | 1.080 | 66.18 | $20.51 \pm 0.45$ | 9.66 |
| 1140* | $5.395 \times 10^{-17}$ | $5.616 \times 10^{-14}$ | $1.156 \times 10^{-14}$ | $2.361 \times 10^{-14}$ | 52.9 | 1.084 | 73.35 | $20.58 \pm 0.53$ | 9.26 |
| 1170* | $5.578 \times 10^{-17}$ | $3.019 \times 10^{-14}$ | $6.188 \times 10^{-15}$ | $2.052 \times 10^{-14}$ | 32.3 | 1.075 | 77.19 | $20.41 \pm 0.74$ | 9.30 |
| 1450 | $8.630 \times 10^{-16}$ | $1.723 \times 10^{-13}$ | $3.678 \times 10^{-13}$ | $3.020 \times 10^{-13}$ | 20.5 | 1.685 | 100.00 | $31.90 \pm 1.69$ | 8.93 |
| Total | $1.542 \times 10^{-15}$ | $7.006 \times 10^{-14}$ | $1.612 \times 10^{-13}$ | $5.953 \times 10^{-13}$ | | 1.243 | | $23.58 \pm 0.72$ | |

**VP94-129, biotite, $J = 1.0692 \times 10^{-2}$**

| Step | | | | | | | | | |
|---|---|---|---|---|---|---|---|---|---|
| 0.2 W* | $6.416 \times 10^{-18}$ | $1.592 \times 10^{-17}$ | $2.911 \times 10^{-15}$ | $3.987 \times 10^{-15}$ | 51.7 | 0.708 | 11.79 | $13.61 \pm 0.90$ | $1.04\text{E}-2$ |
| 0.4 W* | $4.664 \times 10^{-18}$ | $5.955 \times 10^{-18}$ | $5.803 \times 10^{-15}$ | $5.403 \times 10^{-15}$ | 73.4 | 0.683 | 35.29 | $13.13 \pm 0.42$ | $1.95\text{E}-3$ |
| 0.6 W* | $3.918 \times 10^{-18}$ | $5.870 \times 10^{-17}$ | $4.874 \times 10^{-15}$ | $4.456 \times 10^{-15}$ | 73.0 | 0.667 | 55.02 | $12.83 \pm 0.32$ | $2.29\text{E}-2$ |
| 4.0 W | $9.417 \times 10^{-17}$ | $9.956 \times 10^{-17}$ | $1.111 \times 10^{-14}$ | $9.828 \times 10^{-15}$ | 96.1 | 0.850 | 100.00 | $16.32 \pm 0.19$ | $1.70\text{E}-2$ |
| Total | $1.594 \times 10^{-17}$ | $1.801 \times 10^{-16}$ | $2.470 \times 10^{-14}$ | $2.367 \times 10^{-14}$ | | 0.758 | | $14.56 \pm 0.35$ | |

**Table 4.** (continued)

| T (°C) [or watts (W)] | $^{36}Ar$ (mol) | $^{37}Ar$ (mol) | $^{39}Ar$ (mol) | $^{40}Ar$ (mol) | $^{40}Ar^*$ (%) | $^{40}Ar^*/^{39}Ar(K)$ | Cumulative $^{39}Ar$ (%) | Calculated age (Ma ± 1σ) | Ca/K |
|---|---|---|---|---|---|---|---|---|---|
| **VP94-129, biotite, J = 1.0692 × 10$^{-2}$** | | | | | | | | | |
| 0.3 W | $7.643 \times 10^{-19}$ | $4.428 \times 10^{-17}$ | $1.499 \times 10^{-15}$ | $1.715 \times 10^{-15}$ | 86.1 | 0.985 | 35.83 | 18.91 ± 0.59 | 5.61E−2 |
| 0.6 W | $1.825 \times 10^{-18}$ | $5.847 \times 10^{-18}$ | $1.615 \times 10^{-15}$ | $1.646 \times 10^{-15}$ | 66.2 | 0.675 | 74.41 | 12.98 ± 0.82 | 6.88E−3 |
| 0.8 W | $6.783 \times 10^{-21}$ | $7.498 \times 10^{-20}$ | $5.743 \times 10^{-16}$ | $6.604 \times 10^{-16}$ | 98.8 | 1.136 | 88.13 | 21.78 ± 0.21 | 2.48E−4 |
| 4.0 W | $6.783 \times 10^{-21}$ | $7.498 \times 10^{-20}$ | $4.970 \times 10^{-16}$ | $6.488 \times 10^{-16}$ | 98.9 | 1.291 | 100.00 | 24.73 ± 0.26 | 2.87E−4 |
| Total | $2.603 \times 10^{-18}$ | $5.027 \times 10^{-17}$ | $4.185 \times 10^{-15}$ | $4.670 \times 10^{-15}$ | | 0.923 | | 17.71 ± 0.59 | |
| **VP94-132, hornblende, J = 1.0766 × 10$^{-2}$** | | | | | | | | | |
| 0.5 W | $4.864 \times 10^{-18}$ | $1.226 \times 10^{-15}$ | $3.158 \times 10^{-16}$ | $1.531 \times 10^{-15}$ | 12.9 | 0.629 | 11.56 | 12.18 ± 4.16 | 7.40 |
| 0.75 W | $3.738 \times 10^{-18}$ | $6.649 \times 10^{-15}$ | $1.252 \times 10^{-15}$ | $1.800 \times 10^{-15}$ | 70.4 | 1.017 | 57.34 | 19.65 ± 0.79 | 1.01E+1 |
| 1.0 W | $3.449 \times 10^{-18}$ | $4.384 \times 10^{-15}$ | $8.392 \times 10^{-15}$ | $1.205 \times 10^{-15}$ | 46.7 | 0.673 | 88.04 | 13.02 ± 1.71 | 9.96 |
| 4.0 W | $3.657 \times 10^{-18}$ | $1.754 \times 10^{-15}$ | $3.270 \times 10^{-16}$ | $6.267 \times 10^{-16}$ | −48.4 | 0.001 | 100.00 | 0.019 ± 4.209 | 1.02E+1 |
| Total | $1.571 \times 10^{-17}$ | $1.401 \times 10^{-14}$ | $2.734 \times 10^{-15}$ | $5.162 \times 10^{-15}$ | | 0.745 | | 14.41 ± 1.87 | |
| **VP94-132, hornblende, J = 1.0766 × 10$^{-2}$** | | | | | | | | | |
| 800 | $4.408 \times 10^{-16}$ | $5.711 \times 10^{-15}$ | $4.811 \times 10^{-15}$ | $1.482 \times 10^{-13}$ | 12.4 | 3.835 | 3.01 | 72.99 ± 3.30 | 2.26 |
| 900 | $3.131 \times 10^{-17}$ | $5.748 \times 10^{-15}$ | $2.977 \times 10^{-15}$ | $1.268 \times 10^{-14}$ | 30.8 | 1.311 | 4.88 | 25.29 ± 0.82 | 3.67 |
| 980 | $4.902 \times 10^{-17}$ | $3.677 \times 10^{-14}$ | $8.340 \times 10^{-15}$ | $2.141 \times 10^{-14}$ | 47.0 | 1.211 | 10.09 | 23.38 ± 0.59 | 8.40 |
| 1010* | $7.952 \times 10^{-17}$ | $1.111 \times 10^{-13}$ | $2.098 \times 10^{-14}$ | $3.720 \times 10^{-14}$ | 62.5 | 1.113 | 23.20 | 21.49 ± 0.37 | 1.01E+1 |
| 1040* | $1.235 \times 10^{-16}$ | $2.543 \times 10^{-13}$ | $4.708 \times 10^{-14}$ | $6.515 \times 10^{-14}$ | 77.6 | 1.078 | 52.60 | 20.82 ± 0.35 | 1.03E+1 |
| 1070* | $3.600 \times 10^{-17}$ | $3.458 \times 10^{-14}$ | $6.411 \times 10^{-14}$ | $1.485 \times 10^{-14}$ | 48.4 | 1.126 | 56.61 | 21.73 ± 0.74 | 1.03E+1 |
| 1100* | $7.596 \times 10^{-17}$ | $1.087 \times 10^{-13}$ | $1.932 \times 10^{-14}$ | $3.462 \times 10^{-14}$ | 62.2 | 1.120 | 68.67 | 21.62 ± 0.46 | 1.07E+1 |
| 1130* | $5.647 \times 10^{-17}$ | $5.425 \times 10^{-14}$ | $9.877 \times 10^{-15}$ | $2.297 \times 10^{-14}$ | 47.7 | 1.114 | 74.84 | 21.50 ± 0.60 | 1.05E+1 |
| 1170 | $6.914 \times 10^{-17}$ | $5.070 \times 10^{-14}$ | $9.227 \times 10^{-15}$ | $2.747 \times 10^{-14}$ | 41.5 | 1.241 | 80.61 | 23.95 ± 0.96 | 1.05E+1 |
| 1450 | $6.565 \times 10^{-16}$ | $1.608 \times 10^{-13}$ | $3.105 \times 10^{-15}$ | $2.624 \times 10^{-13}$ | 31.3 | 2.658 | 100.00 | 50.89 ± 0.97 | 9.88 |
| Total | $1.618 \times 10^{-15}$ | $8.227 \times 10^{-13}$ | $1.601 \times 10^{-13}$ | $6.469 \times 10^{-13}$ | | 1.502 | | 28.94 ± 0.66 | |
| **VP94-132, biotite, J = 1.0713 × 10$^{-2}$** | | | | | | | | | |
| 4.0 W | $1.049 \times 10^{-17}$ | $5.085 \times 10^{-16}$ | $5.912 \times 10^{-15}$ | $8.274 \times 10^{-15}$ | 62.3 | 0.872 | 100.00 | 16.78 ± 0.66 | 1.63E−1 |
| Total | $1.049 \times 10^{-17}$ | $5.085 \times 10^{-16}$ | $5.912 \times 10^{-15}$ | $8.274 \times 10^{-15}$ | | 0.872 | | 16.78 ± 0.66 | |
| **VP94-132, biotite, J = 1.0713 × 10$^{-2}$** | | | | | | | | | |
| 0.3 W* | $6.030 \times 10^{-18}$ | $8.281 \times 10^{-17}$ | $4.820 \times 10^{-15}$ | $5.491 \times 10^{-15}$ | 66.8 | 0.761 | 25.93 | 14.64 ± 0.43 | 3.26E−2 |
| 0.5 W* | $5.634 \times 10^{-18}$ | $1.838 \times 10^{-16}$ | $5.059 \times 10^{-15}$ | $5.431 \times 10^{-15}$ | 68.7 | 0.737 | 53.15 | 14.19 ± 0.27 | 6.90E−2 |
| 0.7 W* | $4.640 \times 10^{-18}$ | $1.314 \times 10^{-16}$ | $4.168 \times 10^{-15}$ | $4.271 \times 10^{-15}$ | 69.6 | 0.713 | 75.58 | 13.73 ± 0.56 | 5.99E−1 |
| 4.0 W | $3.171 \times 10^{-18}$ | $1.934 \times 10^{-15}$ | $4.540 \times 10^{-15}$ | $4.615 \times 10^{-15}$ | 82.4 | 0.837 | 100.00 | 16.11 ± 0.45 | 8.10E−1 |
| Total | $1.947 \times 10^{-17}$ | $3.515 \times 10^{-15}$ | $1.859 \times 10^{-14}$ | $1.981 \times 10^{-14}$ | | 0.762 | | 14.67 ± 0.42 | |
| **VP94-135, hornblende, J = 1.0597 × 10$^{-2}$** | | | | | | | | | |
| 4.0 W | $4.025 \times 10^{-17}$ | $4.111 \times 10^{-14}$ | $3.655 \times 10^{-15}$ | $1.537 \times 10^{-14}$ | 45.9 | 1.947 | 100.00 | 36.85 ± 1.76 | 2.15E+1 |
| Total | $4.025 \times 10^{-17}$ | $4.111 \times 10^{-14}$ | $3.655 \times 10^{-15}$ | $1.537 \times 10^{-14}$ | | 1.947 | | 36.85 ± 1.76 | |

**VP94-135, hornblende, $J = 1.0597 \times 10^{-2}$**

| Step | | | | | | | | | |
|---|---|---|---|---|---|---|---|---|---|
| 0.3 W | $5.440 \times 10^{-18}$ | $9.894 \times 10^{-17}$ | $7.568 \times 10^{-17}$ | $1.335 \times 10^{-15}$ | −19.8 | 0.001 | 2.81 | 0.019 ± 30.293 | 2.49 |
| 0.6 W | $2.447 \times 10^{-18}$ | $3.938 \times 10^{-15}$ | $6.840 \times 10^{-16}$ | $1.660 \times 10^{-15}$ | 76.9 | 1.875 | 28.11 | 35.49 ± 2.34 | 1.10E+1 |
| 1.0 W | $5.892 \times 10^{-18}$ | $7.009 \times 10^{-15}$ | $1.177 \times 10^{-15}$ | $1.977 \times 10^{-15}$ | 42.5 | 0.717 | 71.65 | 13.66 ± 0.57 | 1.14E+1 |
| 4.0 W | $4.898 \times 10^{-18}$ | $5.012 \times 10^{-15}$ | $7.667 \times 10^{-16}$ | $1.333 \times 10^{-15}$ | 23.9 | 0.418 | 100.00 | 7.97 ± 3.08 | 1.25E+1 |
| Total | $1.868 \times 10^{-17}$ | $1.606 \times 10^{-14}$ | $2.703 \times 10^{-15}$ | $6.305 \times 10^{-15}$ | | 0.905 | | 17.22 ± 2.57 | |

**VP94-135, hornblende, $J = 1.0597 \times 10^{-2}$**

| Step | | | | | | | | | |
|---|---|---|---|---|---|---|---|---|---|
| 800 | $3.311 \times 10^{-16}$ | $5.574 \times 10^{-15}$ | $5.278 \times 10^{-15}$ | $1.103 \times 10^{-13}$ | 11.6 | 2.435 | 5.11 | 45.95 ± 2.21 | 2.01 |
| 900 | $6.338 \times 10^{-17}$ | $6.073 \times 10^{-15}$ | $3.532 \times 10^{-15}$ | $2.168 \times 10^{-14}$ | 15.9 | 0.976 | 8.52 | 18.56 ± 1.34 | 3.27 |
| 980 | $7.942 \times 10^{-17}$ | $6.184 \times 10^{-14}$ | $1.196 \times 10^{-14}$ | $3.310 \times 10^{-14}$ | 45.2 | 1.254 | 20.06 | 23.82 ± 0.84 | 9.86 |
| 1020* | $1.914 \times 10^{-16}$ | $2.472 \times 10^{-13}$ | $3.763 \times 10^{-14}$ | $7.695 \times 10^{-14}$ | 54.3 | 1.115 | 56.31 | 21.20 ± 0.39 | 1.25E+1 |
| 1040* | $4.977 \times 10^{-17}$ | $4.063 \times 10^{-14}$ | $6.714 \times 10^{-15}$ | $1.786 \times 10^{-14}$ | 37.3 | 0.996 | 62.78 | 18.95 ± 0.99 | 1.15E+1 |
| 1070* | $5.235 \times 10^{-17}$ | $3.485 \times 10^{-14}$ | $5.273 \times 10^{-15}$ | $1.783 \times 10^{-14}$ | 30.1 | 1.024 | 67.86 | 19.47 ± 0.83 | 1.26E+1 |
| 1100* | $8.627 \times 10^{-17}$ | $6.586 \times 10^{-14}$ | $9.651 \times 10^{-15}$ | $3.038 \times 10^{-14}$ | 34.8 | 1.102 | 77.16 | 20.94 ± 0.86 | 1.30E+1 |
| 1130* | $1.068 \times 10^{-16}$ | $5.335 \times 10^{-14}$ | $7.869 \times 10^{-15}$ | $3.598 \times 10^{-14}$ | 25.1 | 1.155 | 84.74 | 21.95 ± 0.53 | 1.29E+1 |
| 1170* | $7.910 \times 10^{-17}$ | $4.804 \times 10^{-14}$ | $7.111 \times 10^{-15}$ | $2.697 \times 10^{-14}$ | 28.8 | 1.096 | 91.59 | 20.83 ± 1.37 | 1.29E+1 |
| 1450 | $1.893 \times 10^{-16}$ | $5.606 \times 10^{-14}$ | $8.727 \times 10^{-15}$ | $6.709 \times 10^{-14}$ | 23.8 | 1.840 | 100.00 | 34.84 ± 0.83 | 1.23E+1 |
| Total | $1.229 \times 10^{-15}$ | $6.195 \times 10^{-13}$ | $1.037 \times 10^{-13}$ | $4.381 \times 10^{-13}$ | | 1.243 | | 23.61 ± 0.79 | |

**VP94-135, biotite, $J = 1.0734 \times 10^{-2}$**

| Step | | | | | | | | | |
|---|---|---|---|---|---|---|---|---|---|
| 0.4 W | $1.069 \times 10^{-17}$ | 7.647E-20 | $6.905 \times 10^{-15}$ | $8.381 \times 10^{-15}$ | 61.4 | 0.746 | 33.44 | 14.38 ± 0.23 | 2.10E−5 |
| 0.7 W | $2.356 \times 10^{-18}$ | $3.162 \times 10^{-17}$ | $7.073 \times 10^{-15}$ | $6.693 \times 10^{-15}$ | 88.5 | 0.838 | 67.69 | 16.15 ± 0.23 | 8.49E−3 |
| 4.0 W | $3.123 \times 10^{-18}$ | $1.239 \times 10^{-16}$ | $5.671 \times 10^{-15}$ | $6.230 \times 10^{-15}$ | 84.2 | 0.787 | 100.00 | 15.17 ± 0.28 | 3.53E−2 |
| Total | $1.617 \times 10^{-17}$ | $1.556 \times 10^{-16}$ | $2.065 \times 10^{-14}$ | $2.130 \times 10^{-14}$ | | 0.790 | | 15.24 ± 0.24 | |

**VP94-135, biotite, $J = 1.0734 \times 10^{-2}$**

| Step | | | | | | | | | |
|---|---|---|---|---|---|---|---|---|---|
| 0.25 W | $1.153 \times 10^{-18}$ | $1.944 \times 10^{-17}$ | $1.007 \times 10^{-15}$ | $1.698 \times 10^{-15}$ | 79.4 | 1.338 | 16.91 | 25.73 ± 0.80 | 3.67E−2 |
| 0.4 W | $6.783 \times 10^{-21}$ | $6.793 \times 10^{-20}$ | $1.040 \times 10^{-15}$ | $1.197 \times 10^{-15}$ | 98.9 | 1.139 | 34.36 | 21.93 ± 0.15 | 1.24E−4 |
| 0.6 W | $6.783 \times 10^{-21}$ | $6.793 \times 10^{-20}$ | $1.242 \times 10^{-15}$ | $1.370 \times 10^{-15}$ | 98.9 | 1.091 | 55.21 | 21.01 ± 0.10 | 1.04E−4 |
| 4.0 W* | $2.525 \times 10^{-18}$ | $6.022 \times 10^{-17}$ | $2.668 \times 10^{-15}$ | $2.623 \times 10^{-15}$ | 70.7 | 0.695 | 100.00 | 13.40 ± 0.70 | 4.29E−2 |
| Total | $3.691 \times 10^{-18}$ | $7.980 \times 10^{-17}$ | $5.957 \times 10^{-15}$ | $6.888 \times 10^{-15}$ | | 0.964 | | 18.57 ± 0.50 | |

**VP94-138, biotite, $J = 1.0755 \times 10^{-2}$**

| Step | | | | | | | | | |
|---|---|---|---|---|---|---|---|---|---|
| 0.25W | $1.130 \times 10^{-17}$ | $2.119 \times 10^{-17}$ | $9.297 \times 10^{-15}$ | $1.168 \times 10^{-14}$ | 70.6 | 0.887 | 30.21 | 17.13 ± 0.38 | 4.33E−3 |
| 0.35W | $1.589 \times 10^{-18}$ | $3.972 \times 10^{-17}$ | $4.102 \times 10^{-15}$ | $4.217 \times 10^{-15}$ | 87.9 | 0.904 | 43.54 | 17.45 ± 0.33 | 1.84E−2 |
| 0.6W | $4.010 \times 10^{-18}$ | $1.668 \times 10^{-16}$ | $9.990 \times 10^{-15}$ | $9.708 \times 10^{-15}$ | 86.9 | 0.844 | 76.00 | 16.30 ± 0.26 | 3.17E−2 |
| 4.0W | $2.739 \times 10^{-18}$ | $6.281 \times 10^{-17}$ | $7.387 \times 10^{-15}$ | $7.189 \times 10^{-15}$ | 87.7 | 0.854 | 100.00 | 16.49 ± 0.27 | 1.62E−2 |
| Total* | $1.964 \times 10^{-17}$ | $2.905 \times 10^{-16}$ | $3.078 \times 10^{-14}$ | $3.280 \times 10^{-14}$ | | 0.867 | | 16.75 ± 0.31 | |

**VP94-138, biotite, $J = 1.0755 \times 10^{-2}$**

| Step | | | | | | | | | |
|---|---|---|---|---|---|---|---|---|---|
| 0.2 W | $1.325 \times 10^{-17}$ | $1.035 \times 10^{-16}$ | $8.513 \times 10^{-15}$ | $1.120 \times 10^{-14}$ | 64.3 | 0.846 | 16.45 | 16.35 ± 0.29 | 2.31E−2 |
| 0.5 W | $5.564 \times 10^{-18}$ | $1.609 \times 10^{-16}$ | $1.723 \times 10^{-14}$ | $1.659 \times 10^{-14}$ | 89.1 | 0.858 | 49.73 | 16.57 ± 0.23 | 1.77E−2 |
| 0.7 W | $5.365 \times 10^{-18}$ | $1.987 \times 10^{-16}$ | $1.550 \times 10^{-14}$ | $1.472 \times 10^{-14}$ | 88.2 | 0.838 | 79.67 | 16.19 ± 0.10 | 2.44E−2 |
| 4.0 W | $4.034 \times 10^{-18}$ | $6.666 \times 10^{-17}$ | $1.052 \times 10^{-14}$ | $9.935 \times 10^{-15}$ | 86.9 | 0.821 | 100.00 | 15.86 ± 0.13 | 1.20E−2 |
| Total* | $2.821 \times 10^{-17}$ | $5.298 \times 10^{-16}$ | $5.176 \times 10^{-14}$ | $5.244 \times 10^{-14}$ | | 0.843 | | 16.27 ± 0.18 | |

Table 4. *(continued)*

| $T$ (°C) [or watts (W)] | $^{36}$Ar (mol) | $^{37}$Ar (mol) | $^{39}$Ar (mol) | $^{40}$Ar (mol) | $^{40}$Ar* (%) | $^{40}$Ar*/$^{39}$Ar(K) | Cumulative $^{39}$Ar (%) | Calculated age (Ma ± 1σ) | Ca/K |
|---|---|---|---|---|---|---|---|---|---|
| **VP94-140a, hornblende, $J = 1.0609 \times 10^{-2}$** | | | | | | | | | |
| 0.3 W | $8.222 \times 10^{-17}$ | $1.504 \times 10^{-16}$ | $8.121 \times 10^{-17}$ | $4.663 \times 10^{-15}$ | 48.2 | 27.689 | 0.69 | 464.65 ± 14.04 | 3.52 |
| 0.7 W | $9.622 \times 10^{-18}$ | $1.493 \times 10^{-15}$ | $4.239 \times 10^{-14}$ | $1.817 \times 10^{-14}$ | 91.3 | 3.926 | 36.59 | 73.62 ± 0.96 | 6.71 |
| 4.0 W* | $1.229 \times 10^{-17}$ | $2.711 \times 10^{-14}$ | $7.488 \times 10^{-14}$ | $2.909 \times 10^{-14}$ | 95.5 | 3.718 | 100.00 | 69.80 ± 1.50 | 6.90 |
| Total | $3.014 \times 10^{-17}$ | $4.220 \times 10^{-14}$ | $1.181 \times 10^{-14}$ | $5.193 \times 10^{-14}$ | | 3.958 | | 74.21 ± 1.39 | |
| **VP94-140a, hornblende, $J = 1.0609 \times 10^{-2}$** | | | | | | | | | |
| 0.3 W | $1.206 \times 10^{-16}$ | $3.703 \times 10^{-16}$ | $2.730 \times 10^{-15}$ | $7.754 \times 10^{-15}$ | 54.4 | 15.472 | 1.68 | 274.20 ± 5.77 | 2.58 |
| 0.4 W | $4.807 \times 10^{-18}$ | $1.455 \times 10^{-15}$ | $3.810 \times 10^{-15}$ | $2.104 \times 10^{-15}$ | 38.4 | 2.126 | 4.01 | 40.24 ± 4.97 | 7.28 |
| 0.7 W | $1.855 \times 10^{-17}$ | $4.196 \times 10^{-14}$ | $1.181 \times 10^{-14}$ | $4.675 \times 10^{-14}$ | 95.9 | 3.806 | 76.46 | 71.41 ± 0.37 | 6.77 |
| 4.0 W* | $6.124 \times 10^{-18}$ | $1.378 \times 10^{-14}$ | $3.839 \times 10^{-15}$ | $1.471 \times 10^{-14}$ | 95.7 | 3.675 | 100.00 | 69.00 ± 0.31 | 6.84 |
| Total | $4.155 \times 10^{-17}$ | $5.757 \times 10^{-14}$ | $1.631 \times 10^{-14}$ | $7.132 \times 10^{-14}$ | | 3.932 | | 73.72 ± 0.55 | |
| **VP94-140a, biotite, $J = 1.0775 \times 10^{-2}$** | | | | | | | | | |
| 0.2 W | $2.004 \times 10^{-17}$ | $2.260 \times 10^{-16}$ | $7.866 \times 10^{-15}$ | $1.069 \times 10^{-14}$ | 44.0 | 0.598 | 47.20 | 11.59 ± 0.25 | 5.46E − 2 |
| 0.6 W | $9.717 \times 10^{-18}$ | $1.647 \times 10^{-16}$ | $6.688 \times 10^{-15}$ | $9.671 \times 10^{-15}$ | 69.7 | 1.008 | 87.34 | 19.50 ± 0.21 | 4.68E − 2 |
| 4.0 W | $5.640 \times 10^{-18}$ | $1.376 \times 10^{-16}$ | $2.111 \times 10^{-15}$ | $3.185 \times 10^{-15}$ | 47.4 | 0.715 | 100.00 | 13.84 ± 0.24 | 1.24E − 1 |
| Total* | $3.540 \times 10^{-17}$ | $5.283 \times 10^{-16}$ | $1.666 \times 10^{-14}$ | $2.355 \times 10^{-14}$ | | 0.778 | | 15.05 ± 0.23 | |
| **VP94-140a, biotite, $J = 1.0775 \times 10^{-2}$** | | | | | | | | | |
| 0.3W | $8.561 \times 10^{-18}$ | $8.854 \times 10^{-17}$ | $5.004 \times 10^{-15}$ | $6.586 \times 10^{-15}$ | 60.9 | 0.802 | 45.02 | 15.52 ± 0.44 | 3.36E − 2 |
| 0.5W | $6.054 \times 10^{-18}$ | $1.238 \times 10^{-17}$ | $3.295 \times 10^{-15}$ | $4.259 \times 10^{-15}$ | 57.2 | 0.739 | 74.67 | 14.32 ± 0.58 | 7.14E − 3 |
| 0.7W | $1.685 \times 10^{-18}$ | 7.207E-20 | $1.186 \times 10^{-15}$ | $1.617 \times 10^{-15}$ | 68.4 | 0.933 | 85.34 | 18.05 ± 0.73 | 1.15E − 4 |
| 4.0W | $1.871 \times 10^{-18}$ | $7.347 \times 10^{-17}$ | $1.630 \times 10^{-15}$ | $2.038 \times 10^{-15}$ | 72.3 | 0.905 | 100.00 | 17.50 ± 0.82 | 8.56E − 2 |
| Total* | $1.817 \times 10^{-17}$ | $1.744 \times 10^{-16}$ | $1.111 \times 10^{-14}$ | $1.450 \times 10^{-14}$ | | 0.812 | | 15.72 ± 0.57 | |
| **VP94-140b, hornblende, $J = 1.0622 \times 10^{-2}$** | | | | | | | | | |
| 0.3 W | $5.023 \times 10^{-18}$ | $5.849 \times 10^{-17}$ | $1.403 \times 10^{-16}$ | $2.848 \times 10^{-15}$ | 48.0 | 9.752 | 2.76 | 177.81 ± 10.52 | 7.93E − 1 |
| 0.5 W | $5.891 \times 10^{-19}$ | $4.421 \times 10^{-17}$ | $3.437 \times 10^{-16}$ | $1.575 \times 10^{-15}$ | 89.0 | 4.077 | 9.51 | 76.48 ± 6.09 | 2.44E − 1 |
| 0.85 W | $8.863 \times 10^{-18}$ | $1.073 \times 10^{-15}$ | $2.689 \times 10^{-15}$ | $6.002 \times 10^{-15}$ | 71.6 | 1.603 | 62.24 | 30.47 ± 0.69 | 7.60 |
| 4.0 W* | $6.624 \times 10^{-18}$ | $7.611 \times 10^{-15}$ | $1.926 \times 10^{-15}$ | $3.591 \times 10^{-15}$ | 63.6 | 1.189 | 100.00 | 22.64 ± 1.24 | 7.53 |
| Total | $2.110 \times 10^{-17}$ | $1.844 \times 10^{-14}$ | $5.099 \times 10^{-15}$ | $1.402 \times 10^{-14}$ | | 1.839 | | 34.90 ± 1.54 | |
| **VP94-140b, hornblende, $J = 1.0622 \times 10^{-2}$** | | | | | | | | | |
| 0.3W | $7.720 \times 10^{-18}$ | $1.420 \times 10^{-16}$ | $1.557 \times 10^{-16}$ | $3.819 \times 10^{-15}$ | 40.6 | 9.955 | 3.05 | 181.33 ± 13.39 | 1.73 |
| 0.5W | $6.393 \times 10^{-18}$ | $1.220 \times 10^{-15}$ | $3.865 \times 10^{-16}$ | $2.246 \times 10^{-15}$ | 20.5 | 1.195 | 10.62 | 22.75 ± 12.87 | 6.01 |
| 0.65W | $4.144 \times 10^{-18}$ | $4.023 \times 10^{-15}$ | $1.016 \times 10^{-15}$ | $2.659 \times 10^{-15}$ | 66.9 | 1.756 | 30.49 | 33.34 ± 1.91 | 7.55 |
| 0.85W | $4.977 \times 10^{-18}$ | $6.634 \times 10^{-15}$ | $1.703 \times 10^{-15}$ | $3.552 \times 10^{-15}$ | 74.5 | 1.559 | 63.81 | 29.63 ± 0.87 | 7.42 |
| 4.0W | $5.352 \times 10^{-18}$ | $7.405 \times 10^{-15}$ | $1.850 \times 10^{-15}$ | $3.384 \times 10^{-15}$ | 72.0 | 1.320 | 100.00 | 25.12 ± 0.59 | 7.63 |
| Total | $2.859 \times 10^{-17}$ | $1.942 \times 10^{-14}$ | $5.111 \times 10^{-15}$ | $1.566 \times 10^{-14}$ | | 1.740 | | 33.05 ± 2.27 | |

**VP94-141, biotite, $J = 1.0796 \times 10^{-2}$**

| Step | | | | | % | | % | Age ± err | |
|---|---|---|---|---|---|---|---|---|---|
| 0.3 W | $7.999 \times 10^{-18}$ | $4.021 \times 10^{-17}$ | $3.342 \times 10^{-15}$ | $3.978 \times 10^{-14}$ | 94.0 | 11.187 | 40.96 | $205.71 \pm 2.69$ | 2.29E−2 |
| 0.5 W | $1.162 \times 10^{-18}$ | $2.325 \times 10^{-17}$ | $1.331 \times 10^{-15}$ | $5.310 \times 10^{-15}$ | 93.3 | 3.723 | 57.28 | $71.09 \pm 0.56$ | 3.32E−2 |
| 0.7 W | $6.783 \times 10^{-21}$ | $3.502 \times 10^{-17}$ | $1.623 \times 10^{-15}$ | $6.420 \times 10^{-15}$ | 99.8 | 3.947 | 77.16 | $75.29 \pm 0.85$ | 4.10E−2 |
| 4.0 W | $3.735 \times 10^{-18}$ | $1.071 \times 10^{-16}$ | $1.863 \times 10^{-15}$ | $7.316 \times 10^{-15}$ | 84.8 | 3.329 | 100.00 | $63.70 \pm 1.08$ | 1.09E−1 |
| Total | $1.290 \times 10^{-17}$ | $2.056 \times 10^{-16}$ | $8.159 \times 10^{-15}$ | $5.883 \times 10^{-14}$ | | 6.735 | | $126.63 \pm 1.61$ | |

**VP94-141, biotite, $J = 1.0796 \times 10^{-2}$**

| Step | | | | | % | | % | Age ± err | |
|---|---|---|---|---|---|---|---|---|---|
| 0.2 W | $7.279 \times 10^{-18}$ | $1.257 \times 10^{-16}$ | $4.410 \times 10^{-15}$ | $1.735 \times 10^{-14}$ | 87.4 | 3.438 | 15.89 | $65.76 \pm 0.61$ | 5.42E−2 |
| 0.4 W* | $3.795 \times 10^{-18}$ | $1.315 \times 10^{-16}$ | $8.002 \times 10^{-15}$ | $3.074 \times 10^{-14}$ | 96.1 | 3.692 | 44.72 | $70.52 \pm 0.60$ | 3.12E−2 |
| 0.6 W* | $3.702 \times 10^{-18}$ | $1.216 \times 10^{-16}$ | $5.324 \times 10^{-15}$ | $2.055 \times 10^{-14}$ | 94.5 | 3.647 | 63.90 | $69.67 \pm 0.72$ | 4.34E−2 |
| 4.0 W* | $5.206 \times 10^{-18}$ | $6.053 \times 10^{-16}$ | $1.002 \times 10^{-14}$ | $3.839 \times 10^{-14}$ | 95.9 | 3.672 | 100.00 | $70.14 \pm 0.45$ | 1.15E−1 |
| Total | $1.998 \times 10^{-17}$ | $9.841 \times 10^{-16}$ | $2.776 \times 10^{-14}$ | $1.070 \times 10^{-13}$ | | 3.636 | | $69.47 \pm 0.57$ | |

**VP94-142, biotite, $J = 1.0818 \times 10^{-2}$**

| Step | | | | | % | | % | Age ± err | |
|---|---|---|---|---|---|---|---|---|---|
| 0.3 W | $7.564 \times 10^{-18}$ | $1.307 \times 10^{-16}$ | $2.782 \times 10^{-15}$ | $1.197 \times 10^{-14}$ | 81.2 | 3.494 | 27.18 | $66.92 \pm 1.53$ | 8.93E−2 |
| 0.5 W | $1.697 \times 10^{-18}$ | 7.067E−20 | $3.037 \times 10^{-15}$ | $1.300 \times 10^{-14}$ | 95.9 | 4.104 | 56.86 | $78.36 \pm 0.83$ | 4.42E−5 |
| 0.7 W | $7.587 \times 10^{-19}$ | 7.067E−20 | $1.850 \times 10^{-15}$ | $8.653 \times 10^{-15}$ | 97.2 | 4.545 | 74.94 | $86.59 \pm 1.35$ | 7.26E−5 |
| 4.0 W | $4.325 \times 10^{-18}$ | $1.800 \times 10^{-17}$ | $2.565 \times 10^{-15}$ | $2.763 \times 10^{-14}$ | 95.3 | 10.265 | 100.00 | $189.97 \pm 3.09$ | 1.33E−2 |
| Total | $1.434 \times 10^{-17}$ | $1.489 \times 10^{-16}$ | $1.023 \times 10^{-14}$ | $6.125 \times 10^{-14}$ | | 5.562 | | $105.40 \pm 1.68$ | |

**VP94-142, biotite, $J = 1.0818 \times 10^{-2}$**

| Step | | | | | % | | % | Age ± err | |
|---|---|---|---|---|---|---|---|---|---|
| 0.3 W | $3.813 \times 10^{-17}$ | $4.101 \times 10^{-16}$ | $2.461 \times 10^{-14}$ | $9.564 \times 10^{-14}$ | 88.0 | 3.420 | 34.94 | $65.53 \pm 0.66$ | 3.17E−2 |
| 0.5 W* | $5.181 \times 10^{-18}$ | $1.454 \times 10^{-16}$ | $1.820 \times 10^{-14}$ | $6.894 \times 10^{-14}$ | 97.5 | 3.694 | 60.78 | $70.70 \pm 0.28$ | 1.52E−2 |
| 0.75 W* | $4.202 \times 10^{-18}$ | $4.066 \times 10^{-17}$ | $1.210 \times 10^{-14}$ | $4.585 \times 10^{-14}$ | 97.0 | 3.677 | 77.96 | $70.37 \pm 0.27$ | 6.39E−3 |
| 4.0 W* | $4.744 \times 10^{-18}$ | 8.781E−20 | $1.552 \times 10^{-14}$ | $5.900 \times 10^{-14}$ | 97.3 | 3.701 | 100.00 | $70.82 \pm 0.31$ | 1.08E−5 |
| Total | $5.226 \times 10^{-17}$ | $5.962 \times 10^{-16}$ | $7.042 \times 10^{-14}$ | $2.694 \times 10^{-13}$ | | 3.597 | | $68.87 \pm 0.42$ | |

**VP94-143, biotite, $J = 1.0839 \times 10^{-2}$**

| Step | | | | | % | | % | Age ± err | |
|---|---|---|---|---|---|---|---|---|---|
| 0.3 W | $1.685 \times 10^{-17}$ | $9.225 \times 10^{-18}$ | $4.797 \times 10^{-15}$ | $2.583 \times 10^{-14}$ | 80.5 | 4.336 | 42.45 | $82.84 \pm 1.03$ | 3.65E−3 |
| 0.5 W | $2.475 \times 10^{-18}$ | $1.916 \times 10^{-17}$ | $2.253 \times 10^{-15}$ | $9.475 \times 10^{-15}$ | 92.1 | 3.872 | 62.39 | $74.17 \pm 1.37$ | 1.62E−2 |
| 4.0 W | $7.579 \times 10^{-18}$ | $4.418 \times 10^{-17}$ | $4.250 \times 10^{-15}$ | $1.801 \times 10^{-14}$ | 87.3 | 3.700 | 100.00 | $70.94 \pm 1.86$ | 1.98E−2 |
| Total | $2.691 \times 10^{-17}$ | $7.256 \times 10^{-17}$ | $1.130 \times 10^{-14}$ | $5.331 \times 10^{-14}$ | | 4.004 | | $76.64 \pm 1.41$ | |

**VP94-143, biotite, $J = 1.0839 \times 10^{-2}$**

| Step | | | | | % | | % | Age ± err | |
|---|---|---|---|---|---|---|---|---|---|
| 0.2 W | $2.770 \times 10^{-17}$ | $1.495 \times 10^{-16}$ | $4.039 \times 10^{-15}$ | $1.929 \times 10^{-14}$ | 57.4 | 2.741 | 7.91 | $52.82 \pm 0.37$ | 7.03E−2 |
| 0.5 W* | $9.107 \times 10^{-18}$ | $1.323 \times 10^{-16}$ | $1.655 \times 10^{-14}$ | $6.134 \times 10^{-14}$ | 95.3 | 3.534 | 40.32 | $67.82 \pm 0.18$ | 1.52E−3 |
| 0.7 W* | $5.669 \times 10^{-18}$ | $1.409 \times 10^{-16}$ | $1.219 \times 10^{-14}$ | $4.500 \times 10^{-14}$ | 96.0 | 3.545 | 64.20 | $68.02 \pm 0.55$ | 2.20E−2 |
| 4.0 W* | $6.130 \times 10^{-18}$ | $1.175 \times 10^{-16}$ | $1.828 \times 10^{-14}$ | $6.734 \times 10^{-14}$ | 97.0 | 3.575 | 100.00 | $68.58 \pm 0.34$ | 1.22E−2 |
| Total | $4.860 \times 10^{-17}$ | $4.211 \times 10^{-16}$ | $5.105 \times 10^{-14}$ | $1.930 \times 10^{-13}$ | | 3.489 | | $66.96 \pm 0.34$ | |

**VP94-144, hornblende, $J = 1.0757 \times 10^{-2}$**

| Step | | | | | % | | % | Age ± err | |
|---|---|---|---|---|---|---|---|---|---|
| 0.3 W | $7.085 \times 10^{-17}$ | $6.341 \times 10^{-16}$ | $1.686 \times 10^{-16}$ | $2.399 \times 10^{-15}$ | 15.0 | 2.137 | 2.76 | $41.01 \pm 9.94$ | 7.16 |
| 0.6 W* | $7.966 \times 10^{-18}$ | $1.661 \times 10^{-14}$ | $4.525 \times 10^{-15}$ | $1.924 \times 10^{-14}$ | 95.1 | 4.055 | 76.73 | $77.03 \pm 1.05$ | 6.99 |
| 0.8 W | $2.621 \times 10^{-19}$ | $2.530 \times 10^{-15}$ | $6.059 \times 10^{-16}$ | $2.697 \times 10^{-15}$ | 105.2 | 4.695 | 86.63 | $88.90 \pm 1.90$ | 7.96 |
| 4.0 W | $2.182 \times 10^{-18}$ | $3.338 \times 10^{-15}$ | $8.180 \times 10^{-16}$ | $3.516 \times 10^{-15}$ | 89.8 | 3.870 | 100.00 | $73.58 \pm 1.27$ | 7.78 |
| Total | $1.749 \times 10^{-17}$ | $2.311 \times 10^{-14}$ | $6.117 \times 10^{-15}$ | $2.785 \times 10^{-14}$ | | 4.041 | | $76.76 \pm 1.41$ | |

**Table 4.** (*continued*)

| $T$ (°C) [or watts (W)] | $^{36}Ar$ (mol) | $^{37}Ar$ (mol) | $^{39}Ar$ (mol) | $^{40}Ar$ (mol) | $^{40}Ar^*$ (%) | $^{40}Ar^*/^{39}Ar(K)$ | Cumulative $^{39}Ar$ (%) | Calculated age (Ma ± 1σ) | Ca/K |
|---|---|---|---|---|---|---|---|---|---|
| **VP94-144, hornblende, $J = 1.0757 \times 10^{-2}$** | | | | | | | | | |
| 0.3 W | $8.437 \times 10^{-18}$ | $1.140 \times 10^{-16}$ | $2.956 \times 10^{-16}$ | $2.555 \times 10^{-15}$ | 6.2 | 0.540 | 2.84 | $10.46 \pm 4.67$ | 7.35 |
| 0.5 W | $1.554 \times 10^{-17}$ | $9.281 \times 10^{-15}$ | $2.565 \times 10^{-15}$ | $1.127 \times 10^{-14}$ | 66.3 | 2.920 | 27.52 | $55.79 \pm 1.11$ | 6.89 |
| 0.7 W | $1.151 \times 10^{-17}$ | $1.927 \times 10^{-14}$ | $5.149 \times 10^{-15}$ | $2.156 \times 10^{-14}$ | 91.8 | 3.855 | 77.06 | $73.31 \pm 0.39$ | 7.13 |
| 4.0 W | $1.214 \times 10^{-17}$ | $9.645 \times 10^{-15}$ | $2.385 \times 10^{-15}$ | $1.018 \times 10^{-14}$ | 72.9 | 3.120 | 100.00 | $59.55 \pm 0.88$ | 7.70 |
| Total | $4.762 \times 10^{-17}$ | $3.934 \times 10^{-14}$ | $1.039 \times 10^{-14}$ | $4.556 \times 10^{-14}$ | | 3.361 | | $64.08 \pm 0.80$ | |
| **VP94-144, biotite, $J = 1.0850 \times 10^{-2}$** | | | | | | | | | |
| 0.3 W | $1.097 \times 10^{-18}$ | $1.895 \times 10^{-16}$ | $7.234 \times 10^{-15}$ | $2.651 \times 10^{-14}$ | 87.6 | 3.208 | 32.05 | $61.74 \pm 0.88$ | 4.98E−2 |
| 0.5 W | $3.657 \times 10^{-18}$ | $1.798 \times 10^{-16}$ | $6.835 \times 10^{-15}$ | $2.386 \times 10^{-14}$ | 95.2 | 3.325 | 62.34 | $63.94 \pm 0.95$ | 5.00E−2 |
| 0.7 W* | $6.026 \times 10^{-18}$ | $2.075 \times 10^{-16}$ | $4.302 \times 10^{-15}$ | $1.853 \times 10^{-14}$ | 90.2 | 3.887 | 81.40 | $74.53 \pm 0.65$ | 9.16E−2 |
| 4.0 W | $4.583 \times 10^{-18}$ | $3.739 \times 10^{-16}$ | $4.199 \times 10^{-15}$ | $1.670 \times 10^{-14}$ | 91.8 | 3.653 | 100.00 | $70.12 \pm 1.19$ | 1.69E−1 |
| Total | $2.523 \times 10^{-17}$ | $9.508 \times 10^{-16}$ | $2.257 \times 10^{-14}$ | $8.561 \times 10^{-14}$ | | 3.456 | | $66.41 \pm 0.91$ | |
| **VP94-146, biotite, $J = 1.0850 \times 10^{-2}$** | | | | | | | | | |
| 0.3 W | $5.103 \times 10^{-18}$ | $9.764 \times 10^{-17}$ | $2.953 \times 10^{-15}$ | $1.028 \times 10^{-14}$ | 85.1 | 2.963 | 31.21 | $57.08 \pm 1.03$ | 6.28E−2 |
| 0.5 W | $3.506 \times 10^{-18}$ | $7.852 \times 10^{-17}$ | $2.493 \times 10^{-15}$ | $8.316 \times 10^{-15}$ | 87.3 | 2.912 | 57.56 | $56.12 \pm 0.96$ | 5.98E−2 |
| 0.7 W | $4.579 \times 10^{-18}$ | $1.508 \times 10^{-16}$ | $1.308 \times 10^{-15}$ | $4.439 \times 10^{-15}$ | 69.5 | 2.360 | 71.39 | $45.61 \pm 0.50$ | 2.19E−1 |
| 4.0 W | $2.043 \times 10^{-18}$ | $1.067 \times 10^{-16}$ | $2.707 \times 10^{-15}$ | $9.091 \times 10^{-15}$ | 93.1 | 3.128 | 100.00 | $60.22 \pm 0.95$ | 7.49E−2 |
| Total | $1.523 \times 10^{-17}$ | $4.337 \times 10^{-16}$ | $9.461 \times 10^{-15}$ | $3.212 \times 10^{-14}$ | | 2.913 | | $56.14 \pm 0.91$ | |
| **VP94-146, hornblende, $J = 1.0634 \times 10^{-2}$** | | | | | | | | | |
| 0.5 W | $2.303 \times 10^{-18}$ | $4.546 \times 10^{-18}$ | $1.599 \times 10^{-15}$ | $1.158 \times 10^{-14}$ | 97.4 | 7.072 | 52.23 | $130.81 \pm 2.01$ | 5.41 |
| 1.0 W | $3.933 \times 10^{-18}$ | $3.120 \times 10^{-18}$ | $1.092 \times 10^{-15}$ | $4.705 \times 10^{-15}$ | 80.9 | 3.491 | 87.92 | $65.77 \pm 1.66$ | 5.44 |
| 4.0 W | $1.754 \times 10^{-18}$ | $1.001 \times 10^{-18}$ | $3.696 \times 10^{-16}$ | $1.523 \times 10^{-15}$ | 71.5 | 2.953 | 100.00 | $55.78 \pm 3.53$ | 5.15 |
| Total | $7.990 \times 10^{-18}$ | $8.666 \times 10^{-18}$ | $3.060 \times 10^{-15}$ | $1.781 \times 10^{-14}$ | | 5.297 | | $98.86 \pm 2.07$ | |
| **VP94-146, hornblende, $J = 1.0634 \times 10^{-2}$** | | | | | | | | | |
| 0.2 W | $4.285 \times 10^{-18}$ | $1.362 \times 10^{-16}$ | $1.552 \times 10^{-16}$ | $6.137 \times 10^{-15}$ | 79.5 | 31.465 | 2.64 | $520.71 \pm 7.93$ | 1.67 |
| 0.5 W | $3.701 \times 10^{-18}$ | $5.673 \times 10^{-16}$ | $2.097 \times 10^{-15}$ | $1.043 \times 10^{-14}$ | 94.1 | 4.687 | 38.22 | $87.74 \pm 0.70$ | 5.15 |
| 0.8 W | $3.289 \times 10^{-18}$ | $6.145 \times 10^{-16}$ | $2.186 \times 10^{-15}$ | $9.111 \times 10^{-15}$ | 95.0 | 3.968 | 75.31 | $74.57 \pm 0.56$ | 5.35 |
| 4.0 W* | $2.700 \times 10^{-18}$ | $4.069 \times 10^{-16}$ | $1.455 \times 10^{-15}$ | $5.290 \times 10^{-15}$ | 91.4 | 3.329 | 100.00 | $62.76 \pm 0.52$ | 5.32 |
| Total | $1.397 \times 10^{-17}$ | $1.602 \times 10^{-15}$ | $5.894 \times 10^{-15}$ | $3.096 \times 10^{-14}$ | | 4.791 | | $89.65 \pm 0.80$ | |
| **VP94-144, biotite, $J = 1.0841 \times 10^{-2}$** | | | | | | | | | |
| 0.2 W | $7.888 \times 10^{-18}$ | $1.199 \times 10^{-16}$ | $2.274 \times 10^{-15}$ | $4.442 \times 10^{-15}$ | 47.2 | 0.922 | 13.17 | $17.95 \pm 1.63$ | 1.00E−1 |
| 0.4 W | $5.535 \times 10^{-18}$ | $1.631 \times 10^{-16}$ | $4.291 \times 10^{-15}$ | $5.853 \times 10^{-15}$ | 71.5 | 0.976 | 38.01 | $18.98 \pm 0.89$ | 7.22E−2 |
| 0.6 W* | $5.733 \times 10^{-18}$ | $1.850 \times 10^{-15}$ | $8.340 \times 10^{-15}$ | $1.246 \times 10^{-14}$ | 87.0 | 1.300 | 86.31 | $25.26 \pm 0.38$ | 4.21E−1 |
| 4.0 W | $6.796 \times 10^{-18}$ | $3.239 \times 10^{-16}$ | $2.365 \times 10^{-15}$ | $3.504 \times 10^{-15}$ | 42.8 | 0.634 | 100.00 | $12.35 \pm 0.67$ | 2.60E−1 |
| Total | $2.595 \times 10^{-17}$ | $2.457 \times 10^{-15}$ | $1.727 \times 10^{-14}$ | $2.626 \times 10^{-14}$ | | 1.079 | | $20.97 \pm 0.71$ | |

**VP94-146, biotite, $J = 1.0841 \times 10^{-2}$**

| | | | | | | | | | | |
|---|---|---|---|---|---|---|---|---|---|---|
| 0.25 W* | $5.794 \times 10^{-18}$ | 7.067E-20 | $3.465 \times 10^{-15}$ | $6.445 \times 10^{-15}$ | 72.9 | 1.356 | 20.08 | 26.32 | 1.24 | 3.88E-5 |
| 0.35 W* | $3.175 \times 10^{-18}$ | 7.067E-20 | $2.969 \times 10^{-15}$ | $5.021 \times 10^{-15}$ | 80.7 | 1.365 | 37.29 | 26.50 | 0.43 | 4.52E-5 |
| 0.6 W* | $3.837 \times 10^{-18}$ | $3.271 \times 10^{-17}$ | $6.384 \times 10^{-15}$ | $9.496 \times 10^{-15}$ | 87.4 | 1.300 | 74.29 | 25.25 | 0.39 | 9.74E-3 |
| 4.0 W | $1.193 \times 10^{-18}$ | $1.254 \times 10^{-16}$ | $4.436 \times 10^{-15}$ | $7.357 \times 10^{-15}$ | 94.7 | 1.571 | 100.00 | 30.47 | 0.57 | 5.37E-2 |
| Total | $1.400 \times 10^{-17}$ | $1.583 \times 10^{-16}$ | $1.725 \times 10^{-14}$ | $2.832 \times 10^{-14}$ | | 1.392 | | 27.02 | 0.61 | |

Note: $^{40}K\lambda = 5.5430 \times 10^{-10}$; * = step used in plateau, isochron, weighted mean or maximum age determinations; $^{40}Ar/^{39}Ar$ analyses were performed at La Trobe University. Samples and flux monitor GA1550 biotite (97.9 Ma: McDougall & Roksandic 1974) were wraped in Al foil and placed in a quartz vial. Flux monitors were spaced every 0.5–1.0 cm over the length of the tube. Vials were sealed and irradiated at Soreq Nuclear Research Centre, IRR-1 reactor, Israel (Heimann et al. 1992). Biotite samples, single grains or small numbers of grains, were loaded into milled pits in a Cu plate and placed into a gas extraction line fitted with a glass viewport. Gas was extracted using a defocused 6 W Ar ion laser. Step-wise heating was accomplished by varying the power output of the laser. Gas extraction from hornblende was performed in a computer-controlled double-vacuum resistance furnace. Extrated gas was expanded into a stainless steel clean-up line and purified for c. 10 min with Zr–Ti getters. Ar isotopes were measured using a VG3600 mass spectrometer operating at a sensitivity of $1 \times 10^{-3}$ A/torr$^{-1}$, using a Daly photomultiplier set at a gain of 100 over the Faraday cups. Data was corrected for machine background, line blank and mass discrimination. Extraction line blanks of atmospheric composition were typically $9 \times 10^{-17}$–$1 \times 10^{-16}$ mole $^{40}Ar$ (laser) and $2 \times 10^{-16}$ (furnace). Mass discrimination, determined by analysing atmospheric Ar, was 1.0105 for 1 amu. Correction factors for interfering isotopes were determined by analysing $K_2SO_4$ and $CaF_2$ salts irradiated with the samples, and the following values were used: $(^{40}Ar/^{39}Ar)K = 0.0105$, $(^{36}Ar/^{37}Ar)Ca = 0.000298$ and $(^{39}Ar/^{37}Ar)Ca = 0.000712$.

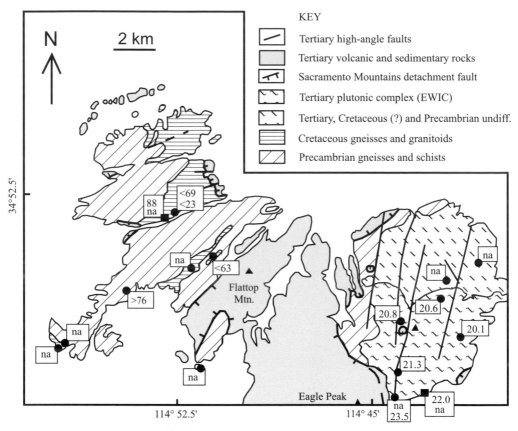

**Fig. 6.** Distribution of hornblende $^{40}Ar/^{39}Ar$ ages.

reproducible degassing behaviour between sample aliquots was a factor in determining weighted means. Biotite ages are similar at 13–14 Ma (Fig. 7), with the exception of VP94-069b, whose release spectra behaviour was not reproducible and therefore uninterpretable. The discordance observed in some of the biotite analyses is probably due to the presence of minor intergrown chlorite related to greenschist facies deformation after crystallization of the EWIC. Chlorite intergrowths in biotite commonly lead to variable amounts of $^{39}Ar$ recoil during neutron irradiation (Lo & Onstott 1989) and in these cases the total fusion age is generally accepted as not representing the cooling age of the sample. This is because the recoil is predominantly within the grain and little, if any, leaves the system for relatively coarse grained material.

Hornblende and biotite samples from the northwest Sacramento Mountains were analysed by the laser step-heating and fusion techniques, displaying concave and convex spectra. Where

replicate analyses show reproducibility, data have been averaged to produce weighted mean ages. In the case of curved spectra, maximum or minimum ages are assigned on the basis of the shape of the release spectra, i.e. where the concave curve of saddle-shaped spectra result from the presence of excess Ar, maximum ages are derived from the curve minimum; where the convex curve of release spectra result from slow cooling or Ar loss, minimum ages are derived from the curve maximum. The resulting ages from the northwest vary from *c.* 50 to 80 Ma (Table 4 and Fig. 6). Biotite samples from this region (all analysed by the laser-dating technique) display degassing patterns similar to those from the northeast, with flat and variable release spectra. Biotite ages range from *c.* 15 to 75 Ma (Table 4 and Fig. 7). Replicate analyses of sample VP94–144 (biotite) did not produce the same results (within $2\sigma$ errors). The hornblende separate from this sample yields an age of $\geq77$ Ma, has a release curve indicative of slow cooling or Ar loss and suggests that, because of

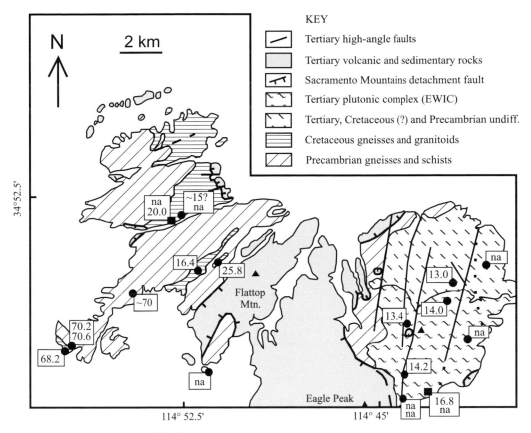

**Fig. 7.** Distribution of biotite $^{40}$Ar/$^{39}$Ar ages. Note transition to Miocene ages in the northwest.

the lower $T_c$ associated with biotite, the biotite age should be <77 Ma. Therefore, an imprecise but reasonable estimate for this biotite age is 65–75 Ma and this is supported by one of the analyses. A younging trend to the northeast is not well defined in the hornblende ages from the northwest part of the range (Fig. 6). This is due to the relatively high closure temperature of hornblende and suggests that all of the foot-wall was at a temperature below hornblende closure prior to intrusion of the EWIC. Biotite ages in the northwest Sacramento Mountains, however, young in the direction of tectonic trans-port (Fig. 7), with a sharp transition between Late Cretaceous ages and Tertiary (mostly Miocene) ages.

## Fission-track analyses

*Analytical methods.* Fission-track (FT) ana-lyses of zircon and apatite were performed at La Trobe University, Melbourne, Australia,

following the procedures described in Foster & Gleadow (1992) and Fitzgerald & Gleadow (1988). FT ages were calculated using the zeta calibration method (Hurford & Green 1983) and errors were calculated using the conventional method of Green (1981), which assumes a Poisson distribution. The 'chi-squared ($\chi^2$)' sta-tistic is used to determine the probability that all grains counted belong to a single population of ages (Galbraith 1981). A probability of <5% is evidence of an asymmetric, i.e. non-Poisson, spread of single-grain ages. Central ages have been reported where the pooled data result in a marginal $\chi^2$ probability, i.e. 10–15%, which may indicate a non-homogeneous age popula-tion. Zircon and apatite grains were mounted in epoxy resin on glass slides, and were ground and polished to reveal internal surfaces of grains. Zircon samples were etched in a mixture of KOH and NaOH, at 200–220°C for up to 4 h, to reveal fossil fission tracks. The amount of time in the etchant bath is dependent upon the amount of radiation damage in the grains,

**Table 5.** *Summary of fission-track analytical results*

| Sample No. | Elevation/ distance (m/km) | No. of grains | Spontaneous track density (×10⁵ cm⁻²) | Induced track density (×10⁶ cm⁻²) | Standard track density (×10⁶ cm⁻²) | Mean track length (μm) | Standard deviation (μm) | Uranium (ppm) | $\chi^2$ probability (%) | Fission-track age (Ma ± 1σ) | |
|---|---|---|---|---|---|---|---|---|---|---|---|
| **Zircons** | | | | | | | | | | | |
| VP94-129 | 700/17.0 | 13 | 35.45 (1397) | 4.284 (1688) | 0.5541 (2266) | – | – | 402 | 93.6 | 20.1 | 1.1 |
| VP94-132 | 729/12.5 | 15 | 15.63 (1005) | 2.292 (1474) | 0.556 (2266) | – | – | 214.4 | 76.3 | 16.6 | 1.0 |
| VP94-135 | 802/14.25 | 15 | 8.129 (588) | 1.203 (870) | 0.5579 (2266) | – | – | 112.1 | 98.6 | 16.5 | 1.1 |
| VP94-138 | 775/8.7 | 15 | 34.58 (1781) | 7.371 (3796) | 0.5204 (2007) | – | – | 566.6 | 99.7 | 15.3 | 0.7 |
| VP94-140a | 780/7.6 | 12 | 15 (1147) | 3.017 (2307) | 0.5178 (2007) | – | – | 233.1 | 94.2 | 16.1 | 0.8 |
| VP94-141 | 655/0.5 | 12 | 43.96 (1444) | 2.033 (668) | 0.5598 (2266) | – | – | 188.9 | 92.9 | 52.9 | 3.3 |
| VP94-143 | 675/0.4 | 12 | 20.93 (1130) | 1.104 (596) | 0.5617 (2266) | – | – | 102.2 | 12.1 | 47.5 | 3.3 |
| VP94-144 | 750/4.3 | 15 | 31.67 (513) | 3.654 (592) | 0.5636 (2266) | – | – | 337.1 | 28.8 | 21.5 | 1.5 |
| VP94-145 | 750/7.0 | 10 | 25.29 (446) | 1.06 (187) | 0.5655 (2266) | – | – | 97.5 | 92 | 58.9 | 5.7 |

**Apatites**

| | | | | | | | | | | | |
|---|---|---|---|---|---|---|---|---|---|---|---|
| DF88-021 | 610/12.25 | 23 | 0.9908 (78) | 1.734 (1365) | 14.1 (100) | 1.519 (5835) | 1.2 | 14.3 | 83.7 | 16.6 | 2.0 |
| VP94-057b | 583/16.5 | 25 | 0.9639 (156) | 1.301 (2105) | 14.2 (100) | 1.403 (2736) | 1.1 | 10.3 | 57.2 | 19.7 | 1.7 |
| VP94-127 | 802/16.75 | 25 | 0.8944 (152) | 1.43 (2431) | 13.8 (100) | 1.403 (2736) | 1.3 | 11.3 | 95.3 | 16.6 | 1.4 |
| VP94-132 | 729/12.5 | 25 | 0.7012 (87) | 1.131 (1403) | 14 (20) | 1.403 (2736) | 1.4 | 10.1 | 83.2 | 16.6 | 1.9 |
| VP94-135 | 802/14.25 | 25 | 0.5252 (90) | 0.8754 (1500) | 14.3 (9) | 1.403 (2736) | 0.93 | 6.9 | 48.5 | 15.9 | 1.8 |
| VP94-138 | 775/8.7 | 25 | 0.8393 (141) | 2.228 (3743) | 14.5 (100) | 1.403 (2736) | 0.95 | 17.6 | 96.6 | 10.0 | 0.9 |
| VP94-140a | 780/7.6 | 25 | 0.9236 (161) | 1.618 (2820) | 14.4 (90) | 1.403 (2736) | 1.1 | 14.4 | 63.1 | 15.2 | 1.3 |
| VP94-140b | 780/7.6 | 20 | 0.6186 (85) | 1.084 (1489) | 14.3 (54) | 1.403 (2736) | 1.07 | 9.7 | 97.9 | 15.2 | 1.7 |
| VP94-141 | 655/0.5 | 25 | 0.3605 (62) | 0.9088 (1563) | 14 (78) | 1.403 (2736) | 1.15 | 8.1 | 12.1 | 10.7 | 1.6 |
| VP94-142 | 655/0.5 | 20 | 0.3469 (51) | 0.4483 (659) | 14.2 (2) | 1.403 (2736) | 0.85 | 4 | 47.7 | 20.6 | 3.0 |
| VP94-143 | 675/0.4 | 25 | 1.348 (245) | 2.426 (4410) | 13.8 (38) | 1.403 (2736) | 1.16 | 21.6 | 56.3 | 14.8 | 1.0 |
| VP94-144 | 750/4.3 | 25 | 0.7792 (130) | 0.995 (1660) | 14 (36) | 1.403 (2736) | 1.56 | 8.9 | 71.3 | 20.8 | 1.9 |
| VP94-146 | 690/10.3 | 20 | 1.114 (139) | 1.41 (1759) | 14 (74) | 1.403 (2736) | 1.32 | 12.6 | 94.3 | 21.0 | 1.9 |

Age calculation using the zeta method [after Green (1981)]: dosimeter glass U3 for zircon (zeta 88) and glass CN5 for apatite (379 or 383). Central ages are reported where pooled data result in low probability from the $\chi^2$ test, i.e. c. 12%, which may suggest a non-homogeneous age population. Number of tracks counted in parentheses. Standard and induced track densitites measured on external detectors ($g = 0.5$), spontaneous (fossil) track densities on internal mica surfaces. Distance = distance down-dip from an arbitrary reference line (the northwest–southeast steel tower transmission line road), perpendicular to the transport direction (Fig. 2).

which is a function of the U content and the sample age. Apatite samples were etched in 5N HNO$_3$ at room temperture for 20 s to reveal fossil fission tracks. All samples were irradiated in the X-7 position of the Australian HIFAR Research Reactor. Thermal neutron fluences were monitored using a muscovite detector.

FT dating of zircon and apatite provides a measure of rock thermal histories $c. <260°C$. This method relies on the temperature-dependent annealing of fission tracks that results in complete erasure of tracks in zircon and apatite at temperatures of $c. 220-260$ and $c. 110°C$, respectively, for $10^6-10^7$ year timescales (Naeser 1979; Gleadow *et al.* 1986; Green *et al.* 1989; Brandon & Vance 1992; Yamada *et al.* 1995; Foster *et al.* 1996*a*). Below these temperatures, track annealing rates decrease progressively with lower temperatures. The temperature interval bounded by the point of total annealing and the point of effective track stability where annealing occurs at very slow rates is known as the partial annealing zone, PAZ (e.g. Fitzgerald & Gleadow 1988). This interval for common compositions

of apatite spans temperatures of $c. 110-60°C$ (Green *et al.* 1989), while that for zircon is less well constrained at $c. 200-260°C$ (Brandon & Vance 1992). The temperatures recorded by apparent zircon and apatite ages are a function of the cooling rate through, and the subsequent time–temperature path within, the PAZ. The relatively simple histories recorded by the FT system in these phases (see Table 5), however, suggests that the effective closure temperatures are $c. 245°C$ for zircon and $c. 110°C$ for apatite, for rapid cooling rates (Brandon & Vance 1992; Foster *et al.* 1996*a*).

*FT results.* FT analytical results are presented in Table 5. Confined track lengths in apatite grains are generally $14 \pm 1$ ($1\sigma$) $\mu$m, providing some degree of confidence that these samples have not resided in the PAZ for a significant period of time and that data associated with these samples, in spite of the relative dispersion in the data set, reflect geologically meaningful ages. Most samples pass the $\chi^2$ test, suggesting age distributions represent a single

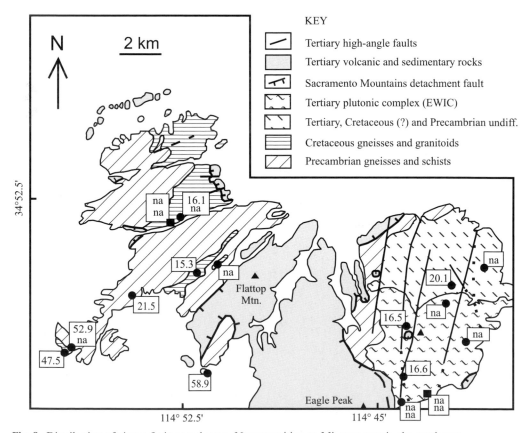

**Fig. 8.** Distribution of zircon fission-track ages. Note transition to Miocene ages in the northwest.

population. Analytical results from other workers are reported for comparison (e.g. apatite FT data from Foster *et al.* 1991), sample number SCM11.

The three zircon FT dates from the northeast Sacramento Mountains (VP94-129, VP94-132 and VP94-135) can be regarded as similar at *c.* 18 Ma (within $2\sigma$ error). There are too few data to determine whether a correlation exists between age and geographic distribution, i.e. along the tectonic transport direction (Fig. 8). Apatite dates in the northeast Sacramento Mountains are similar, having a mean Miocene age of $17.2 \pm 1.6$ Ma (within $2\sigma$ error), and no correlation between age and geographic distribution (Fig. 9).

Zircon FT dates from the northwest Sacramento Mountains show younging in the tectonic transport direction (N60E). Ages vary systematically from *c.* 50 to 15 Ma, with a sharp transition between the Eocene and Miocene ages (Fig. 8). Some of the zircon FT ages in the northwest Sacramento Mountains have poor

analytical precision (e.g. VP94-145, Precambrian upper-plate basement). This appears to be the result of zircon grains being strongly coloured. Grain opacity obscures tracks and dislocations. This problem is further compounded because zircon from the older basements also tend to have higher U contents, generally resulting in higher degrees of radiation damage. Any bias introduced due to zircon coloration is nonsystematic.

The apatite FT data from the northwest Sacramento Mountains (Fig. 2), unlike the zircon FT results, define no correlation between age and geographic distribution. Only Miocene ages are recorded, ranging from *c.* 21 to 10 Ma (Fig. 9). Dispersion within the apatite FT data from the northwest Sacramento Mountains is in excess of experimental error and the variation in ages from a single location is geologically untenable (e.g. VP94-141 and VP94-142, both samples of Precambrian metasediment). FT retention is a function of time, temperature and composition. FT lengths of *c.* 14 $\mu$m suggest that these samples

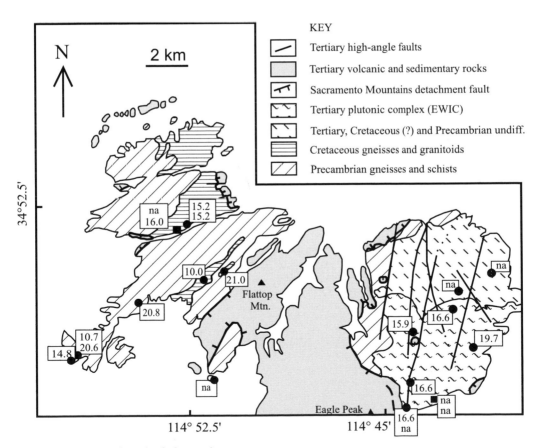

**Fig. 9.** Distribution of apatite fission-track ages.

have not been above apatite $T_c$ since the Miocene, therefore mineral composition was considered. FT in Cl-rich apatites are more resistant to track annealing (shortening) than are Fl-rich apatites (Gleadow & Duddy 1981; Crowley *et al.* 1991; O'Sullivan & Parrish 1995). F-rich and Cl-poor apatites, being more susceptible to track annealing, may have less tracks, producing systematically younger FT ages. The critical Cl content below which tracks in apatites will anneal relatively quickly appears to be *c.* 0.2–0.3 wt%. Electron microprobe apatite analyses of samples from the northwest show that 60% of the apatites analysed are F-apatites (Pease 1997). Consequently, it seems probable that the range in apatite compositions contributes to the unexpected variation in apatite FT ages. It is also possible that counting statistics contribute to this scatter because of the relatively low spontaneous track densities arising from the relatively young age and low U concentrations.

## Pressure and temperature associated with the emplacement of the EWIC

Although contact metamorphic rocks may contain mineral assemblages appropriate for emplacement barometry, applying the various techniques may be difficult in pervasively intruded terranes because of extensive thermal overprinting. In contrast, refractory igneous phases are commonly less affected by younger thermal events. Even in mylonites, porphyroclasts can retain compositions related to igneous crystallization (Anderson 1988). Consequently, samples from the EWIC were used for thermobarometric determinations (Fig. 2); analytical results from other workers are reported for comparison (e.g. a single analysis from Anderson 1988, indicated by "A., '88").

*Thermobarometric analytical methods.* Mineral compositions of hornblende and plagioclase were determined using a Cambridge Instruments Microscan 9 electron microprobe. Operating conditions during analyses were: beam diameter, 0.5 $\mu$m; beam current, 40 nA; accelerating potential, 20 kV. Elemental compositions are generally precise to better than 1% ($2\sigma$), so long as elemental abundance exceeds the detection limit of the microprobe (e.g. Potts 1987). The Fortran program *RECAMP* of Spear & Kimball (1984) was used to recalculate electron microprobe analyses of amphiboles into amphibole structural formulae, using only the 13-cation output (Cosca *et al.* 1991). The chosen normalization scheme does not significantly affect calculated

$Al_{tot}$ or pressure, but it does affect calculated [4]Al and [6]Al, and this can affect the calculated temperatures.

Crystallization temperatures were calculated, using the equations of Holland & Blundy (1994), by assuming a given pressure; temperatures were then substituted into the revised equation of Anderson & Smith (1995) in order to calculate pressures. The resulting, calculated, pressures were then used to solve for the crystallization temperatures, and this iterative process was continued until convergence in the value of both temperature and pressure was achieved. Both equation A ($T_a$) and equation B ($T_b$) of Holland & Blundy (1994) were used to calculate the crystallization temperature of quartz-bearing and marginally quartz-bearing samples (CIPW normative quartz of *c.* 0–5%), whereas only $T_b$ was used for non-quartz bearing samples.

The error associated with these temperature determinations is ±35–50°C ($1\sigma$; Holland & Blundy 1994), including the analytical uncertainties associated with electron microprobe determinations of silica in amphibole and anorthite contents. For the compositions in this study ($X_{ab} = 0.6$–0.8 and Si per formula unit in amphibole $= 6.5$–7.2), the $2\sigma$ uncertainty is ±75°C (Blundy & Holland 1990). It should be noted, however, that the associated error may be larger for samples whose compositions vary from that of the calibration data set, i.e. $T_a$: $T = 400$–900°C, amphiboles with [A]Na > 0.02 per formula unit (pfu), [6]Al < 1.8 pfu, Si = 6.0–7.7 pfu and plagioclases with $X_{an} < 0.90$; $T_b$: $T = 500$–900°C, amphiboles with [M4]Na > 0.03 pfu, [6]Al < 1.8 pfu, Si = 6.0–7.7 pfu and plagioclases with $0.1 < X_{an} < 0.9$.

The Al-in-hornblende barometer assumes that the phase compositions represent near-emplacement conditions, i.e. that crystallization occurs near the $H_2O$-saturated solidus, thus rim compositions in contact with quartz are most likely to reflect emplacement conditions. This barometer is commonly applied to rocks with the restricted mineral assemblage used in the experimental calibration: quartz + alkali feldspar + plagioclase + hornblende + biotite + iron titanium oxide + sphene + liquid. The granodioritic phases of the EWIC usually contain the full buffering assemblage, though quartz and potassium feldspar can be absent in the more mafic phases. There is some evidence to suggest that the full buffering assemblage may not be necessary: Anderson & Smith (1995) found that the absence of K-feldspar and sphene did not significantly affect pressure determinations. It is also common practice to apply the barometer to restricted mineral compositions. Anderson & Bender

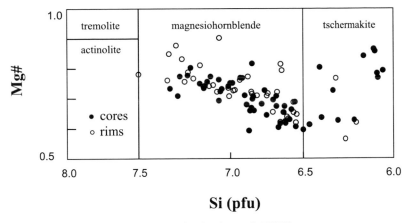

Fig. 10. EWIC amphibole compositions [fields after Leake *et al.* (1997)].

(1988) showed that the barometer fails for Proterozoic anorogenic granitoids in which Fe contents in the hornblendes are higher than those used in the original calibration, therefore use of the barometer is restricted to amphiboles with $Fe^{tot}/(Fe^{tot} + Mg)$ between 0.4 and 0.65 and $Fe^{3+}/(Fe^{3+} + Fe^{2+}) > 0.2$. Also, plagioclase compositions of $c.\,An_{25-35}$ were used in the original calibration, thus the application of the barometer in this study is similarly restricted.

Anderson & Smith (1995) estimate a total error of $\pm 1.5\,kbar$ $(2\sigma)$ for their revised calibration of this barometer. This figure includes the $2\sigma$ variation associated with regression of the calibration data sets ($\pm 0.6\,kbar$), the propagation of $2\sigma$ uncertainties associated with temperature determinations ($\pm 50°C \Rightarrow 0.8\,kbar$), as well as an analytical precision associated with electron microprobe analyses of $Al_2O_3$ ($\pm 1\% \rightarrow 0.1\,kbar$). The analytical precision associated with the determination of $Al_2O_3$ is well within the range of analyses from this investigation and its contribution to the total error is relatively minor, but the contribution from the temperature determination is underestimated. The error contribution from the temperature determination should be evaluated using $\pm 75°C$ $(2\sigma)$. This increases the error contribution associated with temperature determination to $\pm 1.1\,kbar$. Consequently, the best figure for uncertainty associated with this barometer is $1.8\,kbar$ $(2\sigma)$.

*Amphibole Classification.* The Al-in-hornblende barometer assumes that emplacement conditions reflect crystallization near the $H_2O$-saturated solidus. Amphiboles are known to be susceptible to subsolidus alteration. Consequently, amphibole compositions from granodioritic phases of

the EWIC were evaluated, prior to the application of the barometer, to determine whether subsolidus alteration had occurred. Using the calcic amphibole classification of Leake *et al.* (1997), amphibole compositions were determined for the EWIC (Fig. 10). The majority of calcic amphiboles are magnesiohornblende, with $c.\,15\%$ being tschermakitic or edenitic. Using the empirical criterion of Fleet & Barnett (1978), the $^{[4]}Al/^{[6]}Al$ ratio can distinguish unaltered igneous calcic amphiboles from low- and high-pressure metamorphic calcic amphiboles. Amphiboles with $^{[4]}Al/^{[6]}Al > 3.3$ are considered to be the products of crystallization from a melt. It is clear that the majority of amphiboles

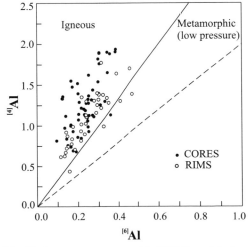

Fig. 11. Igneous v. metamorphic EWIC amphibole compositions [fields after Fleet & Barnett (1978)].

from the granodioritic phases of the EWIC, both cores and rims, represent calcic, igneous compositions (Fig. 11). Consequently, the thermobarometric results which follow are likely to represent the emplacement temperature and pressure of the EWIC at solidus conditions.

*Thermobarometric Results.* The results of thermobarometric calculations are shown in Table 6. Attrition due to unacceptable microprobe analysis, non-stoichiometric hornblende compositions or mineral compositions outside the calibration data set, resulted in generally three to five grain-pairs per thin section comprising the sample means quoted in Table 6. Only $T_b$ was applied to non-quartz-bearing samples and because quartz was not present in the buffering assemblage, pressures calculated using these temperatures should be evaluated with caution (see discussion below). Microprobe analyses and calculated CIPW normative chemistry can be found in Pease (1997).

Two samples require special mention, VP94-124 and VP94-119b. The various causes of attrition (refer to analytical section) resulted in only a single acceptable analysis of VP94-124. The result is reported in Table 6 but is not statistically representative and, consequently, is not used in the final calculations. Though the temperature results from $T_a$ and $T_b$ are the same (within $2\sigma$ error), VP94-119b has a consistently higher $T_a$ than $T_b$, combined with a consistently lower $P_a$ than $P_b$. This is typical for samples which are silica undersaturated (Holland & Blundy 1994). This sample proved to be only nominally silica saturated at *c.* 4–6% normative quartz, whereas the other samples in this investigation are clearly quartz normative (>10%) or lacking in normative quartz (Pease 1997). This suggests that only equation B (Holland & Blundy 1994) should be applied to this sample. Subsequently, only $T_b$ from VP94-119b is used in the final calculations.

There is little difference ($\leq 8°C$) between the calculated temperatures, when compositional criteria allow both equation A and B to be applied (Holland & Blundy 1994; Table 6; samples VP94-061, VP94-130b and VP94-246). The within-sample temperature variation from these analyses is generally between 30 and 70°C, within the $2\sigma$ error of $\pm 75°C$ quoted for the technique. (The notable exception is VP94-130b, where grain-pair-4 is outside the grouping of analyses and is regarded as an outlier, and is therefore not included in the sample mean.) The individual grain analyses are averaged to produce **sample means** for $T_a$ and $T_b$ (Table 6). Sample means for $T_a$ and $T_b$ are within $2\sigma$

analytical error, and can be averaged together to represent an **overall mean** for the sample. The overall sample means show rim temperature variations within the suite of *c.* 30°C, from 660 to 690°C, and the weighted average of the overall means for the temperature of emplacement ($T_1$) is $676 \pm 19°C$.

When compositional criteria allow only equation B (Holland and Blundy 1994) to be used, the calculated temperature of emplacement ($T_2$) is nearly identical to $T_1$ above (within $2\sigma$ analytical error, using VP94-011c, VP94-019, VP94-119b, VP94-123b, VP94-123c and VP94-125). Sample mean temperatures (using equation B only; Holland & Blundy 1994) vary *c.* 75°C, from 655 to 730°C, with a resulting weighted average for $T_2 = 683 \pm 31°C$. These two temperatures are the same, within error, thus the temperature of emplacement of the granodioritic phase of the EWIC is taken to be $680 \pm 19°C$.

By comparing the results of Si-saturated samples ($P_a$) with non-Si-saturated samples ($P_b$), the validity of using non-quartz-bearing, EWIC rocks to calculate pressures using the Al-in-hornblende barometer can be assessed. The pressures calculated using Si-saturated samples (VP94-061, VP94-130b and VP94-246) show excellent agreement between $P_a$ and $P_b$, allowing the average of these sample means to represent the pressure of emplacement for each sample (the overall mean). The resulting weighted average of the overall means ($P_1$) is $3.7 \pm 1.4$ kbar.

$P_b$ from non-saturated samples show a greater range of pressure variation than those from saturated samples. Using the six reliable non-saturated samples, an overall mean of $3.1 \pm 1.2$ kbar ($P_2$) is generated. This value is indistinguishable (within error) from the Si-saturated mean of $3.7 \pm 1.4$ kbar ($P_1$ above). Over 50% of the non-saturated samples yield results in excellent agreement with $P_1$ and it is interesting to note that VP94-125 yields one of the highest pressures calculated for the EWIC (4.5 kbar). Si-undersaturated minerals are entirely absent from the non-saturated samples; it may be that the activity of Si ($a_{Si}$) of these samples, though low, is close to unity, and the pressure calculation continues to yield accurate results. Consequently, it would seem that, in some instances, samples without normative quartz can be used to determine pressure with the Al-in-hornblende barometer. This kind of application, however, requires an 'external' pressure determination to verify the results or an independent check on $a_{Si}$ on a 'per sample' basis. Finally, because $P_1$ and $P_2$ are within error, and can be combined, the pressure of emplacement for the

**Table 6.** *Results from amphibole–plagioclase thermometry and Al-in-hornblende barometry*

| Sample ID | Rim | | | |
|---|---|---|---|---|
| | $T_a$ | $T_b$ | $P_a$ | $P_b$ |
| VP94-011c 1-r | | 660 | | 4.9 |
| VP94-011c 2-r | | 633 | | 3.5 |
| VP94-011c 3-r | | 674 | | 3.4 |
| VP94-011c 4-r | | 683 | | 2.9 |
| VP94-011c 5-r | | 656 | | 1.3 |
| **Mean** | | **661 ± 34** | | **3.2 ± 0.8** |
| VP94-019 4-r | | 691 | | 1.9 |
| VP94-019 5-r | | 677 | | 1.8 |
| VP94-019 6-r | | 646 | | 0.6 |
| **Mean** | | **671 ± 43** | | **1.4 ± 1.0** |
| VP94-061 1-c* | 680 | 652 | 3.6 | 3.8 |
| VP94-061 3-c* | 648 | 687 | 1.8 | 1.6 |
| VP94-061 4-c* | 725 | 691 | 3.9 | 4.4 |
| VP94-061 4-r * | 707 | 697 | 3.4 | 3.5 |
| VP94-061 5-c* | 723 | 718 | 3.9 | 4.0 |
| VP94-061 5-r * | 670 | 669 | 3.6 | 3.6 |
| **Mean** | **692 ± 31** | **685 ± 31** | **3.4 ± 0.7** | **3.5 ± 0.7** |
| **Overall mean** | | **689 ± 31** | | **3.4 ± 0.7** |
| VP94-119b 1-c* | 740 | 707 | 3.7 | 4.2 |
| VP94-119b 2-c* | 724 | 709 | 3.5 | 3.7 |
| VP94-119b 3-c* | 768 | 733 | 2.5 | 3.2 |
| VP94-119b 5-c* | 776 | 735 | 2.4 | 3.2 |
| VP94-119b 5-r* | 782 | 718 | 2.3 | 3.5 |
| **Mean** | **758 + 34** | **720 ± 34** | **2.9 ± 0.8** | **3.6 ± 0.8** |
| VP94-123b 1-r | | 667 | | 2.6 |
| VP94-123b 2-r | | 646 | | 1.9 |
| VP94-123b 3-r | | 644 | | 1.2 |
| VP94-123b 4-r | | 655 | | 2.5 |
| VP94-123b 5-r | | 653 | | 3.0 |
| **Mean** | | **653 ± 34** | | **2.2 ± 0.8** |
| VP94-123c 3-r | | 726 | | 3.8 |
| VP94-123c 4-r | | 699 | | 4.1 |
| VP94-123c 5-r | | 759 | | 3.7 |
| **Mean** | | **728 ± 43** | | **3.9 ± 1.0** |
| (VP94-124 1-r | | 641 | | 1.6) |
| VP94-125 1-r | | 685 | | 5.0 |
| VP94-125 2-r | | 654 | | 4.7 |
| VP94-125 4-r | | 663 | | 4.0 |
| VP94-125 5-r | | 666 | | 4.4 |
| **Mean** | | **667 ± 38** | | **4.5 ± 0.9** |
| VP94-130b 1-r | 638 | 647 | 3.2 | 3.2 |
| VP94-130b 2-r | 681 | 677 | 3.4 | 3.5 |
| VP94-130b 3-r | 681 | 656 | 3.4 | 3.6 |
| VP94-130b 4-r | (787) | (734) | (4.7) | (5.9) |
| VP94-130b 5-r | 651 | 646 | 3.0 | 3.0 |
| **Mean** | **663 ± 38** | **660 ± 38** | **3.3 ± 0.9** | **3.3 ± 0.9** |
| **Overall mean** | | **662 ± 38** | | **3.3 ± 0.9** |
| VP95–246 1-r | 714 | 704 | 4.5 | 4.6 |
| VP95–246 2-r | 727 | 707 | 4.3 | 4.6 |
| VP95–246 3-r | 685 | 689 | 4.9 | 4.8 |
| VP95–246 4-r | 690 | 683 | 4.2 | 4.3 |
| VP95–246 5-r | 708 | 702 | 4.3 | 4.4 |
| **Mean** | **705 ± 34** | **697 ± 34** | **4.4 ± 0.8** | **4.6 ± 0.8** |
| **Overall mean** | | **671 ± 34** | | **4.5 ± 0.8** |

$T_a$ and $T_b$ (temperature in °C) for silica-saturated and unsaturated samples, respectively (Holland & Blundy 1994). $P_a$ and $P_b$ (pressure in kbar) calculated using $T_a$ and $T_b$ respectively (Anderson & Smith 1995). Error ($2\sigma$) of individual analyses: $T \pm 75$°C, $P \pm 1.8$ kbar. Sample ID includes field sample number and grain-pair number (cf. Pease 1997). Numbers in parentheses are not included in the mean (see text). $P_b$ should be evaluated with caution, as compositional criteria may not be met.
* Fine grained, no real distinction between rim and core.

granodioritic phase of the EWIC is calculated to be $3.3 \pm 0.8$ kbar.

Anderson (1988) used the same techniques on a single sample of EWIC from the quarry along Eagle Pass Road: the same location as the biotite–hornblende granodiorite, VP94-119b (Fig. 2). The results of his four rim pairs, recalculated (Anderson 1996) using Anderson & Smith (1995) for pressure and Holland & Blundy (1994) for temperature, are $700 \pm 9°C$ and $4.3 \pm 0.5$ kbar, respectively. The $2\sigma$ errors associated with Anderson's data can also be recalculated and compare well to those from VP94-119b ($T_b = 720 \pm 34°C$ and $P_b = 3.6 \pm 0.8$ kbar). These results verify the reproducibility of the technique and the combined data yield a mean of $711 \pm 25°C$ and $3.9 \pm 0.6$ kbar for the quarry site in the northeast Sacramento Mountains. Although at the upper limit of the temperature variation, these values are in good agreement with the overall results defining the initial conditions related to the intrusion of the EWIC: i.e. depths of 7.5–12 km ($3.3 \pm 0.8$ kbar) and a crystallization temperature of $680 \pm 19°C$ (Table 6).

## Discussion

### Thermal evolution of the northwest Sacramento Mountains

There is a well-defined, contrasting thermal history between the northeast and northwest regions of the Sacramento Mountains (Figs 12 and 13). Younging in the direction of tectonic transport is obvious in the northwest Sacramento Mountains, notably in the biotite and zircon (FT) data (Fig. 12), consistent with cooling related to exhumation via normal faulting. This region also has a pronounced Cretaceous thermal signature (Figs 12 and 13) which may be associated with Late Cretaceous plutonism and/or be derived from earlier metamorphism and crustal thickening associated with Cretaceous thrusting. Cretaceous greenschist- to amphibolite-grade metamorphism, resulting from Laramide compression (thrusting and magmatism), is well documented in the region (Burchfiel & Davis 1981; John & Mukasa 1990; Foster *et al.* 1992). Argent (1996) identified a regional amphibolite-facies deformation event of

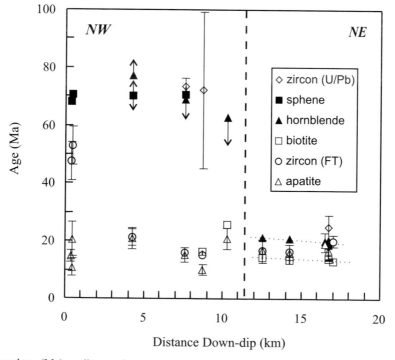

**Fig. 12.** Mineral age (Ma) vs. distance down-dip (km). Ages given with $2\sigma$ error bars (arrows represent maximum or minimum ages for hornblende). Note the break in slope associated with both zircon (FT) and biotite data in the northwest.

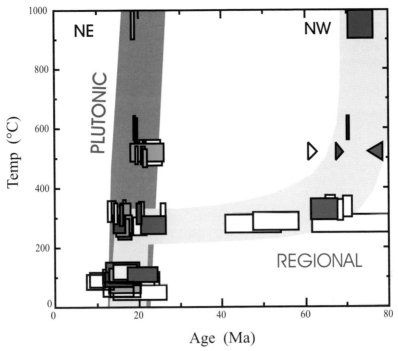

**Fig. 13.** Cooling history of the northern Sacramento Mountains. Closure temperature (°C) vs. mineral age (Ma) show contrasting histories between footwalls of the northwest and northeast regions. The vertical dimension of error boxes represents the range in closure temperature for each mineral system (refer to text), while the horizontal dimension represents $2\sigma$ errors (Ma). Triangles represent maximum and minimum ages for hornblende analyses, where the orientation of the triangle suggests < or > as shown. To identify trends, mineral suites from the same sample have been colour-coded when three or more dating techniques have been applied. The distribution of apatite fission track ages has been exaggerated in the vertical dimension in order to show all of the analyses; the correct range is 90–120°C.

possible Cretaceous age throughout much of the northwest Sacramento Mountains.

Late Cretaceous ages in the northwest Sacramento Mountains associated with the $T_c$ of zircon and sphene ($\geq$600°C), as well as with the $T_c$ of biotite (*c.* 315°C), require cooling rates of the order of 100°C Ma$^{-1}$ (Fig. 13). This figure is consistent with cooling rates associated with Late Cretaceous batholith emplacement in the Old Woman Mountains area 10 km west of the Sacramento Mountains (Foster *et al.* 1989, 1990*b*, 1992), where the Ar systematics in hornblende from Proterozoic rocks have been reset in the Late Cretaceous, suggesting that the thermal profile of the northwest Sacramento Mountains may also be dominated by Late Cretaceous plutonism. Rapid Late Cretaceous cooling in the Old Woman Mountains area contrasts with slower rates of 1–10°C Ma$^{-1}$, both immediately prior to and following this thermal event (Foster *et al.* 1990*b*). Post-Late Cretaceous cooling in the northwest Sacramento Mountains slowed to comparable figures, where the

difference between biotite and zircon (FT) ages in the southwesternmost region indicate cooling of *c.* 1–6°C Ma$^{-1}$. Finally, the northeasternmost region of the northwest Sacramento Mountains records very rapid cooling rates of 38–53°C Ma$^{-1}$ [through the $T_c$ associated with biotite, zircon (FT), and apatite to ambient surface temperatures] between *c.* 19 and 14 Ma. The high-temperature portion of the cooling curve for the northwest Sacramento Mountains is consistent with cooling of the crust by thermal conduction (Carslaw & Jaeger 1959; Harrison & Clarke 1979; Harrison & McDougall 1980; Foster *et al.* 1992), due to the relatively high thermal contrast between the intrusion and the country rock (especially in igneous bodies intruded into the upper crust). The sudden acceleration of the cooling rate to 38–53°C Ma$^{-1}$ between 19 and 14 Ma, however, again suggests that tectonic denudation (unroofing) as a result of detachment faulting was the driving mechanism for the final cooling of the northwest Sacramento Mountains.

## Thermal evolution of the northeast Sacramento Mountains

The thermal regime in the northeast Sacramento Mountains is dominated by the intrusion of the syntectonic EWIC. The initial conditions relating to the intrusion of the EWIC were depths of 7.5–12 km ($3.3 \pm 0.8$ kbar) and a crystallization temperature of $680 \pm 19°C$. After intrusion of the EWIC at $c.$ 19 Ma, very rapid cooling from magmatic temperatures to $c. <110°C$ ($T_c$ of apatite; Fig. 13) and exposure of the EWIC at the surface occurred by 14 Ma, as indicated by the extrusion of the Eagle Peak rhyodacite onto the fault surface. This requires cooling rates $>100°C$ Ma$^{-1}$ and exhumation rates of 1.5–3 km Ma$^{-1}$. A tabular (sheeted) pluton cooling solely by thermal conduction *in situ* will define a cooling curve with a strongly concave upward appearance (Carslaw & Jaeger 1959; Harrison & Clarke 1979; Harrison & McDougall 1980; Foster *et al.* 1992). Such a cooling path is not defined by the EWIC in the northeast Sacramento Mountains, requiring an additional mechanism of convective heat transfer and/or uplift. Regardless of whether or not convective heat transfer occurred, the exhumation rate of 1.5–3 mm y$^{-1}$ clearly necessitates invoking tectonic denudation. Given the contemporaneity of detachment faulting with the age of the EWIC, tectonic denudation associated with Miocene detachment faulting is most likely to be the cause of exhumation and must have made a significant contribution to the cooling of the EWIC. Accelerated exhumation from 0.3 to 1.3–2 mm y$^{-1}$ in the Whipple Mountains (south) and Santa Catalina Mountains (southeast Arizona) has also been correlated with tectonic decompression coincident with mid-Tertiary detachment faulting (Anderson *et al.* 1988).

## Palaeoisotherms

Using the 'palaeoisotherms' derived from the closure temperatures associated with each of the mineral systems, the thermal profile of the footwall in the northern Sacramento Mountains, just before faulting at 20–21 Ma, can be determined (Fig. 14). Inasmuch as these boundaries are likely to be diffuse, given the range in $T_c$ associated with each system, some latitude exists with respect to their locations. For example, all of the apatite ages are Miocene, suggesting that the entire region was above the $T_c$ of fission tracks in apatite ($c. 110°C$) prior to this time. Given that the various mineral data suggest that deeper, i.e. hotter, portions of the crust are being exposed to the northeast, the $c. 110°C$ palaeoisotherm must be located somewhere to the southwest of the sampling area. Without knowledge of the geothermal gradient at 20–21 Ma, the location of the $c. 110°C$ palaeoisotherm is somewhat unconstrained. The $c. 245°C$ palaeoisotherm must be located between VP94-141 and VP94-144. All of the zircon samples northeast of, and including, VP94-144 yield Miocene FT ages and must have been at temperatures above that required for total track annealing in zircon prior to this time. It follows that all of the zircon samples southwest of, and including, VP94-141 record much older FT ages and would have been below, or within, the given PAZ prior to the Miocene. Using the same logic, the palaeoisotherm associated with $c. 315°C$ ($T_c$ of biotite) can also be relatively placed. These palaeoisotherms have been oriented along a northwesterly trend because: (1) some lateral control on their orientation is provided by the spatial distribution of sample locations, and the zircon (FT) and biotite data have similar ages (within each data set) along-strike of this trend; (2) this orientation is perpendicular to the Miocene tectonic transport direction.

This thermal profile suggests that the northwest Sacramento Mountains represents a tilted block of crust in which deeper levels of the footwall were exposed as Miocene detachment faulting progressed. The inferred tilting of this block is top-to-the-southwest, consistent with the overall dip of fault blocks within the extensional corridor. Block rotation of the northwest Sacramento Mountains could have been facilitated by a relatively older imbricate splay of the detachment fault system, or by isostatic rebound (Spencer 1984). Merey & Ticken (1989) identified a gently east dipping reflector at 0.9–1.2 s (two-way traveltime) on a seismic reflection profile adjacent to the northwest Sacramento Mountains. This feature is consistent with the existence of an underlying detachment fault splay. As the palaeoisotherms increase in temperature to the northeast, the cross-sectional elevation also increases. From geometric constraints, assuming that the palaeoisotherms were originally horizontal, the present-day configuration of palaeoisotherms in the northwest Sacramento Mountains requires 15–20° of tilt. The rotation of this block must have occurred prior to dyke intrusion (as dykes in the region are currently near vertical) but while the block was still relatively hot in its deeper portion, as dykes in the northeasternmost region exhibit mylonitic fabrics (Pease & Argent 1999). These dykes are thought to be 19–17 Ma (Spencer 1985).

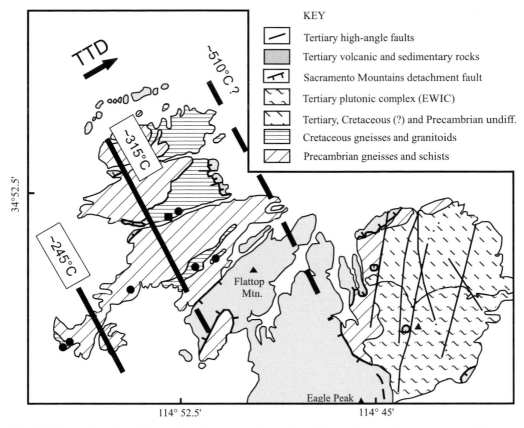

**Fig. 14.** Thermal profile of the northern Sacramento Mountains. The northwest region represents ambient conditions of upper crustal levels prior to the onset of detachment faulting at 21–20 Ma.

Inasmuch as ages young to the northeast, it appears that faulting continued towards that direction. The southwest portion of the northwest Sacramento Mountains would have been at lower temperatures, i.e. shallower depths, than the deeper, down-dip, northeastern region. This suggests that differential uplift and cooling, related to detachment faulting, occurred in the region. The thermal contribution from synchronous intrusion of the EWIC would have resulted in a localized, and transient, elevated geotherm in the northeast. Previous investigations utilizing more limited data sets (Foster *et al.* 1990*b*, 1991) were unable to reveal this detail and complexity in the thermal profile of the northern Sacramento Mountains.

## Angle of detachment faulting

To calculate the angle of faulting, the simple trigonometric relationship is used whereby an angle ($\theta$) is expressed

$$\theta = \sin^{-1}(o/h) \qquad (1)$$

where $o$ is the opposite limb and $h$ the hypotenusal limb. Using the nomenclature of Wernicke (1995), this relationship can be expressed as a function of the palaeogeothermal gradient and field palaeothermal gradient, with $\theta$ as the angle of the fault (Fig. 15). The field paleothermal gradient (the palaeothermal gradient of the footwall measured along the fault) between two points with a temperature difference ($\Delta T$) is related to the paleogeothermal gradient by the average dip of the fault ($\theta$), which from equation 1 is

$$\theta = \sin^{-1}((dT/dw)/(dT/dz)) \qquad (2)$$

where $dT/dz$ is the geothermal gradient just prior to unroofing and $dT/dw$ is the measured field palaeothermal gradient. In this case, $\Delta T$

$$\theta = \sin^{-1} \frac{z}{w}; \quad \text{where} \quad z = \frac{\Delta T}{dT/dz} \quad \text{and} \quad w = \frac{\Delta T}{dT/dw}$$

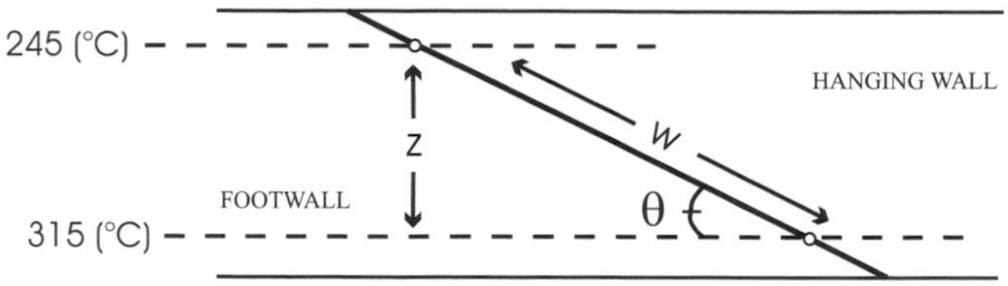

**Fig. 15.** Geometric relationship between the angle of faulting and the geotherm [after Wernicke (1995)]. In this case, the palaeogeothermal gradient $(z) = 20–30°C\,km^{-1}$ and the field palaeothermal gradient $(w)$ is calculated: $\Delta T = 70°C$ and the distance separating these isotherms along the detachment fault is 6 km.

is defined by the relevant palaeoisotherms (the $c.315$ and $c.245°C$ palaeoisotherms in the northwest Sacramento Mountains).

Several assumptions are inherent in combining palaeotemperature variations with depth in order to derive the angle of faulting: (1) the palaeogeothermal gradient must be known; (2) palaeoisotherms should be horizontal and approximate a steady state; and (3) samples proximal to the fault surface should have maintained their relative positions during faulting. Each of these assumptions is evaluated below with respect to the northern Sacramento Mountains.

(1)  The palaeogeothermal gradient must be known. The ambient geothermal gradient prior to Tertiary unroofing has been determined in several areas of the Basin-and-Range Province. In these investigations, the time–temperature history has been determined from rocks with independently estimated palaeodepths. In the region of this investigation, the eastern Mojave Desert, the average geothermal gradient in the upper 10 km of the crust 18 Ma ago has been estimated to be $c.50°C\,km^{-1}$ (Foster *et al.* 1991), 30–50°C km$^{-1}$ (John & Foster 1993) and 16–35°C km$^{-1}$ prior to $c.22$ Ma (Howard & Foster 1996). For other regions in the Basin-and-Range Province: east-central Nevada (upper 10 km of crust), $c.20°C\,km^{-1}$ at 35 Ma (Dumitru *et al.* 1991) and 25–35°C km$^{-1}$ at 8–10 Ma (Holm & Wernicke 1990; Holm *et al.* 1992); in southern Nevada (upper 5 km of the crust), 20–24°C km$^{-1}$ at 15 Ma (Howard & Foster 1996); and in south-central Arizona (upper 12 km of crust), $c.17°C\,km^{-1}$ at $c.25$ Ma (Howard & Foster 1996). The paleogeothermal gradient $(dT/dz)$ prior to detachment faulting in the northern Sacra-

mento Mountains (20–21 Ma) probably ranged from 20 to 30°C km$^{-1}$.

(2)  Palaeoisotherms should approximate a steady-state condition. It seems likely that the palaeogeotherm in the northwest Sacramento Mountains had reached a steady state prior to the onset of detachment faulting, given the absence of magmatotectonic events since Late Cretaceous time (Fig. 13) in this region. Ketchum (1996) has shown that palaeoisotherms from shallow crustal levels (<10 km in the northern Sacramento Mountains) retain their original configuration (i.e. are not compressed), merely shifting to higher crustal levels in response to superposition against a cold hanging wall. Intrusion of the EWIC, as a localized heat source, might have affected the palaeogeothermal gradient in the northwest Sacramento Mountains. Extension in the region, however, was already underway when the EWIC was intruded at $c.19$ Ma, and it is apparent that the rapid exhumation and cooling of the footwall (Fig. 13) would have outpaced thermal perturbations resulting from the intrusive complex.

(3)  Samples in the footwall are assumed to have maintained their relative positions during faulting. Geologic mapping suggests that this is true, as the high-angle and secondary low-angle faults in the range have only minor offsets (10–100 m; Pease & Argent 1999).

Within the limits of these assumptions, the angle of faulting and the slip rate associated with the SMDF can be calculated. This is done by using the apparent distance between palaeoisotherms in the northwest Sacramento Mountains, and by combining exhumation rates with slip rates on detachment faults across the extensional corridor.

*Spacing of isotherms.* In this case, the field palaeothermal gradient ($dT/dw$) is derived from the distance between the two observed palaeoisotherms in the northwest Sacramento Mountains. The distance between the $c.\,245$ and $315°C$ palaeoisotherms (zircon fission track and biotite, respectively) is not well constrained due to the range in closure temperature associated with each mineral system. Nevertheless, the apparent distance separating these two palaeoisotherms can be reasonably determined to be $c.\,6$ km. (A maximum separation of 8 km is achieved if the 245°C palaeoisotherm is located near VP94-141 and the 315°C paleoisotherm is located near VP94-138 but, as previously discussed, this is regarded as implausible). A distance of $c.\,6$ km separating $c.\,70°C$ ($\Delta T$) results in a field palaeothermal gradient ($dT/dw$) of $c.\,12°C$ km$^{-1}$. Using Equation 2, a palaeogeothermal gradient ($dT/dz$) of 20–30°C km$^{-1}$, and a field paleothermal gradient ($dT/dw$) of 12°C km$^{-1}$, the angle of faulting is calculated to be 23–35°.

*Exhumation/Slip rates.* If tectonic denudation associated with Tertiary detachment faulting is the dominant mechanism responsible for the exhumation of Miocene plutonic rocks, then the angle of faulting can be determined using Equation 1. Combining the exhumation rate ($o$) with the slip rate along the fault ($h$), $\theta$ can be calculated. Several investigations have determined the time-averaged exhumation rates of Miocene plutonic rocks in the extensional corridor using geobarometry: 1.5–2.4 mm yr$^{-1}$, this study; 1.4 mm yr$^{-1}$, Chemehuevi Mountains (Foster *et al.* 1990*a*, *b*); 1.3–2.3 mm yr$^{-1}$, Whipple Mountains (Anderson *et al.* 1988); 1.3 mm yr$^{-1}$, Santa Catalina Mountains (Anderson *et al.* 1988). These estimates define comparable and relatively rapid rates of uplift within the extensional corridor from plutons emplaced early in the history of detachment faulting.

Slip rates associated with the early history of detachment faulting in the extensional corridor have been determined using geologic constraints (offset markers, opening and closing of syntectonic basins, etc.): 7–9 mm yr$^{-1}$, Harcuvar/Buckskin–Rawhide Mountains (Spencer & Reynolds 1991), and 7–8 mm yr$^{-1}$, Whipple Mountains (Davis 1988). There is good agreement between slip rates determined by geologic constraints and those derived using thermochronologic data: 7–8 mm yr$^{-1}$, Harcuvar/Buckskin–Rawhide Mountains (Foster *et al.* 1993; Foster & John 1999), and $c.\,8$ mm yr$^{-1}$, Whipple Mountains (Foster *et al.* 1994). The agreement between these techniques verifies the accuracy of

slip rates derived using geologic constraints and is not unexpected, as these footwall blocks were unroofed from beneath the same detachment fault system at about the same time. Data from the central part of the extensional corridor, where extension began later in time and the total amount of slip is less, reveal consistently lower slip rates (3–4 mm yr$^{-1}$, Foster & John 1999; 3–6 mm yr$^{-1}$, this study).

Assuming that tectonic denudation was the dominant mechanism responsible for exhumation in the extensional corridor, using an exhumation rate of 1.3–3.0 mm yr$^{-1}$, and a slip rate of 7–9 mm yr$^{-1}$, the angle of detachment faulting within the extensional corridor is calculated to be 8–25°. This determination reflects the angle of faulting early in the extensional history of the region. Although this calculation is by no means rigorous, it combines data from different authors, using different techniques, working in different mountain ranges. Consequently, some degree of confidence can be placed in the final result: *an active, low-angle fault system*.

## Conclusions

The northern Sacramento Mountains preserve both a Cretaceous–post-Cretaceous cooling record interpreted as the ambient, regional thermal profile in the northwest, as well as a localized thermal regime related to syntectonic, Miocene plutonism in the northeast. This plutonism, represented by the EWIC, occurred in the northeast and central Sacramento Mountains $c.\,19$ Ma ago, recording emplacement at 7.5–12 km ($3.3 \pm 0.8$ kbar) and $680 \pm 19°C$. Thermal profiles in both the northwest and northeast Sacramento Mountains record rapid Miocene exhumation and cooling (to temperatures $c.\,{<}110°C$ by 14 Ma ago), suggesting that tectonic denudation associated with Tertiary detachment faulting was the dominant mechanism by which the crust was exhumed and cooled.

Two techniques have been used to determine the angle of detachment faulting associated with crustal extension in Sacramento Mountains region and yield similar results: 23–35° in the northwest Sacramento Mountains and 8–25° across the extensional corridor. The combined results of these two calculations suggests that the SMDF in the northern Sacramento Mountains was active at $c.\,25°$. These new data contribute to a handful of studies (Foster *et al.* 1990*b*; Richard *et al.* 1990; Dokka 1993; Holm & Dokka 1993; John & Foster 1993) which suggest that detachment faults initiate and are active at low angles, i.e. <30.

This research was funded in part by NERC grants GT4/92/240/G and GT4/93/234/G, the Australian Research Council, the US Geological Survey, the Bureau of Land Management, and the Australian Institute of Nuclear Science and Energy. We thank the personnel at the Soreq Reactor, M. J. Whitehouse for ion-probe analytical assistance (VP94–129), and Stan Szczepanski for help with argon analyses. Comments by M. Krabbendam and an anonymous reviewer improved the manuscript. The authors also wish to thank Amerada Hess Ltd. (UK Exploration) for their generous funding of the colour figure.

# References

ANDERSON, J. 1988. Core complexes of the Mojave–Sonaran desert: Conditions of plutonism, mylonitization, and decompression. *In*: ERNST, W. G. (ed.) *Metamorphic and Crustal Evolution of the Western Conterminous, U.S. Rubey Volume VII.* Prentice-Hall, 503–525.

——1996. Status of thermobarometry in granitic batholiths. *Transactions of the Royal Society of Edinburgh*, **87**, 125–138.

—— & BENDER, E. 1988. Nature and origin of Proterozoic A-type granitic magmatism in the southwestern United States. *Lithos*, **23**, 19–52.

—— & CULLERS, R. 1990. Middle to upper crustal plutonic construction of a magmatic arc; an example from the Whipple Mountains metamorphic core complex. *In*: ANDERSON, J. (ed.) *The Nature and Origin of Cordilleran Magmatism.* Geological Society of America Memoir, **174**, 47–69.

—— & SMITH, D. 1995. The effects of temperature and *f*O₂ on the Al-in-hornblende barometer. *American Mineralogist*, **80**, 549–559.

——, BARTH, A. & YOUNG, E. 1988. Mid-crustal Cretaceous roots of Cordilleran metamorphic core complexes. *Geology*, **16**, 366–369.

ARGENT, J. 1996. *The geology and Tertiary structural evolution of the northwest Sacramento Mountains footwall core complex, southeast California, USA.* DPhil Thesis, University of Oxford.

ARMSTRONG, R. L. 1982. Cordilleran metamorphic core complexes – from Arizona to southern Canada. *Annual Review of Earth and Planetary Sciences*, **10**, 129–154.

BLUNDY, J. & HOLLAND, T. 1990. Calcic amphibole equilibria and a new amphibole–plagioclase geothermometer. *Contributions to Mineralogy and Petrology*, **104**, 208–224.

BRANDON, M. & VANCE, J. 1992. Tectonic evolution of the Cenozoic Olympic subduction complex, Washington State, as deduced from fission-track ages for detrital zircons. *American Journal of Science*, **292**, 565–636.

BUCK, W. R. 1988. Flexural rotation of normal faults. *Tectonics*, **7**, 959–975.

BURCHFIEL, B. & DAVIS, G. 1981. Mojave Desert and environs. *In*: ERNST, W. (ed.) *The Geotectonic Development of California Rubey Volume I.* Prentice-Hall, 217–252.

BROOKS, W. E. & MARTIN, R. F. 1985. Discordant isotopic ages and potassium metasomatism in volcanic rocks, Yavapai County, Arizona. *Geological Society of America Abstracts with Programs*, **17**, 344.

CARSLAW, H. S. & JAEGER, J. C. 1959. *Conduction of Heat in Solids.* Oxford University Press.

CHERNIAK, D. 1993. Lead diffusion in titanite and preliminary results on the effects of radiation damage on Pb transport. *Chemical Geology*, **110**, 117–194.

COMPSTON, W., WILLIAMS, I. & MEYER, C. 1984. U–Pb geochronology of zircons from lunar breccia 73217 using a sensitive high mass-resolution ion microprobe. *Journal of Geophysical Research*, **98**, Suppl, B525–B534.

——, KIRSCHVINK, J., ZHANG, Z. & MA, C. 1992. Zircon U–Pb ages for the Early Cambrian time scale. *Journal of the Geological Society, London*, **149**, 171–184.

CONEY, P. J. 1980. Cordilleran metamorphic core complexes. An overview. *In*: CRITTENDEN, M. D., JR, CONEY, P. J. & DAVIS, G. H. (eds) *Cordilleron Metamorphic Core Complexes.* Geological Society of America Memoir, **153**, 7–31.

COPELAND, P., PARRISH, R. & HARRISON, T. 1988. Identification of inherited radiogenic Pb in monazite and implications for U–Pb systematics. *Nature*, **333**, 760–763.

COSCA, M., ESSENE, E. & BOWMAN, J. 1991. Complete chemical analyses of metamorphic hornblendes: Implications for normalizations, calculated H₂O activities, and thermobarometry. *Contributions to Mineralogy and Petrology*, **108**, 472–484.

CRITTENDEN, M. D., JR, CONEY, P. J. & DAVIS, G. H. (eds) 1980. *Cordilleran Metamorphic Core Complexes.* Geological Society of America Memoir, **153**.

CROWLEY, K., CAMERON, M. & SCHAEFER, R. 1991. Experimental studies of annealing of etched fission tracks in fluorapatite. *Geochimica et Cosmochimica Acta*, **55**, 1449–1465.

DALRYMPLE, G. & LANPHERE, M. 1971. ⁴⁰Ar/³⁹Ar technique of K–Ar dating: A comparison with the conventional technique. *Earth and Planetary Science Letters*, **12**, 300–308.

DAVIS, G. A. 1988. Rapid upward transport of midcrustal mylonitic gneisses in the footwall of a Miocene detachment fault, Whipple mountains, southeastern California. *Geologische Rundschau*, **77**, 191–209.

—— & LISTER, G. S. 1988. Detachment faulting in continental extension: Perspectives from the southwestern, U.S. Cordillera. *Geological Society of America Special Paper*, **218**, 135–159.

——, ANDERSON, J. L., FROST, E. G. & SHACKELFORD, T. S. 1980. Mylonitization and detachment faulting in the Whipple–Buckskin–Rawhide Mountains terrane, southeastern California and western Arizona. *In*: CRITTENDEN, M. D., JR, CONEY, P. J. & DAVIS, G. H. (eds) *Cordilleran Metamorphic Core Complexes.* Geological Society of America Memoir, **153**, 79–130.

——, ——, MARTIN, D., KRUMMENACHER, D., FROST, E. & ARMSTRONG, R. 1982. Geologic and geochronologic relations in the lower plate of the Whipple detachment fault, Whipple Mountains, southeastern California: a progress report. *In:* FROST, E. & MARTIN, D., (eds) *Mesozoic-Cenozoic Tectonic Evolution of the Colorado River Region California, Arizona and Nevada.* Cordilleran Publishers, San Diego, California, 409–432.

DODSON, M. H. 1973. Closure temperature in cooling geochronological and petrological systems. *Contributions to Mineralogy and Petrology,* **40,** 259–274.

DOKKA, R. K. 1993. Original dip and subsequent modification of a Cordilleran detachment fault, Mojave extensional belt, California. *Geology,* **21,** 711–714.

DUMITRU, T., GANS, P., FOSTER, D. & MILLER, E. 1991. Refrigeration of the western Cordilleran lithosphere during Laramide shallow-angle subduction. *Geology,* **19,** 1145–1148.

FITZGERALD, P. & GLEADOW, A. 1988. Fission-track geochronology, tectonics and structure of the Transantarctic Mountains in northern Victoria Land, Antarctica. *Chemical Geology,* **73,** 169–198.

FLECK, R., SUTTER, J. & ELLIOT, D. 1977. Interpretation of discordant $^{40}Ar/^{39}Ar$ age spectra of Mesozoic tholeiites from Antarctica. *Geochimica et Cosmochimica Acta,* **41,** 15–32.

FLEET, M. & BARNETT, R. 1978. $Al^{[4]}/Al^{[6]}$ partitioning in calciferous amphiboles from the Frood mine, Sudbury, Ontario. *Canadian Mineralogist,* **16,** 527–532.

FOSTER, D. A. & FANNING, C. M. 1997. Geochronology of the northern Idaho-Bitterroot batholith and Bitterroot metamorphic core complex: magmatism preceding and contemporaneous with extension. *Geological Society of America Bulletin,* **109,** 379–394.

—— & GLEADOW, A. 1992. The morphotectonic evolution of rift-margin mountains in central Kenya, Constraints from apatite-fission track thermochronology. *Earth and Planetary Science Letters,* **113,** 157–171.

—— & JOHN, B. 1999. Quantifying tectonic exhumation in an extensional orogen with thermochronology: examples from the southern Basin and Range province. *In:* RING, U., BRANDON, M. & LISTER, G. (eds) *Exhumation Processes: Normal Faulting, Ductile Flow, and Erosion.* Geological Society, London, Special Publications, in press.

——, HARRISON, T. M. & MILLER, C. 1989. Age, inheritance, and uplift of the Old Woman–Piute batholith, California and implications for K-feldspar age spectra. *Journal of Geology,* **97,** 232–243.

——, HOWARD, K. A. & JOHN, B. E. 1994. Thermochronologic constraints on the development of meta-morphic core complexes in the lower Colorado River area. *US Geological Survey Circular,* **1107.**

——, JOHN, B. & GLEADOW, A. 1996a. Sphene and zircon fission track closure temperatures revisited: empirical calibrations from $^{40}Ar/^{39}Ar$ diffusion studies of K-feldspar and biotite. International Workshop on Fission Track Dating, August 26–30, University of Gent, Gent, 37.

——, MILLER, D. S. & MILLER, C. F. 1991. Tertiary extension in the Old Woman Mountains area, California: Evidence from apatite fission track analysis. *Tectonics,* **10,** 875–886.

——, GLEADOW, A. J., REYNOLDS, S. J. & FITZGERALD, P. G. 1993. The denudation of metamorphic core complexes and the reconstruction of the Transition Zone, west-central Arizona: Constraints from apatite fission-track thermochronology. *Journal of Geophysical Research,* **98,** 2167–2185.

——, HARRISON, T., COPELAND, P. & HEIZLER, T. 1990a. Effects of excess argon within large diffusion domains on K-feldspar age spectra. *Geochimica et Cosmochimica Acta,* **54,** 1699–1708.

——, ——, MILLER, C. F. & HOWARD, K. A. 1990b. $^{40}Ar/^{39}Ar$ thermochronology of the eastern Mojave Desert, California, and adjacent western Arizona: With implications for the evolution of metamorphic core complexe. *Journal of Geophysical Research,* **95,** 20005–20024.

——, JOHN, B., CAMPBELL, E. & FANNING, C. 1996b. The role of syntectonic plutonism in development of the Colorado River Extensional Corridor: Sacramento Mountains example. *Geological Society of America Abstracts with Programs,* **28,** A450.

——, MILLER, C., HARRISON, T. & HOISCH, T. 1992. The $^{40}Ar/^{39}Ar$ thermochronology and thermobarometry of metamorphism, plutonism, and tectonic denudation in the Old Woman Mountains area, California. *Geological Society of America Bulletin,* **104,** 176–191.

GABER, J., FOLAND, K. & CORBATO, C. 1988. On the significance of argon release from biotite and amphibole during $^{40}Ar/^{39}Ar$ vacuum heating. *Geochimica et Cosmochimica Acta,* **10,** 2457–2465.

GALBRAITH, R. 1981. On statistical models for fission track counts. *Mathematical Geology,* **13,** 471–488.

GLEADOW, A. & DUDDY, I. 1981. A natural long term annealing experiment for apatite. *Nuclear Tracks,* **5,** 169–174.

——, ——, GREEN, P. & LOVERING, J. 1986. Confined fission track lengths in apatite: A diagnostic tool for thermal history analysis. *Contributions to Mineralogy and Petrology,* **94,** 405–415.

GREEN, P. 1981. A new look at statistics in fission track dating. *Nuclear Tracks,* **5,** 77–86.

——, DUDDY, I., LASLETT, G., HEGARTY, K., GLEADOW, A. & LOVERING, J. 1989. Thermal annealing of fission tracks in apatite 4. Quantitative modelling techniques and extension to geological time scales. *Chemical Geology (Isotope Geoscience Section),* **79,** 155–182.

GROVE, M. & HARRISON, T. 1996. 40Ar? diffusion in Fe-rich biotite. *American Mineralogist,* **81,** 940–951.

HAMILTON, W. 1988. Detachment in the Death Valley region, California. *US Geological Survey Bulletin,* **1790,** 763–771.

HAMMARSTROM, J. & ZEN, E. 1986. Aluminum in hornblende: An empirical igneous geobarometer. *American Mineralogist*, **71**, 1297–1313.

HARRISON, T. M. 1981. Diffusion of $^{40}$Ar in hornblende. *Contributions to Mineralogy and Petrology*, **78**, 324–331.

——1983. Some observations on the interpretation of $^{40}$Ar/$^{39}$Ar age spectra. *Isotope Geoscience*, **1**, 319–338.

—— & CLARKE, G. 1979. A model of the thermal effects of igneous intrusion and uplift as applied to Quottoon Pulton, British Columbia. *Canadian Journal of Earth Science*, **16**, 411–420.

—— & MCDOUGALL, I. 1980. Investigations of an intrusive contact, northwest Nelson, New Zealand – I. Thermal, chronological and isotopic constraints. *Geochimica et Cosmochimica Acta*, **44**, 1985–2003.

——, DUNCAN, I. & MCDOUGALL, I. 1985. Diffusion of $^{40}$Ar in biotite: Temperature, pressure and compositional effects. *Geochimica et Cosmochimica Acta*, **52**, 2461–2468.

——, MCKEEGAN, K. & LEFORT, P. 1995. Detection of inherited monazite in the Manaslu leucogranite by $^{208}$Pb/$^{232}$Th ion microprobe dating: Crystallization age and tectonic implications. *Earth and Planetary Science Letters*, **133**, 271–282.

HEIMANN, A., STEINITZ, G. & ZAFRIR, H. 1992. Irradiation of samples for $^{40}$Ar/$^{39}$Ar dating using the Soreq Nuclear Research Center IRR-1 reactor. *Israelly Nuclear Geophysics*, **6**, 273–286.

HEIZLER, M. & HARRISON, T. 1988. Multiple trapped argon isotope components revealed by $^{40}$Ar/$^{39}$Ar isochron analysis. *Geochimica et Cosmochimica Acta*, **52**, 1295–1303.

HOLLAND, T. & BLUNDY, J. 1994. Non-ideal interactions in calcic amphiboles and their bearing on amphibole-plagioclase thermometry. *Contributions to Mineralogy and Petrology*, **116**, 433–447.

HOLLISTER, L., GRISSOM, G., PETERS, E., STOWELL, H. & SISSON, V. 1987. Confirmation of the empirical correlation of Al in hornblende with pressure of solidification of calc-alkaline plutons. *American Mineralogist*, **72**, 231–239.

HOLM, D. & DOKKA, R. 1993. Interpretation and tectonic implications of cooling histories: An example from the Black Mountains, Death Valley extended terrane, California. *Earth and Planetary Science Letters*, **116**, 63–80.

—— & WERNICKE, B. 1990. Black Mountains crustal section, Death Valley extended terrain, California. *Geology*, **18**, 520–523.

——, SNOW, J. & LUX, D. 1992. Thermal and barometric constraints on the intrusive and unroofing history of the Black Mountains: Implications for timing, initial dip, and kinematics of detachment faulting in the Death Valley region, California. *Tectonics*, **11**, 507–522.

HOWARD, K. A. & FOSTER, D. A. 1996. Thermal and unroofing history of a thick, tilted Basin and Range crustal section in the Tortilla Mountains, Arizona. *Journal of Geophysical Research*, **101**, 511–522.

—— & JOHN, B. E. 1987. Crustal extension along a rooted system of imbricate low-angle faults, Colorado River extensional corridor, California

and Arizona. *In*: COWARD, M. P., DEWEY, J. F. & HANCOCK, P. L. (eds) *Continental Extensional Tectonics*. Geological Society, London, Special Publications, **28**, 299–311.

——, STONE, P., PERNOKAS, M. A. & MARVIN, R. F. 1982. Geologic and geochronologic reconnaissance of the Turtle Mountains area, California: West border of the Whipple detachment terrane. *In*: FROST, E. G. & MARTIN, D. L (eds) *Mesozoic–Cenozoic Tectonic Evolution of the Colorado River Region, California, Arizona, and Nevada (Anderson–Hamilton volume)*. Cordilleran Publishers, 341–354.

——, WOODEN, J., SIMPSON, R. & PEASE, V. 1996. Extension-related plutonism along the Colorado River extensional corridor. *Geological Society of America Abstracts with Programs*, **28**, A450.

——, JOHN, B., DAVIS, G., ANDERSON, J. & GANS, P. 1994. *A guide to Miocene extension in the lower Colorado River region, Nevada, Arizona, and California*. US Geological Survey Open-File Report 94–246.

HURFORD, A. & GREEN, P. 1983. The zeta age calibration of fission track dating. *Chemical Geology (Isotope Geoscience Section)*, **1**, 258–317.

JOHN, B. E. 1982. Geologic framework of the Chemehuevi Mountains, southeastern California. *In*: FROST, E. G. & MARTIN, D. L. (eds) *Mesozoic–Cenozoic Tectonic Evolution of the Colorado River Region, California, Arizona, and Nevada (Anderson–Hamilton Volume)*. Cordilleran Publishers, 317–325.

—— & FOSTER, D. A. 1993. Structural and thermal constraints on the initiation angle of detachment faulting in the southern Basin and Range: The Chemehuevi Mountains case study. *Geological Society of America Bulletin*, **105**, 1091–1108.

—— & MUKASA, S. 1990. Footwall rocks of the mid-Tertiary Chemehuevi detachment fault: A window into the middle crust in the southern Cordillera. *Journal of Geophysical Research*, **95**, 463–475.

—— & WOODEN, J. 1990. Petrology and geochemistry of the metaluminous to peraluminous Chemehuevi Mountains Plutonic Suite, southeastern California. *In:* ANDERSON, J. (ed.) *The Nature and Origin of Cordilleran Magmatism*. Geological Society of America Memoir, **174**, 71–98.

KETCHUM, R. 1996. Thermal models of core-complex evolution in Arizona and New Guinea: Implications for ancient cooling paths and present-day heat flow. *Tectonics*, **15**, 993–951.

LANPHERE, M. & DALRYMPLE, G. 1976. Identification of excess $^{40}$Ar by the $^{40}$Ar/$^{39}$Ar age spectrum technique. *Earth and Planetary Science Letters*, **32**, 141–148.

LEAKE, B., WOOLLEY, A., ARPS, C. *et al.* 1997. Nomenclature of amphiboles: report of the Subcommittee on Amphiboles of the International Mineralogical Association, commission on new minerals and mineral names. *American Mineralogist*, **82**, 1019–1037.

LEE, J., WILLIAMS, I. & ELLIS, D. 1997. Pb, U and Th diffusion in natural zircon. *Nature*, **390**, 159–162.

Lo, Ch-H. & Onstott, T. 1989. $^{39}$Ar recoil artefacts in chloritized biotite. *Geochimica et Cosmochimica Acta*, **53**, 2697–2711.

Lucchitta, I. & Suneson, N. 1993. Stratigraphic section of the Castaneda Hills – Signal area, Arizona. *US Geological Survey Bulletin*, **2053**, 139–144.

McClelland, W. 1982. Structural geology of the central Sacramento Mountains, San Bernardino County, California. *In*: Frost, E. G. & Martin, D. L. (eds) *Mesozoic–Cenozoic Tectonic Evolution of the Colorado River Region, California, Arizona, and Nevada (Anderson–Hamilton Volume)*. Cordilleran Publishers, 401–406.

——1984. *Low-angle Cenozoic faulting in the central Sacramento Mountains, San Bernardino County, California*. MSc Thesis, University of Southern California.

McDougall, I. & Harrison, T. 1988. *Geochronology and Thermochronology by the $^{40}$Ar/$^{39}$Ar Method*. Oxford University Press.

—— & Roksandic, Z. 1974. Total fusion $^{40}$Ar/$^{39}$Ar ages using HIFAR reactor. *Journal of the Geological Society of Australia*, **21**, 81–89.

Merey, C. & Ticken, E. 1989. Geophysical investigation, California low-level waste disposal project, Ward Valley, California., US Ecology: California low-level waste disposal facility Licence Application, Appendix 2320B.

Miller, C., Wooden, J., Bennett, V., Wright, J., Solomon, G. & Hurst, R. 1990. Petrogenesis of the composite peraluminous–metaluminous Old Woman–Piute Range batholith, southeastern California; Isotopic constraints. *In:* Anderson, J. (ed.) *The Nature and Origin of Cordilleran Magmatism*. Geological Society of America Memoir, **174**, 99–109.

Molnar, P. & England, P. 1990. Temperatures, heat flux, and frictional stress near major thrust faults. *Journal of Geophysical Research*, **95**, 4833–4856.

Naeser, C. 1979. Fission-track dating and geologic annealing of fission tracks. *In*: Jäger, E. & Hunziker, J. (eds) *Lectures in Isotope Geology*. Springer-Verlag, New York, 145–169.

Nielson, J. E. & Beratan, K. K. 1995. Stratigraphic and structural synthesis of a Miocene extensional terrane, southeast California and west-central Arizona. *Geological Society of America Bulletin*, **107**, 241–252.

O'Sullivan, P. & Parrish, R. 1995. The importance of apatite composition and single-grain ages when interpreting fission track data from plutonic rocks: a case study from the Coast Ranges, British Columbia. *Earth and Planetary Science Letters*, **132**, 213–224.

Parrish, R. 1990. U–Pb dating of monazite and its application to geological problems. *Canadian Journal of Earth Sciences*, **27**, 1431–1450.

Pease, V. 1997. *Geology of the northeast Sacramento Mountains, California*. DPhil Thesis, University of Oxford.

—— & Argent, J. 1999. The northern Sacramento Mountains, southwest United States. Part I: Structural profile through a crustal extensional detachment. *This volume*.

——, Foster, D., Wooden, J., O'Sullivan, P. & Argent, J. 1995. Tertiary Plutonism and extension in the Sacramento Mountains, SE California. *EOS, Transactions of the American Geophysical Union*, **76**, F639.

Potts, P. 1987. *Handbook of Silicate Rock Analysis*. Blackie.

Reynolds, S. J. & Lister, G. S. 1990. Folding of mylonitic zones in Cordilleran metamorphic core complexes: Evidence from near the mylonitic front. *Geology*, **18**, 216–219.

Richard, S. M., Fryzell, J. E. & Sutter, J. F. 1990. Tertiary structure and thermal history of the Harquahala and Buckskin Mountains, west-central, Arizona: Implications for denudation by a major detachment fault. *Journal of Geophysical Research*, **95**, 19 973–19 988.

Roddick, J. 1978. The application of isochron diagrams in $^{40}$Ar/$^{39}$Ar dating: A discussion. *Earth and Planetary Science Letters*, **41**, 233–244.

Schweitzer, J. 1991. *Structural evolution of crystalline lower plate rocks, central Sacramento Mountains, southeastern California*. PhD Thesis, Virginia Polytechnic Institute and State University.

Simpson, C., Schweitzer, J. & Howard, K. A. 1991. A reinterpretation of the timing, position, and significance of part of the Sacramento Mountains detachment fault, southeastern California. *Geological Society of America Bulletin*, **103**, 751–761.

Spear, F. & Kimball, K. 1984. *RECAMP* – A Fortran IV program for estimating $Fe^{3+}$ contents in amphiboles. *Computers and Geosciences*, **10**, 317–325.

Spencer, J. E. 1984. The role of tectonic denudation in the warping and uplift of low-angle normal faults. *Geology*, **12**, 95–98.

——1985. Miocene low-angle normal faulting and dike emplacement, Homer Mountain and surrounding areas, southeastern California and southernmost Nevada. *Geological Society of America Bulletin*, **96**, 1140–1155.

—— & Reynolds, S. J. 1991. Tectonics of mid-Tertiary extension along a transect through west-central Arizona. *Tectonics*, **10**, 1204–1221.

—— & Turner, R. 1983. *Geologic map of part of the northwestern Sacramento Mountains, southeastern California*. US Geological Survey Open-File Report 83–614.

Steiger, R. & Jäger, E. 1977. Subcommission on geochronology: Conventions on the use of decay constants in geo- and cosmochronology. *Earth and Planetary Science Letters*, **36**, 359–362.

Tera, R. & Wasserburg, E. 1972. U-Th-Pb systematics on lunar rocks and inferences about lunar evolution and the age of the moon. Proceedings of the 5th Lunar Congress, *Geochimica Cosmochimica Acta Supplement 5*, **2**, 1571–1599.

Wernicke, B. 1995. Low-angle normal faults and seismicity: A review. *Journal of Geophysical Research*, **100**, 20 159–20 174.

Wernicke, B. & Axen, G. 1988. On the role of isostasy in the evolution of normal fault systems. *Geology*, **16**, 848–851.

WHITEHOUSE, M., STACEY, J. & MILLER, F. 1992. Age and nature of the basement in northeastern Washington and northern Idaho: Isotopic evidence from Mesozoic and Cenozoic granitoids. *The Journal of Geology*, **100**, 691–701.

WILLIAMS, I. & CLAESSON, S. 1987. Isotopic evidence for the Precambrian provenance and Caledonian metamorphism of high grade paragneisses from the Seve Nappes, Scandinavian Caledonides. II. Ion microprobe zircon U–Th–Pb. *Contributions to Mineralogy and Petrology*, **97**, 205–217.

YAMADA, R., TAGAMI, T., NISHIMURA, S. & ITO, H. 1995. Annealing kinetics of fission tracks for zircon: An experimental study. *Chemical Geology*, **122**, 249–258.

# Evolution and development of the Levant (Dead Sea Rift) Transform System: a historical–chronological review of a structural controversy

## Z. R. BEYDOUN‡

*American University of Beirut, Beirut, Lebanon*

**Abstract:** The Levant Fracture System is a transform plate boundary that connects the extensional Red Sea spreading centre, in the south, with the collisional Tauride region, in the north, across the Levant. The debate regarding its origin, evolution and development is reviewed, with proposed models ranging from a primordial geosuture, along which only vertical movements have occurred, to a Neogene transform, along which the Arabian Plate has moved 107 km left-laterally northwards in two stages, 62 km in the Miocene and 45 km in the Pliocene–Recent. Overwhelming field evidence supports this degree of motion south of Lebanon but in Lebanon, where the system branches into a braided fault system generally trending north-northeast, evidence for such motion is lacking. Proponents of the transform model have explained away this lack of evidence by invoking division of the total horizontal motion to the south amongst the main fault strands across Lebanon and ascribing the uplift/depression of the Lebanon–Palmyride structural units to transpressional–transtensional deformation by this horizontal motion. The inversion of the Lebanon–Palmyride belts had, in fact, been largely achieved by the Late Oligocene as a consequence of oblique collision between the Eurasian–Arabian Plates during closure of the Neo-Tethys. The Neogene initiation of the Levant Fracture System, resulting from the opening of the Gulf of Aden–Red Sea, could only, therefore, have modified existing structures, not been responsible for them. Pliocene bypassing of the principal Lebanese Yammouneh Fault requires that active motion in the south be transmitted through other faults. The currently active north–south Roum Fault has been proposed as the principal candidate and its trace northwards appears identifiable on satellite images. Field checks suggest it has broken into short left-stepping segments and it probably cuts the coast obliquely just south of Beirut. Current research to establish the geometry and magnitude of horizontal motion on this fault is discussed.

The Levant Fracture System, or the Dead Sea Rift Transform, is generally regarded by Earth scientists as a major left-lateral (sinistral) strike-slip fault forming the northwest boundary of the Arabian Plate. The system has a right step-over trend oscillating between north-north-east–south-southwest at its two extremities and through Lebanon, with north–south trending connecting segments. It extends from the Gulf of Aqaba and Wadi Araba in the south (north-northeast–south-southwest), through the Dead Sea–Jordan Valley (north–south), through the Bekaa Depression in Lebanon (north-northeast–south-southwest), the Ghab Depression in Syria (north–south) and finally through the Kara Su Valley of the Amanus region of north-west Syria–Iskenderun (north-northeast–south-southwest) be-fore abutting against the Taurus Mountains of southern Turkey in the extreme north (Figs 1 and 2). The northern segment may link-up with the sinistral East Anatolian Transform Fault as it arcs to the northeast and east. The total amount of horizontal displacement along the Levant Fracture System as a whole is a major subject of controversy (Figs 1 and 2).

## Early views

Dubertret (1932), following the ideas of Lartet (1869), developed the working hypothesis of sinistral movement along the Levant Fracture System to explain the evolution of the mountain ranges of Syria and Lebanon. He considered the Arabian block to be fixed relative to the Sinai–Levant maritime block, the latter moving some 160 km southwards, with the African block rotating 6.4° clockwise around a rotation centre situated in the Ionian Sea. However, having completed detailed geological mapping of most of Lebanon, especially within the Bekaa region, he could find no evidence in either the stratigraphy or the relationship between the Libano–Syrian structural units for large-scale horizontal

‡ Deceased, reprint requests should be addressed to the editors.

*From*: Mac Niocaill, C. & Ryan, P. D. (eds) 1999. *Continental Tectonics*. Geological Society, London, Special Publications, **164**, 239–255. 1-86239-051-7/99/$15.00 © The Geological Society of London 1999.

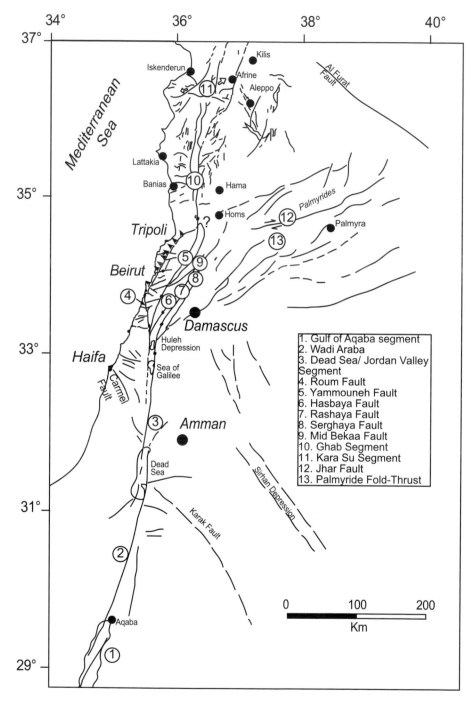

**Fig. 1.** Simplified map of the Levant Fracture Transform System. Only principal faults are depicted. The map is compiled from Dubertret (1955, 1967, 1971–72), Quenell (1984), Muehlberger (1981), McBride *et al.* (1990), Chaimov & Barazangi (1990), Searle *et al.* (1993) and Khair *et al.* (1997).

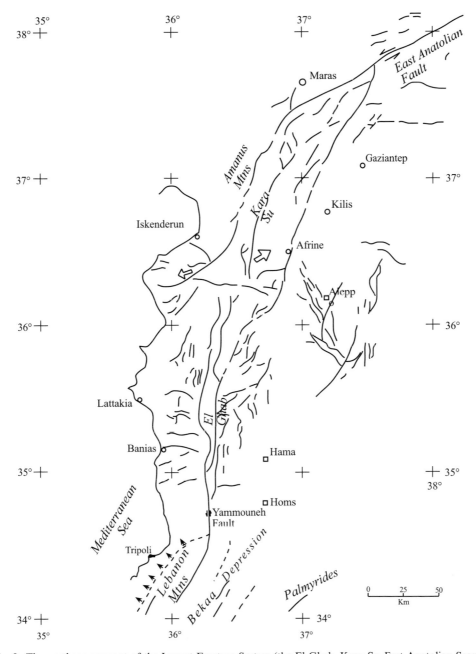

**Fig. 2.** The northern segment of the Levant Fracture System (the El Ghab–Kara Su–East Anatolian Segment). Simplified from Muehlberger (1981), and based on satellite imagery and field checks.

movement and abandoned his ideas. He concluded that the influence of the fracture, which he regarded as primordial, did not extend northwards beyond the Huleh Depression at the north end of the Jordan Valley adjacent to the Golan Heights (Dubertret 1947). It is at this point that the fracture system diverges into a horsetail of branching faults.

Quennell (1951, 1958, 1959) revived Dubertret's (1932) hypothesis and further developed the idea of sinistral strike-slip motion, proposing a model that utilized stratigraphic, geomorphological and palaeontological evidence from both sides of the rift in the Aqaba–Dead Sea–Jordan Valley segments. He quantified the major shift along this segment as having taken place in two stages; 62 km in the Miocene and 45 km in the Pleistocene. This motion moved the eastern or trans-Jordanian (Arabian) block some 107 km northwards relative to the western cis-Jordanian block with 5.5° of anticlockwise rotation of Arabia with respect to the Sinai–Levant. He also recognized the Dead Sea area as a rhomb-graben resulting from transtensional stresses. His faults were shown as arcs with a centre of rotation in the Mediterranean at 33°N 24°E. Girdler (1989) referred to Quennell as the 'father of transform faults and poles of rotation'; however, Dubertret (1932), must be credited as being the first to observe the concept of a centre of rotation. Quennell recognized that the rift extended north to the Huleh Depression, beyond which movement required invoking other factors. He assumed, as others were doing, that the Dead Sea–Jordan Valley fault path would be found to be the faults, folds and uplifts in Lebanon. The 25 km crustal gap left as a result of the recognized movements was believed to be accounted for by shortening in the Palmyrides, and folding and upthrusting of the Lebanon and Syrian horsts, and by their bulk displacement to the northeast.

Wetzel & Morton (1959), contradicting Quennell, saw the Dead Sea Rift as a marginal graben. They observed that regional isopach lines ran parallel to it, merely demonstrating a deep basin to the west and a shoreline to the east, the site of the rift having acted as a basin hinge line. Although the isopach lines in the south come in from the west, swing north and then to the east this sinusoidal pattern, according to the authors, was mistakenly taken for horizontal offset. Bender (1968; and in Freund *et al.* 1970) carried out detailed geological mapping in the Jordan–Dead Sea rift area and agreed with the views of Wetzel & Morton (1959), believing that horizontal movement was unsubstantiated by the mapping. He regarded the rift zone as the site of an old geosuture along which vertical movement had taken place during the Precambrian, as well as in later times, playing the role of a hinge zone which controlled facies changes, trends and thicknesses. Picard (1967) also supported the hinge zone view and disagreed with the postulated horizontal movement. However, Bender (1983) later revised some of his earlier views, admitting to direct evidence of lateral movement in the unconsolidated Quaternary fill of the rift where left-lateral displacement of several hundred metres is seen. He concluded that this

movement may have resulted in horizontal displacement totalling 30–35 km.

Interestingly enough, at this same later date, Bahat & Rabinovitch (1983) still questioned that the Levant Fracture had initiated as a shear transform associated with continental collision. They considered that fracture initiation, propagation and bifurcation along the rift are best explained by extensile processes. There has been recent confirmation of transitional properties of the crust–mantle boundary along the rift coupled with considerable thinning of the crust on the west side between the Dead Sea and the Sea of Galilee (Folkman & Bein 1978; Ginzburg & Folkman 1980) which can be interpreted as indications of an active mantle diapir having initiated the fracture. Propagation north and south reached conditions of bifurcation in several localities along the rift; these areas showing characteristic local uplifts and small grabens which are oblique to the rift. Fracture branching and bilateral bifurcation associated with a transitional crust–mantle boundary in the area of crustal thinning implies an active rather than a passive upper mantle below the rift, and volcanism suggests relative hotspot migration north or northeast on the east side in the Late Cenozoic. The evolution of the Levant Rift is thus considered by these authors as beginning with an early stage of extensional fracturing which only later evolved into the known major sinistral strike-slip fracture system.

de Sitter (1962), however, confirmed evidence of movement along the rift fracture in pre- and post-Turonian times. Freund (1965) took this into account when proposing his model which moved the Arabian block northwards *c.* 60–80 km since the Middle Cretaceous as part of the drift of Arabia northwards away from Africa during the opening of the Gulf of Aden and the Red Sea. Freund (1965) suggested that the main structural elements of the southeastern Mediterranean region may be explained as secondary effects of strike-slip movement along the Dead Sea Fault (Levant Fracture), with the high structures of Lebanon and the synclinal shape of the Bekaa Valley accounting for the 25 km of shortening. Drag along the contact between the two blocks would give rise to folds and thrusts on either side of the fracture, especially on the convex parts of the blocks where secondary dextral wrenching probably occurred, forming a rift open from the Gulf of Aqaba to the Huleh Depression, narrowing and shallowing northwards.

The northward propogation model was further developed by Freund & Raab (1969) and Freund *et al.* (1970), who cited the offset of ophiolite outcrops on both sides of the northwest Syria–Iskenderun as supporting evidence. Freund *et al.* (1970) refined the amount and timing of the horizontal movement to a total of 106 km in the south (Dead Sea area) with the Arabian block shifting northwards in two stages, Late Miocene and Plio-Pleistocene. Dubertret (1976), rebutted, claiming that the apparent offset of the ophiolite in the Kara Su region is entirely an orographic effect. He also reiterated that, if this movement is carried into the northern part, it would leave a large crustal gap where the Levant Fracture System changes direction from south–north to south-southwest–north-northeast along the Yammouneh and other faults of Lebanon. Detailed gravity surveys in Lebanon, including the Bekaa, show no evidence for such a gap nor is there support for other than small-scale horizontal movement along the fracture (Tiberghien 1974). Freund *et al.* (1970) suggested that accommodation of all or most of such a crustal gap is possible by a complex combination of sinistral, dextral and vertical movements in the Lebanon–Anti-Lebanon Ranges, in the intervening Bekaa Depression and in the Galilee area to the south plus over a wide zone on either side of the rift. Freund & Tarling (1979) extended on this idea by suggesting that part of the motion along the Lebanon segment of the Levant Fracture System is accommodated by internal rotation of small fault blocks on the west side of the rift.

The importance of the role of vertical movements in determining the structure of the northern Levant states (Syria and Lebanon) was stressed by Dubertret (1967, 1969, 1970, 1971–72), who suggested that to invoke large scale horizontal movement is totally inconsistent with the details of the stratigraphy and structural block relationships of those areas. He also reiterated his conviction regarding the great antiquity of the Levant Fracture System, perhaps going as far back in origin as the Late Precambrian. He cited field evidence to indicate that horizontal movement had occurred along the westernmost of its Lebanese branches, the Roum Fault, as far back as the Late Jurassic to the medial Cretaceous. Dubertret (1971–72) suggested that the opening of the Red Sea could perhaps be linked to horizontal movements along the Najd Fault System, the northwest–southeast trending strike-slip faults dissecting the Arabian Shield, rather than along the length of the whole of the Levant Fracture System. He did admit to the difficulty of reconciling the ancient nature of the Najd system with the recent nature of the movements that led to the opening of the Red Sea. His main objective still remained, however,

to discount the occurrence of large scale horizontal movement along the Lebanese segments of the Levant Fracture System, especially along the Yammouneh Fault. He envisaged minor horizontal displacement, shared amongst the other branches of the Lebanese faults that make up the Levant Fracture System, especially the Serghaya Fault (Figs 1 and 3) where he had noted horizontal slickensides on the fault plane dissecting the Jurassic succession (Dubertret 1969). He appeared rather ambivalent regarding the role of the Roum Fault, at times remarking that the line of separation between the Arabian and Levant–Sinai Plates should be looked for in the Roum Fault while later remarking that to the north of Roum, towards Beirut, the fault trend is taken up by the western flexure that bounds Mount Lebanon without sign of any surface dislocation (Dubertret 1967).

Hancock & Atiya (1979) concurred that left-lateral displacement had taken place along the northern part of the Yammouneh Fault which is of Neogene age, but that it is limited and does not exceed 10 km. They attributed deformation of the Mount Lebanon Range and the Bekaa Depression as having developed within a transpressive regime related to oblique convergence of the Arabia and the Sinai–Levant Plates across a north-northeast trending boundary which took place mainly in the Late Palaeogene. This was at variance with the generally accepted timing for the initiation of the movement of the Arabian Plate along the Levant Fracture System transform in the Late Miocene but not with the initial uplift of the Lebanon Ranges in the Late Eocene and Oligocene. They invoked pressure solution as a mechanism for shortening which, they believed, may have achieved substantial shortening (locally as much as 50%), thus accommodating the left lateral movement.

Neev (1975) invoked a somewhat novel approach questioning the position of the boundary of the Arabian Plate. He considered the Sinai and the Levant as continuous with a substantial portion of northeast Africa, referring to this region as the Central Plate, divisible into the Arabian and northeast African Sub-plates. He referred to the western boundary of the Central Plate as the Pelusium Line which he envisaged as a transcontinental shear, trending northeast across much of Africa, entering the Mediterranean at the eastern edge of the Nile Delta and then following an offshore subparallel course along the Levant coast up to the Bay of Iskenderun. To the east of the Central Plate is the Iranian Plate and to the west is the northwest African Plate, which includes the Levantine Basin and Cyprus. He suggested that all

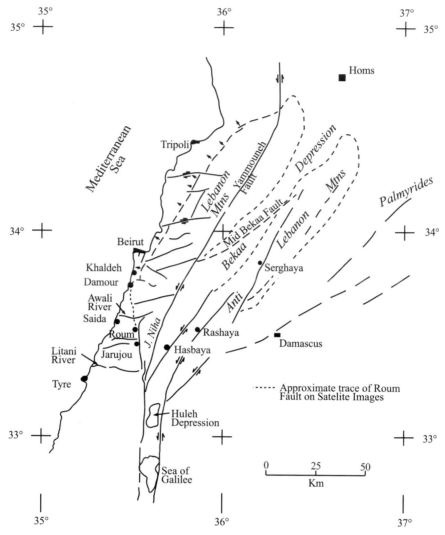

**Fig. 3.** The braided fault system of Lebanon. Simplified from several sources including Dubertret (1955, 1967), Quenell (1984) and Butler *et al.* (1997).

three plates are moving northwards, with the Zagros Front (thrust or crush zone) and the Pelusium Line functioning as transcurrent faults along which the three plates have differentially moved, the Central Plate moving northward faster than the other two and, therefore, being wedged between them. The resulting oblique collision energies (with Anatolia) would create internal transpressional shear parallel to the transcurrent faults, elongating the Central Plate parallel to the transcurrent faults and giving rise to northeast trending compressional features and northwest trending tensional features. This stretching of the Central Plate, according to the author, has probably gone on since the Late Palaeozoic time. An analogous mechanism is proposed to account for the development of the Mesopotamian and Gulf fold systems.

The above model, although embodying some interesting arguments, rests on the identification of the Pelusium Line running across Africa. This is, at best, tenuous at this time and its antiquity open to question. Whether the western boundary of the Levant–Sinai Plate is along the Pelusium Line, or there is simply a subsiding hinge zone between the Levant mainland and the Levantine Basin or the Arabia–Levant interface is a transtensional one are the debatable questions.

Dubertret (1955, 1963, 1966, 1974–1975) demonstrated, through meticulous field mapping and stratigraphic structural measurements, observations and correlations, that the emergence and break-up of Lebanon into its tripartite megastructural components of Mount Lebanon in the west, the Bekaa Depression in the centre and the Anti-Lebanon Range in the east, had already taken place prior to the Miocene. Upper Eocene and Oligocene deposits are absent throughout the country; therefore, inversion (whether due to vertical forces or to compression) commenced after the deposition of the Middle Eocene but prior to any Miocene sedimentation. In the coastal area near Beirut at Nahr el Kalb, marine Miocene carbonates with a basal clastic unit (Middle to Upper) onlap, with a sharp angularity, to the steeply dipping Cretaceous succession of western Mount Lebanon, cutting a terrace into the flexural beds. In the Bekaa Depression, lacustrine and coarse alluvial deposits are evidence of erosion and deposition into an enclosed inland basin from bordering uplifts (Fig. 3). Beydoun (1977), following on Dubertret's earlier work discussed above, summarized these events which predate the inception of the Levant Fracture System Transform and any horizontal movement along it.

## Refinement of concepts

Several authors have contributed to specific aspects of the debate. Arthaud et al. (1978) suggested that the Libano–Syrian portion of the Levant Fracture System may not represent a simple linear rigid plate boundary but rather an extensive deformed and fractured zone. The problem of northward propagation of the Red Sea Rift and sinistral movement along the Jordanian segment of the fracture must not be envisaged as merely accommodated by a simple rotation. Part of the shortening is probably taken up by deformation within the Lebanon, Anti–Lebanon and Palmyride belts. Moreover, this displacement could be absorbed across a very wide zone, which could include the north–south Roum Fault in southern Lebanon (Fig. 3). Masson et al. (1982), utilizing numerical processing of Landsat images over Lebanon and northwest Syria, identified previously unrecognized lineaments in the Beirut region (south and southeast of Beirut). These appeared to coincide with deformation associated with the main northerly trending faults. Two main lineament directions were established – north-northwest–south-southeast and northwest–southeast. The

first, parallel to the Roum Fault, would allow for a connection of the main Levant trend with the Roum Fault and the coast (not seen on geological maps). These lineaments appear not to coincide with either morphological or lithological boundaries. Earthquake epicentres have, however, been noted in the region. The second northwest–southeast trend of lineaments were noted between the Roum Fault and Beirut, where they cut major north trending structures. This second trend is similar to the orientation of a string of Quaternary volcanic cones situated to the south of the Palmyrides in the Jebel Druze area of Syria. Independent microtectonic analyses have identified three main phases of deformation associated with this area: a mid–upper Miocene compressional phase (shortening west-northwest–east-southeast); a Late Miocene–Pliocene extensional phase; and a Pliocene–Quaternary strike-slip phase (shortening north-northwest–south-southeast). If the structural nature of the second lineament trend is confirmed, it could correspond to the normal fault trend that affected the area during the Plio-Pleistocene strike-slip movement phase (Fig. 3).

Beydoun (1981), following Dubertret, remarked on the predominance of vertical tectonics in Lebanon, nevertheless, he concluded that the transverse and en echelon folds of the Bekaa Depression and the faults and asymmetric sharp folds of the Jabal Niha Wedge, should be regarded as representing first-order deformations in local response to the sinistral strike-slip motion along the Levant Fracture. In the Jabal Niha Wedge, a young recumbent overfold in competent Cenomanian carbonates can be seen above Niha village. On the nearby Roum fault plane, in the vicinity of Jarjou' c. 14 km south of the Awali River (Fig. 3), Beydoun (1974, unpublished) observed well-preserved horizontal slickensides in Cretaceous carbonates, indicative of recent horizontal motion. The prevailing karstic conditions of the area would have obliterated these had they represented an old phase of motion. The recumbent overfold, in combination with the evidence of horizontal motion, could represent a flower structure that developed as a result of transpression (Sengor, pers. comm.).

Garfunkel (1981) suggested that crustal separations of varying, but small, magnitudes occurred along many segments of the Levant Fracture System (Transform) so that, in part, it should be leaky. Crustal separation became important during the second slip phase (Plio-Pleistocene), as a result of a change in relative plate motions, resulting in the development of rhomb-shaped grabens which grew by becoming longer rather than wider. Visible igneous activity, however, is generally absent in the interior of the basins, probably because of their small width, slow rates of widening, and thick sediment fill. He noted that the Levant Transform was activated after the abandonment of northwest extensional crack formation, the initial mode of yielding.

Further evidence confirming large scale horizontal motion along the southern segment of the Levant Fracture System has been provided by Hatcher *et al.* (1981) and is based on aeromagnetic and magnetic data from both sides of the Levant Fracture. East–west trending basement structural anomalies, previously unrecognized, are identified on the eastern side and compared to similar anomalies on the western side, the comparison supporting a sinistral offset movement of slightly >100 km. Bandal (1981) provided similar evidence for the same amount of displacement based on Mesozoic stratigraphic correlations, particularly of the Jurassic succession in southern Jordan. He reviewed structural, stratigraphic, lithological and magmatic features of Mesozoic rocks, east and west of the rift, to conclude that there was nothing in the stratigraphic column to indicate activity along the Levant Fracture System before Tertiary time. Walley (1983) also reached a similar conclusion regarding horizontal displacement along the southern segment of the fracture. His conclusion was based on correlation of Callovian–Oxfordian strata and fauna from the southern side of Mount Hermon where Tethyan faunal connections indicate a northwest–southeast directed trough open to the Tethys, which is now displaced, on the east (Mount Hermon) side, some 100 km northwards. Abed (1985) concluded, on the basis of new field studies in Jordan, that previous arguments, utilizing the premise that the occurrence of dykes and Infra-Cambrian and Cambrian basal conglomerates and coarse clastics are more abundant in the rift, that the rift as a site of a long-lived zone of crustal weaknesses was invalid. He concluded that future arguments in favour of important ancient (vertical) tectonics along it will have to rely on other evidence.

Quennell (1983, 1984) modified his early models to account for the lack of any significant horizontal movement along the Lebanese (particularly the Yammouneh) segment of the fracture system. He suggested that during the Lower Miocene the Afro–Arabian crust was ruptured by a first phase of movement on the Levant Fracture System, probably along a geosuture in the continental plate which was originally a simple arc [see also Bender (1983) who believed

that the southern transform zone overlies a Late Proterozoic suture, intermittently active during the Phanerozoic]. A common 'root zone', in the form of a single principal fracture at great depth, has also been suggested by Kashai (1988) on the basis of the en echelon nature and short overlaps of the master faults. The first phase of northward movement of the Arabian Plate, due to the opening of the Red Sea, according to Quennell (1983, 1984), was a 62 km displacement along the southern segment in the Miocene and Lower Pliocene. This rift was a diastrophically initiated, terminating, cross-fault of the Red Sea spreading zone, its northerly course being influenced by preferred basement trends and rotation of the Arabian Platform on a small circle transform. In the Upper Pliocene, the Syrian Platform was forced eastwards by oblique compression across the East Anatolian Transform, with dextral distortion and shear acting on the Lebanon–Palmyride belt, leading to uplift, deformation and faulting. The Lebanon–Palmyride fold belt appears to have formed in the weakest zone of the Arabian Plate between stable and unstable shelves with differing thicknesses of sedimentary cover. A second phase of movement was accompanied by consumption of the edge of the Arabian Stable Platform beneath the Lebanon–Palmyride fold belt. Thus, by the Upper Pliocene, the once continuous Levant Fracture System was divided into a southern transform system south of the southern Lebanese border and a northern transform system north of the northern Lebanese border. Both of these are sinistral strike-slips but now act independently, being separated by the belt of oblique folding, faulting and thrusting of the Lebanon–Palmyride zone. The Yammouneh Fault, a young feature, probably lies above the geosuture. Unfortunately, unequivocal evidence for plate consumption in this region is not seen on published gravity and magnetic maps (Ponikarov *et al.* 1967), nor do regional seismic refraction lines across the area from the Arabian Stable Shelf into the Palmyride Basin or 'aulacogene' appear to substantiate more than the presence of a thickened sedimentary succession (>8 km) resting on continental basement (Beydoun 1991).

The Levant Fracture's southern section consists of a number of morphotectonic pull-apart depressions, the best known being the Dead Sea Graben, and the Gulf of Aqaba and the Sea of Galilee–Huleh Depressions. These have generally been referred to as rhomb-grabens. Interpretation and analyses of reflection–seismic, gravity and magnetic surveys, and deep-drilling results have shown that the Dead Sea Depression (and several of the others) is bounded along its length on either side by steep faults with large, vertical throws, down-stepping from the high shoulders to the deep, sunken floor. The graben is *c*. 15 km wide and the steep, eastern boundary fault juxtaposes Palaeozoic or Precambrian of the Jordan Plateau, against Cenozoic of the graben floor. The steep, stepped, boundary faults belong to the earlier of the two phases of faulting, thought to be Middle Miocene. It is bound at its southern end (within the Wadi Araba region) by a cross-fault which is downthrown to the north. Its northern end is characterized by a steadily rising graben floor without any major cross-faults (Kashai & Croker 1987; Ben-Avraham & ten Brink 1989). Multiple minor cross faults do occur within the Dead Sea Graben and others, dissecting each graben into a series of subsidiary basins. These grabens are therefore not typical 'leaky' pull-aparts, since they exhibit neither significant gravimetric, magnetic or high heat flow anomalies, nor unusual magmatism. On the contrary, the Dead Sea Graben, exhibits low–normal heat-flow, and seismic surveys and other geophysical data indicate the presence of considerable thicknesses of sedimentary cover above a continental crystalline basement.

The sinistral, en echelon strike-slip master faults of the southern segment characteristically die out to the north by bending to the east. Horizontal shortening is taken up by the next master fault located to the west. Where two such master faults overlap, narrow and very deep grabens (such as the Dead Sea Graben) have developed; the narrowness dictated by the closeness of the overlapping segments. Width appears to be an indicator of depth (Kashai 1988).

The Dead Sea Graben contains syn-rift strata >12 000 m. Miocene deposition initially comprised continental clastics with limited carbonates. During quiescent periods, sedimentation was dominated by Pliocene evaporites with subordinate fluviolacustrine deposits. The evaporites, which include considerable amounts of salt, can reach 5000 m in thickness (Powell 1988). Resumption of movement resulted in rapid, major subsidence, so that up to 5000 m of Upper Pliocene–Quaternary clastics (alluvial fans) along the margins interfinger with fluviolacustrine sediments in the centre. Rapid loading resulted in salt mobilization, and diapiric structures and surface piercements and salt-withdrawal troughs are found over much of the basin. The Usdom (Sedom) and Lisan Domes in the southern part are the most prominent. Low-angle, listric growth faults have developed in this young, post-salt fill which sole out at the base of the salt.

## Observations on the northern Ghab-Kara Su Segment (Fig. 2)

Along the northern transform segment [using Quennell's (1984) terminology], considerably less work has been carried out. Muehlberger *et al.* (1979) and Muehlberger (1981) utilized Skylab and Apollo-Soyuz Test Project photographs. These show that the northern system splinters into a number of splays between the Ghab Depression in northwest Syria and the East Anatolian Transform Fault. The Ghab Depression is a rhomb-graben, one of a number of small ones lying between the Syrian–Lebanese border and the Aafrine area of northwest Syria. Other major linear trends associated with closed depressions, such as those near Antioch and Lake Amik, occur in this northern segment. This splintering disrupts modern drainage systems indicating that the area is currently tectonically active. Bending or segmentation of the fracture system is assumed to have started after the first period of slip and to be the result of initial continent–continent collision of the Arabian and Anatolian Plates. As a result of this kinking, the direction of motion along the northern segment during the current episode of slip (commencing *c*. 4.5 Ma) has an east-of-north component which generated the East Anatolian Transform Fault Zone, progressively tightened the Palmyride Zone folds and elongated coastal Lebanon (Muehlberger 1981). This resulted in reducing the 45 km slip measurable south of Lebanon to *c*. 10 km north of Lebanon. The kinking splintered the region south and east of Lake Amik, the southern wedge of the Kara Su Valley, where the northwestern corner of the Arabian Plate is deforming by simple shear and northeast movement, parallel to the East Anatolian Transform Fault Zone (Muehlberger 1981).

Giannerini *et al.* (1988) carried out a structural analysis of the northern part of the Levant Fracture System in Syria and of the Palmyride area using satellite imagery and field investigations. They too concluded that deformation along the Levant Fracture System, and in the Palmyrides, took place during two main episodes in the Miocene and the Plio-Quaternary, separated by a relatively quiet episode during the Upper Miocene. Miocene deformations took place under a principal stress oriented N140–150E, giving rise to N40–50E synsedimentary folds and north–south sinistral and east–west dextral wench faults. Plio-Quaternary deformation was controlled by a submeridian principal stress orientation and gave rise, in the Palmyrides to synsedimentary folds oriented N80E, and to conjugate transverse faults on

N35–60E (sinistral) and N100–120E (dextral) orientations.

During this period of convergence, both tensional and compressional cycles occurred spatially during the phases of deformation. The transform zones and the orientation of the main stresses define the junctions between tensional and compressional areas. In the Miocene, migration of the Arabian Plate was strictly meridional but by the Plio-Quaternary the Arabian Plate moved in a more easterly direction (sinistral motion in the Palmyrides area). This change of direction engendered crustal separations along much of the length of the Levant Fracture System creating the Aqaba, Dead Sea, Al-Ghab Depressions and other pull-apart features. Thus, the Palmyride region behaved as a sinistral intracontinental transform fault.

Matar & Mascle (1993) carried out microfault analysis in the Ghab Depression area to derive the initiating stress fields. They confirmed that the motion in this part of the Levant Fracture System was pure sinistral strike-slip and that the Ghab Depression opened as a pull-apart structure located on a left-lateral relay, with both the compressional and tensional stress directions being horizontal. The Ghab Depression appears to represent a textbook example of a pull-apart (transtensional) basin.

## Recent observations on the Lebanon–Palmyride Segment (Figs 1 and 3)

Returning to the Lebanon–Palmyride segment of the Levant Fracture System, many years of civil strife and war, instability, invasion and occupation of substantial parts of southern Lebanon, have made field investigations in critical areas of Lebanon virtually impossible from 1975 until very recently, apart from occasional quick forays during some extended lulls (e.g. Arthaud et al. 1978). Within Israeli occupied parts of Lebanese (and Syrian) territory, Israeli Earth scientists have had limited access and have published a few papers of a local or theoretical nature (e.g. Ron 1987; Heimann et al. 1990). Structural studies since those of Hancock & Atiyah (1979), (field work carried out prior to 1975) have had to rely on analysis of the published maps of Dubertret [e.g. Dubertret (1955) and more detailed 1:50 000 sheets] and on satellite images. Thus, most research has utilized existing data and reassessed or reinterpreted this in the light of new concepts and ideas. In contrast, research involving the Palmyride area has not been thus constrained.

Rothstein & Kafka (1982) interpreted the Palmyride region as the focus of a new north dipping subduction zone resulting from the collision further north along the Bitlis Suture in southeast Turkey. Lovelock (1984) was also of the opinion that underthrusting of the Palmyride region would account for sufficient shortening to meet the objection of those who do not accept the 105–110 km sinistral displacement along the Dead Sea (Levant Fracture) System. He considered that the folding of the Palmyride zone appears to be coeval with the onset of oceanic ridge push on the eastern (Arabian) block of the Levant Fracture System, with a possible sequence of events being as follows: failure of the Gulf of Suez Rift in the Early Miocene, closely followed by formation of the Levant Transcurrent (Transform) Fault. Continental collision along the Bitlis Suture in southeast Turkey followed in Early–Middle Miocene times, with underthrusting and box folding in the Palmyride zone in the Middle–Late Miocene times and, finally, strong uplift in the north of Turkey with overthrusting along the Bitlis Suture in Late Miocene–Pliocene times. This led to a reactivation of lineaments within the Arabian Plate in Late Pliocene–Recent times and transcurrent motion along the Levant Transcurrent (Transform) Fault, giving rise to the present-day structures and accentuating the underthrusting in the Palmyrides.

Ron (1987) developed a model to predict the deformational mechanism and magnitude along the Yammouneh Fault which, he felt, represents a restraining bend of the Levant (Dead Sea) Transform. He concluded that the plate margin had been deformed by north-northeast folding parallel to the restraining bend and by east–west trending dextral strike-slip faults. Palaeomagnetic rotational data and fault kinematics suggest that the mechanism which accommodated sinistral shear was simultaneous dextral strike-slip faulting on secondary faults and block rotation. According to this model, the magnitude of deformation is directly controlled by the geometry of the restraining bend and the cumulative displacement along the transform fault. The author has tested this idea along the plate margin of the Yammouneh restraining bend but existing detailed geological maps of the area (Dubertret 1951) do not unequivocally substantiate the conclusions.

Walley (1988) reviewed the nature of the fault systems in Lebanon and southwest Syria utilizing published papers and detailed maps supplemented by personal observations where possible. He concluded that the Yammouneh Fault is only one of a number of large-scale strike-slip faults in

Lebanon and should not be regarded as constituting the principal prolongation of the Dead Sea Transform. The arrangement of the Lebanese faults suggests that the Levant Fracture System Transform anastomoses to give a braided strike-slip complex which is related to the eastward flexing of the Levant Fracture System in this area, allowing the crust of the Arabian Plate to bypass the resultant locked zone. The effects of lateral motion on the various fault branches (all sinistral) resulted in large-scale transpressive uplifts in the region; the main structural features such as Mount Lebanon, the Anti-Lebanon, the Palmyrides and the Bekaa Depression are explicable in terms of this braided strike-slip model. The data available are in harmony with two phases of motion during the Lower Miocene and Pliocene–Recent phases of the opening of the Red Sea, with the main lateral motion taking place during the first phase along the various faults. Small pull-apart basins are present along the Yammouneh Fault (Lake Yammouneh) and the Serghaya Fault (the Zabdani Basin).

Heimann *et al.* (1990) found one of these small rhomb-grabens, the Barahta, along the Rachaya Fault which crosses Mount Hermon on a northeast–southwest trend. It is formed between two left-stepping segments which gave rise to the graben. The authors attribute its formation to a local stress field arising from the fault configuration and not to a regional stress field, as dimensions are in the order of a few kilometres length and 1 km width. The structure can largely be explained by rotation rather than by the classic pull-apart model, the step-over of strike-slip faults being essential for the development of rhomb grabens. The mode of formation differs in each case, being dependent on the fault configuration and direction of block motion (Heimann *et al.* 1990).

Girdler (1990) utilized satellite imagery, seismicity, existing geological maps and bathymetric charts to come up with a new map showing the major features of the Dead Sea (Levant Fracture) Transform into Lebanon, with special attention being given to the possible northward continuation of the transform system beneath the Mediterranean near Damour south of Beirut. The map shows the system to be a series of offset, overlapping, left-lateral transform faults with rhomb grabens between each pair. He concluded that, in considering the transform system in its regional setting, namely as extending from the Red Sea spreading centre in the south to the Eurasian collision zone in the north, the possibility that it may intersect the latter somewhere east of Cyprus is strongly suggested, making that area the northernmost termination of the Dead

Sea Transform System. He projected the sinistral and well documented Roum Fault, which he incorrectly renamed the Damur (Damour Fault), beyond its earlier northern mappable end as a fault to the coast partly across some Recent sediments into the Mediterranean, utilizing satellite imagery and a certain degree of generalized long-distance interpretation.

In the Palmyride region, considerable surface and subsurface work has been underway since the mid 1970s, throughout the 1980s and into the 1990s. As a result, much more data has been available to researchers than from the Lebanese area to its west. Some of these data had been released in earlier publications on the area (e.g. Ponikarov *et al.* 1967), while more recent information had to be searched for between the lines (e.g. Beydoun 1977, 1981, 1988). More recently, publications involving considerable subsurface geophysical and borehole data has appeared (e.g. Mc Bride *et al.* 1990; Chaimov & Barazangi 1990; Beydoun 1991; Chaimov *et al.* 1992; Barazangi *et al.* 1993; Saber *et al.* 1993; Searle 1993).

The Palmyride Basin has been described as an aulacogene or a failed rift (Ponikarov *et al.* 1967; Beydoun 1981). It constitutes a $350 \times 100$ km, northeast–southwest oriented area which originated during the Levantine margin rifting events of the Late Permian–Middle Triassic. It was inverted in the Cenozoic, being sandwiched between two relatively stable blocks, the Aleppo and Rutbah Uplifts to its north and south, respectively. These uplifts are, in turn, inverted Palaeozoic intracratonic basins which only received a relatively thin veneer of younger sediments (Beydoun 1991). McBride *et al.* (1990) estimated at least 5000 m of post-Palaeozoic sediments were laid down in the Palmyride Basin, thickening abruptly into the basin from the flanking uplifts. These authors considered that the Rutbah Uplift was active during the early formation of the basin with down-flexuring of the lower Palaeozoic platform sequence beneath the southern front along a basement hinge zone. The Aleppo High, to the north, acted passively. The Palmyride Basin is divisible into two main parts: (1) a southwestern to southern portion characterized by short wavelength (5–10 km) folds; (2) a northeastern to northern portion characterized by a northeast plunging anticlinorium. The southern area folds are developed on regional low angle detachments within Triassic evaporites, with small Jurassic and Early Cretaceous normal faults, whose displacement was reversed during basin inversion mainly during the Cenozoic (Chaimov *et al.* 1992). The northeastern to northern portion is characterized by a northeast plunging anticlinorium whose outer flanks are marked by smaller superimposed asymmetrical anticlines associated with outward verging thrusts to give this region a rough symmetry (McBride *et al.* 1990). The approximately east–west trending Jhar Fault separates the two regions. The folding and thrusting in the Palmyride belt is restricted to the narrow 100 km wide swath, in contrast to the relatively undeformed Phanerozoic strata of the platforms on either side. Palinspastically restored cross-sections across the southwestern segment demonstrate a minimum northwest–southeast shortening of *c.* 20 km in the southwest, diminishing to 1–2 km in the northeast (Chaimov & Barazangi 1990). A first inversion phase is noted in the succession during the Upper Cretaceous (Beydoun 1991; Chaimov *et al.* 1992), which was followed by tectonic quiescence throughout the Palaeogene, with the main inversion taking place by Late Oligocene–Early Miocene (Chaimov *et al.* 1992). The first inversion phase is associated with the collision and ophiolite obduction along the Arabian Plate margin to the north during the Late Cretaceous, whilst the main inversion and south directed thrusting is due to the oblique collision between the Anatolian and Arabian Plates, starting in the Middle Eocene, leading to suturing of the two in the Neogene (Hempton 1987) – this predates development of the Levant Fracture System Transform. Only restricted deformations that modified an already established fold belt can be attributed to the Levant (Dead Sea) Fault System, the belt being too wide and too extensive to be deformed only by left-lateral motion (Lovelock 1984). Saber *et al.* (1993) and Barazangi *et al.* (1993) recorded depth to metamorphic basement beneath the Palmyrides and the thickness of Phanerozoic rocks decreasing from *c.* 12 km in the southwest to 9 km in the central segment, whilst basement depth beneath the Aleppo High is *c.* 6 km and beneath the Rutbah Uplift is *c.* 8 km. Shortening along the Palmyrides has been insufficient to invert the previously extended basement morphology under the basin, and estimates of crustal shortening in the Palmyride belt indicate that the northern segment of the Levant Fracture System is younger (Pliocene) than the southern part (Barazangi *et al.* 1993).

Searle (1993), who carried out field mapping in the eastern part of both segments of the Palmyrides, south and north of the Jhar Fault, confirmed that shortening amounts are very small in the eastern part (*c.* 1 km) but increase towards the west to *c.* 20 km. Tight box folds and slightly asymmetric folds in the southeast

have localized basal detachments within the Triassic evaporite horizon which, in some cases, flows into the fold core.

A Mid-Bekaa Fault is identified which bifurcates from the Yammouneh Fault about halfway along its north-northeast–south-southwest length and strikes northeast under the Bekaa alluvium to join the Serghaya Fault extension of the Anti-Lebanon in Syria. Khair *et al.* (1997) confirmed and refined the delineation of this Mid-Bekaa Fault, using multiple source Werner deconvolution estimates for Bouguer gravity anomalies. The authors calculated the thickness of the Phanerozoic succession in the southwestern part of the Palmyrides as *c.* 13 km with the shortening possibly >30 km. They suggested that an appreciable amount of the total left-lateral horizontal displacement, measurable in the southern part of the Levant Fracture System Transform south of the Lebanese border, could have been accounted for by movement on the Serghaya–Mid-Bekaa Fault without transmission to the northern (Ghab) segment of the system, and they explained the splitting of the Dead Sea Transform south of the Lebanese border into the Serghaya, Mid-Bekaa, Yammouneh and Roum Faults by rotation of the detached post-Triassic succession over a stable, deep sinistral fracture of the Dead Sea Fault in the underlying crust.

Within Lebanon, Khair *et al.* (1993), utilizing Bouguer gravity anomalies to determine crustal structure, have confirmed that crustal thickness decreases from *c.* 35 km under the Anti-Lebanon to *c.* 27 km beneath the coast. Depth-to-basement ranges between 3.5 and 6 km below sea level under the mountain ranges to *c.* 8–10 km below sea level under the Bekaa Depression. These figures are not far off calculations made by Goedicke [in Beydoun (1977)] confirming a marked crustal break.

Westway (1995) developed a quantitative model incorporating deformation which he applied to step-overs in strike-slip fault zones and used the north-northeast–south-southwest Yammouneh Fault of Lebanon as a case study. Step-overs are defined as areas where left-lateral faults step rightward or right-lateral faults step leftward. A step-over is regarded as containing an internal strike-slip fault (here the Yammouneh), whose ends link to transform faults (here the north–south Dead Sea Transform to the south and north–south Ghab Transform to the north). The kinematics are governed by the ratio of slip rates on the internal fault and transform faults, and the angle between these faults. The convergence direction across the deformation is an azimuth around which vectors do not rotate around vertical axes. In the Yammouneh step-over, the time-averaged deformation since 15 Ma accounts for 30% of anticlockwise rotation of Cretaceous magnetization vectors in the Lebanon Mountains and c. 50% rotation of Jurassic vectors. There is evidence of a recent evolution from low to high slip-rate ratios.

## Current research on the Lebanon Segment

Field work has recently become much more achievable in Lebanon, except in the occupied areas in the south and some critical areas adjacent to these occupied. Accordingly, researchers from Leeds University and the American University of Beirut have reopened reconnaissance field based investigations concerning the role of the principal faults of the braided Levant Fracture System in Lebanon and the amount of horizontal displacement, if any, that is currently taking place along each (Fig. 3).

Tabet (1993) remarked, quoting seismicity plots, on the probable presence of a seismic gap covering the northern 40 km of the Yammouneh Fault and a diffuse band of activity over the southern portion the fault trace. Considering that only one active monitoring station, located in central Lebanon, is the source for recording seismic activity for northern Lebanon with no operational monitors providing records from northern Syria and southern Turkey at the present time, the lack of activity along the northern portion of the Yammouneh Fault could be an artifact rather than a seismic gap. Alternatively, it could be due to the presence of a bypass fault that is now inactive.

Butler et al. (1997) established that the northern segment of the Yammouneh Fault (where it changes direction from north-northeast–south-southwest to north–south, to become the Ghab Segment) has been totally inactive for the past 5 Ma or so. This was achieved by dating undeformed basalts in Wadi Chadra flanking Jabal Akroum in northernmost Lebanon, where highly fractured and deformed Cretaceous limestones occur where the Yammouneh Fault runs through them. The upper volcanic flows gave ages of $5.2 \pm 0.2$ Ma while the pillow lavas at the base gave $5.7 \pm 0.5$ Ma. None of these flows are cut by major faults and the whole sequence appears undeformed, suggesting that deformation of the carbonates, by fault activity, preceded the eruption of the adjacent basalts. The apparent offset of the basalts on either side of the Yammouneh Fault is due purely to orographic effects, the basalts flowing around the elevated eastern side

where the Jabal Akroum Ridge occurs to infill the low-lying topography on the west side.

Since the southern segment of the Levant Fracture System Transform (the Dead Sea Transform) is active and displacement estimates in the past 5 Ma (Pliocene on) are 45 km (Quennell 1958, 1959, 1983, 1984; Freund et al. 1970), this displacement must either be accommodated by distributed crustal shortening through Lebanon and the Palmyrides, as has been proposed by various authors discussed earlier, or else by a bypass offshore along the Roum Fault into the eastern Mediterranean, as has been suggested by Girdler (1990) (Fig. 3).

The Roum Fault is currently active and recent earthquakes in 1956 and 1997 (magnitudes c. 6 and 5, respectively) have occurred on it with epicentres in the Awali River Sector (Plassard & Kogoj 1981; Beydoun 1997). Although Dubertret (e.g. 1969) had in the past denied the Roum Fault's continuity northwards beyond the Awali River (35 km south of Beirut), crossing the mountainous west dipping Cenomanian carbonates that form the western flank and flexure of the Mount Lebanon Uplift, because of the apparent lack of field evidence, this trace is evident on satellite imagery. Butler et al. (1997) have established from random field checks along the prolonged Roum Trend north of the Awali River, that the Roum Fault continues as a left-stepping northward prolongation through the Cretaceous carbonates and that it has been apparently broken up into short segments of no more than a few kilometres in length and a few hundred metres in width. These authors mapped two strands across the Damour River (21 km south of Beirut) which show left-lateral movement, the fault here forming the eastern tectonic boundary of a Miocene marine terrace abutted against the main flank of the Lebanon Uplift (Fig. 3). Another 10 km of Cretaceous carbonates separate this locality from the 10 km long (by 4 km wide) region south of Beirut covered by Quaternary deposits. Thus, it is uncertain if the trace of the fault obliquely heads out to sea just across the southern edge of this Quaternary area at Khaldeh, as depicted by Girdler (1990), or continues under the Quaternary loose cover of sediment under the airport runways and southern suburbs of Beirut to join either the Nahr Beirut Fault on the eastern edge of Beirut or the Harbour Fault in the town centre (Beydoun 1997). Much of the town centre area between these two faults has a thick infill of Quaternary deposits faulted against Cretaceous carbonates on the west and Miocene carbonates on the east. Evidence of sediment liquefaction and of great earthquake destruction have been uncovered in current archaeological excavations associated with post early Roman to early Byzantine structures in the town centre (Beydoun 1997).

Butler *et al.* (1997) remarked that it is the Roum Fault and not the Yammouneh Fault which has been the principal active strand of the Dead Sea Fault System through Lebanon (as seems to be the case in consequence of the earlier discussion). They estimated that the Roum Fault has moved left laterally some 30 km since the Messinian on the basis of displacement of major east–west flowing river gorges and catchment areas across the fault. The transform plate boundary would then presumably join the East Anatolian Fault and the Cyprus Arc at a triple junction in the northeast Mediterranean. They further considered the Mid–Late Miocene as having been the main transpressive phase, with the Yammouneh Fault being the principal transcurrent structure. Transpression is indicated by major left-lateral faults and regional on-line folds. These features infer that transpression was partitioned into transcurrent faulting and fault-perpendicular compression. The later Miocene to the present stage of movement was characterized by displacement bypass of the transpressive bend with bulk transcurrent displacement approximate to the total relative plate motion across the transform, being accommodated on the Roum Fault.

## Conclusions

The Levant (Dead Sea) Fracture System is a Neogene–Recent transform system that forms the northwestern boundary of the Arabian Plate. It came into being in response to the opening of the Gulf of Aden–Red Sea Rifts and the separation of Arabia (with anticlockwise rotation) from Africa. It links the extensional Red Sea spreading centre in the south with the collisional Tauride collisional region in the north and consists of segments that follow north-northeast–south-southwest to south-north orientations. The fracture system is, thus, a young structure which may or may not preferentially follow ancient lines of weakness in the crust.

The Levant Fracture System consists principally of three portions. A southern portion, south of the Lebanese border, and a northern portion, north of Lebanon, essentially consisting of north–south trending left-stepping strike-slip faults. These are separated by a north-northeast–south-southwest braided strike-slip system cutting across Lebanon and into the Palmyride fold-thrust belt (Fig. 1).

The initiation of the Levant Fracture System post-dates the inversion, folding and uplift of Lebanon and of the Palmyride belt, inversion essentially being completed prior to the Miocene, especially in Lebanon. This was as a consequence of oblique collision between the Arabian Plate and the Anatolian Plate, and closure of the Neo Tethys to the north which had commenced in Late Middle Eocene times and was mainly complete by the Late Miocene. Thus, the initiation of horizontal movement along the Levant Fracture in Middle–Late Miocene time could not have had more than a local effect in the enhancement of existing structures through transpression and transtension, and could not have been the cause of this main structuration, especially within Lebanon.

Prior to the development of the restraining bend on the Yammouneh Fault (end Miocene), the braided system in Lebanon would have absorbed most or all of the 62 km of early lateral motion measurable south along the Dead Sea–Jordan, with the Yammouneh, Serghaya and probably Roum Faults taking the lion's share. In the northern (Ghab) segment, the reduced total northern displacement through Lebanon would have been transmitted through the Ghab system.

From the Pliocene, the Yammouneh Fault in Lebanon appears to have been bypassed and the 45 km horizontal motion measurable south of Lebanon must be transmitted northwards through Lebanon and the Palmyrides in some other way. Some of this would be through deformations along the other branches of the braided fault system in Lebanon, and by local modification and adjustments to the existing structures in Lebanon and the Palmyrides. How much excess to this accommodation is then available to be transmitted to the Ghab Segment remains an open question but the splintering of this latter segment over the northern portion in the Kara Su Valley demonstrates current activity. Is this essentially due to westward tectonic escape of the Anatolian wedge, between the East Anatolian and North Anatolian Transform Faults (Dewey *et al.* 1986), or is there still lateral activity along the Ghab Transform transmitted independently by that segment as suggested by Quennell (1984)?

Current research indicates that the Roum Fault has probably taken up all activity since the bypassing of the Yammouneh Fault (Butler *et al.* 1997) and that Recent horizontal movement south of Lebanon is being transmitted along this fault as it heads towards Beirut and thence offshore into the Mediterranean. Any horizontal displacement not transmitted by the Roum Fault is probably shared by transpressive–transtensional movement along other branches in Lebanon and the Palmyrides. Fortunately, the Roum Fault appears to have become fragmented into a series of short left-stepping segments along which the total energy release during rupture is probably not likely to reach magnitude 7 on the Richter Scale. In addition to the current attempts at enhancing the recording of seismicity across Lebanon, and in the Beirut region in particular, through the establishment of a network of seismic monitors and seismographs, a separate detailed field mapping programme along the whole satellite-identified northern trace of the Roum Fault is planned for rapid execution. This will run across the well-bedded and evenly dipping monotonous Cretaceous carbonates to delineate details of any segmentation similar to that mapped near Damour, establish the geometry of each strand of the fault and resolve this interesting question.

# References

ABED, A. M. 1985. On the supposed Precambrian palaeosuture along the Dead Sea Rift, Jordan. *Journal of the Geological Society, London*, **142**, 527–531.

ARTHAUD, F., DUBERTRET, L., MASSON, P. & MERCIER, J. L. 1978. Etats de contraintes successifs post-eocenes le long des failles libano–syriennes au Nord de Beyrouth. 6ᵉ Reunion, *Annales des Science de la Terre*, Orsay, Avril 1978 (en depot a la Societe Geologique de France).

BAHAT, D. & RABINOVITCH, A. 1983. The initiation of the Dead Sea Rift. *Journal of Geology*, **91**, 317–332.

BANDEL, K. 1981. New stratigraphical and structural evidence for lateral dislocation in the Jordan Rift valley connected with a description of the Jurassic rock column in Jordan. *Neues Jahrbuch fur Geologie und Palaontologie. Abhandlungen*, **161**, 271–308.

BARAZANGI, M., SABER, D., CHAIMOV, T., BEST, J., LITAK, R., AL SAAD, D. & SAWAF, T. 1993. Tectonic evolution of the northern Arabian plate in western Syria. *In*: BOSCHI, E. *et al.* (eds) *Recent Evolution and Seismicity of The Mediterranean Regions*. Kluwer, 117–140.

BEN-AVRAHAM, Z. & TEN BRINK, U. 1989. Transverse faults and segmentation of basins within the Dead Sea Rift. *Journal of African Earth Sciences*, **8**, 603–616.

BENDER, F. 1968. Uber das Alter und die Entstehungsgeschichte des Jordangrabens am Beispiel Sudabschnittes (Wadi Araba, Jordanian). *Geologische Jahrbuch*, **86**, 117–196.

——1983. On the evolution of the Wadi Araba-Jordan Rift. *In*: ABED, A. M. & KHALED, K. M. (eds)

*1st Jordanian Geological Conference Proceedings*. Jordanian Geologists Association, 415–445.

BEYDOUN, Z. R. 1977. The Levantine countries: The geology of Syria and Lebanon (Maritime regions). *In*: NAIRN, K. E. M., KANES, W. H. & STEHLI, F. G. (eds) *The Ocean Basins and Margins. Volume 4A The Eastern Mediterranean*. Plenum, 319–353.

——1981. Some open questions relating to the petroleum prospects of Lebanon. *Journal of Petroleum Geology*, **3**, 303–314.

——1988. *The Middle East: Regional Geology and Petroleum Resources*. Scientific Press.

——1991. *Arabian Plate Hydrocarbon Geology and Potential – A Plate Tectonic Approach*. AAPG Studies in Geology, **33**.

——1997. Earthquakes in Lebanon: an overview. *Lebanese Science Bulletin*, **10**.

BUTLER, R. W. H., SPENCER, S. & GRIFFITHS, H. 1997. Structural evolution of a transform restraining bend – transpression on the Lebanese sector of the Dead Sea Fault System. *In*: HOLDSWORTH, R. E., STRACHAN, R. A. & DEWEY, J. F. (eds) *Continental Transpressional and Transtensional Tectonics*. Geological Society, London, Special Publications, **135**, 81–106.

——, SPENCER, S. & GRIFFITHS, H. M. 1997. Transcurrent fault activity on the Dead Sea Transform in Lebanon and its implications for plate tectonics and seismic hazard. *Journal of the Geological Society, London*, **154**, 757–760.

CHAIMOV, T. A. & BARAZANGI, M. 1990. Crustal shortening in the Palmyride fold belt, Syria, and implications for movement along the Dead Sea fault system. *Tectonics*, **9**, 1369–1383.

——, ——, AL SAAD, D., SAWAF, T. & GEBRAN, A. 1992. Mesozoic and Cenozoic deformation inferred from seismic stratigraphy in the southwestern intracontinental Palmyride fold-thrust belt, Syria. *Geological Society of America Bulletin*, **104**, 704–715.

DEWEY, J. F., HEMPTON, M. R., KIDD, W. S. F., SAROGLU, F. & SENGOR, A. M. C. 1986. Shortening of continental lithosphere: The neotectonics of Eastern Anatolia – a young collision zone. *In*: COWARD, M. P. & RIES, A. C. (eds) *Collision Tectonics*. Geological Society, London, Special Publications, **19**, 3–36.

DUBERTRET, L. 1932. Les formes structurales de la Syrie et de la Palestine, *Comptes rendu de l'Academie des Sciences (Paris)*, **195**, 66.

——1947. Problemes de la geologie du Levant. *Bulletin de la Societe Geologique de France*, **17**, 1–31.

——1951. *Carte geologique an 50,000e, feuille de Merdjayoun avec notice explicative*. Ministere des Travaux Publics, Republique Libanaise, Beyrouth.

——1955. *Carte geologique du Liban au 1/200 000e avec notice explicative*. Ministere des Travaux Publics, Republique Libanese, Beyrouth.

——1963. Liban, Syrie: chaine des grands massif cotiers et confins a l'est. *In*: DUBERTRET, L. (eds) *Lexique Stratigraphique International, Vol. III Asie*. Centre National de la Recherches Scientifique, Paris, **10**, 9–103.

——1966. Liban, Syrie et bordure des pays voisins. *Notes et Memoires sur la Moyen-Orient*, **8**, 251–358.

——1967. Remarques sur le fosse de la Mer Morte et ses prolongements au nord jusqu'au Taurus. *Revue de Geographie Physique et Geologie Dynamique*, **9**, 3–16.

——1969. Le Liban et la derive des continents. *Hannon, revue Libanaise de Geographic, Universite Libanaise*, **4**, 53–61.

——1970. Revue of structural geology of the Red Sea and surrounding areas. *Philosophical Transactions of the Royal Society, London*, **A267**, 9–20.

——1971–72. Sur la dislocation de l'ancienne plaque sialique Afrique–Sinai–Peninsule Arabique, proposition d'un nouvelle modele. *Notes et Memoires sur le Moyen-Orient*, **12**, 227–243.

——1974–1975. Introduction a la carte geologique a 1/50000e du Liban. *Notes et Memoire sur la Moyen-Orient*, **12**, 227–243.

——1976. *La peninsule Arabique: explanatory text to sheet 16, Tectonic Map of Europe 1 : 2 500 000 scale*. International Geological Congress Sub Commission of Tectonic Maps, Chapter 6, section 4.

FOLKMAN, Y. & BEIN, A. 1978. Geophysical evidence for a pre-late Jurassic fossil continental margin oriented east-west under Central Israel. *Earth and Planetary Science Letters*, **39**, 335–340.

FREUND, R. 1965. A model of the structural development of Israel and adjacent areas since Upper Cretaceous times. *Geological Magazine*, **102**, 189–205.

—— & RAAB, M. 1969. Lower Turonian ammonites from Israel. *Tectonophysics*, **2**, 457–474.

—— & TARLING, D. H. 1979. Preliminary Mesozoic palaeomagnetic results from Israel and inferences for a micro plate structure in Lebanon. *Tectonophysics*, **60**, 189–205.

——, GARFUNKEL, Z., ZAK, I., GOLDBERG, M., WEISSBROD, T. & DERIN, B. 1970. The shear along the Dead Sea Rift. *Philosophical Transactions of the Royal Society, London*, **A267**, 107–130.

GARFUNKEL, Z. 1981. Internal structure of the Dead Sea leaky transform (Rift) in relation to plate kinematics. *Tectonophysics*, **80**, 81–108.

GIANNERINI, G., CAMPREDON, R., FERAUD, G. & ABOU ZAKHEM, B. 1988. Deformations intraplaques et volcanisme associe: exemple de la bordure NW de la plaque Arabique au Cenozoique. *Bulletin de la Societe Geologique de France*, **4**, 937–947.

GINZBURG, A. & FOKLKMAN, Y. 1980. The crustal structure between the Dead Sea Rift and the Mediterranean. *Earth and Planetary Science Letters*, **51**, 181–188.

GIRDLER, R. W. 1989. A. M. Quennell father of transform faults and poles of rotation? *EOS (Transactions of the American Geophysical Union)*, **70**.

——1990. The Dead Sea transform fault system. *Tectonophysics*, **180**, 1–13.

HANCOCK, P. C. & ATIYA, M. S. 1979. Tectonic significance of mesofracture systems associated with the Lebanese segment of the Dead Sea transform fault. *Journal of Structural Geology*, **1**, 143–153.

HATCHER, R. D. JR, ZIETZ, I., REGAN, R. D. & ABU-AJAMIEH, J. 1981. Sinistral strike-slip motion on the Dead Sea Rift: Confirmation from new magnetic data. *Geology*, **9**, 458–462.

HEINMANN, A., EYAL, M. & EYAL, Y. 1990. The evolution of the Barahta rhomb-shaped graben, Mount Hermon, Dead Sea Transform. *Tectonophysics*, **180**, 101–110.

HEMPTON, M. R. 1987. Constraints on Arabian plate motion and extensional history of the Red Sea. *Tectonics*, **6**, 687–705.

KASHAI, E. L. 1988. A review of the relation between the tectonics, sedimentation and petroleum occurrences of the Dead Sea–Jordan rift system. *In*: MANSPEITZER, W. (ed.) *Triassic–Jurassic Rifting, Continental Break-up and the Origin of the Atlantic Ocean and Passive Margins*. Elsevier, 883–909.

—— & CROKER, P. E. 1987. Structural geometry and evolution of the Dead Sea–Jordan rift system as deduced from new subsurface data. *Tectonophysics*, **141**, 33–60.

KHAIR, K., TSOKAS, G. N. & SAWAF, T. 1997. Crustal structure of the northern Levant region: multiple source Werner deconvolution estimates for Bouguer gravity anomalies. *Geophysical Journal International*, **128**, 605–616.

——, KHAWLIE, M., HADDAD, F., BARAZANGI, M., SEBER, D. & CHAIMOV, T. 1993. Bouguer gravity and crustal structure of the Dead Sea transform fault and adjacent mountain belts in Lebanon. *Geology*, **21**, 739–742.

LARTET, L. 1869. Essai sur la geologie de la Palestine. *Annales des Sciences Geologiques, Paris*.

LOVELOCK, P. E. R. 1984. A review of the tectonics of the northern Middle East region. *Geological Magazine*, **121**, 577–587.

McBRIDE, J. E., BARAZANGI, M., BEST, J., AL-SAAD, O., SAWAF, T., AL-OTRI, M. & GEBRAN, A. 1990. Seismic reflection structure of intracratonic Palmyride fold-thrust belt and surrounding Arabian platform, Syria. *AAPG Bulletin*, **74**, 238–259.

MASSON, P., CHEVEL, P., EQUILBY, S. & MARION, A. 1982. Apports du traitement numerique d'images Landsat a l'etude des failles Libano–Syrienne. *Bulletin de la Societe Geologique de France*, **24**, 63–71.

MATAR, A. & MASCLE, G. 1993. Cinematique de la faille du Levant au nord de la Syrie: analyse microtectonique du fosse d'Alghab. *Geodynamica Acta*, **6**, 153–160.

MUEHLBERGER, W. R. 1981. The splintering of the Dead Sea fault zone in Turkey. *Yerbilimleri (Bulletin of the Institute of Earth Sciences of Hacetepe University, Ankara)*, **8**, 125–130.

——, GOETZ, L. K. & BELCHER, R. C. 1979. Analysis of Skylab and ASTP photographs of the Levantine (Dead Sea) fault zone. *In*: EL-BAZ, F. & WARNER, D. M. (eds) *Apollo–Soyuz Test Project, Summary Science Report, Vol. II. Earth Observations and Photography*. NASA, Washington, DC, 45–62.

NEEV, D. 1975. Tectonic evolution of the Middle East and the Levantine basin (easternmost Mediterranean). *Geology*, **3**, 683–686.

PICARD, L.,1967. Thoughts on the graben system in the Levant. *Geological Survey Canada Professional Paper*, **66–14**, 22–32.

PLASSARD, J. & KOGOJ, B. 1981. *Seismicite du Liban. Annales-Memoires de l'Observatoire de Ksara, V.4 (seismologie) cahier 1*, 3rd edn. Centre National de la Recherche Scientifique, Beyrouth.

PONIKAROV, V. P., KAZMIN, V. G., MIKHAILOV, *et al.* 1967. *The geology of Syria: explanatory notes on the geological map of Syria, scale 1:500 000. Part I Stratigraphy, igneous rocks and tectonics.* Ministry of Industry, Damascus.

POWELL, J. H. 1988. *The geology of the Karak area. Explanatory notes and geological map on 1:50 000 scale.* Natural Resources Authority Bulletin 8, Amman, Jordan.

QUENNELL, A. M. 1951. Geology and mineral resources of (former) Transjordan. *Colonial Geology and Mineral Resources*, **2**, 85–115.

——1958. The structural and geomorphic evolution of the Dead Sea Rift. *Quarterly Journal of the Geological Society, London*, **114**, 1–24.

——1959. Tectonics of the Dead Sea Rift. *20th International Geological Congress Proceedings*, Mexico, ASGA, 385–403.

——1983. Evolution of the Dead Sea Rift: A review. *In*: ABED, A. M. & KHALED, K. M. (eds) *1st Jordanian Geological Conference Proceedings.* Jordanian Geologists Association, 460–482.

——1984. The Western Arabia rift system, *In*: DIXON, J. E. & ROBERTSON, A. H. F. (eds) *The Geological Evolution of the Eastern Mediterranean.* Geological Society, London, Special Publications, **17**, 775–778.

RON, H. 1987. Deformation along the Yammouneh, the restraining bend of the Dead Sea Transform. Paleomagnetic data and kinematic implications. *Tectonics*, **6**, 6553–6566.

ROTHSTEIN, Y. & KAFKA, A. L. 1982. Seismotectonics of the southern boundary of Anatolia, eastern Mediterranean region: Subduction, collision and arc jumping. *Journal of Geophysical Research*, **87**, 7694–7706.

SABER, D., BARAZANGI, M., CHAIMOV, T. A., AL-SAAD, D., SAWAF, T. & KHADDOUR, M. 1993. Upper crustal velocity structure and basement morphology beneath the intracontinental Palmyride fold-thrust belt and north Arabian platform, Syria. *Geophysical Journal International*, **113**, 752–766.

SEARLE, M. P. 1993. Structure of the intraplate eastern Palmyride fold belt, Syria. *Geological Society of America Bulletin*, **106**, 1332–1350.

SITTER, L. U. DE 1962. Structural development of the Arabian Shield in Palestine. *Geologie en Mijnbouw*, **41**, 116–124.

TABET, C. A. 1993. Possible seismic gap along the northern segment of the Yammouneh Fault in Lebanon (Abstract). *Seminar, Cooperative Program for Reducing Earthquake Losses in the Eastern Mediterranean Region (RELEMR)*, Cairo, Oct. 1993, UNESCO–USGS.

TIBERGHIEN, V. 1974. Le champ de la pesanteur au Liban et ses interpretations. *Publications Technique et Scientific de l'Ecole Superieure d'Ingenieurs de Beyrouth*, **26**.

WALLEY, C. D. 1983. The palacoccology of the Callovian–Oxfordian strata of Majdal Shams (Syria) and its implications for Levantine palaeogeography and tectonics. *Palaeogeography, Palaeoclimatology, Palaeoecology*, **42**, 323–340.

——1988. A braided strike-slip model for the northern confirmation of the Dead Sea Fault and its implications for Levantine tectonics. *Tectonophysics*, **145**, 63–72.

WESTWAY, R. 1995. Deformation around stepovers in strike-slip fault zones. *Journal of Structural Geology*, **17**, 831–846.

WETZEL, R. & MORTON, D. M. 1959. Contribution a la geologie de la Transjordanie. *Notes et Memoires sur la Moyen-Orient*, **7**, 95–191.

# The lithospheric structure of the Kenya Rift as revealed by wide-angle seismic measurements

M. A. KHAN,[1] J. MECHIE,[2] C. BIRT,[1] G. BYRNE,[3] S. GACIRI,[4] B. JACOB,[5]
G. R. KELLER,[6] P. K. H. MAGUIRE,[1] O. NOVAK,[5] I. O. NYAMBOK,[4]
J. P. PATEL,[7] C. PRODEHL,[3] D. RIAROH,[8] S. SIMIYU,[8] & H. THYBO[9]

[1] *Department of Geology, University of Leicester, University Road, Leicester LEI 7RH, UK*
[2] *GeoForschungsZentrum Potsdam (GFZ), Department 2 Telegrafenberg A17,
14473, Potsdam, Germany*
[3] *Geophysikalisches Institut, Universitaet Karlsruhe, Hertzstrasse 16,
76187 Karlsruhe, Germany*
[4] *Department of Geology, University of Nairobi, PO Box 14576, Nairobi, Kenya*
[5] *Dublin Institute for Advanced Studies, School of Cosmic Physics, 5 Merrion Square,
Dublin 2, Ireland*
[6] *Department of Geological Sciences, The University of Texas at El Paso, El Paso,
TX79968–0555, USA*
[7] *Department of Physics, University of Nairobi, PO Box 14576, Nairobi, Kenya*
[8] *Ministry of Energy, Nyayo House, PO Box 30582, Nairobi, Kenya*
[9] *Institute of Geology, University of Copenhagen, Oester Voldgade 10,
1350 Copenhagen K, Denmark*

**Abstract:** The Kenya Rift International Seismic Project (KRISP) seismic refraction–wide-angle reflection experiments carried out between 1985 and 1994 show abrupt changes in Moho depths and $P_n$ phase velocities as the rift boundaries are crossed. Beneath the rift flanks, normal $P_n$ phase velocities of $8.0$–$8.3 \, \mathrm{km \, s^{-1}}$ are observed, except for the Chyulu Hills volcanic field, east of the rift, where it is $7.9$–$8.0 \, \mathrm{km \, s^{-1}}$. Also to the east, some of the thickest crust (38–44 km) encountered so far beneath Kenya has been observed over a distance of *c.* 300 km. However, beneath the surface expression of the rift itself, the uppermost mantle velocity of the $P_n$ phase is anomalously low at $7.5$–$7.8 \, \mathrm{km \, s^{-1}}$ throughout its length.

Beneath the rift itself, there are major differences in crustal thickness, extension and upper mantle velocity structure between the north and the south. Beneath the section from the centre of the Kenya Dome southwards, where the extension is estimated to be 5–10 km, the crust is thinned by *c.* 10 km to a thickness of 35 km, and the narrow low-velocity zone in the mantle extends to a depth of at least 65 km. However, in the north beneath Turkana, where the extension is 35–40 km, the crust is only *c.* 20 km thick and two layers with velocities of 8.1 and $8.3 \, \mathrm{km \, s^{-1}}$ are embedded in the low velocity mantle material at depths of 40–45 km and 60–65 km. This mantle velocity structure indicates that the depth to the onset of melting is at least 65 km beneath the northern part of the rift and is thus not shallower than the corresponding depth (45–50 km) in the south. These results, taken together with those from teleseismic studies, petrology and surface geology, have been used to deduce that anomalously hot mantle material appeared below the present site of the Kenya Rift *c.* 20–30 Ma ago. This led to widespread volcanism along the whole length of the rift and modification of the underlying crust by mafic igneous underplating and intrusion.

The Kenya Rift, which runs north–south through central Kenya (Fig. 1), has fascinated geologists and geophysicists since it was first described just over 100 years ago by J. W. Gregory, after whom the rift is often named. Since then the East African Rift System, extending from Ethiopia in the north to the Zambezi in the south, has become recognized as the classical example of an active continental rift zone. The Kenyan segment, which has experienced widespread

*From*: MAC NIOCAILL, C. & RYAN, P. D. (eds) 1999. *Continental Tectonics.* Geological Society, London, Special Publications, **164**, 257–269. 1-86239-051-7/99/$15.00 © The Geological Society of London 1999.

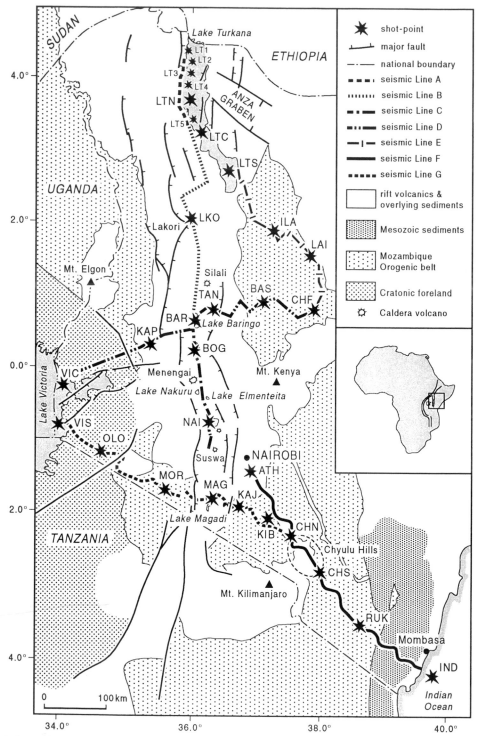

**Fig. 1.** Location map of the KRISP'90 and KRISP'94 seismic refraction–wide-angle reflection surveys. The 1985 refraction–wide-angle reflection lines extended from Lake Baringo to Lake Magadi and across the rift just north of the Susua Volcano.

volcanicity as well as normal faulting, became the focus of extensive geological investigations. The shallow geological features therefore became well known but their relation to the deeper structure was not. It was recognized that a knowledge of the deep structure and dynamics of this unique rift system is fundamental to the understanding of the processes leading to the break-up of continents and the development of ocean basins (Burke & Dewey 1973). The primary need was for high-resolution, long-range seismic refraction data to provide information on the velocity–depth structure beneath the rift and surrounding areas.

This led to the Kenya Rift International Seismic Project (KRISP) which carried out seismic refraction–wide-angle reflection investigations along profiles within, across and outside the rift, at the locations shown in Fig. 1 in 1985, 1990 and 1994. The results of these experiments have been published in a series of papers (e.g. KRISP Working Group 1987, 1995; Henry et al. 1990; KRISP Working Party 1991), and in two special issues of Tectonophysics (Prodehl et al. 1994a; Fuchs et al. 1997) devoted to the crustal and upper mantle structure of the Kenya Rift. During the major experiment in 1990, five deployments were carried out (seismic lines A–E, Fig. 1). Lines A–C make up a 750 km long axial rift profile, line D a 450 km long cross-rift profile, north of the Kenya Dome, and line E a 300 km long profile outside the rift to the northeast. During the experiment in 1994, two further profiles, F and G, were completed. Line F is a 420 km long profile crossing the Chyulu Hills Quaternary volcanic field, on the southeastern flank of the Kenya Rift, and line G is a 430 km long cross-rift profile south of the Kenya Dome. During the earlier test phase of the project in 1985, a line extending from Lake Baringo to Lake Magadi was completed as was a short cross-rift profile just north of the Suswa Volcano (Fig. 1).

This contribution summarizes the lithospheric structure and evolution as revealed by the seismic data combined with the geological and other geophysical data.

## Seismic data

Representative examples of the 1990 and 1994 seismic data, displayed in the form of distance v. reduced-time record sections, are shown for the 1990 shot-point LTN along the rift (Fig. 2a and b) and for the 1994 shot-point VIS across the rift (Fig. 3a and b). Figure 2a shows mainly the crustal phases while Fig. 2b displays mainly the mantle phases. At distances out to 90 km, the first arrivals are from the $P_g$ phase, the refracted phase through the Precambrian upper crystalline crust. At distances >100 km, the first arrivals are from the $P_n$ phase, the refracted phase from just below the crust-mantle boundary. Both the $P_g$ and $P_n$ phases have highly variable apparent velocities, and the topography of the base of the various rift basins can be easily recognized in the travel-time advances and delays of both (Fig. 2a). In addition to these first arrivals, several second-arrival phases can be identified. The most prominent of these include the $P_M P$ phase, the reflection from the crust–mantle boundary, and $d_1$, a reflected phase from a boundary within the uppermost mantle. Although the intracrustal phase $P_{i1}P$, the reflection from the top of the lower crust, cannot be identified with any certainty on the LTN record section, it can be recognized on the nearby LT4 record section, recorded to the south along line A (Gajewski et al. 1994). At extreme wide-angle distances, the strong crustal reflection has been correlated as $P_{i2}P$, the reflected phase from the top of the 6.8 km s$^{-1}$ basal crustal layer. In the northern part of the rift there is no evidence on either the LTN or LTC record sections that the $P_{i2}P$ phase occurs between the $P_{i1}P$ and $P_M P$ phases at distances <100 km. Also, there is no evidence on either of these record sections for the $P_{i2}$ phase, the refraction through the basal crustal layer, and the strong crustal reflection at distances >150 km can be better fitted with apparent velocities nearer 6.4–6.5 than 6.7–6.9 km s$^{-1}$. Thus, in the northern part of the rift, the 2 km thickness with which the 6.8 km s$^{-1}$ layer has been modelled is the maximum thickness it can have, otherwise at least the $P_{i2}P$ phase should have been recognizable between the $P_{i1}P$ and $P_M P$ phases at distances <100 km.

At a distance of c. 300 km on the LTN record section, the $P_n$ phase is overtaken by the first arrival of a faster phase, $d_1'$ (Fig. 2b). As argued by Keller et al. (1994a), the $d_1'$ phase is the diving wave (head wave) turning in the layer whose top surface produces the $d_1$-reflected phase. On the record sections from both the LTN and LTC shot-points, a later reflection, $d_2$, can also be recognized. As in the case of the $d_1$ reflection, the polarity of the $d_2$ phase is the same as that of the first arrivals at the same distance, indicating that the velocity below the $d_2$ reflector is greater than that above it (Keller et al. 1994a).

Record sections showing seismograms recorded in 1994, along the southern cross-rift profile G, are shown on Fig. 3a and b. As before, Fig. 3a mainly shows the crustal phases while

**(a)**

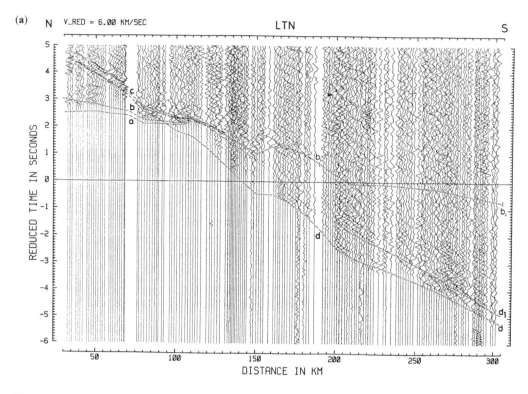

**(b)**

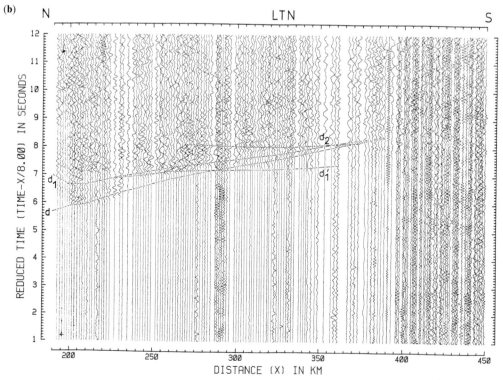

Fig. 3b displays mainly the mantle phases. At distances <150 km, the $P_g$ and $P_1$ phases form the first arrivals, while at distances >150 km the $P_n$ phase forms the first arrivals. The prominent secondary arrival phases between 70 and 200 km distance have been correlated as two reflected phases; the $P_MP$ phase from the crust–mantle boundary and the $P_{i_3}P$ phase from the top of the basal layer deep within the crust. Between the $P_g$ and $P_{i_3}P$ phases, between 40 and 100 km distance, two additional reflected phases have been recognized. These are $P_{i_1}P$ from the top of the lower crust and $P_{i_2}P$ from within the lower crust. Beyond $c.$ 200 km distance (Fig. 3b), reflected phases $D_1$, $D_2$ and $D_3$, from boundaries within the upper mantle, can be recognized.

## Structure of the crust and uppermost mantle under the Kenya rift

Based on the phase correlations on the record sections, 2D earth models were derived for each line using ray tracing (Cerveny et al. 1977). In a first step, the models were iterated by trial and error, or by using the inversion scheme of Zelt & Smith (1992), until a satisfactory match between the travel times of the observed and theoretical phases was obtained. In a second step, further iteration was carried out to model the general features of the relative amplitudes curves. This was done either by calculating ray-theoretical seismograms (Cerveny et al. 1977) or by calculating 2D finite-difference synthetic seismograms (Kelly et al. 1976; Sandmeier 1990).

The major lithospheric structures derived from the KRISP'85, KRISP'90 and KRISP'94 data are presented in the model in Fig. 4. The most important features are the crustal thickness variations both along and across the rift, and the distribution of upper-mantle velocities beneath the rift. Crustal thickness varies dramatically along the rift by 15 km from $c.$ 35 km in the south, beneath and to the south of the Kenya Dome, to $c.$ 20 km in the north beneath the Turkana region. This is accomplished to a major extent by the thinning of the 6.8 km s$^{-1}$ basal crustal layer from $c.$ 9 km in the south to 2 km in the north. There is also a correlation between crustal thickness and topography, rift width, surface geological estimates of crustal extension and Bouguer anomaly. Beneath the Kenya Dome, the crust is thick (35 km), the elevation of the rift floor is high (2–3 km), the rift is narrow and well defined (50–70 km wide), estimates of extension are small (5–10 km) (Baker & Wohlenberg 1971; Strecker 1991) and the Bouguer anomaly is low ($-200 - -250$ mgal). Towards the north, beneath Turkana, the crust thins to 20 km, the elevation of the rift floor decreases to $c.$ 400 m the rift widens to 150–200 km, surface estimates of extension increase to 35–40 km (Morley et al. 1992) and the Bouguer anomaly increases to $c. -50$ mgal. The regional 150–200 mgal change in Bouguer anomaly from north to south can be completely explained by the change in crustal thickness along the rift axis (Mechie et al. 1994b). Along the northeastern flank, profile E between Archers Post (CHF on Fig. 1) and Lake Turkana (LTS), the crustal thickness decreases from 35 km (Fig. 4) to $c.$ 29 km just east of the rift (Prodehl et al. 1994b), halfway between LTS and ILA (Fig. 1). At LKO, 100 km to the west in the middle of the rift, the crustal thickness is only 20 km. As the northeastern flank profile lies between the Anza Graben to the northeast and the Kenya Rift to the southwest, the thinning beneath the axial rift profile with respect to that beneath the northeastern flank profile must be due to the Tertiary rifting episode.

Beneath the 750 km long axial rift profile, the uppermost mantle velocity, $P_n$, is 7.5–7.8 km s$^{-1}$, whereas beneath the Kenya Dome, in the southern part of the rift, these low velocities continue uninterrupted down to 60–65 km depth. Beneath the northern part of the rift, two layers with velocities $\gg$7.8 km s$^{-1}$ alternate with the low-velocity material between 40 and 60 km depth (Keller et al. 1994a). The shallower of the two layers at 40–45 km depth has a velocity of 8.1 km s$^{-1}$, while the deeper layer at 60–65 km depth has a velocity of $c.$ 8.3 km s$^{-1}$. South of the Kenya Dome, along the cross-rift profile G two uppermost mantle reflectors at 45–50 and 60–65 km depth have been identified under both the rift itself and the rift flanks (Byrne et al. 1997). Below the northeastern flank (profile E), a reflector at $c.$ 45 km depth has been detected

Fig. 2. (a) Trace-normalized, low-pass filtered (12 Hz) record section for shot-point LTN recorded to the south along the axial rift profile, with travel times, calculated from the model shown in Fig. 4, drawn in. Reduction velocity, 6 km s$^{-1}$. Phase notation: a, $P_g$; b, $P_{i_1}P$; b$_1$, $P_{i_2}P$; c, $P_mP$; d, $P_n$; d$_1$, upper mantle reflection. Square brackets indicate where b$_1$ can no longer be seen in the observed data (after Mechie et al. 1994b). (b) Trace-normalized, low-pass filtered (12 Hz) record section for shot-point LTN recorded to the south along the axial rift profile, with travel times for the mantle phases calculated from the model shown in Fig. 4, drawn in. Reduction velocity, 8.0 km s$^{-1}$. Phase notation: d, $P_n$; d$_1$, reflection from first upper-mantle reflector; d$_1'$, refraction from the layer beneath the d$_1$ reflector; d$_2$, reflection from the second upper-mantle reflector.

**(a)**

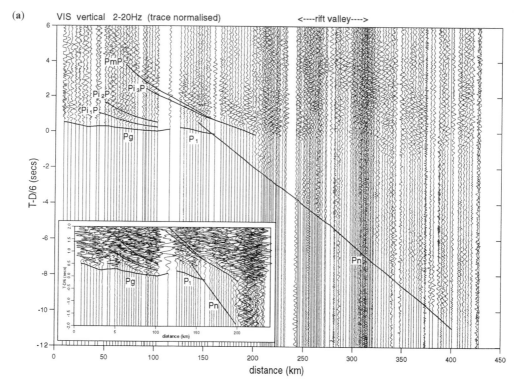

VIS vertical  2-20Hz (trace normalised)                <----rift valley---->

**(b)**

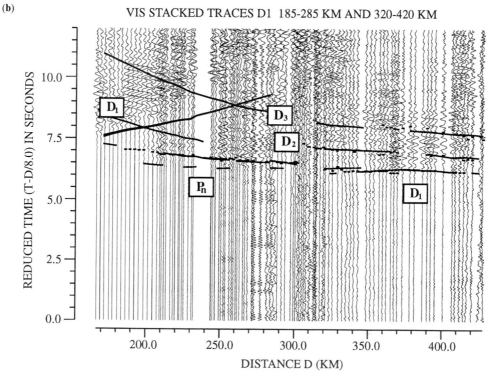

VIS STACKED TRACES D1  185-285 KM AND 320-420 KM

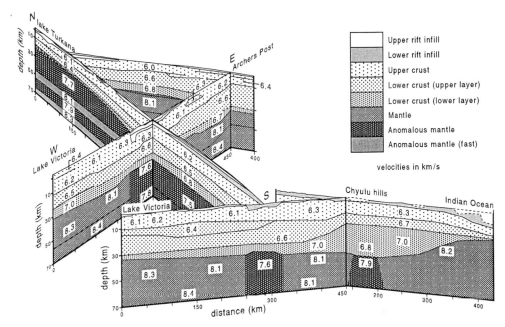

**Fig. 4.** Fence diagram showing crustal and uppermost mantle structure of the Kenya Rift from Lake Turkana to Lake Magadi and beneath the neighbouring flanks from Lake Victoria to the Indian Ocean. P-wave velocities are shown in $km\,s^{-1}$ [modified after Keller *et al.* (1994b)].

(Prodehl *et al.* 1994b), while beneath the cross-rift profile D north of the Kenya Dome, reflectors at 55–60 km depth have been identified outside the rift below both the eastern and western flanks (Maguire *et al.* 1994; Masotti *et al.* 1996).

Across the rift, at the latitudes of both cross-rift profiles D and G crustal thinning and the low uppermost mantle $P_n$ velocities of 7.5–7.8 $km\,s^{-1}$ are restricted to below the surface expression of the rift itself. As soon as the rift boundaries at the surface are crossed, at depth the crustal thickness and the $P_n$ velocities increase abruptly. Beneath the rift flanks, except for the Chyulu Hills area, the $P_n$ velocities have normal values of 8.0–8.3 $km\,s^{-1}$. The abrupt change in upper mantle velocities, as the rift boundaries are crossed, persists down to 100–150 km depth (Achauer *et al.* 1994). Beneath the Quaternary volcanic field of the Chyulu Hills the $P_n$ velocity is 7.9–8.0 $km\,s^{-1}$. Along the cross-rift profile D at the latitude of Lake Baringo, *c.* 100 km north

of the Kenya Dome, the crustal thinning below the rift is 5–10 km. In contrast, along the cross-rift profile G at the latitude of Lake Magadi, *c.* 100 km south of the dome, it is somewhat less at *c.* 2–3 km. Furthermore, the crustal thinning beneath the Lake Magadi portion of the rift south of the dome is flanked by thicker crust to the east (38–44 km) than to the west (<35 km), where the major Nguruman Fault bounding the rift is located (Fig. 4). This is in contrast to the situation below Lake Baringo, north of the dome, where the crust on either side of the rift tends to be more equal in thickness, or to be thicker beneath the western flank (Braile *et al.* 1994; Maguire *et al.* 1994), where the major rift-bounding Elgeyo Fault Rift is located. Nyblade & Pollack (1992) used the pattern of gravity anomalies in this region to suggest that some of the crustal variations just west of the rift valley are due to presence of the suture between the Mozambique Belt and the Tanzania Craton. Below the cross-rift profile south of the Kenya

---

**Fig. 3.** (a) Trace-normalized, band-pass filtered (2–20 Hz) record section for shot-point VIS recorded to the east across the rift south of the Kenya Dome along line G. Travel times, calculated from the model shown in Fig. 4, are drawn in. Reduction velocity, 6.0 $km\,s^{-1}$ [after Birt *et al.* (1997)]. (b) Trace-norrnalized, low-pass filtered (12 Hz) record section showing the mantle phases $D_1$, $D_2$ and $D_3$ recorded from shot-point VIS to the east across the rift south of the Kenya dome along line G. Reduction velocity, 8.0 $km\,s^{-1}$. Each trace in the distance range from 185 to 285 km, 305–420 km is the average of itself and its two nearest neighbours for the $D_1$ phase [after Byrne *et al.* (1997)].

Dome, relatively thin crust (<35 km) was found west of the rift all the way to Lake Victoria (Birt *et al.* 1997). East of the rift, some of the thickest crust observed so far beneath Kenya (38–44 km) extends eastwards for over 300 km, from the Lake Magadi region to beyond the Chyulu Hills–Kilimanjaro region (Novak *et al.* 1997a). In addition, there is a difference between the basement velocity which is 6.0 km s$^{-1}$ in the Archaean to the west and 6.35 km s$^{-1}$ in the Proterozoic to the east of the rift, supporting the idea that the rift is developing along the old suture.

Major structural variations occur not only at and below the level of the crust–mantle boundary but also within the crust. Although along the axial and cross-rift profiles the rift infill generally varies in thickness between 3 and 5 km, large variations occur. At the crests of steeply tilted fault blocks, Precambrian basement is exposed, as in the case of the structural high between the Kerio and Lokichar basins at LKO (Fig. 1), *c.* 250 km from the northern end of the axial profile. In contrast, *c.* 180 km from the western end of the northern cross-rift profile D the rift infill is 8–9 km thick in the Elgeyo basin, at the base of the Elgeyo escarpment (Fig. 1).

The crystalline crust is divided into three principal layers. The upper crystalline crust constitutes one of these principal layers, except beneath the southeastern flank (profile F) where a subdivision into two layers can be recognized. The upper crystalline crust generally has velocities of 6.0–6.3 km s$^{-1}$. However, limited 'blocks' or 'layers' with velocities up to 0.2 km s$^{-1}$ greater than the background values occur; e.g. beneath the rift flanks along the northern cross-rift and northeastern flank profiles (Maguire *et al.* 1994; Prodehl *et al.* 1994b), and within the crystalline crust under the central part of the rift valley (Ritter & Achauer 1994). The velocity jump to 6.4–6.7 km s$^{-1}$ at 9–18 km depth marks the top of the lower crust. Smaller depths to the top of the lower crust are found beneath the axial rift profile in the north where extension has been greatest, beneath the western part of the southern cross-rift profile, towards Lake Victoria, and under the flank profiles, with the exception of those towards the Indian Ocean. Larger depths to the top of the lower crust are encountered beneath the southern part of the axial profile, beneath the northern cross-rift profile, under the eastern part of the southern cross-rift profile and towards the Indian Ocean. The basal crustal layer is characterized by velocities of 6.7–7.0 km s$^{-1}$. Below the northern part of the axial rift profile, the basal crustal layer is too thin for the reflection from it to be distinguished from the reflection from the

crust–mantle boundary (Moho), and thus it is represented as a 2–3 km thick zone beneath the northernmost 250 km of the axial rift profile. Beneath the western end of the southern cross-rift profile, and towards the Indian Ocean, the basal crustal layer is also only *c.* 2 km thick. Under most of the rest of the profiles the basal crustal layer is *c.* 10 km thick. However, in the vicinity of the Chyulu Hills, below the southeastern flank profile F the basal crustal layer is *c.* 20 km thick (Fig. 4) and includes a low-velocity body with a maximum thickness of *c.* 10 km, just above the crust–mantle boundary (Novak *et al.* 1997b). The high velocity layer at the base of the crust is unlikely to be ancient lower crustal material at the high temperatures to be expected at depths of *c.* 30 km. It has therefore been argued (Keller *et al* 1994a) that it is mantle derived mafic igneous material at the base of the crust.

## Discussion and conclusions

The fact that volcanism precedes faulting, in both the northern and southern parts of the rift, by a few million years (Morley *et al.* 1992; Smith 1994), plus the fact that the stretching ($\beta$) factor does not exceed 1.7 anywhere along the rift, provides evidence that much of the volcanism has been generated by melting of the mantle at anomalously high temperatures (Hendrie *et al.* 1994). Although active uprising of anomalously hot mantle material is required both in the north and the south, the amount of stretching and extension in the northern part of the rift due to the Tertiary rifting episode is greater than in the southern part. In the northern part, where the crustal thickness is 20 km, Hendrie *et al.* (1994) have estimated for the Tertiary rifting episode a stretching factor of 1.55–1.65 associated with 35–40 km of total surface extension. In the southern part of the rift, where the crustal thickness is 35 km, estimates of total surface extension range from 5 to 10 km (Baker & Wohlenberg 1971; Strecker 1991) and are thus smaller than those for the northern part of the rift. For the southern part of the rift, estimates of the stretching factor, based on the ratio of crystalline crustal thickness outside the rift to crystalline crustal thickness within the rift, are always <1.4. However, volcanism and rifting started early in the Tertiary in the northern part of the rift and get progressively younger southwards (Dawson 1992). This may be the reason why stretching and extension in the north is greater than in the south. Of the 35–40 km of surface extension which took place in the north, 25–30 km took place before the Upper Miocene

(10 Ma ago) (Morley *et al.* 1992; Hendrie *et al.* 1994). In the south, faulting began in the period 12–7 Ma ago (Baker *et al.* 1988), and so it is possible that the surface extension in both the north and the south over the past 10 Ma is about 10 km and thus of similar magnitude.

The petrological interpretation of the uppermost mantle seismic velocities indicates that the depth to the onset of melting below the Kenya Dome, in the southern part of the rift, is $50 \pm 10$ km (Mechie *et al.* 1994*a*). Below the northern part of the Kenya Rift the depth to the onset of melting must be >65 km, as velocities of 8.3 km s$^{-1}$ are found at *c.* 60–65 km depth (Fig. 4). This interpretation also indicates that the influence of hot mantle material actively uprising from below is greater under the Kenya Dome than under the northern part. Beneath the eastern shoulder, mantle xenoliths brought to the surface during the Quaternary volcanism have been recovered. Geothermobarometric measurements have been carried out on these xenoliths, and their pressure and temperature conditions of formation have been established (Henjes-Kunst & Altherr 1992). The observed seismic velocities are compatible with the xenolith-derived geotherms established in this way. For the eastern shoulder, the $P_n$ velocities beneath the KRISP'90 flank line E are compatible with an average undepleted spinel peridotite composition (50% olivine, 30% orthopyroxene, 18% clinopyroxene and 2% spinel) at depths of 30–35 km and temperatures of 300–400°C. The Pollack & Chapman (1977) geotherms show that these correspond to surface heat flows of 35–50 mW m$^{-2}$, in agreement with the observed average values from heat flow measurements (Morgan 1983). The $P_n$ velocities along line E are also compatible with a composition containing 10–15% more olivine than undepleted spinel peridotite at 30–35 km depth and temperatures of 600–670°C, corresponding to a surface heat flow of *c.* 75 mW m$^{-2}$, as has been proposed for certain areas; e.g. the Merille–Nyambeni Hills region near the southern part of line E (Fuchs *et al.* 1990). For the Chyulu Hills area, the $P_n$ velocity of 7.9–8.0 km s$^{-1}$ is compatible with an average undepleted spinel peridotite composition (see above) at a depth of *c.* 45 km and temperatures of 650–825°C which, using the Pollack & Chapman (1977) geotherms, correspond to surface heat flows of 60–75 mW m$^{-2}$. For the Chyulu Hills area, the xenolith-derived geotherm of Henjes-Kunst & Altherr (1992) corresponds to a surface heat flow of 60 mW m$^{-2}$.

The above results from the KRISP'85, KRISP'90 and KRISP'94 seismic refraction–wide-angle reflection experiments, together with those from the teleseismic studies, petrology and surface geology, provide the basis for the following evolutionary and present-day structural summary. Anomalously hot mantle material migrating southwards from the mantle plume centred under Afar (northern Ethiopia), began to rise below northern Kenya and gave rise to the volcanism *c.* 30 Ma ago in the northern part of the rift, in the Turkana region. Between 30 and 25 Ma ago, faulting, extension and associated crustal thinning began in the northern part of the rift, and by the end of the Palaeogene (23.5 Ma) 12–13 km of surface extension had taken place (Hendrie *et al.* 1994). During the Lower and Middle Miocene faulting, extension and crustal thinning continued in the northern portion of the rift and essentially by 10 Ma ago 25–30 km of surface extension had taken place in the north. The magmatic event which occurred between 23 and 14 Ma ago, which produced a phase of alkaline basaltic volcanism in the northern part of the rift, was probably responsible for a phase of magmatic underplating and intrusion in the southern part of the rift (Hay *et al.* 1995*a*). This magmatically underplated and intruded material was, in turn, the source of the extensive flood phonolites which erupted in the southern part of the rift between 14 and 11 Ma (Hay & Wendlandt 1995; Hay *et al.* 1995*b*). In the south, volcanism began between 16 and 20 Ma ago, probably associated with the active uprising of mantle plume material below the Nyanza (Tanzanian) Craton and extending to beneath the present site of the Kenya Dome. Faulting, extension and crustal thinning began in the southern part of the rift at *c.* 10 Ma ago. Around 10 km of surface extension has occurred in both the northern and southern parts of the rift in the past 10 Ma. Thus, a total of 35–40 km of surface extension has occurred in the northern part of the rift where the crust is now only 20 km thick. In contrast, the much smaller amount of surface extension in the southern part of the rift is associated with a present-day 35 km thick crust.

The active uprising of anomalously hot mantle material beneath the Kenya Rift for the past 20–30 Ma, and the associated faulting, surface extension and crustal thinning, has given rise to the following present-day structure (Fig. 5). Within the mantle, below the surface expression of the rift, a region of anomalously low P-wave velocities exists due to the uprising of the anomalously hot mantle material. The boundary between the region of low P-wave velocities under the rift and normal mantle P-wave velocities beneath the rift flanks is sharp, but it is consistent with the anomalously hot mantle

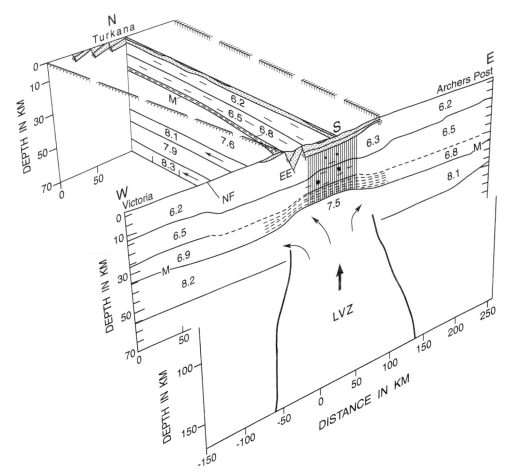

**Fig. 5.** Idealized model of the present-day geological structure of the Kenya Rift from Lake Turkana to Lake Baringo and beneath the neighbouring shoulders from Lake Victoria to Archers Post. The rift widens from *c.* 100 km at the latitude of Lake Baringo to *c.* 175 km at the latitude of shot-point LTN. The 40 km wide dyke-injection zone in the centre of the rift is illustrated, as is the significant amount of mafic igneous material intruding the basal crustal layer. Crustal igneous intrusions small enough so that the refraction–wide-angle reflection experiment will not detect any individual body are also shown. The boundary of the mantle low-velocity zone (LVZ), derived from the teleseismic delay-time studies, is shown, together with arrows indicating the directions of possible flow of hot mantle material caused by shearing radially away from the Kenya Dome. P-wave velocities are in km s$^{-1}$. Stippled areas, rift infill; 6.2–6.3 km s$^{-1}$, upper crystalline crust; 6.5 km s$^{-1}$, upper lower crustal layer; 6.9 km s$^{-1}$, basal crustal layer; 7.5 km s$^{-1}$ and greater, uppermost mantle; M, Moho; EE, Eigeyo Escarpment; NF, Nandi Fault separating Nyanza (Tanzanian) Archaean Craton to the west from Mozambique Proterozoic belt to the east [after Mechie *et al.* (1997)].

material having been present beneath the rift for the past 20–30 Ma. Petrological modelling (Keller *et al.* 1994*a*) indicates that the low P-wave velocities of 7.5–7.8 km s$^{-1}$ under the rift can be explained by 3–5% of basaltic melt which, under the Kenya Dome in the southern part of the rift, probably exists as *in situ* partial melt below *c.* 45–50 km depth. In contrast, below the northern part of the rift in the Turkana region, a

layer with a high velocity of 8.3 km s$^{-1}$ indicates that the depth to the onset of melting is at least 65 km in this portion of the rift and is, if anything, greater than under the Kenya Dome. It is envisaged that, at the top of the anomalous region in the mantle beneath the Kenya Dome, radial flow away from the region, possibly caused by shearing, occurs. This could explain the preferred orientation of olivine which is necessary to explain the high velocities of 8.1 and $8.3\,km\,s^{-1}$ in the two mantle layers beneath the northern part of the rift.

Magmatic activity associated with the rifting process has also modified the crust. For example, of the 9 km thick layer at the base of the crust beneath the Kenya Dome, it is estimated that at least 3–4 km is due to mafic igneous underplating and intrusion during the rifting process. In addition, a dyke injection zone in the central 40 km of the rift, penetrating both the upper and lower crust, has been proposed to explain the axial gravity high (Swain 1992). The zone, however, is too diffuse to have been identified by the refraction–wide-angle reflection experiments carried out so far.

The picture of the present-day structure indicates that the upper crust is thinned primarily by simple shear along normal faults, in some cases leading to asymmetric basins with maximum depths of almost 10 km (e.g. the Elgeyo Basin at the foot of the Elgeyo Escarpment in Figs 4 and 5), and tilted fault blocks in which Precambrian crystalline basement rocks are sometimes exposed in the crests. The model also demonstrates that crustal thinning and anomalous mantle material are essentially restricted to, and symmetrically centred beneath, the surface expression of the present-day rift. This is consistent with the mantle, and the deeper levels of the crust below the rift, deforming by pure shear and is inconsistent with models requiring that the major lithospheric thinning is asymmetrically offset beneath one of the rift flanks (e.g. Bosworth 1987).

Although the presence of anomalously hot mantle material giving rise to the low velocity zone under the Kenya Dome in the southern part of the rift seems to dominate the scene today, this was not always so in the past history of the Kenya Rift. The evidence points to the southwards migration of material from the large plume centred under Afar (northern Ethiopia), giving rise to the upwelling of anomalously hot mantle material, initial volcanism and subsequent rifting 30–25 Ma ago in the Turkana region in the northern part of the rift. However, the broad zone of extension in the north, overlying thin crust, suggests that passive lithospheric stretching has also played an important role in this region. As volcanism and rifting propagated southwards with time, the presence of the plume centred under the Nyanza (Tanzanian) Craton began to be felt. The narrow zone of extension over the culmination of the Kenya Dome, overlying thick crust and a mantle low-velocity zone extending down to 150–200 km depth, suggests that active uprising of anomalously hot mantle material, with thinning of the lithosphere from below, is the dominant rifting mechanism beneath the Kenya Dome in the southern part of the rift.

A great debt of gratitude is owed to all the people who took part in the organization, fieldwork, data processing and interpretation associated with the KRISP project. Thanks are due to Ulrich Achauer, Rainer Altherr and Karl Fuchs for the many stimulating discussions on KRISP scientific matters. Special thanks are due to the Government of Kenya, and the many officials and organizations in the country who helped to make the experiment possible. The project was financed by the DFG through the SFB 108 at Karlsruhe University, and the state of Baden-Wuertternberg in Germany; NSF in the USA; NERC and British Petroleum in the UK; and the EEC through SCI contract 00064, and the Human Capital and Mobility Program contract CHX-CT93 0308 (DG12 COMA). Many of the calculations were carried out on the Siemens S600 super-computer and other computers of the Computer Centre at Karlsruhe University.

## References

ACHAUER, U. & THE KRISP TELESEISMIC WORKING GROUP. 1994. New ideas on the Kenya rift based on the inversion of the combined dataset of the 1985 and 1989–90 seismic tomography experiments. *In*: PRODEHL, C., KELLER, G. R. & KHAN, M. A. (eds) *Crustal and Upper Mantle Structure of the Kenya Rift*. Tectonophysics, **236**, 305–329.

BAKER, B. H. & WOHLENBERG, J. 1971. Structure and evolution of the Kenya Rift Valley. *Nature*, **229**, 538–542.

——, MITCHELL, J. G. & WILLIAMS, L. A. J. 1988. Stratigraphy, geochronology and volcano-tectonic evolution of the Kedong–Naivasha–Kinangop region, Gregory Rift Valley. Kenya. *Journal of the Geological Society, London*, **145**, 107–116.

BIRT, C. S., MAGUIRE, P. K. H., KHAN, M. A., THYBO, H., KELLER, G. R. & PATEL, J. 1997. The influence of pre-existing structures on the evolution of the southern Kenya Rift Valley – evidence from seismic and gravity studies. *In*: FUCHS, K., ALTHERR, R., MÜLLER, B. & PRODEHL, C. (eds) *Structure and Dynamic Processes in the Lithosphere of the Afro–Arabian Rift System*. Tectonophysics, **278**, 211–242.

BOSWORTH, W. 1987. Off axis volcanism in the Gregory rift, East Africa: implications for models of continental rifting. *Geology*, **15**, 397–400.

BRAILE, L. W., WANG, B., DAUDT, C. R., KELLER, G. R. & PATEL, J. P. 1994. Modelling the 2-D velocity structure across the Kenya rift. *In*: PRODEHL, C., KELLER, G. R. & KHAN, M. A. (eds) *Crustal and Upper Mantle Structure of the Kenya Rift*. Tectonophysics, **236**, 251–269.

BURKE, K., & DEWEY, J. F. 1973. Plume generated triple junctions: key indicators in applying Plate Tectonics to old rocks. *Journal of Geology*, **81**, 406–433.

BYRNE, G. F., JACOB, A. W. B., MECHIE, J. & DINDI, E. 1997. Seismic structure of the upper mantle beneath the southern Kenya Rift from wide-angle data. *In*: FUCHS, K., ALTHERR, R. MÜLLER, B. & PRODEHL, C. (eds) *Structure and Dynamic Processes in the Lithosphere of the Afro-Arabian Rift System.* Tectonophysics, **278**, 243–260.

CERVENY, V., MOLOTKOV, I. A. & PSENEIK, I. 1977. *Ray Method in Seismology.* University of Karlova, Prague.

DAWSON, J. B. 1992. Neogene tectonics and volcanicity in the North Tanzania sector of the Gregory rift valley; contrast with the Kenya sector. *Tectonophysics*, **204**, 81–92.

FUCHS, K., ALTHERR, R. & STRECKER, M. 1990. Spannung und Spannungsumwandlung in der Lithosphere. *Geowissenschaften*, DFG, Mitteilung **XVIII**, 143–179.

——, MÜLLER, B. & PRODEHL, C. (eds) 1997. *Structure and Dynamic Processes in the Lithosphere of the Afro—Arabian Rift System.* Tectonophysics, **278**.

GAJEWSKI, D., SCHULTE, A. & RIAROH, D. 1994. Deep seismic sounding in the Turkana Depression, northern Kenya Rift. *In*: PRODEHL, C., KELLER, G. R. & KHAN, M. A. (eds) *Crustal and Upper Mantle Structure of the Kenya Rift.* Tectonophysics, **236**, 165–178.

HAY, D. E. & WENDLANDT, R. F. 1995. The origin of Kenya rift plateau-type flood phonolites: Results of high-pressure/high-temperature experiments in the systems phonolite–$H_2O$ and phonolite-$H_2O$–$CO_2$. *Journal of Geophysical Research*, **100**, 401–410.

——, —— & KELLER, G. R. 1995*a*. The origin of Kenya rift plateau-type flood phonolites: Integrated petrologic and geophysical constraints on the evolution of the crust and upper mantle beneath the Kenya rift. *Journal of Geophysical Research*, **100**, 10 549–10 557.

—— & WENDLANDT, E. D. 1995*b*. The origin of Kenya rift plateau-type flood phonolites: Evidence from geochemical studies for fusion of lower crust modified by alkali basaltic magmatism. *Journal of Geophysical Research*, **100**, 411–422.

HENDRIE, D. B., KUSZNIR, N. J., MORLEY, C. K. & EBINGER, C. J. 1994. Cenozoic extension in northern Kenya: a quantitative model of rift basin development in the Turkana region. *In*: PRODEHL, C., KELLER, G. R. & KHAN, M. A. (eds) *Crustal and Upper Mantle Structure of the Kenya Rift.* Tectonophysics, **236**, 409–438.

HENJES-KUNST, F. & ALTHERR, R. 1992. Metamorphic petrology of xenoliths from Kenya and northern Tanzania and implications for geotherms and lithospheric structures. *Journal of Petrology*, **33**, 1125–1156.

HENRY, W. J., MECHIE, J., MAGUIRE, P. K. H., KHAN, M. A., PRODEHL, C., KELLER, G. R. & PATEL, J. 1990. A seismic investigation of the Kenya Rift Valley. *Geophysical Journal International*, **100**, 107–130.

KELLER, G. R., MECHIE, J., BRAILE, L. W., MOONEY, W. D. & PRODEHL, C. 1994*a*. Seismic structure of the uppermost mantle beneath the Kenya Rift. *In*:

PRODEHL, C., KELLER, G. R. & KHAN, M. A. (eds) *Crustal and Upper Mantle Structure of the Kenya Rift.* Tectonophysics, **236**, 201–216.

——, PRODEHL, C., MECHIE, J. *et al.* 1994*b*. The East African rift system in the light of KRISP 90. *In*: PRODEHL, C., KELLER, G. R. & KHAN, M. A. (eds) *Crustal and Upper Mantle Structure of the Kenya Rift.* Tectonophysics, **236**, 465–483.

KELLY, K. R., WARD, R. W., TREITEL, S. & ALFORD, R. M. 1976. Synthetic seismograms: A finite-difference approach. *Geophysics*, **41**, 2–27.

KRISP WORKING GROUP. 1987. Structure of the Kenya rift from seismic refraction. *Nature*, **325**, 239–242.

——1995. A new look at the lithosphere underneath southern Kenya. *EOS Transactions of the American Geophysical Union*, **76**, 81–82.

KRISP WORKING PARTY. 1991. Large-scale variation in lithospheric structure along and across the Kenya rift. *Nature*, **354**, 223–227.

MAGUIRE, P. K. H., SWAIN, C. J., MASOTTI, R. & KHAN, M. A. 1994. A crustal and uppermost mantle cross-sectional model of the Kenya rift derived from seismic and gravity data. *In*: PRODEHL, C., KELLER, G. R. & KHAN, M. A. (eds) *Crustal and Upper Mantle Structure of the Kenya Rift.* Tectonophysics, **236**, 217–249.

MASOTTI, R., MAGUIRE, P. K. H. & MECHIE, J. 1996. On the upper mantle beneath the Kenya Rift. *Geophysical Journal International*, **126**, 579–592.

MECHIE, J., FUCHS, K. & ALTHERR, R. 1994*a*. The relationship between seismic velocity, mineral composition and temperature and pressure in the upper mantle – with an application to the Kenya Rift and its eastern flank. *In*: PRODEHL, C., KELLER, G. R. & KHAN, M. A. (eds) *Crustal and Upper Mantle Structure of the Kenya Rift.* Tectonophysics, **236**, 453–464.

——, KELLER, G. R., PRODEHL, C., KHAN, M. A. & GACIRI, S. J. 1997. A model for the structure, composition and evolution of the Kenya rift. *In*: FUCHS, K., ALTHERR, R., MÜLLER, B. & PRODEHL, C. (eds) *Structure and Dynamic Processes in the Lithosphere of the Afro–Arabian Rift System.* Tectonophysics, **278**, 95–119.

——, ——, —— *et al.* 1994*a*. Crustal structure beneath the Kenya Rift from axial profile data. *In*: PRODEHL, C., KELLER, G. R. & KHAN, M. A. (eds) *Crustal and Upper Mantle Structure of the Kenya Rift.* Tectonophysics, **236**, 179–200.

MORGAN, P. 1983. Constraints on rift thermal processes from heat flow and uplift. *In*: MORGAN, P. & BAKER, B. H. (eds) *Processes of Continental Rifting.* Tectonophysics, **94**, 277–298.

MORLEY, C. K., WESCOTT, W. A., STONE, D. M., HARPER, R. M., WIGGER, S. T. & KARANJA, F. M. 1992. Tectonic evolution of the northern Kenya Rift. *Journal of the Geological Society, London*, **149**, 333–348.

NOVAK, O., PRODEHL, C., JACOB, A. W. B. & OKOTH, W. 1997*a*. Crustal structure of the southeastern flank of the Kenya rift deduced from wide-angle P-wave data. *In*: FUCHS, K., ALTHERR, R., MÜLLER, B. &

PRODEHL, C. (eds) *Structure and dynamic processes in the lithosphere of the Afro-Arabian rift system*. Tectonophysics, **278**, 171–186.

——, RITTER, J. R. R., ALTHERR, R. *et al.* 1997*b*. An integrated model for the deep structure of the Chyulu Hills volcanic field, *Kenya*. *In*: FUCHS, K., ALTHERR, R., MÜLLER, B. & PRODEHL, C. (eds) *Structure and Dynamic Processes in the Lithosphere of the Afro-Arabian Rift System*. Tectonophysics, **278**, 187–209.

NYBLADE, A. & POLLACK, H. N. 1992. A gravity model for the lithosphere in western Kenya and northeastern Tanzania. *Tectonophysics*, **212**, 257–267.

POLLACK, H. N. & CHAPMAN, D. S. 1977. On the regional variation of heat flow, geotherms, and lithospheric thickness. *Tectonophysics*, **38**, 279–296.

PRODEHL, C., KELLER, G. R. & KHAN, M. A. (eds) 1994*a*. *Crustal and Upper Mantle Structure of the Kenya Rift*. Tectonophysics, **236**.

——, JACOB, A. W. B., THYBO, H., DINDI, E. & STANGL, R. 1994*b*. Crustal structure on the northeastern flank of the Kenya rift. *In*: PRODEHL, C., KELLER, G. R. & KHAN, M. A. (eds) *Crustal and Upper Mantle Structure of the Kenya Rift*. Tectonophysics, **236**, 271–290.

RITTER, J. R. R. & ACHAUER, U. 1994. Crustal tomography of the central Kenya Rift. *In*: PRODEHL, C., KELLER, G. R. & KHAN, M. A. (eds) *Crustal and Upper Mantle Structure of the Kenya Rift*. Tectonophysics, **236**, 291–304.

SANDMEIER, K.-J. 1990. *Untersuchung der Ausbreitungseigenschaften seismischer Wellen in geschichteten und streuenden Medien*. PhD Thesis, Karlsruhe University.

SMITH, M. 1994. Stratigraphic and structural constraints on mechanisms of active rifting in the Gregory Rift, Kenya. *In*: PRODEHL, C., KELLER, G. R. & KHAN, M. A. (eds) *Crustal and upper mantle structure of the Kenya rift*. *Tectonophysics*, **236**, 3–22.

STRECKER, M. R. 1991. *Das zentrale und sadliche Kenia-Rift unter besonderer BerAcksichtigung der neotektonischen Entwicklung. Habil. Schrift*, Karlsruhe University.

SWAIN, C. J. 1992. The Kenya rift axial gravity high: a re-interpretation. *Tectonophysics*, **204**, 59–70.

ZELT, C. A. & SMITH, R. B. 1992. Seismic traveltime inversion for 2-D crustal velocity. *Geophysical Journal International*, **108**, 16–34.

# Style, timing and distribution of tectonic deformation across the Exmouth Plateau, northwest Australia, determined from stratal architecture and quantitative basin modelling

GARRY D. KARNER[1] & NEAL W. DRISCOLL[2]

[1] *Lamont-Doherty Earth Observatory of Columbia University, Palisades, NY 10964, USA.*
*(e-mail: garry@ldeo.columbia.edu)*
[2] *Department of Geology and Geophysics, Woods Hole Oceanographic Institution,*
*Woods Hole, MA 02543, USA*

**Abstract:** By combining a kinematic and flexural model for the deformation of the lithosphere with sequence stratigraphy, the distribution amplitude, depth-partitioning and interaction of the tectonic events responsible for the formation of the Exmouth Plateau, northwest Australia, have been determined. The style of deformation varied continually as a function of space and time. Initially, the deformation was characterized by a broadly distributed late Permian event. During late Triassic–middle Jurassic time, deformation became more localized and formed a series of sub-basins; the Exmouth, Barrow and Dampier Sub-basins. This localized phase of deformation was followed by a substantially more regional deformation event in the Tithonian–Valanginian that generated large, post-Valanginian regional subsidence across the Exmouth Plateau, with only minor accompanying brittle deformation and erosion. After the initiation of seafloor spreading, a phase of inversion occurred, which correlated with the reorganization of the Indo-Australian plates during magnetic anomaly M0–34 time. This inversion induced minor reactivation of fault systems within the sub-basins and adjacent areas.

The style and distribution of each tectonic event are recorded by a diagnostic stratal architecture. For example, the late Permian event is recorded by the deposition of the broadly distributed and monotonous sequences of the Locker Shales and Mungaroo Formation. These units are more appropriately described as intracratonic in character. In contrast, the localized early Triassic and Callovian rifting events are characterized by regressive packages of thick shales (the upper and lower Dingo claystones). Crucial in the development of the upper and lower Dingo claystones was the segmented nature of the border faults that delineated the Exmouth, Barrow and Dampier Sub-basins. En echelon fault patterns and a basin synform structure, within which the pre-rift stratigraphy dips subparallel to or away from the fault trace, strongly implies a wrench influence in the development of the northern sub-basins (central Barrow and Dampier Sub-basins). This wrenching is consistent with the difference in regional trend of both the central Barrow and Dampier Sub-basins relative to the spreading fabric within either the Argo and/or Cuvier oceanic crust.

The regional distribution and amplitude of the post-Valanginian subsidence is not consistent with the minor amounts of Tithonian–Valanginian brittle upper crustal extension observed on the margin. Facies and microfossil analyses suggest that, prior to extension, large portions of the platform were emergent, or at shallow water depths. To match the distribution and magnitude of the post-Valanginian thermal-type subsidence requires significant lower crustal and mantle extension across the Exmouth Plateau Basin. Such a distribution of extension implies the existence of an intracrustal detachment having a ramp–flat–ramp geometry that effectively thinned the lower crust and lithospheric mantle. A large injection of heat should accompany the Tithonian–Valanginian extension, being governed by the geometry of the detachment, and the distribution and amplitude of the lower plate extension. Implications for the generation of excessive tholeiitic magmatism during rifting support the notion that margin magmatism can occur within rift environments in the absence of a plume.

The sedimentary record is punctuated by unconformities and their correlative conformities, the timing, spatial distribution and formation of which are critical in evaluating the importance of both the syntectonic and post-tectonic development of basins and passive margins. During continental extension, the depositional packages and bounding surfaces are a consequence of the

*From*: MAC NIOCAILL, C. & RYAN, P. D. (eds) 1999. *Continental Tectonics*. Geological Society, London, Special Publications, **164**, 271–311. 1-86239-051-7/99/$15.00 © The Geological Society of London 1999.

spatial and temporal distribution of rifting across and along the margin, and the resulting structural and sedimentological interactions. In particular, the structural interplay between multiple rift events can result in complex stratigraphic stacking patterns in response to changing sediment source regions (i.e. basement, pre-rift or earlier synrift sediments), physiography, drainage modification and the exploitation of accommodation zones, and eustasy. Irrespective of this complexity, the preserved sedimentary record, because it is a tape recorder of the vertical motions of the lithosphere, can be used to constrain and map the deformation of the lithosphere. By combining quantitative basin modelling with sequence stratigraphy (termed 'quantitative basin analysis'; Karner et al. 1997; Driscoll & Karner 1998a), it becomes possible to determine the chronostratigraphic framework for the development of a basin system and, thus, the tectonic significance of the preserved stratigraphy. Iterating between the modelled time-line stratigraphy and the observed stratal patterns allows establishment of a quantitative relationship between the tectonic deformation of the lithosphere and the resultant stratigraphic sequences.

The large regional post-break-up subsidence that characterizes many passive margins, but is also associated with little synrift crustal deformation, has been a particularly difficult observation to explain in terms of causative mechanisms. Given that lithospheric extension involves both the crust–upper crust and the mantle lithosphere, the amplitude and distribution of the brittle deformation necessarily needs to balance with the mantle thinning responsible for the post-rift or thermal subsidence. Driscoll & Karner (1998a, b) have termed this generation of large regional subsidence with little attendant brittle deformation as the 'upper plate' paradox, i.e. both sides of many conjugate margins appear to be the 'upper plate'. Despite this apparent basin symmetry, studies have tended to concentrate on the variance between crustal and lithospheric mantle extension and the resulting asymmetric development of rift basins. By default, depth-dependent extension implies the existence of detachment faults that mechanically decouple the upper extensional system (the upper plate) from the lower system (the lower plate). 'Detachment tectonics' was rapidly embraced because geological mapping of large extensional regions, such as the Basin-and-Range Province, demonstrated that low-angle normal faults, or detachments, were an important part of the process by which the continental crust thinned during extension (e.g. Wernicke & Burchfiel 1982; Wernicke 1985). Nevertheless, cogent evidence for the existence of intercrustal detachments remains limited. Even when inferred detachment faults are imaged seismically, it is difficult to demonstrate that differential displacement has occurred across these surfaces. The northwest Australian margin is no exception. Tithonian–Valanginian rifting generated large post-Valanginian regional subsidence across the Exmouth Plateau with only minor synrift brittle deformation and erosional truncation. Extensive volcanism is also associated with the Tithonian–Valanginian break-up. Shallow intrusive and extrusive tholeiitic basalts and seaward dipping reflector sequences exist on the Scott and Exmouth Plateaux, and the Gascoyne and Cuvier margins (Crawford & von Rad, 1994). In contrast with the White & McKenzie (1989) model, which suggests that margin magmatism is a direct consequence of mantle plumes, none of these basalts show a plume signature (Crawford & von Rad 1994), nor is there any evidence for the arrival or departure of a plume from the Exmouth Plateau region. These geophysical and geochemical observations strongly support the notion that excessive margin magmatism can occur within rift environments in the absence of a plume. It is postulated here that the mechanism responsible for the development of large,

**Fig. 1.** Bathymetry map of the Exmouth Plateau region showing the relationship between the plateau and the Argo, Gascoyne and Cuvier Abyssal Plains. Contour interval is 200 m. The yellow box represents the region shown in Fig. 2. Note the marked difference in margin architecture along the northwest Australian region in response to the north to south propagation of Mesozoic rifting, post-extensional volcanic constructions and the Mesozoic–Recent margin progradation. The Argo and Cuvier Abyssal Plains represent narrow margins, whereas the Exmouth Plateau represents a wide margin. The Exmouth, Barrow and Dampier Sub-basins have no bathymetric expression and are located beneath the present-day continental shelf. The large transform faults that separate the Exmouth Plateau from the Argo Abyssal Plain to the north and the Cuvier Abyssal Plain to the south are well imaged by the bathymetry data. The grid of AGSO, GECO and Digicon seismic reflection data made available to the study is shown by the fine red lines, the bold blue lines pertaining to seismic lines for which examples are presented in the text. The onset of seafloor spreading propagated from north to south along the northwest margin: Callovian in the Argo Abyssal Plain, and Valanginian in the Gascoyne and Cuvier Abyssal Plains. Flow lines and age are shown by rose-coloured arrows and numbers: 1, Callovian, 2, Valanginian.

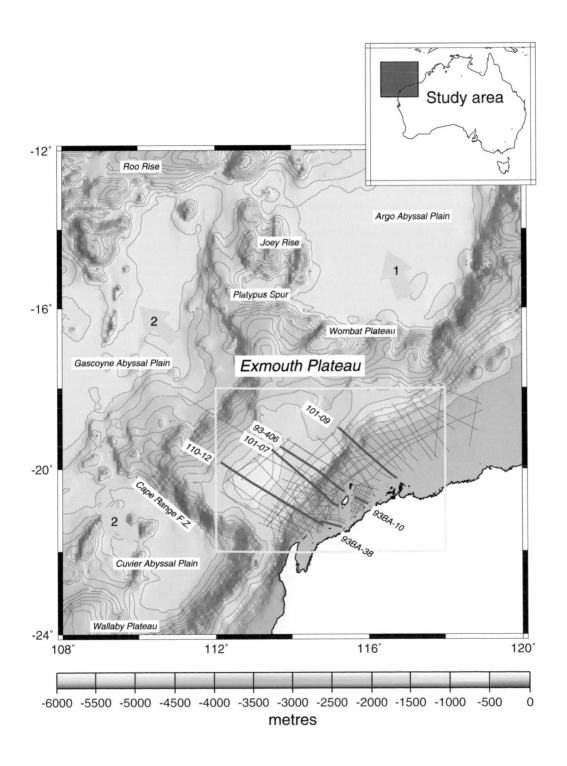

regional, post-rift subsidence, with little attendant brittle synrift deformation and the occurrence of excess volcanism, may be intimately
related, and may help to explain the excess
volcanism along non-volcanic margins (e.g. US
east coast; Holbrook & Kelemen 1993).

Using an integrated approach, incorporating
both kinematic basin modelling and seismic
sequence stratigraphy to constrain the first-
order processes responsible for both the deformation of the lithosphere and the transport and
deposition of sediments, the objectives of this
paper are to: (1) determine the amplitude, depth-
partitioning and interaction of the tectonic
events responsible for the crustal architecture
of the Exmouth Plateau, and the geometry of the
various sub-basins that comprise the plateau;
(2) define how the style of deformation changes
through the margin history; (3) define the role of
detachments in the thinning of the Exmouth
Plateau continental crust; and (4) investigate the
thermal consequences of detachment-controlled
continental extension and its role in the generation of late-stage, synrift tholeiitic basalts.

## Regional Geology of the Exmouth Plateau

The northwestern Australian region consists of
several geologic terranes: Archean and early
Proterozoic cratons (Kimberley and Pilbara Cratons), Proterozoic basins (Hamersley and Ban-
gemall Basins), Palaeozoic basins (Canning and
Carnarvon Basins), Mesozoic oceanic basins
(Argo, Gascoyne and Cuvier Basins), and a
series of Mesozoic and Cenozoic basins. The
onset of seafloor spreading propagated from
north to south along the northwest Australian
margin (Fig. 1); in the Argo Basin, seafloor
spreading began during Callovian time (marine
magnetic anomaly M25), whereas in the Gascoyne, Cuvier, and Perth Basins, seafloor spreading began in Valanginian time (anomaly M10;
Exon et al., 1982; Boote & Kirk 1989; von Rad
et al. 1992). A marked change in spreading

direction from north-northwest to west-northwest occurred between the Callovian spreading
in the Argo Basin and the Valanginian spreading in the Gascoyne, Cuvier and Perth Basins
(Fig. 1). The flow lines (fracture zones) observed
in the Argo and Gascoyne abyssal plains are one
of the best indicators of the extensional transport direction during the transition between
continental extension and the onset of seafloor
spreading. Note that the north-northeast trend
of the tectonic fabric observed across the
Exmouth Plateau is highly oblique to the Callovian, and inferred pre-Callovian extensional
transport direction and is more orthogonal to the
post-Callovian transport direction. The angle
between the pre-existing structural trend and the
extensional transport direction plays an important role in controlling the structural style of the
deformation and sediment patterns in the sub-
basins through time (e.g. Tron & Brun 1991).

The Exmouth Plateau is a thinned and sub-
sided continental block with a crustal thickness
of $c.20$ km and an average water depth of
$c.1000$ m (Fig. 1; Exon & von Rad 1994). The
Exmouth Plateau encompasses the Barrow,
Dampier and Exmouth Sub-basins, as well as
the Peedamullah and Lambert Shelves and the
Kangaroo Trough (Figs 1–3; Copp 1994; Stagg
& Colwell 1994). Previous seismic reflection and
exploratory well studies across the Exmouth
Plateau indicate that the preserved sedimentary
succession consists of 5000 m of Palaeozoic
rocks, an average of 3000 m of Triassic sedi-
ments, and 1000 m of younger Cretaceous and
Cenozoic sediments (Willcox & Exon 1976).
Four main Palaeozoic basins appear to have
influenced the Mesozoic and Cenozoic development of the northwest Australian margin: the
Perth Basin (Permian only), the Carnarvon
Basin (Ordovician–Permian), the Canning Basin
(Ordovician–Permian) and the Bonaparte Basin
(Cambrian–Permian). The offshore equivalents
of these basins collectively form the Westralian
Superbasin (Yeates et al. 1987). The Palaeozoic
stratigraphy of the superbasin consists of three
megasequences that correlate in time with the

**Fig. 2.** Crustal Bouguer gravity anomaly map of the Exmouth Plateau. Contour interval is 10 mgal. The crustal
Bouguer gravity anomaly map was generated by removing the gravity effect of the bathymetry from the GEOSAT
free-air gravity anomaly and filtered using a least-squares bicubic trend surface that approximates the east to
west crustal thinning across the Exmouth Plateau. The resulting anomaly tends to accentuate sedimentary basins
and flexurally compensated features such as rift flanks. The most prominent anomalies are: (1) the positive
anomalies associated with the northwestern location of the ocean–continent boundary, the Rankin Trend and the
Alpha Arch; and (2) the negative anomalies that correlate with the eroded and underplated southwest margin
of the Exmouth Plateau, the Exmouth, Barrow and Dampier Sub-basins, and the Kangaroo Trough. Shown also
are the three modelled cross-sections (bold blue lines). From south to north, seismic lines AGSO110-12 and
93BA-38, GPCT93-406 and 93BA-10, and AGSO101R-09. Critical wells used to constrain the quantitative basin
modelling are also located.

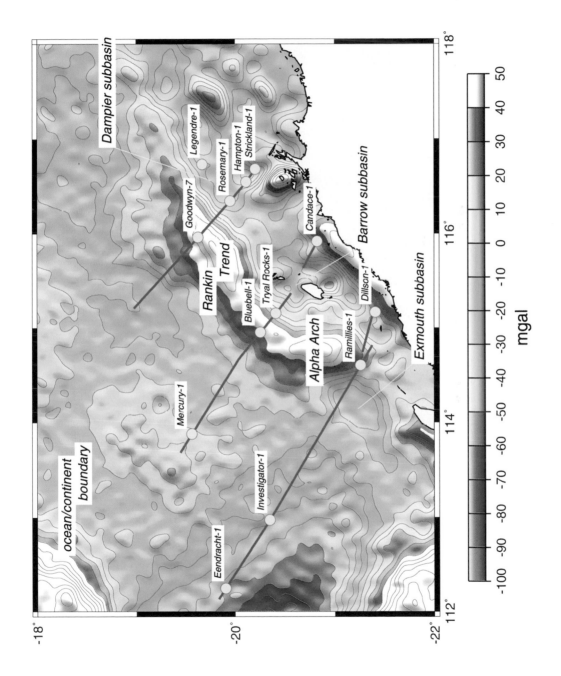

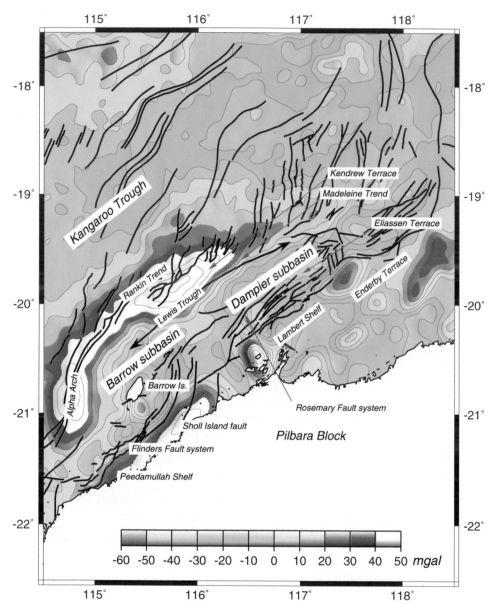

**Fig. 3.** Structural detail of the Barrow and Dampier Sub-basins, and the Kangaroo Trough, eastern Exmouth Plateau. Key structural elements from the AGSO North West Shelf Study Group (1994) have been superposed onto the crustal Bouguer gravity anomaly map. Contour interval is 10 mgal.

cessation of major compressional orogenic events within Gondwanaland (Warris 1993). The spatial dimensions of the Westralian Superbasin (e.g. Cockbain 1982), its monotonous and consistently parallel stratigraphy over very large areas (e.g. Veenstra 1985), and evidence for long-lived subsidence (Cambrian–Triassic) suggest

that the Westralian Superbasin is an intracratonic basin.

The major tectonic and structural elements comprising the Exmouth Plateau are illustrated in Figs 2 and 3, superposed onto the crustal Bouguer gravity anomaly. The gravity map was constructed by determining the Bouguer gravity

C) Southern transect - Exmouth & Barrow subbasins

AGSO 110-12

93BA-38

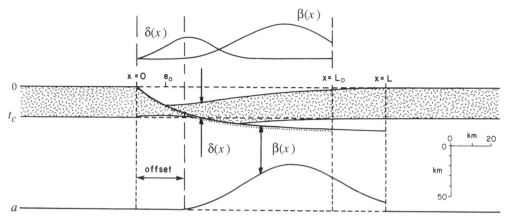

**Fig. 4.** Kinematic description of instantaneous extension of the lithosphere involving simple slip with heave $e_0$ along a detachment (equal to the amount of extension $L - L_0$) combined with pure shear $\delta(x)$ in the upper plate or hanging-wall block, and pure shear $\beta(x)$ in the lower plate or footwall block. The deformed configuration of the kinematic model is found by using the extension parameters $\delta$ and $\beta$ as mapping functions. The initial shape of the detachment is denoted by the dotted curve while its shape in the deformed configuration is the solid curve. The initial crustal and lithosphere thicknesss are $t_c$ and $a$, respectively. The lithosphere thickness is defined as the depth where temperature reaches $T_m$, the asthenosphere temperature. The relative distribution and amplitude of the estimated $\delta(x)$ and $\beta(x)$ ultimately determine if some form of detachment was operative during the extension process.

passive rise of the lithosphere–asthenosphere boundary and, in so doing, introduces heat to the base of the thinned lithosphere. The brittle faults often detach into a region of plasticity. Following Driscoll & Karner (1998*a*), a detachment represents a diffuse zone that separates (detaches) the brittle deformation in the upper crust from the plastic deformation in the lower crust and lithospheric mantle. The more restrictive definition of detachments proposed by Wernicke (1981), and subsequently modified and refined by Lister & Davis (1989), that define detachments as narrow zones of relative movement that pass through the entire crust and/or lithosphere and are recorded by 'mylonitic detachment terranes' (or metamorphic core complexes), is not adopted here.

Referring to Fig. 4, the response of the lithosphere to extension can be described by considering the distribution and magnitude of the various kinematically produced loads. Starting with a pre-rift lithosphere of thickness $a$ and crustal thickness $t_c$, simple slip along a basin bounding fault ($e_o$), and bulk thinning of the hanging-wall block, produces a surficial hole by collapse of the block. The ratio of the pre-rift to the post-rift hanging-wall block thickness is termed $\delta(x)$. The delta function determines the distribution of the rift-phase subsidence and ultimately the maximum thickness of sediment, both rift and post-rift, that can accumulate within a basin. Similarly, the ratio of the pre-rift to

the post-rift footwall block thickness is termed $\beta(x)$. The beta function principally controls the distribution of the post-rift subsidence. Both $\delta(x)$ and $\beta(x)$ determine the amplitude of the post-rift subsidence. The relative distribution and amplitude of the estimated $\delta(x)$ and $\beta(x)$ functions, obtained by matching the observed distribution and thickness of syn- and post-rift sequences, will ultimately determine if some form of detachment was operative during the extension process, rather than an *a priori* modelling assumption. The modelling parameters and constants are summarized in Table 1.

During rifting, the partial removal of the hanging-wall block unloads the lithosphere, inducing a flexural rebound (e.g. Weissel & Karner 1989; Braun & Beaumont 1989). The form of this rebound is controlled by the flexural strength of the lithosphere during rifting. The flexural strength of the lithosphere, or, equivalently, its effective elastic thickness, is important in controlling the regional architecture of the rift and post-rift basins. The effective elastic thickness varies spatially and temporally both during and after rifting, and tracks the changing thermal structure of the extended lithosphere. In addition to the flexural rebound induced by crustal unloading, thinning of the lithosphere adds heat to the base of the lithosphere causing density variations that represent a buoyant load (Fig. 4). With increasing time after rifting, the perturbed lithospheric temperature structure

**Table 1.** *Modelling parameters for the Exmouth Plateau*

| | | |
|---|---|---|
| $x, z$ | horizontal and vertical coordinates | |
| $t$ | time since rifting/re-rifting event | |
| $\Delta t_{rift}$ | finite rifting interval | 5–20 Ma |
| $\delta(x)$ | upper plate extension factor | |
| $\beta(x)$ | lower plate thinning factor | |
| $\delta_i(x)$ | upper plate compression (inversion) factor | |
| $\beta_i(x)$ | lower plate compression (inversion) factor | |
| $T(x, z, t)$ | lithospheric temperature structure | |
| $T_m$ | asthenosphere temperature | 1333°C |
| $\alpha$ | thermal expansion coefficient | $3.28 \times 10^{-5}\,\mathrm{K}^{-1}$ |
| $\kappa$ | thermal diffusivity | $8 \times 10^{-7}\,\mathrm{m^2\,s^{-1}}$ |
| $C_p$ | specific heat | $1.05\,\mathrm{J\,gm^{-1}\,K^{-1}}$ |
| $q_0$ | background heat flow | $42\,\mathrm{mW\,m^{-2}}$ |
| $A$ | Foucher *et al.* (1982) | 0.4 |
| $B$ | | $3.65 \times 10^{-3}\,\mathrm{K}^{-1}$ |
| $C$ | | $3 \times 10^{-10}\,\mathrm{K\,km}^{-1}$ |
| $D$ | | 1100°C |
| $G$ | adiabatic temperature gradient | $0.3\,\mathrm{K\,km}^{-1}$ |
| $L$ | latent heat of fusion | $334\,\mathrm{J\,g}^{-1}$ |
| $t_c$ | pre-rift crustal thickness | 34 km |
| $t'_c$ | post-rift crustal thickness | |
| $t_d(x)$ | intracrustal detachment geometry | |
| $a$ | equilibrium lithospheric thickness | 125 km |
| $a'$ | post-rift lithospheric thickness | |
| $\rho_a$ | density of asthenosphere at 0°C | $3330\,\mathrm{kg\,m^{-3}}$ |
| $\rho'_a$ | density of asthenosphere at $T_m$ | $3179\,\mathrm{kg\,m^{-3}}$ |
| $\rho_c$ | density of crust at 0°C | $2800\,\mathrm{kg\,m^{-3}}$ |
| $\rho_s(z)$ | sediment density | |
| $\rho_a$ | sediment grain density | $2650\,\mathrm{kg\,m^{-3}}$ |
| $\rho_{melt}$ | density of melted asthenosphere | $2.6\,\mathrm{g\,cm^{-3}}$ |
| $\Phi(z)$ | bulk sediment porosity | $\Phi_0\,e^{-k_p z}$ |
| $\Phi_0$ | surface porosity | 60% |
| $k_p^{-1}$ | porosity decay constant | 2.5 km |
| $D(x, t)$ | lithospheric rigidity | $ET_e^3(x, t)/12(1 - \nu^2)$ |
| $T_e(x, t)$ | effective elastic thickness | |
| $\nu$ | Poisson ratio | 0.25 |
| $E$ | Young modulus | $6.5 \times 10^{10}\,\mathrm{Pa}$ |
| $g$ | gravitational acceleration | $9.82\,\mathrm{m\,s^{-2}}$ |
| $Z_c$ | controlling isotherm for $T_e$ | 450°C |
| $h(x, t)$ | topographic relief | |
| $\Delta h(x, t)$ | topographic load removed by erosion | |
| $k_e^{-1}$ | erosional time constant | |
| | crust | 82 Ma |
| | sediment | 40 Ma |
| $\Delta S_L$ | maximum first-order eustatic sea-level variation | 250 m |

re-equilibrates, allowing the lithosphere to cool, contract and subside. Thus, the rapid initial subsidence associated with the brittle deformation of the rift phase is replaced by a slower subsidence associated with the conductive cooling of the rifted lithosphere.

A major controversy remains over whether extended continental lithosphere associated with extensional basins and passive continental margins possesses mechanical strength or flexural rigidity. Several recent studies of gravity anomalies and subsidence patterns of basins and passive margins have concluded that the flexural rigidity of continental lithosphere remains negligible once it is extended (e.g. Barton & Wood 1984; Watts 1988; Fowler & McKenzie 1989; Marsden *et al.* 1990; Roberts *et al.* 1993; Watts & Marr 1995; Watts & Fairhead 1997; Watts & Stewart 1998). Moreover, other studies have concluded from gravity data that the morphology of rift basins and their flanks are an expression of regional isostatic compensation which, in fact, requires that extended lithosphere possesses finite mechanical strength (e.g. Weissel

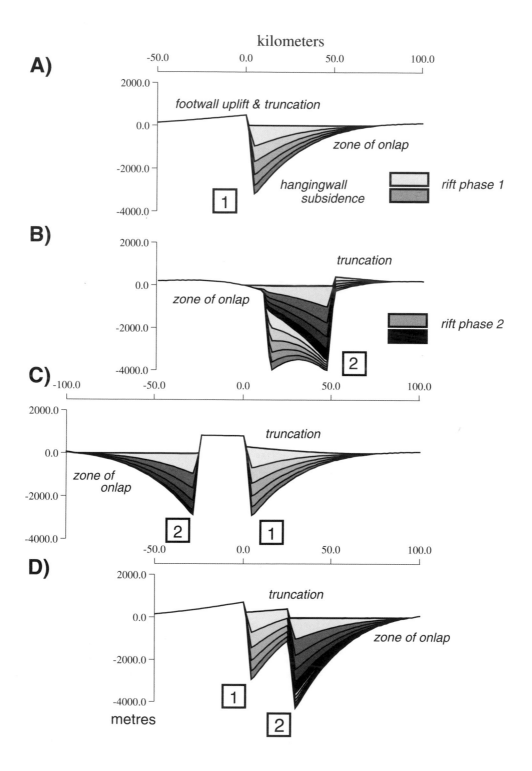

& Karner 1989; Ebinger *et al.* 1991; Kooi *et al.* 1992; Keen & Dehler 1997). Extreme amplitude free-air and Bouguer gravity anomalies (−120 to 140 mgals) over extensional basins (e.g. Tucano Basin of Brazil; Karner *et al.* 1992), disqualifies the concept that rifted lithosphere necessarily loses its flexural strength. Given this, it is important to critically re-examine the assumptions and methods used in the 'low rigidity' papers. Firstly, all of these studies suffer from a paucity of data to adequately define the various rift-induced loads acting on the lithosphere. Secondly, many of these studies were conducted in regions characterized by multiple stages of continental extension (e.g. the North Sea). For these areas, the isostatic state of the lithosphere, either prior to rifting (e.g. Barton & Wood 1984; Fowler & McKenzie 1989) or at the close of rifting (e.g. Watts 1988; Watts & Marr 1995; Watts & Fairhead 1997), was assumed in order to interpret the gravity anomalies associated with the rift phase being studied. It is disconcerting to discover that in all these cases, the authors assumed local isostasy to reconstruct the crustal structure at the end of rifting. Thirdly, the approach adopted by Marsden *et al.* (1990) and Roberts *et al.* (1993), the so-called flexural cantilever model of Kusznir *et al.* (1991), is also problematical. In this modelling scheme, the flexural strength of the rifted lithosphere is assigned based on the curvature of the collapsed hanging-wall block. The fact that the hanging wall tends to be completely fractured by antithetic and synthetic faults attests to its low strength. Consequently, the low strength and high curvature of the hanging-wall block are hardly proxies for the flexural strength of the entire rifted lithosphere. Finally, as pointed out by Weissel & Karner (1989), and applied by Kooi *et al.* (1992) and Keen & Dehler (1997), the depth of necking within the lithosphere profoundly affects the subsidence and gravity anomalies of extensional basins. In particular,

the depth of necking appears to be related to an intrinsic strength maximum within the lithosphere – shallow necking depths are a function of a relatively hot lithosphere, whereas deeper necking depths are a consequence of low extension rates and small degrees of extension (Keen & Dehler 1997). Shallow necking depths should thus be associated with rift systems as break-up is approached. Extension dominated by shallow necking depths will tend to induce subsidence of the footwall region, a result used here to explain the long wavelength, bathymetric rollover of the Exmouth Plateau. Intermediate necking depths avoid the need for low flexural rigidities during rifting (Kooi *et al.* 1992), producing local-type gravity anomalies, even though the lithosphere exhibits finite flexural strength (Keen & Dehler 1997). Given these arguments, it will explicitly be assumed that rifted lithosphere possesses flexural strength both during, and following, extension.

Many rift systems are characterized by multiple fault systems with the dominant faults changing their importance as a function of space and time during extension (e.g. Karner *et al.* 1997; Driscoll & Karner 1998*a, b*). The kinematic and flexural interactions of hanging-wall and footwall blocks helps to explain the complexity of rift basin architecture and the pattern of clastic delivery to the evolving rifts. The incorporation of finite rifting and re-rifting capabilities in this modelling approach is an important aspect in simulating the architecture and geological development of rift systems. To illustrate the flexural interaction of multiple-border fault systems and the effect of rift-induced topography on fluvial network development, the resultant basin and rift flank architecture for a range of fault geometries has been calculated. Extension occurs by multiple slip across an eastward dipping border fault (Fig. 5a). The amplitude and wavelength of the footwall uplift is a function of the heave across the fault

**Fig. 5.** Characteristic responses to multiple faulting associated with synthetic and antithetic border-fault systems. (**a**) Multiple slip across a single fault system. Fault heave is 5 km, fault dip 30° and effective elastic thickness of the lithosphere during rifting is 20 km. Hanging-wall collapse and rotation results in the generation of a series of onlap surfaces and a thickening synrift sediment wedge towards the border fault. Note that during rifting accommodation is both created (collapsed hanging wall) and destroyed (uplifted footwall). Fluvial drainage networks will tend to flow down the back side of the footwall away from the rift basin proper, only gaining access to the basin in structurally low-lying regions between fault segments (i.e. accommodation zones). (**b**) Structural interactions induced by antithetic faulting. The hanging wall of the initial deformation now becomes the footwall to this second phase of deformation. The uplifted footwall is capped by deposits associated with the first rifting phase with the basin depocentre migrating from west to east. (**c**) Structural interactions induced by antithetic faulting. The footwall to the initial deformation, is also the footwall to this second phase of deformation, leading to the formation of a horst block. The sediments associated with the first rifting phase are rotated and uplifted above baselevel as part of the footwall to the second phase of deformation. (**d**) Structural interactions induced by synthetic faulting. The hanging wall to the initial deformation is now part of the footwall to this phase of deformation, and is consequently uplifted and deformed.

($e_0 = 5$ km), the dip of the fault (30°), and the effective elastic thickness of the lithosphere (assumed a constant value of 20 km in this example) during rifting. A characteristic structural and stratigraphic response is generated; rifting induces *both* the creation (i.e. the rift basins) and destruction (i.e. rift flank) of accommodation. This characteristic, asymmetric footwall and hanging-wall geometry is a consequence of the finite flexural strength of the rifted lithosphere during extension. The multiple collapse and rotation of the hanging-wall block results in the generation of a series of onlap surfaces and a thickening synrift sediment wedge towards the border fault. The flexural uplift of the rift flank represents a relative sea-level fall, allowing erosional reworking of pre-rift sediment and/or basement. During the first rift phase, fluvial drainage networks will tend to flow down the back side of the footwall, away from the rift basin proper. These drainage networks can only gain access to the rift basin in structurally low-lying regions between fault segments (i.e. accommodation zones). Because of both the lag in clastic production by the erosion of rift flank topography and the preferential delivery of clastics to the accommodation zones, the facies within the rift basin proper will tend to consist of regressive packages built on top of a condensed section.

To investigate the structural and stratigraphic effects of synthetic and antithetic border fault systems, it is assumed that the first rift phase is associated with an eastward dipping border-fault system. A second rift phase is accommodated by extension across a westward dipping fault located to the east of the previous basin-bounding fault (Fig. 5b). The hanging wall to the initial deformation now becomes the footwall to this phase of deformation. The uplifted footwall is capped by deposits associated with the first rifting phase, with the basin depocentre migrating from west to east. Alternatively, if the second rifting phase is accommodated by extension across a westward dipping fault, located significantly to the west of the first border fault (Fig. 5c), then the footwall to the initial

deformation is also the footwall to this second phase of deformation, leading to the formation of a horst block. The sediments associated with the first rifting phase are rotated and uplifted above baselevel as part of the footwall to the second phase of deformation. Finally, the case of a second rifting phase in which extension is accommodated across an eastward dipping fault located to the east of the original basin-bounding fault (Fig. 5d), is explored. The hanging wall to the initial deformation is now part of the footwall to this phase of deformation, and is consequently uplifted and deformed.

This alternating footwall–hanging wall role of border-fault systems is an important factor determining the architecture of the various sub-basins that comprise the Exmouth Plateau and the facies development of the Jurassic and early Cretaceous sequences. Understanding this interplay between rift-induced topography and fluvial networks allows predictions to be made concerning source and reservoir quality along and across the evolving rift basins. As stated earlier, during the initial rift phase, fluvial drainage networks flowed down the back side of the footwall away from the rift basin proper. These drainage networks can only gain access to the rift basin in structurally low-lying regions, which are controlled by the location of fault segments (i.e. accommodation zones). For instance, the deposition of the lower Dingo claystones was the result of river drainage systems being diverted away from the rift by footwall uplift; the fluvial networks gain access into the basin towards the north, along several large accommodation zones, resulting in the deposition of the fluvial-dominated Legendre Formation. This formation represents a major southward prograding system that infilled the Barrow and Dampier Sub-basins longitudinally, parallel to the long axis of the basin. Renewed extension, depending on the synthetic–antithetic nature of the second border-fault system, will either promote major capture of the footwall river systems, or uplift, structuring, and reworking of the earlier rift sediments. As will be discussed later, the capture of these stream networks,

**Fig. 7.** Generalized Mesozoic and Tertiary chronostratigraphy for the northwest Australian margin [modified from Kopsen & McGann (1985)] showing the relationship between extensional and inversion events across the margin and the various stratigraphic packages. Four major episodes of extensional deformation and one minor inversion event affected the northwest Australian margin: (1) Late Permian extension; (2) Rhaetian extension associated with the onset of rifting in the Argo Basin; (3) Callovian extension that culminated in the onset of seafloor spreading in the Argo Basin (marine magnetic anomaly M25); (4) Tithonian–Valanginian extension associated with rifting and the onset of seafloor spreading in the Gascoyne and Cuvier Basins (marine magnetic anomaly M10). Note that the base of each tectonic megasequence is defined by a flooding surface and marine onlap within the depocentre. (5) Reorganization of the Indo-Australian Plates is roughly synchronous with the Turonian (Gearle horizon) compressive reactivation of fault systems within the Exmouth, Barrow and Dampier Sub-basins.

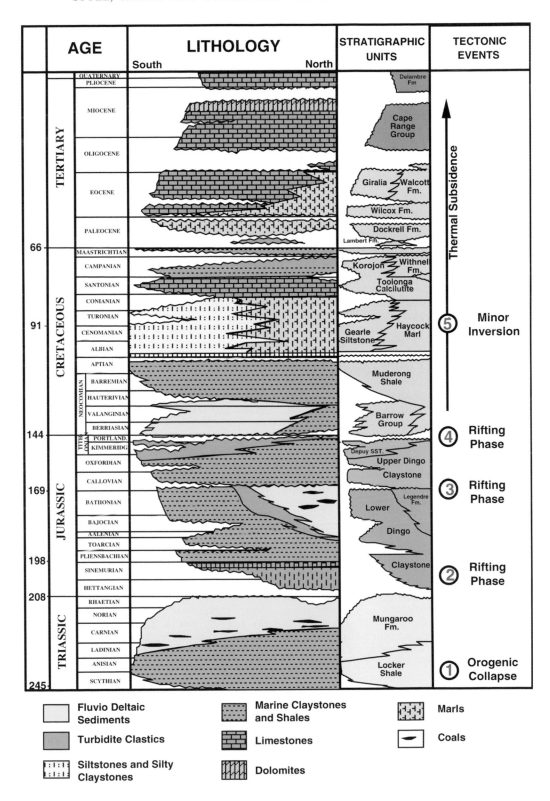

and sediment progradation across the hanging wall, is consistent with the development of the Angel Sands.

## Structural and stratigraphic development of the Exmouth Plateau

Three representative cross-sections have been selected to model the tectonic and stratigraphic development of the Exmouth Plateau. These cross-sections are coincident with the following seismic lines, from south to north: AGSO110-12 and 93BA-38; GPCT93-406 and 93BA-10; and AGSO101-09 (Figs 1 and 2). These cross-sections were selected because they capture the along-margin structural and stratigraphic variability associated with the development of the Exmouth, Barrow and Dampier Sub-basins. These data (Fig. 6), coupled with age and sedimentary facies information from the margin (Fig. 7), form the constraints for the quantitative basin modelling of the margin presented here. Figure 6 presents interpreted depth section and simulated time-line stratigraphy, and basin architecture, for the three Exmouth Plateau cross-sections: the Exmouth and southern Barrow Sub-basins (southern transect; Fig. 6c); the central Barrow Sub-basin (central transect; Fig. 6b); and the Dampier Sub-basin (northern transect; Fig. 6a); the location of the critical exploratory wells used in the analysis, and their respective penetrations, are also shown. Stacking velocities were used to convert two-way travel-time (TWTT) to depth. The lithologic colour scheme is keyed to the modelled cross-sections.

Seismic line AGSO110-12 crosses the Exmouth and southern Barrow Sub-basins. The Exmouth Sub-basin is delineated by the Alpha Arch to the east and the Exmouth Plateau to the west (Figs 2 and 6c). The northern extent of the Exmouth Sub-basin overlaps with the southern portion of the Barrow Sub-basin and the termination of these basins is controlled by major accommodation zones (Fig. 2). The southern Barrow subbasin is bounded to the east by the Peedamullah Shelf and to the west by the Alpha Arch. Seismic lines GPCT93-406 and

93BA-10 cross the central Barrow Sub-basin, bounded by the Rankin Trend to the west and the Peedamullah Shelf to the east (Figs 2, 3 and 6b). The observed data gap between the GECO and Digicon seismic lines is associated with the shoals surrounding Barrow Island. Seismic line AGSO101R-09 crosses the Dampier Sub-basin, delineated by the Rankin Trend to the west and the Lambert Shelf to the east (Figs 2, 3 & 6a).

### Late Permian Extension

Late Permian extension, while not structurally well-represented across the Exmouth Plateau, is required to explain the accumulation of the Locker Shales and Mungaroo Formation. The basal member of the Triassic is a transgressive marine sheet sand resting unconformably on Palaeozoic sediments (Kopsen & McGann 1985; Figs 6 and 7). A northwesterly prograding fluviodeltaic system forms the overlying regressive component, the lower part of which is represented by the pro-delta Locker Shales (Boote & Kirk 1989; Fig. 7). Towards the east, sand-rich units become more prevalent within the Locker Shale because of the proximity to the delta front (Kopsen & McGann 1985). The Locker Shale passes upward into the Mungaroo Formation, which is a deltaic system that grades vertically into a thick unit of stacked fluvial channel deposits (Barber 1982; Exon et al. 1992). The overall thickness of the Triassic section systematically increases toward the west with the depocentre occurring along the western edge of the Exmouth Plateau (Fig. 6; Cockbain 1982). Minor, but regional, lateral thickness variations of the Locker Shale and Mungaroo Formation are observed across the Exmouth Plateau, Rankin Platform, and the Peedamullah and Lambert Shelves. For example, the Triassic section across the Exmouth Plateau in the Dampier Sub-basin region is c. 1 km thinner than the equivalent sections west of the Barrow and Exmouth Sub-basins, as evidenced by the accumulation of relatively thin lower Dingo claystones. The Carnian–Norian section of the Mungaroo Formation is dominated by

---

**Fig. 8.** Interpreted seismic reflection profile GPCT93-419, illustrating the oblique clinoforms within the Barrow Delta. The oblique nature of the clinoforms indicates that there was little to no accommodation for aggradation and, consequently, the sediment prograded laterally from east to west. This stratal pattern suggests that the top of the oblique clinoforms was at, or near, sea level during the time of deposition. The present-day eastward dip of the top of the clinoforms, and the underlying Tithonian and late Triassic horizons, are a function of differential subsidence that has occurred across the Exmouth Plateau since Tithonian time. Note the parallel stratal geometry of the Mungaroo Formation and the Locker shales beneath the late Triassic horizon, as well as the lateral variation in reflector amplitude. Seismic line GPCT93-419 is coincident with AGSO110-12.

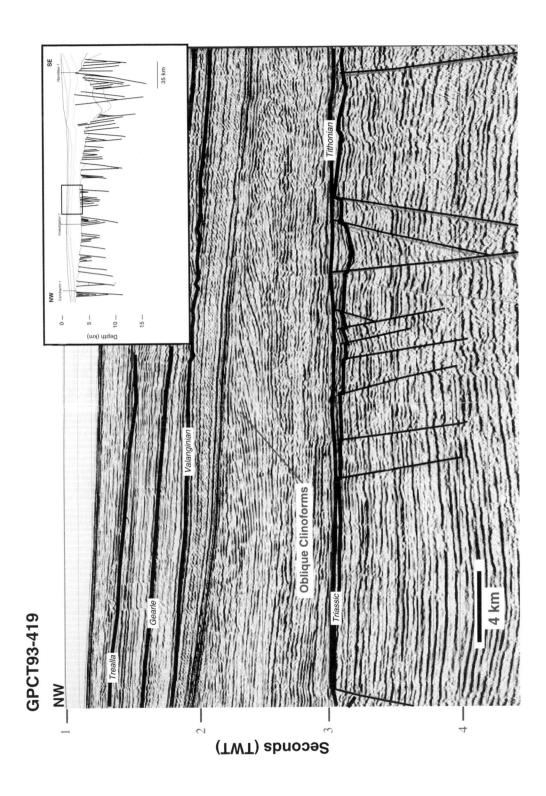

pro-delta, distributary–fluvial channel complexes and coal swamps with deltaic progradation in a general northwesterly direction (Cook *et al.* 1985; Fig. 7). The Mungaroo Formation is capped by a thin transgressive succession of nearshore and shelf fine-grained clastics and carbonates of Hettangian–Sinemurian age. This marine transgression occurred in a general southeasterly direction. The great thickness of fluviodeltaic deposits (*c.* 3000 m; Fig. 6) indicates that basin subsidence across the region was occurring during most of the middle–late Triassic (Cockbain 1982).

Within the various sub-basins, only the top of the Mungaroo Formation is seismically identified (Fig. 8). The base of the Locker Shale is difficult to locate because the data quality diminishes with depth and there is no seismic criteria to define the base of section (i.e. change in stratal geometry and/or distribution of the sequence). Within the southern Barrow Sub-basin, the deepest reflector identified is an intra-Jurassic reflector, recording an early–middle Jurassic (e.g. Pliensbachian) phase extension. The top of the Mungaroo Formation is not imaged because the Digicon seismic data are only recorded to 6 s TWTT (Fig. 6c). Nevertheless, the modelled thickness and distribution of the Triassic section is consistent with seismic refraction work across the basin (Stagg & Colwell 1994) and regional isopach maps for depth to magnetic basement (Cockbain 1982). Seismic data in the central Barrow and Dampier Sub-basins successfully image the base of the Locker Shales as determined by ties to the Candace-1 and Hampton-1 Wells, respectively (Fig. 6b). The Candace-1 Well is located just west of the Sholl Island Fault, which defines the eastern edge of the Barrow Sub-basin. The Candace-1 Well recovered Palaeozoic rocks (i.e. Kennedy and Lyons Groups) underlying the Triassic deposits (Mungaroo Formation and Locker Shales) on the Lambert Shelf (Bentley 1988), while the Hampton-1 Well recovered Palaeozoic rocks (i.e. Kennedy and Lyons Groups) underlying the Triassic deposits (Mungaroo Formation and Locker Shales) on the Lambert Shelf. Because the Hampton-1 Well penetrated the footwall cut-off, it failed to recover the entire Locker and Mungaroo sequences (cf. Fig. 6; Stein 1994).

Major hiatuses separate the Mungaroo Formation below (late Triassic) from the Mardie Greensands and Birdrong Sandstones above (Hauterivian–Barremian) in the central Barrow Sub-basin, and the lower Dingo claystones below (early–middle Jurassic) from the Mardie Greensands and Birdrong Sandstones above

(Hauterivian–Barremian) in the Dampier Sub-basin (Fig. 7). For both sub-basins, the Lambert Shelf is relatively uneroded, suggesting that the hiatus primarily records a period of non-deposition during the Jurassic and early Cretaceous in response to the westward migration of extension. For the central Barrow Sub-basin, extension migrated from the Sholl Island Fault in the Permian to the Flinders Fault Zone in the late Triassic–early Jurassic, while for the Dampier Sub-basin, this same hiatus indicates rift migration away from the Eliasson Fault System (eastern boundary of the Eliassen Terrace) in the Permian towards the Rosemary Fault Zone (western boundary of the Eliassen Terrace; Figs 3 and 6a).

The seismic data gap due to Barrow Island (e.g. Figs 1 and 2) and adjacent shoals necessitated using a technique like Werner deconvolution to estimate the depth to basement. This technique uses the magnetic anomaly to determine the depth to magnetic source using either a dyke or interface approximation (Stagg *et al.* 1989; Karner *et al.* 1991). Gridded AGSO total magnetic data were projected along the GPCT93-406 and 93BA-10 seismic lines. The Werner deconvolution depth estimates shown in Fig. 9 are marked by solid circles, the circle diameter being scaled to the intensity of magnetization. If the deconvolution is successful in defining a 'real' magnetic body, then depth estimates should define the depth range of an interface or its upper boundary. Because Werner deconvolution only identifies depth to magnetic source, it cannot discern between basement and intruded dykes and sills. For example, the deep reflectors shown in seismic line GPCT93–406 are highlighted by the technique (Fig. 9). These high-amplitude, deep reflectors are interpreted as sills emplaced during late Permian and Tithonian–Valanginian extension. The generation of these volumetrically significant volcanics are, in turn, a consequence of decompression melting induced by the relatively large degree of Permian ($\beta = 1.5$) and late Jurassic ($\beta = 2.65$) extension. In addition to their strong magnetic signature, the reflectors dramatically shoal toward the east near the Rankin Trend, having *c.* 10 km relief from west to east (Fig. 9). It is proposed here that these reflectors are sills that exploited crustal weaknesses associated with previous orogenic events (e.g. thrusts of the Alice Springs Orogeny) because it is difficult to reconcile 10 km of uplift with extensional-type processes. Similar high-amplitude, deep reflectors are observed in the Browse Basin and have been correlated to Permian volcanics (Symonds *et al.* 1994). Permian tholeiites have also been

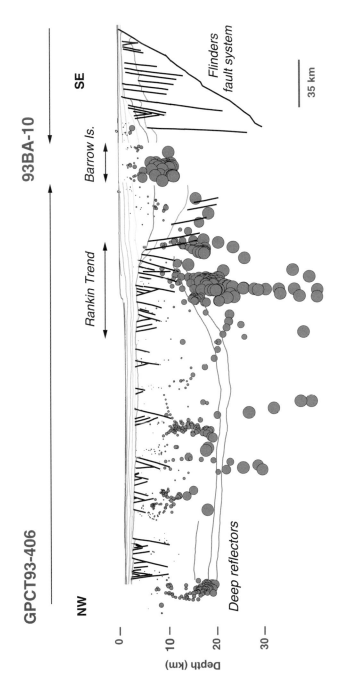

**Fig. 9.** Interpreted depth section across the central Barrow Sub-basin (seismic lines GPCT93-406 and 93BA-10) with depth to magnetic source estimates from the Werner deconvolution technique superposed (grey circles). Circle diameter is scaled to the intensity of magnetization. The Werner estimates define the depth range of either the interface or the upper boundary of magmatic intrusions, or magnetic basement. For example, depth estimates across the Rankin Trend correlate with a set of deep reflectors which are interpreted as sills emplaced during late Permian and Tithonian extension. In contrast, the estimates located toward the east of Barrow Island (i.e. within the seismic data gap) appear to be imaging the edge of a large Palaeozoic(?) basement block. This would suggest that the Flinders Fault System comprises a series of down-to-the-west fault blocks with Barrow Island located on the collapsed hanging wall to this fault system.

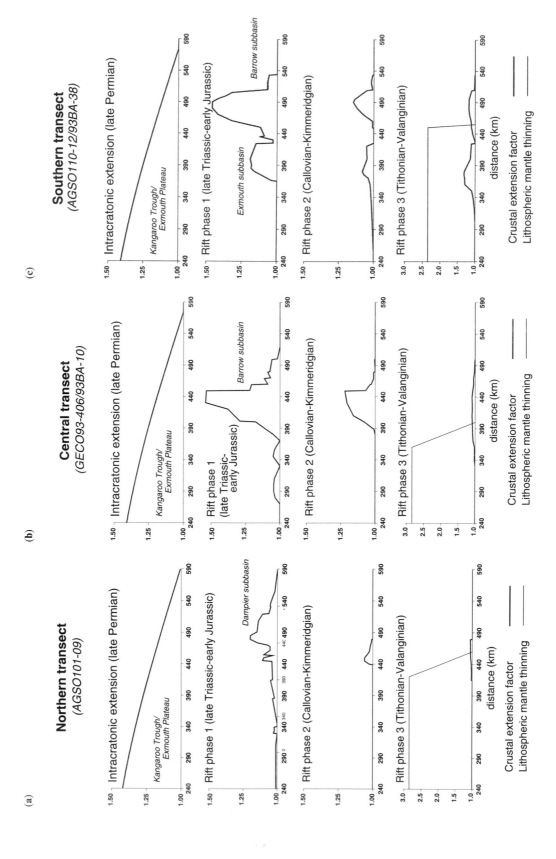

documented in the Western Canning Basin (Colwell *et al.* 1994). Although most of the magnetic signatures identified by the Werner deconvolution technique are associated with dykes and sills, the large magnetic signature located toward the east of Barrow Island (i.e. within the seismic data gap) appears to be imaging the edge of a basement block (Fig. 9). If correct, this suggests that the Flinders Fault System comprises a series of down-to-the-west fault blocks, with Barrow Island located on the collapsed hanging wall to this fault system.

## Model Predictions for late Permian Extension

The distribution and thickness of the synrift Triassic Locker Shale and the post-rift Mungaroo Formation across the margin are a direct consequence of the distribution and magnitude of late Permian rifting. In matching the subsidence history of the early Permo-Carboniferous basins of the northwest Australian margin, we required that the pre-rift lithosphere thickness be 125 km and crustal thickness 34 km. The thickness and distribution of the Locker Shales and the Mungaroo Formation suggest that they represent the late but subaerial stages of Westralian Superbasin (intracratonic) development.

To model the thickness and distribution of the Locker Shales requires broadly distributed, depth-independent extension across the region during the Permian. Permian extension increased toward the west, with a maximum value of 1.5 for all three transects (Fig. 10a–c). Matching the present-day distribution and thickness of the Permo-Triassic sequences across the Peedamullah and Lambert Shelves constrains the spatial extent and magnitude of the late Permian extension (Figs 6 and 10a–c). For example, if the lateral distribution of Permian extension is increased further towards the east, then substantial thicknesses of Permo-Triassic sediments are predicted to occur across the region of the high-standing Pilbara Craton, contrary to

observation. The subsidence history across the Exmouth Plateau also constrains the magnitude of the late Permian extension. Increasing the amplitude of Permian extension across the region increases the amplitude of the post-rift or thermal subsidence. If too large a value is assumed for the Permian extension, then significant post-rift accommodation is generated across the margin into the Jurassic. The observed localized Jurassic depocentres suggest that their accommodation was generated predominantly by restricted brittle (i.e. fault- induced) deformation, rather than by regional thermal subsidence associated with the late Permian extensional event. That is, late Permian induced thermal subsidence needs to be essentially complete prior to the early Jurassic extension. Even though similar Permian stretching factors were obtained for the three transects, the palaeotopographic relief for the Dampier Sub-basin was lower (600 m v. 325 m), which was required in order to create the necessary accommodation for the regionally deposited lower Dingo claystones. Based upon the model results, it is proposed that the depocentre for the Triassic sequence is located on the Exmouth Plateau west of the central Barrow Sub-basin (Fig. 6).

The observed minor lateral thickness variations of the Mungaroo Formation and Locker Shales predominantly reflect post-depositional processes due to rift-induced uplift and erosion associated with the formation of the Exmouth, Barrow and Dampier Sub-basins, and to differential loading and compaction by overlying sediment packages rather than structural control (Fig. 6). The formation of large, discrete rift basins and their associated footwall topography during late Permian rifting are not observed across the interior of the Exmouth Plateau in the seismic reflection data (e.g. Fig. 8). In contrast, rift flank topography (*c.* 750 m) rimming the large intracratonic basin (i.e. Westralian Superbasin) appears to have been sufficient to uplift and rotate existing fluvial networks. Consequent fluvial regrading and erosion supplied large amounts of clastic material to the evolving

**Fig. 10.** Modelled distribution, amplitude and timing of extension necessary to simulate the general basin architecture and preserved stratigraphy of the sub-basins and troughs comprising the Exmouth Plateau for: (a) the northern transect (Dampier Sub-basin); (b) the central transect (central Barrow Sub-basin); and (c) the southern transect (Exmouth and Barrow Sub-basins). The amount of crustal thinning determines the distribution of rift-phase subsidence. In turn, crustal thinning is a combination of the degree of upper plate thinning, defined by $\delta(x)$, and the degree of lower plate thinning, defined by $\beta(x)$. Intracrustal detachments determine the geometry of the upper and lower plates. The $\beta$ function also defines the degree of lithospheric mantle thinning, and critically determines the amount of thermal uplift and heat input into the lithosphere during rifting, and the amplitude and distribution of post-rift subsidence. Intracrustal detachments appear to have played a role only during Tithonian–Valaginian extension.

AGSO101-09

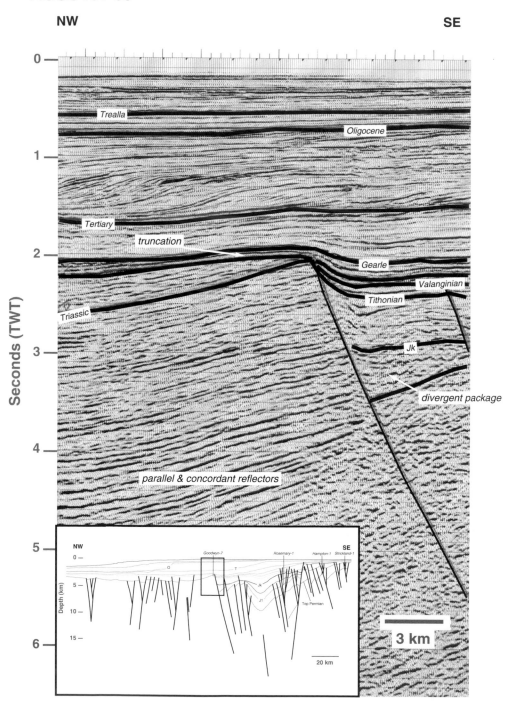

basin. Note that it is extremely difficult to predict how the non-marine accommodation responded to base level perturbations (i.e. the equilibrium profile; Schumm 1993). Poor age constraints and thickness determinations for the Mungaroo Formation and Locker Shales have forced the use of regional isopach maps to constrain the thickness of the Triassic deposits (Cockbain 1982), except along portions of the Peedamullah and Lambert Shelves where Digicon seismic reflection data and exploratory well data can be used with confidence (Fig. 6).

## Late Triassic Extension

The deposition of continental siliciclastics at the end of the Triassic supercycle contrasts with the marine and marginal-marine siliciclastics of the lower and upper Jurassic supercycles (e.g. rift phases 2 and 3; Fig. 7). The rapid formation of accommodation, as evidenced by the abrupt transition from progradational fluviodeltaic clastics of the Mungaroo Formation to the open-marine silts and clays of the lower Dingo claystone within the Exmouth, Barrow and Dampier Sub-basins, marks the onset of renewed extension beginning in late Triassic–early Jurassic time (Kopsen & McGann 1985; Veenstra 1985; Boote & Kirk 1989).

The divergence and rotation of seismic reflectors above the late Triassic–early Jurassic unconformity across the Exmouth Plateau are indicative of differential subsidence and block rotation (e.g. Figs 11 and 12). With the higher resolution seismic data in the Dampier sub-basin, at least four extensional events are recognized; late Triassic–early Jurassic, Pleinsbachian(?), Callovian and Kimmeridgian (Figs 11 and 12). The depocentre for the lower Dingo claystone is located in the Barrow, Dampier and Exmouth Sub-basins (Fig. 6; Boote & Kirk 1989). A defendable criteria has now been developed to define the boundary between the Mungaroo Formation and the overlying lower Dingo claystones within the depocentres by recognizing that the Locker Shales and Mun-

garoo Formation are regionally developed with only minor lateral thickness variations, and the lower Dingo claystone is locally developed with large lateral thickness variations. The first package of seismic reflectors that diverge and onlap onto the underlying parallel, and roughly concordant, reflectors records the boundary between the top of the Mungaroo Formation and the base of the lower Dingo Formation (Figs 11 and 12). Along the structural highs, well data corroborate this interpretation. Nevertheless, in the basin depocentres, where there is no well control, the criteria used here to identify the base of the lower Dingo claystones predicts that the early Jurassic depocentres are much deeper than previously interpreted (e.g. Boote & Kirk 1989; Figs 6 and 9).

The late Triassic–Callovian interval thickens towards the east in both the Exmouth and southern Barrow Sub-basins (Fig. 6c). For the Dampier Sub-basin, the late Triassic–Callovian interval is thickest in the centre of the Lewis Trough, forming a sag-type geometry, with reflectors dipping away from the basin-bounding faults along the Rosemary Fault System and the Madeleine Trend (also Madeleine Fault System; Figs 6a and 12). Away from the Lewis Trough, a secondary depocentre occurs along the east portion of the sub-basin near the Rosemary Fault System (Fig. 6a). Antithetic faults occur along the Rankin Trend, creating isolated depocentres along the western portion of the sub-basin (i.e. Kendrew Terrace; Fig. 6a; Bint & Marshall 1994). The observed thickness variations suggest that the basin-bounding fault for the Exmouth Sub-basin was located along the western flank of the Alpha Arch (Fig. 6c). For the southern Barrow Sub-basin, the border fault was located along the western edge of the Peedamullah Shelf (Fig. 6c), while for the central Barrow and Dampier Sub-basins, the border fault was located along the Lambert Shelf (Figs 6a and 6b). Rift-induced footwall uplift of the Alpha Arch, and Peedamullah and Lambert Shelves during the second phase of deformation affected the drainage networks across the shelf and thus limited the access of fluvial systems

**Fig. 11.** Interpreted seismic reflection profile AGSO101–09 showing the uplift, rotation and truncation of the western border of the Dampier Sub-basin (Rankin Trend), which was the footwall block to Callovian and Kimmeridgian extension in the Dampier Sub-basin. The alternating footwall–hanging-wall block role of the Eliassen Terrace and Lambert Shelf within the eastern Dampier Sub-basin in the Callovian v. the Kimmeridgian is part of a westward migration of extension as a function of time. This migration allows prediction of the temporal and spatial variability of the sediments infilling the evolving sub-basin. For example, the amount of coarse-grained deposits along the eastern Dampier Sub-basin should systematically increase after the Callovian extensional event, culminate in late Kimmeridgian–early Tithonian time and then abruptly end in the Tithonian. The base of the divergent package is late Triassic–early Jurassic in age.

## AGSO101-09

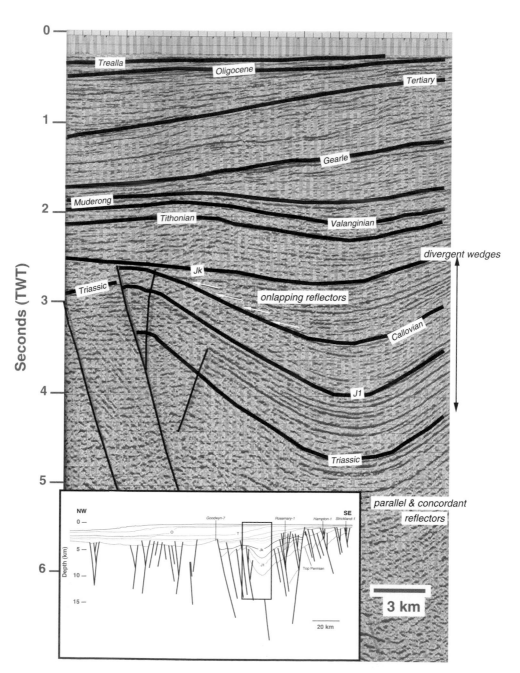

into the rift basin proper (cf. Fig. 5). The onset of the second phase of extension is not well established because the industry wells are located predominantly along structural highs and, consequently, have not sampled the oldest onlapping packages within the sub-basins. Therefore, the age determinations from the well data yield a minimum date for the onset of extensional deformation in the region. For simplicity, a late Triassic–early Jurassic age has been adopted for the onset of the second rift phase. Although deposition of the lower Dingo claystone was predominantly within a relatively deep, open-marine environment (outer neritic–upper bathyal), point sources of sediment delivery gained access into the basin via large offset accommodation zones that resulted in the accumulation of time-equivalent deltaic wedges (the Learmonth delta facies and the Legendre Formation) towards the northeast (Fig. 7).

Overlying the lower Dingo claystone are the Biggada Sands (Fig. 7). Boote & Kirk (1989) regarded the Biggada Sands as recording the onset of renewed tectonic activity and that they were shed directly into the basin from the newly formed rift flanks. However, the spatial distribution of the Biggada Sands within the Barrow–Dampier Sub-basins suggests that they are sourced by sediments entering the basin through various accommodation zones that segment the border-fault systems (Kopsen & McGann 1985). As these structurally controlled deltas prograded across the shelf, infilling the available accommodation, they delivered coarse-grained sediments to the basin. Deposition of the Biggada Sands within the basin resulted from localized deltaic progradation and consequent sediment bypass, delivering sands further basinwards. In this interpretation, the Biggada Sands are recording a period of tectonic quiescence, recording both the destruction of accommodation and flanking topography across the region prior to the next phase of extension in the Callovian. In a tectonostratigraphic sense, the Biggada Sands belong to the lower Dingo claystone rather than the basal unit of the upper Dingo claystone (cf. Boote & Kirk 1989).

## Model Predictions for Late Triassic Extension

Renewed extension in the late Triassic–early Jurassic affected the northwest Australian margin, generating an extensive along-strike network of basins. However, the difference in regional trends of both the central Barrow and Dampier Sub-basins relative to the spreading fabric within either the Argo and/or Cuvier oceanic crust (e.g. Fig. 1) suggests that these sub-basins should contain a significant wrench component. Extension was limited to a narrow portion of the margin, producing the Barrow, Dampier and Exmouth Sub-basins, and presumably the Argo, Gascoyne and Cuvier Rift Basins (i.e. the future sites of the ocean–continent boundary; Fig. 1). Even though the basins and sub-basins were narrow and localized, their position along the edge of the Peedamullah and Lambert Shelves, and the western edge of the Exmouth Plateau, made them efficient traps for clastic sediment derived from the Australian continent to the east and the Greater Indian continent to the west. Consequently, thick synrift depositional wedges were confined within the structurally controlled topography of these evolving basins, thereby starving the Exmouth Plateau of sands and coarse clastic material (Exon et al. 1992; Haq et al. 1992). In addition to the extensional deformation being spatially limited, model results indicate that the brittle and ductile extension was depth independent, with maximum values ranging from 1.46 to 1.54 in the south to 1.15 in the north (Fig. 10a–c). The eastern limit of the late Triassic extensional deformation was shifted towards the west with respect to the Permian extension (Figs 6 and 10a–c).

The Flinders Fault Zone was the basin-bounding fault system to late Triassic–early Jurassic extension, with the southern and central Barrow Sub-basins being the hanging-wall block and the Peedamullah and Lambert Shelves the footwall block (Fig. 6b and c). Rift-induced flexural uplift and erosion of the shelves truncated a large section of the underlying Mungaroo

---

**Fig. 12.** Interpreted seismic reflection line AGSO101-09, illustrating the relationship between the basin structure and stratigraphy in the Dampier Sub-basin (Lewis Trough). The pre-rift stratigraphy dips away from, or subparallel to, the fault trace for both western (Madeleine) and eastern (Rosemary) fault systems. The resulting geometry creates a synform with the axis located near the centre of the Lewis Trough, a pattern characteristic of oblique extension. The depocentre for the overlying synrift sequence is also spatially coincident with the axis of the syncline. The observed pre-rift geometry in the hanging wall is not consistent with the structural and stratigraphic predictions for either rotational or uniform subsidence of a hanging-wall block. While a series of synthetic faults could explain the basinward shift of the depocentre away from the basin-bounding fault, this faulting style cannot account for the dip of the pre-rift stratigraphy. At least four extensional events are recognized; late Triassic–early Jurassic (Triassic), (?)Pliensbachian (JI), Callovian and Kimmeridgian (JK).

Formation. At Dillson-1 Well, for example, the Mungaroo Formation and the Locker Shales are relatively thin (Fig. 6c). However, because the Dillson-1 Well is located on a downdropped block that effectively competed with the rift-induced footwall uplift, the Mungaroo section in this region has been protected from erosion. Furthermore, the thickness of the Locker Shales recovered at the Dillson-1 Well is not truly representative because the well intersected the border fault and the lower section of the sequence has been structurally removed. The asymmetric development of the southern Barrow and Exmouth Sub-basins during late Triassic time is consistent with the location of the lower Dingo depocentre adjacent to the Flinders fault zone and the Alpha Arch/Rankin Trend, respectively (Fig. 6c). Accordingly, the modelling results presented here suggest that the Alpha Arch, which delineates the western boundary of the southern Barrow Sub-basin and the eastern boundary of the Exmouth Sub-basin (e.g. Fig. 2), represents both the collapsed hanging wall to the Flinders Fault and the uplifted footwall to the faults immediately west of the Alpha Arch at this time. In contrast to the Peedamullah and Lambert Shelves, the Alpha Arch was associated with only minor erosion. The Ramillies-1 Well, located on the crest of the Alpha Arch (Figs 2 and 6c), recovered the upper Mungaroo Formation indicating that only minor erosion had occurred across the arch. This is consistent with the tectonic history of the Alpha Arch. In essence, the Alpha Arch had a dual role; it was the hanging wall to the extensional deformation in the east and the foot-wall to the deformation in the west (Fig. 6c). The eastern edge of the Exmouth Plateau also experienced uplift due to extensional unloading, even though it was the hanging-wall block to the late Triassic extension. This uplift occurred because the wavelength of the basin was narrow with respect to the flexural wavelength of the lithosphere ($c$. 60–80 km) and the series of antithetic faults caused minor extensional unloading. Minor faulting across the Exmouth Plateau and the Kangaroo Trough accompanied the late Triassic–early Jurassic development of the Exmouth and southern Barrow Sub-basins (e.g. Figs 6c and 8).

The situation in the central Barrow and Dampier Sub-basins is a little more complicated. During late Triassic rifting, the Rankin Trend was formed as part of the broad regional hanging wall to extension along the Flinders Fault System (Figs 6a, b and 11). Small antithetic faults along the eastern edge of the Rankin Trend accentuated the uplift. Even though the Rankin Trend was the regional hanging wall, it also experienced uplift due to extensional unloading because the wavelength of the basin was narrow with respect to the flexural wavelength of the lithosphere (e.g. Fig. 6a and b). The uplift of the Rankin Trend is greater than that along the Lambert Shelf because the extension deformation along the eastern edge of the sub-basin was distributed across numerous faults within the Flinders Fault System (Fig. 6b). Minor extension across the Kangaroo Trough accompanied the late Triassic formation of the Barrow and Dampier Sub-basins and gave rise to the undulating structure observed in the region (Figs 6a, b and 8).

En echelon fault patterns within the central Barrow and Dampier Sub-basins attest to a significant wrench component (e.g. Fig. 3). In oblique extension along curved faults, the displacement can vary from predominantly dip-slip to strike-slip, depending on the orientation of the fault. Furthermore, strain partitioning occurs to create distinct networks of dominant oblique-slip and dip-slip faults (Tron & Brun 1991). Despite the fact that experimental models have their limitations (e.g. rheological scaling between crust and model material, strain rate, stretching distribution, effects of isostasy, etc.), they are powerful analogues with which to simulate both the geometry and internal infill structure of complex extensional systems. The experimental modelling results of Withjack & Jamison (1986), Tron & Brun (1991) and McClay & White (1995) have been useful in defining the basin architecture and infill geometry for increasing amounts of wrenching. In terms of architecture, the spatial distribution and orientation of the modelled faults strongly depend on the angle of obliquity. For pure extension, the orientation of faults are distributed evenly around the rift trend; i.e. the number of faults that have been rotated clockwise with respect to the rift trend is approximately equal to the number of faults rotated counter-clockwise. As obliquity increases, the fault population becomes more skewed towards faults with a clockwise rotation with respect to the rift trend. If the offset across the rift trend had been left-lateral, then the population of faults would be skewed towards fault populations with counter-clockwise rotations. The geometry of the pre- and syndeformational modelled stratigraphy is of critical importance in deciphering the role of extension v. wrenching. For small degrees of wrenching, the resulting deformation causes the pre-rift stratigraphy in the hanging wall to dip towards the fault trace, an observation that is consistent with pure extensional fault kinematic predictions and observations. In contrast, for increasing wrenching the resulting deformation creates synforms with the pre-rift stratigraphy dipping subparallel to, or away from, the fault trace.

The observed structural and stratigraphic features in the Lewis Trough cannot be reconciled in terms of either simple or complex extensional displacement between the hanging wall and the footwall (Figs 6a and 12). The regional trend of both the central Barrow and Dampier Sub-basins in relation to the spreading fabric within either the Argo and/or Cuvier oceanic crust indicate that these basins have a significant wrench component (c. 30°; Fig. 1). Oblique extensional systems are characterized by the following: (1) en echelon fault patterns; (2) a fault orientation with respect to the rift trend strongly dependent on the degree of wrenching; (3) an average fault trend not perpendicular to the extensional transport direction; and (4) basin synform structure with the pre-rift stratigraphy dipping subparallel to, or away from, the fault trace. As these observations apply to the Lewis Trough region in particular (Figs 3 and 12), oblique extension potentially explains why the pre-rift stratigraphy preserved in the deformed hanging wall dips away from the fault trace (Fig. 12). Even though the major depocentre was created by oblique late Triassic–early Jurassic exension, this modelling indicates that the amount of footwall uplift and accommodation created is nevertheless commensurate with the component of normal extension within the Lewis Trough (Figs 6a and 10a).

Finally, the modelling suggests that the Learmonth prograding clastics are produced by the erosion of the rift flank topography developed adjacent to the Argo Rift Basin (to the north of the Exmouth Plateau) during Rhaetian extension. The drainage networks developed within the eastward dipping Argo Rift Basin footwall gained access in the northern Barrow–Dampier Sub-basins along a series of accommodation zones separating the Argo Abyssal Plain from the Exmouth Plateau (cf. Fig. 5).

## Callovian and Kimmeridgian Extension

Two Mesozoic tectonic events resulted in continental break-up and the formation of oceanic crust (Sager et al. 1992); a Callovian event that generated oceanic crust within the Argo Abyssal Plain (marine magnetic anomaly M26; 163 Ma; Fig. 1), and a Tithonian–Valanginian event that resulted in the break-up of the west and northwest Australian margin, and the generation of oceanic crust in the Gascoyne, Cuvier and Perth Abyssal Plains (marine magnetic anomaly M10; 132 Ma; Fig. 1). These events document the complete continental rupturing and separation of west and northwest Australia from Greater India and related continental fragments (Boote & Kirk 1989). The Callovian extensional episode activated the border faults of the Exmouth, Barrow and Dampier Sub-basins and marked a period of widespread claystone deposition (the upper Dingo claystone) with only minor occurrences of coarse-grained deposits (Fig. 6a–c; Barber 1994). The upper Dingo claystone, which consists of a thick succession of silty claystone deposited below wavebase (water depths of hundreds of metres), records this Callovian episode of fault reactivation (e.g. Figs 11 and 12; Kopsen & McGann 1985; Veenstra 1985; Boote & Kirk 1989). Only minor faulting occurred across the Kangaroo Trough and Exmouth Plateau during the late Triassic–early Jurassic and Callovian reactivation of the sub-basins, evidenced by small wedges of lower and upper Dingo claystone deposition across the region (Fig. 6; Driscoll & Karner 1998a). West of the Dampier Sub-basin, regional subsidence was conducive to the deposition of a thin blanket of lower and upper Dingo claystone (Fig. 6a). The depocentre for the upper Dingo claystones in the southern Barrow Sub-basin remained adjacent to the eastern portion of the basin along the western edge of the Peedamullah Shelf (Fig. 6c). In the Exmouth Sub-basin, the depocentre for the upper Dingo claystones migrated to the west along the eastern edge of the Exmouth Plateau (Fig. 6c), whereas in the Dampier Sub-basin, Callovian sequences are predominantly confined to the Lewis Trough with the depocentre located toward the west along the Madeleine Trend (Fig. 6a).

Along the Peedamullah and Lambert Shelves, the initial pro-delta shales of the upper Dingo claystone were eventually replaced by the Dupuy Sands (Fig. 7). These sands are associated with prograding deltas and shingled coarse-grained turbidites. As accommodation was filled along the eastern margin, the river systems delivered coarse-grained clastics, sourced from the hinterland, to the outer Peedamullah and Lambert Shelves and the adjacent Barrow Sub-basin. The Dupuy Sands prograded from east to west and accumulated on a gently dipping slope that bordered most of the eastern margin of the Barrow and Dampier Sub-basins. The amplitude of the clinoforms observed in the seismic reflection data indicate that minimum water depths at this time were of the order of c. 200 m (Tait

1985; Boote & Kirk 1989). The Dupuy Sands failed to infill completely the rift-induced accommodation in the western regions of the Barrow Sub-basin, and were isolated from the Exmouth Sub-basin and the Kangaroo Trough by the Alpha Arch and Rankin Trend, respectively.

Within the Dampier Sub-basin, the upper Dingo claystone records both a Callovian and a Kimmeridgian episode of fault reactivation (Figs 6a, 11 and 12). During the Kimmeridgian, extension shifted westward from the Madeleine Trend to the Kendrew Terrace Fault System (Fig. 3). This change in extensional deformation was also accompanied by a marked increase in coarse-grained facies across the sub-basin. Along the eastern portion of the sub-basin, near the Rosemary-1 and Legendre-1 Wells (Fig. 2), coarse-grained clastics began accumulating at this time, in contrast to the underlying Callovian deposits that are predominantly claystones. Along the eastern portion of the Rankin Trend, near the Goodwyn-7 Well, the lower and middle Jurassic reflectors are rotated and truncated, indicative of a tectonic event (Fig. 11). Previous workers proposed that this facies change along the eastern Dampier Sub-basin recorded a shift in the tectonic hinge line from the Eliassen to the Rosemary Fault System (Barber 1994). However, such a shift would be recorded by a rapid increase in subsidence west of the Rosemary Fault System and a consequent deepening because the fluvial networks would be diverted away from region by the uplifted footwall. This is the exact opposite of what is observed.

## Model Predictions for Callovian and Kimmeridgian Extension

The Callovian extensional episode, which correlates with the onset of seafloor spreading in the Argo Abyssal Plain, reactivated the border fault systems that delineated the Exmouth, Barrow and Dampier Sub-basins with only minor accompanying deformation across the Kangaroo Trough and Exmouth Plateau (Fig. 6a–c). Similar to the late Triassic–early Jurassic event, the Callovian event appears to be depth independent with a maximum extension factor ranging from 1.15 to 1.22 in the south (Exmouth and Barrow Sub-basins; Fig. 10c) to 1.06 in the north (Dampier Sub-basin; Fig. 10a). Within the Exmouth Sub-basin, the dominant border-fault system appears to have switched from along the western edge of the Alpha Arch (late Triassic extension) to the eastern edge of the

Exmouth Plateau (middle Jurassic extension; Fig. 6c). Because the extensional deformation was distributed along the eastern edge of the Exmouth Plateau, the uplift due to extensional unloading was counteracted by crustal thinning, thus explaining the accommodation created at this time for the deposition of the upper Dingo claystone (Figs 6 and 10a–c). The alternating footwall–hanging-wall role of the eastern edge of the Exmouth Plateau is suggested by the westward migration of the depocentre in the Exmouth Sub-basin. The structural development of the Peedamullah Shelf in the southern Barrow Sub-basin and the Alpha Arch show a similar border fault history. The Peedamullah Shelf was the footwall block during both the late Triassic and middle Jurassic extension. The Alpha Arch had a more complicated deformational history because of its location with respect to the active border faults. In particular, it was part of the collapsed hanging wall to the late Triassic extension in the southern Barrow Sub-basin, as well as the footwall during concomitant extension in the Exmouth Sub-basin (Fig. 6c). During the middle Jurassic extension, the Alpha Arch was the hanging wall to the extensional deformation in both the southern Barrow and Exmouth Sub-basins. It is important to note, however, that because of the segmented nature of normal faults and extensional systems, the structure and development of the Alpha Arch varies along-strike.

For the Barrow Sub-basin, the active fault system remained along the eastern edge of the sub-basin, i.e. the Flinders Fault System along the Peedamullah and Lambert Shelves (Fig. 6b). Because the extensional deformation was accommodated across numerous faults within the Flinders Fault System (Fig. 10b), the uplift due to extensional unloading also tended to be regionally distributed, and therefore subdued with only minor truncation of the lower Dingo claystones across the outer Lambert Shelf. Sediment loading in the central Barrow Sub-basin and thermal subsidence across the Lambert Shelf eventually allowed river access to the sub-basin, and possibly explains the westward progradation of Dupuy Sands along large portions of the eastern margin. The deposition of the Dupuy Sands was interrupted by renewed rift-induced uplift in the Tithonian–Valanginian. The kinematic and flexural modelling presented here allows the sequence of events responsible for the structural development of the Lambert Shelf to be defined as follows (Figs 3, 6a and b): (1) extension was initially accommodated along the Sholl Island Fault during the late Permian; (2) rifting shifted westward to the Flinders Fault

System in the late Triassic, causing portions of the previously collapsed hanging wall to be uplifted; and (3) Callovian extension, also focused along the Flinders Fault System, induced only minor uplift.

Further north in the Dampier Sub-basin, the Callovian stratigraphic interval is confined predominantly to the Lewis Trough, with the depocentre located towards the west along the Madeleine Trend (Figs 3 and 6a). Extension migrated away from the Madeleine Trend, active during Callovian extension, to the Kendrew Terrace Fault System during the Kimmeridgian rifting event. Callovian extension was accommodated along the Rankin Trend with the Lambert Shelf acting as the collapsed hanging wall. This along-strike variation in the deformational style of the Peedamullah and Lambert Shelves from the footwall in the Barrow Sub-basin to the hanging wall in the Dampier Sub-basin plays an important role in governing where and when fluvial networks gained access to the evolving rift. The uplift and erosion of the Rankin Platform removed lower and middle Jurassic sediments (e.g. Fig. 11). The westward migration of the extensional deformation is clearly demonstrated by the depocentres for the lower and upper Dingo claystones (Fig. 6a). The Eliassen Terrace was the hanging wall to both the Callovian and Kimmeridgian extension. The collapse of the hanging wall and the associated westward dip of the block could potentially capture existing fluvial systems along the Lambert Shelf. The deposition of Oxfordian and Tithonian sands within the subbasin is consistent with the Lambert Shelf and Eliassen Terrace being the collapsed hanging wall to both the Callovian and Kimmeridgian phase of deformation. Whereas Callovian extension induced minor faulting across the Exmouth Plateau, lithospheric cooling associated with the late Triassic–early Jurassic extension generated regional accommodation allowing for a thin veneer of both lower and upper Dingo claystones to be deposited.

Similar to the structural and stratigraphic development of the central Barrow and Dampier Sub-basins during late Triassic extension, the Callovian and Kimmeridgian wrench development of the Lewis Trough cannot be reconciled in terms of either simple or complex extensional displacement between the hanging-wall and footwall blocks. Nevertheless, previous workers have attempted to model the development of the Dampier Sub-basin by postulating the existence of non-seismically resolvable faults to control the general morphology of the footwall and hanging-wall blocks (e.g. Baxter et al. 1997;

Baxter 1998). In particular, the synrift basin geometry is mimicked by regionally distributing the deformation over a large number of small faults. In this case, the regional dip of the faulted basin flank will dip towards the basin axis while individual fault blocks will tend to dip towards the bounding faults. While assuming that the existence of these faults may allow the basement geometry to be reproduced, the pre- and synrift stratal relationships are not. For the Dampier Sub-basin, the pre-rift stratigraphy dips either subparallel to, or away from, the presumed border fault trace (Figs 6a and 12), thus requiring the controlling fault to be beneath the basin depocentre.

## The Tithonian–Valanginian Extension

During Tithonian–Valanginian time, the continental lithosphere along the western boundary of the Exmouth Plateau and the southwestern Australian margin was breached, and oceanic lithosphere was emplaced (marine magnetic anomaly M10; Boote & Kirk 1989). Fault reactivation occurred within the Barrow and Exmouth Sub-basins and along the Rosemary Fault System within the Dampier Sub-basin (Fig. 6a–c). Once again, the Eliassen Terrace become the footwall to the extensional deformation within the Dampier Sub-basin, allowing the Angel Sands, which are time equivalent to the Dupuy Formation, to be capped by the Forestier claystones (Fig. 7; Bint & Marshall 1994). AGSO seismic reflection profiles 101-07 and 110-12 (Fig. 1), which cross the Alpha Arch and Exmouth Plateau, imaged a Tithonian angular unconformity that records minor truncation of the underlying Mesozoic stratigraphic successions and onlap of the overlying early Cretaceous sequences (Figs 8 & 13). This truncation is observed across most of the region, suggesting that large portions of the plateau were emergent or at very shallow water depths prior to the development of the overlying Barrow Delta System. A broad shallow-water environment across large portions of the platform is also consistent with facies analysis of the Barrow Delta sediments (the Barrow Group), in addition to the mapped northward and westward distribution of the delta foresets (Tait 1985; Erskine & Vail 1988). The depositional packages that make up the Barrow Group display a systematic evolution from lower turbidite packages across which prograded thin prodelta shales that, in turn, coarsen upward into delta-top sands typical of wave-dominated systems. The clinoform foresets mark the location

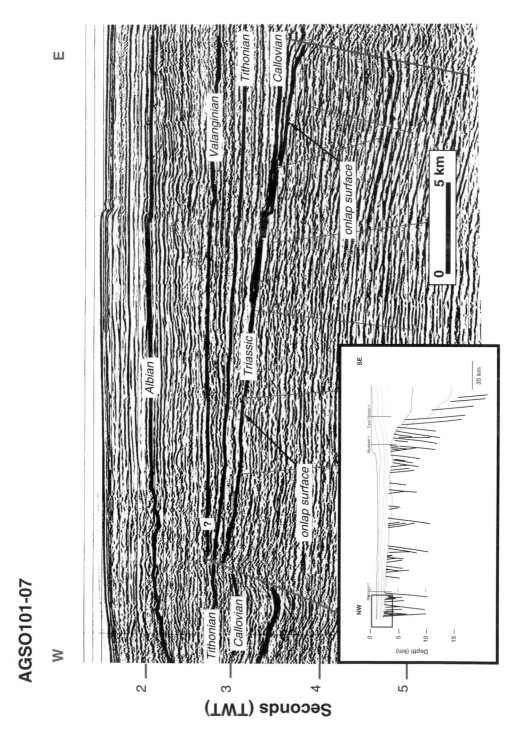

**Fig. 13.** Interpreted seismic reflection line AGSO101-07, illustrating the minor erosional truncation beneath the Tithonian unconformity. The thinning of the Callovian–Valanginian package is primarily controlled by onlap. The truncation associated with this unconformity is observed across most of the Exmouth Plateau, suggesting that large portions of the plateau were emergent, or at very shallow water depths, prior to the deposition of the Barrow Delta. The top of the Barrow succession is marked by the Valanginian unconformity.

of increased palaeowater depth and examination of their amplitude suggests minimum water depths of 200–500 m (Fig. 8; Ross & Vail 1994; Tait 1985). Palaeowater depth estimates for the Valanginian, together with the present-day bathymetry and thickness of the late Cretaceous and Tertiary stratigraphic sequences, indicate that significant subsidence occurred across the northern Exmouth Plateau since Valanginian time. Only the distal facies equivalent of the Barrow Delta (i.e. the Forestier claystones) are observed in the Dampier Sub-basin and, as a result, the Tithonian–Valanginian interval represents a condensed section.

The oblique clinoforms observed within the lower sections of the Barrow Delta indicate that there was little to no accommodation for aggradation and, consequently, the sediment prograded laterally giving rise to an oblique geometry without the development of clinoform topsets (e.g. Fig. 8). Recognition of these stratal patterns in the seismic reflection data is critical because they allow identification of horizons that were at, or near, sea level during the time of deposition. As subsidence outpaced sediment supply, the delta progressively began to aggrade before rapidly backstepping to the east-southeast. The post-Valanginian accommodation appears to be regionally distributed with a slight increase toward the west (i.e. towards the ocean–continent boundary), which is in marked contrast to the localized accommodation associated with the lower and upper Dingo claystones (Fig. 6c). Note that the post-Valanginian accommodation has been amplified along the eastern margin by sediment loading and, taken together with the long wavelength subsidence that increases toward the west, has resulted in an overall arching of the margin (Figs 1, 6a and c). The apex of the arch occurs along the Exmouth Plateau, near Eendracht-1 Well and west of the Mercury-1 Well. Because of the increased amount of sediment loading in the southern Barrow and Exmouth Sub-basins (e.g. AGSO line 110-12), the arching of the margin is more pronounced in the south and diminishes toward the north (e.g. Fig. 1). In fact, because the post-Tithonian sediment wedge is relatively thin in the Dampier Sub-basin region (Fig. 6a), the long wavelength subsidence has played a more dominant role such that the margin is characterized by a westward dipping monocline rather than an arch.

Along the eastern margin of the Barrow Sub-basin, the lower units of the Barrow Group onlap onto the upper Dingo–Dupuy sequence (Boote & Kirk 1989). The observed onlap of the Barrow Delta onto the Dupuy Sands results

from a shift in the sediment source region from the east to the southeast (Tait 1985). These turbidite packages of the Barrow Group are overlain by thin pro-delta shales that coarsen upward into delta-top sands typical of wave-dominated systems. Toward the top of the Barrow Delta, a large unconformity with extensive downcutting is observed, suggesting that large portions of the delta platform were emergent at this time (Fig. 6; Boote & Kirk 1989). The blocky Flag turbidites have been interpreted as lowstand deposits associated with an eustatic sea-level fall (Haq et al. 1987). Alternatively, the Flag turbidites may represent reworked Barrow Delta sediments that record the thermal uplift of the southern boundary of the Exmouth Plateau in response to the migration of the Cuvier spreading centre along the Cape Range Fracture Zone (Exon & Buffler 1992).

There is only minor accompanying Tithonian–Valanginian brittle deformation across the Exmouth Plateau (Figs 6a–c, 8 and 13). The amount of brittle deformation and fault reactivation is small in the central Barrow Sub-basin, decreasing from south to north away from the southern transform margin (i.e. the Cape Range Fracture Zone; Fig. 1). Major late-stage structuring of the margin in the southern Barrow and Exmouth Sub-basins occurred during a possible fourth phase of extension and is responsible for structuring of the Cuvier continental margin. In contrast, Tithonian rifting only induced minor reactivation of the Flinders Fault System in the central Barrow Sub-basin (Fig. 6b). Given the minor Tithonian–Valanginian brittle deformation across the Exmouth Plateau, an interesting paradox results, where regional post-rift subsidence is associated with only minor amounts of upper crustal brittle deformation, i.e. the lithosphere has undergone depth-dependent extension in the rifting phase that progressed to continental break-up.

Extensive break-up volcanism has been identified along the West Australian margin (e.g. Symonds et al. 1998). In general, there are three distinct groups: (1) diverse volcanics, including basalts, hyaloclastite and volcaniclastics, found on the northern Exmouth Plateau and related to middle Jurassic volcanism (Exon & Buffler 1992; Colwell et al. 1994), i.e. the Callovian break-up of the northern northwest margin; (2) a group comprising shallow intrusive and extrusive tholeiitic basalts and seaward-dipping reflector sequences on the Scott and Exmouth Plateaux and on the Gascoyne and Cuvier margins, and having the same age as the Tithonian–Valanginian break-up of the Exmouth Plateau and Cuvier margin (Fig. 9;

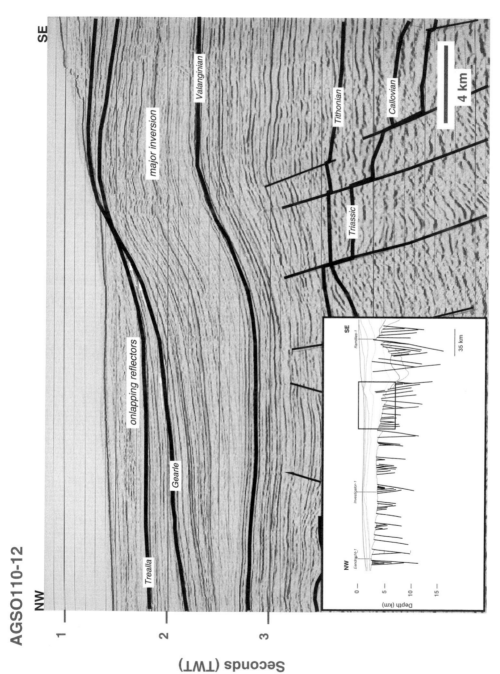

**Fig. 14.** Interpreted seismic reflection profile AGSO110-12, illustrating the forced fold (inversion) along the northwest edge of the Exmouth Sub-basin. The thickness of the individual layers increases into the crest of the anticline. The Turonian inversion observed in the Exmouth, Barrow and Dampier Sub-basins appears to correlate with a major plate reorganization in the Indian Ocean. The inversion appears to be best developed in the large accommodation zones along the margin where sub-basins terminate and the extensional deformation is offset in a right-lateral manner.

Crawford & von Rad 1994); and (3) major volcanic constructs formed either at the spreading centre or emplaced on relatively young oceanic crust are represented by the Wombat Plateau, Platypus Spur, Joey Rise, Roo Rise and Wallaby Plateau (Fig. 1). None of the basalts in (1) and (2) have a plume signature (Crawford & von Rad 1994), although they are clearly related to high degrees of partial melting associated with enhanced thermal gradients below rifted continental lithosphere (Hopper *et al.* 1992; Ludden & Dionne 1992). These observations strongly suggest that excessive margin magmatism can occur within rift environments in the absence of a plume. In contrast, the basalts of (3) have many of the characteristics of other eastern Indian Ocean large igneous provinces, such as the Kerguelen Plateau and Broken Ridge. All appear to have formed some 10–20 Ma after Tithonian–Valanginian break-up by voluminous outpourings of plume-related magmas (Colwell *et al.* 1994; Crawford & von Rad 1994).

Reflector geometries observed in the seismic data indicate the following (Fig. 6): (1) the accommodation generated by brittle deformation in Tithonian time across the Exmouth Plateau was localized and its magnitude was minor in comparison to the regional subsidence; (2) the truncation associated with the Tithonian unconformity was minimal; (3) the Tithonian–Valanginian sequence thins predominantly by onlap onto structural highs; and (4) major portions of the margin were at, or near, sea level with minimal amounts of accommodation for aggradation. Exploratory well data suggests that the Barrow Delta was deposited in a shallow-marine to nearshore environment. In contrast, significant palaeowater depths existed in the eastern Dampier Sub-basin while the Eliassen Terrace and Lambert Shelf were at, or near, sea level. These observations are the critical constraints requiring large, broadly distributed extension across the Exmouth Plateau to accommodate the Barrow Delta sediments, the general arching of the Exmouth Plateau, and the large post-Valanginian subsidence required to explain both the present-day water depths and the post-rift sediment thicknesses observed across the Exmouth Plateau.

Following break-up, the Barrow Delta (including its northerly pro-delta equivalents, i.e. the Forestier claystone) was capped by the diachronous Mardie–Birdsong Sands and the overlying Muderong shales (Figs 6 and 7). The Muderong Shale forms a widespread transgressive marine shale blanketing most of the Exmouth Plateau and the northwest Australian margin. The shift from terrigenous to predominantly carbonate sediments along the Exmouth Plateau resulted from a relative sea-level rise induced by the thermal re-equilibration of the rifted lithosphere, coupled with a systematic change in climatic conditions associated with the northward migration of the Australian continent. The modelling highlighted the need for small amounts of inversion in Turonian time to match the stratal geometry, and temporal and spatial distribution of subsidence (Figs 6 and 14). This inversion resulted in the uplift of Barrow Island, located on the hanging wall to the Flinders Fault System, minor folding along the eastern margin of the Rankin Trend (e.g. Fig. 14), and uplift of the hanging wall immediately west of the Rosemary Fault System (e.g. Rosemary-1 Well) on the eastern edge of the Dampier Sub-basin (Fig. 6a). The inversion appears to be best developed in the large accommodation zones along the margin where sub-basins terminate and the extension is offset in a right-lateral sense. The amount of shortening associated with this reactivation is minor (<500 m). The timing of inversion correlates with the major plate reorganization that occurred between the Greater Indian and Australian Plates at this time. Subsequent to the Turonian inversion, Palaeogene and Neogene carbonate sediments prograded across the shelf, infilling the available accommodation (Figs 6 and 7). It is also recognized that, in addition to the Turonian inversion, a marked change in depositional style occurred along the margin from gravity dominated to current-controlled deposition.

## Model predictions for Tithonian - Valanginian extension

Clearly, the regional distribution and amplitude of the post-Valanginian subsidence is not consistent with the minor amounts of Tithonian–Valanginian brittle crustal extension observed across the sub-basins of the Exmouth Plateau (Fig. 10a–c). The form, distribution and longevity of this subsidence phase is characteristic of the cooling and contraction of extended lithospheric mantle. However, the magnitude of extension needed to generate the regional post-Valanginian subsidence [maximum $\beta(x)$ of 2.8; Fig. 10a–c] precludes it from being the thermal subsidence phase to the minor and laterally restrictive extension that occurred during Callovian time [maximum $\beta(x)$ of 1.15; Fig. 10a–c]. In addition, it is difficult to reconcile the long delay (*c.* 30 Ma) between the Callovian extensional event and the onset of post-Valanginian thermal subsidence. Consequently, to model the

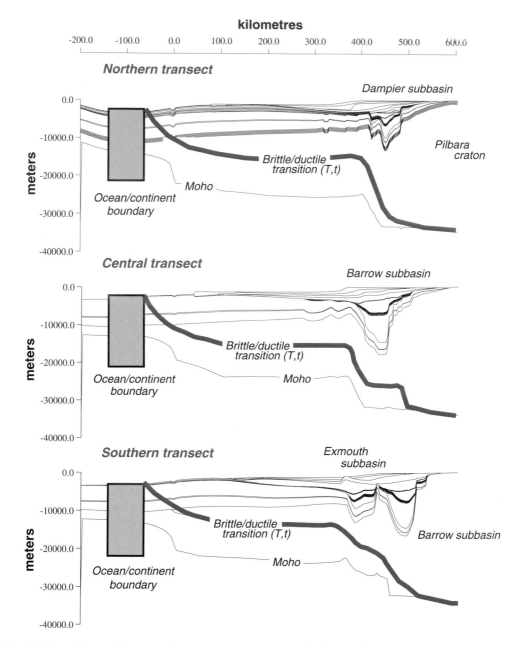

**Fig. 15.** General 'ramp–flat–ramp' detachment geometry responsible for partitioning extension between the upper and lower plates during Tithonian–Valanginian extension. The detachment breached, or shoaled close to, the surface of the crust near the continent–ocean boundary. The 'flat' component of the detachment occurs at mid-crustal depths across the plateau and ramps beneath the Pilbara Craton. The form of the post-Valanginian subsidence constrains the eastward dip of the detachment in this region. The geometry of the detachment beneath the Dampier Sub-basin, the central Barrow Sub-basin, and the southern Barrow and Exmouth Sub-basins is quite similar. However, there are some important differences. For example, the detachment beneath the Dampier Sub-basin is slightly deeper than for either the southern or central Barrow Sub-basins. As a result, the initial and total subsidences are less than observed in the central or southern Barrow Sub-basins. Because the depth of the detachment migrates throughout the history of the rifting, in response to the input of heat, the level of detachment might also be considered the brittle–ductile transition, but one whose depth varies as a function of temperature during the rifting process. Interpretation of seismic refraction data from across the Exmouth Plateau corroborates this interpretation of an intracrustal detachment, suggesting that lower crustal thinning occurred across the plateau with the boundary between the upper and lower crust occurring at *c*. 15 km depth.

distribution and magnitude of the post-Valanginian subsidence required significant lower crustal and mantle extension across the Exmouth Plateau with only minor upper crustal deformation. Such a depth-dependent distribution of extension implies the existence of an intracrustal detachment that effectively thinned the lower crust and lithospheric mantle, generating thermal-type subsidence across the region (Fig. 15). If $\beta(x)$ reflected only mantle thinning, then there would be no net subsidence after the rift-induced heat dissipated from the region because basin formation in an extensional setting is ultimately the consequence of crustal thinning. In contrast to the depth-independent extension during late Permian time, the Tithonian depth-dependent extension required that minor accommodation be created during the rifting event with large amounts of subsidence generated in the post-rift phase (Fig. 6). Because the lower crustal ductile extension and thermal uplift caused by lithospheric mantle thinning are of a comparable wavelength during the Tithonian–Valanginian rifting event, then the thermal uplift approximately balances the rift-induced subsidence, effectively delaying the generation of accommodation until after rifting.

The geometry of the detachment is tightly constrained from the observed stratal sediment facies and palaeowater-depth information from across the Exmouth Plateau. In particular, the basin modelling quantifies the amount of upper crustal and lower crustal–lithospheric mantle extension required to honour the stratal patterns, and the observed palaeowater-depth and sedimentary facies information; i.e. the distribution of extension defines the geometry of the detachment. The detachment is required to have a ramp–flat–ramp geometry so that it breached, or shoaled close to the Gascoyne continent–ocean boundary (Fig. 15). The eastward dip of the detachment is constrained by the eastward decrease in subsidence from the continent–ocean boundary to the Exmouth Arch (Figs 6a–c and 15). The magnitude and wavelength of this subsidence away from the Exmouth Arch cannot be explained by flexural coupling across the continent–ocean boundary (Karner *et al.* 1991). The 'flat' component of the detachment occurs at mid-crustal depths across the plateau and its depth is constrained by the magnitude of the post-Valanginian subsidence (Fig. 15). For example, if the depth of the detachment becomes shallower (Fig. 16), then the magnitude of the post-Valanginian subsidence would increase similarly to that occurring west of the Exmouth Arch (Fig. 15). Likewise, if the detachment was located deeper within the crust (Fig. 16), then

the subsidence would diminish. Figure 16 shows that the thickness of the Barrow Group, the degree of accommodation created across the plateau, the depth to the top of the Mungaroo Formation, and the relationship of the Alpha Arch summit with the Barrow Group and Palaeogene formations all help to constrain the geometry of the detachment. Towards the east, the detachment again ramps down, merging with Moho beneath the Australian continent (i.e. Pilbara Craton; Fig. 15). Its easterly vertical position is constrained by the distribution of Tithonian subsidence, with minimal subsidence occurring east of the ramp. Seismic refraction work across the northwest Australian margin corroborates our interpretation of an intracrustal detachment, suggesting that lower crustal thinning occurred across the Exmouth Plateau with the boundary between the upper and lower crust occurring at *c*. 15 km depth (Exon *et al.* 1992; Stagg & Colwell 1994; Driscoll & Karner 1998*b*).

The depth and shape of the detachment operative beneath the Dampier, central Barrow, and southern Barrow and Exmouth Sub-basins are rather similar (Fig. 15). However, there are some important differences. For example, the detachment beneath the Dampier Sub-basin is slightly deeper than for either the southern or central Barrow Sub basins (Fig. 15). As a result, the initial and total subsidences are less than observed in the central or southern Barrow Sub-basin cf. the Dampier Sub-basin (cf. Fig. 6b and c). Furthermore, because the subsidence caused by the detachment systematically decreases toward the east in the Dampier Sub-basin, then only one ramp was required, similar to the detachment geometry determined for the southern Barrow Sub-basin (Fig. 15). The thickness of the Barrow Group and overlying sediments is much greater in the southern portion of the Barrow Sub-basin because the depth of the detachment is shallower (Fig. 6c). In summary, the shape of the detachment and how it merges with the Moho is determined by the style, distribution and timing of accommodation generated across the margin. From our modelled crustal and lithospheric mantle extension factors, the late Permian, late Triassic–early Jurassic and Callovian extension events were all depth independent. In marked contrast, the extension event that progressed to break-up was characterized by depth-dependent extension. It is proposed here that lower crustal extension appears to be most dominant during the late stages of the rifting phase, just prior to continental break-up, because the upwelling of asthenospheric heat causes the lower crust to

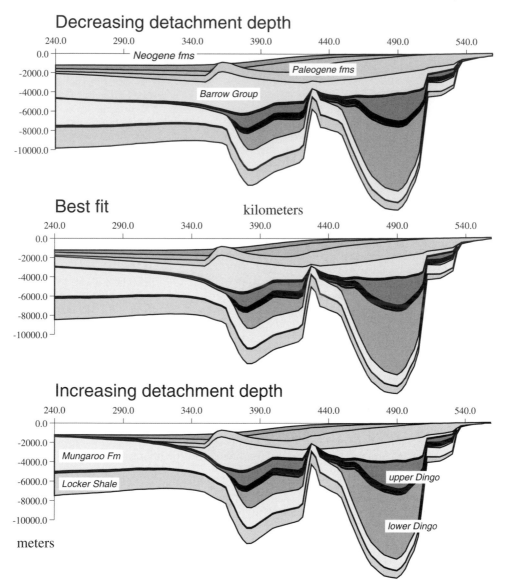

**Fig. 16.** Sensitivity of margin stratigraphy to detachment geometry. The detachment geometry is tightly constrained from the stratal relationships, sediment facies and palaeowater-depth information obtained from across the Exmouth Plateau. With reference to the southern transect, decreasing the detachment depth results in an unacceptable increase in post-Valanginian subsidence and total sediment thickness to the west of the Exmouth Sub-basin. Increasing detachment depth results in increased erosional truncation of the pre-Tithonian section and a general decrease in post-Valanginian subsidence. Although somewhat more subtle, the summit of the Alpha Arch and its relationship to the Barrow Group is also sensitive to detachment depth.

deform plastically. As break-up is approached, a diffuse zone separates the brittle and ductile deformation in the crust (i.e. a detachment) that shoals in the region of maximum heat input. The balancing brittle deformation is focused in a narrow region adjacent to the continent–ocean boundary and soles into the detachment. The deformed continental crust in this region is highly intruded and overprinted by volcanism associated with rift-induced decompression melting. Because the depth of the detachment migrates throughout the history of the rifting, in response to the input of heat, the level of detachment might also be considered the brittle–ductile transition, but one whose depth varies as a function of temperature during the rifting process (Driscoll & Karner 1998b).

An alternative hypothesis for the observed thermal subsidence is to have the northwest Australian margin situated over a hotspot during the earlier rift phases (late Triassic–early Jurassic and Callovian), and subsequently, during the Valanginian, have the region move off the hotspot, thereby allowing the thermally modified lithosphere to cool and subside. In effect, this process would allow thermal subsidence in the absence of crustal extension similar to the subsidence pattern observed at other margins influenced by documented hotspot activity (e.g. Karner et al. 1991; Driscoll et al. 1995). Although geochemical analyses of Indian Ocean crust sampled during ODP drilling Leg 123 in the Argo Abyssal Plain suggest that the crust reflects high degrees of partial melting associated with an enhanced thermal gradient below rifted continental lithosphere (Ludden & Dionne 1992), geologic evidence supporting the existence of a hotspot and, more importantly, the timing of its arrival along the northwest Australian margin, remains equivocal. For example, the emplacement of isolated volcanic constructs along the northwest Australian margin (e.g. Joey Rise and Platypus Spur; Fig. 1) appear insufficient to explain the distribution and timing of the observed widespread subsidence because their emplacement post-dates the onset of seafloor spreading. Furthermore, because the thermal input by a hotspot only delays the re-equilibration of previously extended lithosphere, this delayed subsidence needs to be consistent with the cumulative distribution and magnitude of earlier crustal extensional events. The small magnitude and restrictive distribution of the late Triassic and middle Jurassic crustal extension across the Exmouth Plateau cannot explain the large, regional post-Valanginian subsidence observed across the basin. The lack of a plume control on magma generation in the region is likewise strengthened by the fact that, following Tithonian–Valanginian break-up and formation of the Cuvier Basin, initial seafloor spreading resulted in the generation of thick oceanic crust indicative of relatively high asthenosphere temperatures, while concomitant spreading off the adjacent Exmouth Plateau formed normal thickness oceanic crust (Hopper et al. 1992). Crust of normal thickness was generated in the Cuvier Basin within 3–4 Ma after spreading began (Hopper et al. 1992). Therefore, the observed subsidence across the Exmouth Pla-

teau is best explained in terms of an intracrustal detachment.

Much uncertainty remains concerning the formation of seaward-dipping reflectors along passive continental margins and recent studies (e.g. Holbrook & Kelemen 1993; Keen et al. 1994; Kelemen & Holbrook 1993) question whether the large volume of igneous material observed along passive continental margins, as interpreted from high p-wave velocities, can be explained simply in terms of decompression melting of anomalously hot mantle during extension. Two end-member models have been proposed to explain the occurrence of large volcanic features associated with passive continental margins: (1) continental rifting associated with a hot mantle plume (e.g. White & McKenzie 1989) or with elevated mantle temperatures (Gurnis 1985); and (2) continental rifting with small-scale convection (e.g. Mutter et al. 1989; Anderson 1995). However, the amount of underplated material emplaced at the base of the Exmouth Plateau crust consistent with the modelled upper and lower plate extension factors can be calculated using the approach developed by Foucher et al. (1982) and McKenzie & Bickle (1988). This approach relates magma chemistry and volume to the degree of extension of the continental lithosphere. The physical basis of this theory involves the rise of the lithosphere–asthenosphere boundary and partial melt generation by decompression, as this boundary intersects the mantle solidus. Lithospheric extension generates only minor melt unless $\beta > 2$ and the potential temperature of the asthenospheric mantle ($T_p$) $> 1380°C$ (i.e. the mantle temperature at the surface; McKenzie & Bickle 1988). The maximum amount of magma generation occurs when both the crust and mantle are thinned to zero (i.e. $\beta \rightarrow \infty$). The calculated detachment geometry across the Exmouth Plateau accentuates the effective $\beta$ of the lithosphere so that, with even normal asthenosphere temperatures (i.e. 1333°C), the distribution and magnitude of the modelled Tithonian–Valanginian extension predicts an underplated layer thickness varying from 0.6 km across the central Exmouth Plateau to $> 7$ km towards the ocean–continent boundary (Fig. 17). These predictions are in general agreement with the expanding-spread seismic experiments of Mutter et al. (1989) and underplated thickness estimates from Stagg & Calwell (1994). The regional distribution of predicted tholeiitic melts generated by Tithonian–Valanginian rifting may help to explain the profusion of non-plume related tholeiitic sills and dykes dredged, drilled and imaged from across the

## *Southern transect*

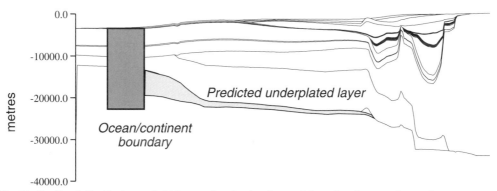

**Fig. 17.** Predicted distribution and thickness of underplated material emplaced across the southern transect during Tithonian–Valanginian rifting. Magma generation, driven by decompressive melting of asthenospheric material, is a function of the upper and lower plate extension factors and the geometry of the detachment. Underplated layer thicknesses vary from 0.6 km across the central Exmouth Plateau to >7 km towards the ocean–continent boundary, and are in general agreement with anomalous lower crustal velocities (7.2–7.4 km s$^{-1}$), presumed to be underplated magmatic material, observed across the northwestern and southwestern boundaries of the Exmouth Plateau. The regional distribution of predicted tholeiitic melts generated by Tithonian–Valanginian rifting may help to explain the profusion of non-plume related tholeiitic volcanics dredged, drilled and imaged from across the Exmouth Plateau.

Exmouth Plateau. Magmatic underplating at the base of the crust represents a negative density contrast with respect to the displaced mantle and will result in flexural uplift of the crust (McKenzie 1984). Thus, similar to the thermal effects of the detachment described above, magmatic underplating will also produce shallowing. However, in marked contrast to a thermal effect, underplating will not induce appreciable post-rift subsidence. The synrift uplift induced by magmatic underplating is simply counteracted by increased crustal extension (Fig. 17).

## Conclusions

The preserved sedimentary record, because it is a tape recorder of the vertical motions of the lithosphere, can be used to constrain and map the deformation history of the lithosphere. Based on quantitative basin modelling results for the geological development of the Exmouth Plateau, northwest Australia, the following conclusions are drawn.

• By combining a kinematic and flexural model for the deformation of the lithosphere with sequence stratigraphy, the distribution amplitude, depth-partitioning and interaction of the tectonic events responsible for the formation of the Exmouth Plateau, northwest Australia, have been determined. The northwest Australian margin was formed as a consequence of four major extension events and one inversion event: (1) a broadly distributed late Permian event; (2) a predominantly localized late Triassic–early Jurassic event, responsible for the formation of the Exmouth, Barrow and Dampier Sub-basins, and the deposition of the lower Dingo claystone; (3) a predominantly localized Callovian fault reactivation within the Exmouth, Barrow and Dampier Sub-basins that led to the deposition of the upper Dingo claystone; and (4) a major Tithonian–Valanginian event that created accommodation space for the Barrow Group but, much more importantly, generated large post-Valanginian regional subsidence. This pattern of subsidence indicates significant lower crustal and lithospheric mantle extension.

- Each tectonic event engendered a characteristic structural, stratigraphic and stratal response. The late Permian event is characterized by the deposition of the broadly distributed and monotonous sequences of the Locker Shales and the Mungaroo Formation. In contrast, the localized early Triassic and Callovian rifting events are characterized by regressive packages of thick shales (the upper and lower Dingo claystones). The stratal patterns associated with the upper and lower Dingo claystones show divergence and rotation of seismic reflectors, indicative of differential subsidence and block rotation. The onlap patterns that define the timing of the rift-onset unconformity record a flooding surface within the developing basin. Because the Locker Shales and the Mungaroo Formation are regionally developed with only minor lateral thickness variations, and the lower Dingo claystones are locally developed with large lateral thickness variations, the first package of seismic reflectors that diverge and onlap onto the underlying parallel, and approximately concordant, reflectors defines the boundary between the top of the Mungaroo Formation and the base of the lower Dingo Formation. Tithonian–Valanginian rifting was characterized by a very different stratigraphic and structural response compared with the earlier rift events. This final rift phase on the northwest Australian margin engendered the development of a major, regional, angular unconformity, that records minor truncation of the underlying Mesozoic stratigraphic successions and onlap of the overlying early Cretaceous sequences. This truncation is observed across most of the region, suggesting that large portions of the plateau were emergent, or at very shallow water depths, prior to the development of the overlying Barrow Delta System. Palaeo-water-depth estimates for the Valanginian, together with the present-day bathymetry and thickness of the late Cretaceous and Tertiary stratigraphic sequences, indicate that significant subsidence has occurred across the northern Exmouth Plateau since Valanginian time. The regional distribution and amplitude of the post-Valanginian subsidence is not consistent with the minor amounts of Tithonian–Valanginian brittle, upper crustal extension observed on the margin.

- Crucial in the development of the synrift depositional packages was the segmented nature of the border faults that delineated the sub-basins. In particular, along-strike alternation of the footwall block in one rift phase to becoming the hanging-wall block in another, played a crucial role in governing the basin architecture and where and when fluvial networks gained access to the sub-basins. It is recognized that the observed structural and stratigraphic features observed in the Lewis Trough of the central Barrow and Dampier Sub-basins cannot be reconciled in terms of either simple or complex extensional displacement between the hanging-wall and footwall blocks. The en echelon fault patterns, sag-type geometry of the Lewis Trough and the pre-rift stratigraphy, dipping subparallel or away from the fault trace, all attest to the role played by wrenching (induced by oblique extension), consistent with the difference in regional trends of both the central Barrow and Dampier Sub-basins relative to the spreading fabric within the Argo and/or Cuvier oceanic crust. This sag-type basin geometry cannot by achieved by regionally distributing the deformation over a large number of smaller faults. While the required rift basin geometry may be reproduced, the pre- and synrift stratal relationships are not.

- The style of deformation varied continually as a function of space and time. Initially, the late Permian deformation event was broadly distributed but depth independent. During late Triassic–middle Jurassic time, the deformation style was localized, remained depth independent and was accompanied by the formation of a series of rift sub-basins; i.e. the Exmouth, Barrow and Dampier Sub-basins. This localized phase of deformation was followed by a regional deformation event in the Tithonian–Valanginian, which was depth dependent and generated large post-Valanginian regional subsidence across the Exmouth Plateau, but with only minor accompanying brittle deformation and erosion. That is, the rift phase subsidence consisted of two components, an initial phase, associated with the localized, brittle deformation of the crust, which was replaced by a rapid, regional subsidence phase induced by the preferential plastic deformation of the lower crust with little involvement of the upper crust.

- To match the distribution and magnitude of the post-Valanginian 'thermal-type' subsidence requires significant lower crustal and mantle extension across the Exmouth Plateau. Such a distribution of extension implies the existence of an intracrustal detachment having a ramp–flat–ramp geometry that effectively thinned the lower crust and lithospheric mantle. The detachment geometry and how it merges with the Moho is mapped from the style, distribution and timing of accommodation generated across the margin. It is proposed that lower crustal extension appears to be most dominant during the late stages of the rifting phase, just prior to continental break-up, because the upwelling of asthenospheric heat causes the lower crust to deform plastically. As break-up is approached, a diffuse zone separates the brittle and ductile deformation in the crust (i.e. a detachment) that shoals in the region of maximum heat input. The balancing brittle deformation is focused in a narrow region adjacent to the continent–ocean boundary and soles into the detachment. The deformed continental crust in this region is highly intruded and overprinted by volcanism associated with rift-induced decompression melting. Because the depth of the detachment migrates throughout the history of the rifting in response to the input of heat, the level of detachment might also be considered the brittle–ductile transition, but one whose depth varies as a function of temperature during the rifting process.

- The generation of excessive tholeiitic magmatism during rifting has generally been explained by the decompressive melting of mantle with elevated temperatures, the high temperatures being a consequence of either plume activity or small-scale convection. In contrast, the calculated detachment geometry across the Exmouth Plateau accentuates the effective $\beta$ of the lithosphere so that with even normal asthenosphere temperatures, the distribution and magnitude of the modelled Tithonian–Valanginian extension predicts an underplaced layer thickness varying from 0.6 km across the central Exmouth Plateau to >7 km towards the ocean–continent boundary, in general agreement with expanding-spread seismic experiments and seismic refraction results. The predicted distribution of tholeiitic melts generated by Tithonian–Valanginian rifting may help explain the profusion of non-plume related tholeiitic sills and dykes dredged, drilled, and imaged from across the Exmouth Plateau. This style of rifting, extension partitioning of the lower crust and lithospheric mantle, supports the notion that excessive margin magmatism can occur within rift enviroments in the absence of a plume or elevated mantle.

We gratefully acknowledge discussions with N. Christie-Blick, J. Colwell, N. Exon, H. Stagg and P. Symonds concerning the behaviour of rifted lithosphere and the geological development of the Australian northwest shelf. We also thank the Australian Geological Survey Organization (AGSO), GECO-PRAKLA and Digicon for granting us access to seismic reflection data, and AGSO for releasing exploratory well, bathymetric and potential field data used in our basin analysis of the northwest Australian margin. The Wessel and Smith GMT software was used extensively in the construction of the figures shown in this paper. This work was supported by the Minerals & Energy Research Institute of Western Australia (MERIWA) grant M260, and by Ampolex, BHP Petroleum, MOBIL, Santos, Texaco, WAPET, Western Mining Corporation Petroleum and Woodside Offshore Petroleum. Additional funding was provided by the Palisades Geophysical Institute and the Office of Naval Research grant N00014-96-1-0026 and National Science Foundation grant OCE 98-00047. Lamont-Doherty Earth Observatory contribution #5867.

# References

ANDERSON, D. L. 1995. Lithosphere, asthenosphere, and perisphere. *Reviews of Geophysics*, **33**, 125–149.

AGSO NORTH WEST SHELF STUDY GROUP. 1994. Deep reflections on the North West Shelf: Changing perceptions of basin formation. *In*: PURCELL, P. G. & PURCELL, R. R (eds) *The Sedimentary Basins of Western Australia*. Proceedings of Petroleum Exploration Society of Australia Symposium, Perth, 63–76.

BARBER, P. M. 1982. Paleotectonic evolution and hydrocarbon genesis of the central Exmouth Plateau. *Australian Petroleum Exploration Association Journal*, **22**, 131–144.

——1994. Sequence stratigraphy and petroleum potential of upper Jurassic–lower Cretaceous depositional systems in the Dampier subbasin, northwest shelf, Australia. *In*: PURCELL, P. G. & PURCELL, R. R. (eds) *The Sedimentary Basins of Western Australia*. Proceedings of Petroleum Exploration Society of Australia Symposium, Perth, 525–542.

BARTON, P. & WOOD, R. 1984. Tectonic evolution of the North Sea basin: crustal stretching and subsidence. *Geophysical Journal of the Royal Astronomical Society*, **79**, 987–1022.

BAXTER, K. 1998. The role of small-scale extensional faulting in the evolution of basin geometries. An example from the late Palaeozoic Petrel Subbasin, northwest Australia. *Tectonophysics*, **287**, 21–41.

——, COOPER, G. T., O'BRIEN, G. W., HILL, K. C. & STURROCK, S. 1997. Flexural isostatic modelling as a constraint on basin evolution, the development of sediment systems, and paleo-heat flow: Application to the Vulcan Sub-basin, Timor Sea. *The Australian Petroleum Production and Exploration Association Journal*, **37**, 137–153.

BENTLEY, J. 1988. The Candace Terrace – A geological perspective. *In*: PURCELL, P. G. & PURCELL, R. R. (eds) *The North West Shelf, Australia*. Proceedings of the North West Shelf Symposium, Perth, 157–172.

BINT, A. N. & MARSHALL, N. G. 1994. High resolution palynostratigraphy of the Tithonian Angel Formation in the Wanaea and Cossack oil field. *In*: PURCELL, P. G. & PURCELL, R. R. (eds) *The Sedimentary Basins of Western Australia*. Proceedings of Petroleum Exploration Society of Australia Symposium, Perth, 543–556.

BOOTE, D. R. D. &, KIRK, R. B. 1989. Depositional wedge cycles on evolving plate margin, western and northwestern Australia. *AAPG Bulletin*, **73**, 216–243.

BOYD, R., WILLIAMSON, P. & HAQ, B. 1992. Seismic Stratigraphy and Passive Margin Evolution of the Southern Exmouth Plateau. *Proceedings of the Ocean Drilling Program, Scientific Results*, **122**, 39–59.

BRAUN, J. & BEAUMONT, C. 1989. A physical explanation of the relationship between flank uplifts and the breakup unconformity at rifted continental margins. *Geology*, **17**, 760–764.

CHRISTIE-BLICK, N. & DRISCOLL, N. W. 1995. Sequence stratigraphy. *Annual Review of Earth and Planetary Sciences*, **23**, 451–478.

COCKBAIN, A. E. 1982. The North West Shelf. *Australian Petroleum Exploration Association Journal*, **29**, 529–545.

COLWELL, J. B., SYMONDS, P. A. &, CRAWFORD, A. J. 1994. The nature of the Wallaby (Cuvier) Plateau and the other igneous provines of the west Australian margin. *AGSO Journal of Australian Geology and Geophysics*, **15**, 137–156.

COOK, A. C., SMYTH, M. & VOS, R. G. 1985. Source potential of Upper Triassic fluvo-deltaic systems of the Exmouth Plateau. *Australian Petroleum Exploration Association Journal*, **25**, 204–215.

COPP, I. A. 1994. *Depth to base Phanerozoic Map of Western Australia explanatory notes*. Geology Survey of Western Australia, Department of Minerals and Energy, Record 1994/9.

COWIE, P. A. & KARNER, G. D. 1990. Gravity effect of sediment compaction: Examples from the North Sea and the Rhine Graben. *Earth and Planetary Science Letters*, **99**, 141–153.

CRAWFORD, A. J. & VON RAD, U. 1994. The petrology, geochemistry and implications of basalts dredged from the Rowley Terrace–Scott Plateau and Exmouth Plateau margins, northwestern Australia. *AGSO Journal of Australian Geology and Geophysics*, **15**, 43–54.

DRISCOLL, N. W. & KARNER, G. D. 1998a. Lower crustal extension along the Northern Carnarvon basin, Australia: Evidence for an eastward dipping detachment. *Journal of Geophysical Research*, **103**, 4975–4992.

—— & ——1998b. The Upper Plate Paradox: The importance of lower crustal extension in conjugate margin development (abst). *EOS Transactions Am. Geophys. Union*, **79**, 5336.

——, HOGG, J. R., KARNER, G. D. & CHRISTIE-BLICK, N. 1995. Extensional tectonics in the Jeanne d'Arc basin: Implications for the timing of break-up between Grand Banks and Iberia. *In*: SCRUTTON, R. A., STOKER, M. S., SHIMMIELD, G. B. & TUDHOPE, A. W. (eds) *The Tectonics, Sedimentation and Palaeoceanography of the North Atlantic Region*. Geological Society, London, Special Publications, **90**, 1–28.

EBINGER, C. J, KARNER, G. D. & WEISSEL, J. K. 1991. Mechanical strength of extended continental lithosphere: Constraints from the Western rift system, East Africa. *Tectonics*, **10**, 1239–1256.

ERSKINE, R. & VAIL, P. R. 1988. Seismic stratigraphy of the Exmouth Plateau. *In*: BALLY, A. W. (ed.) *Atlas of Seismic Stratigraphy (Vol. 2)*. AAPG Studies in Geology, **27**, 163–173.

EXON, N. F. & BUFFLER, R. T. 1992. Mesozoic seismic stratigraphy and tectonic evolution of the western Exmouth Plateau. *In*: VON RAD, U., HAQ, B. U. *et al.* (eds) *Proceedings of the ODP, Scientific Results*, **122**, 61–81.

——, HAQ, B. U. & VON RAD, U. 1992. Exmouth Plateau revisited: scientific drilling and geological framework. *In*: VON RAD, U., HAQ, B. U. *et al.* (eds) *Proceedings of the ODP, Scientific Results*, **122**, 3–20.

——, VON RAD, U. & VON STACKELBERG, U. 1982. The geological development of the passive margins of the Exmouth Plateau off northwest Australia. *Marine Geology*, **47**, 131–152.

—— & ——1994. The Mesozoic and Cainozoic sequences of the northwest Australian margin, as revealed by ODP core drilling and related studies. *In*: PURCELL, P. G. & PURCELL, R. R. (eds) *The Sedimentary Basins of Western Australia*. Proceedings of the Petroleum Exploration Society of Australia Symposium, Perth, 181–199.

FOUCHER, J. P., LE PICHON, X. & SIBUET, J. C. 1982. The ocean–continent transition in the uniform lithospheric stretching model. Role of partial melting in the mantle. *Philosophical Transactions of the Royal Society, London, Series A*, **305**, 27–43.

FOWLER, S. & McKENZIE. D. P. 1989. Flexural studies of the Exmouth and Rockall Plateaux using SEASAT altimetry. *Basin Research*, **2**, 27–34.

FROSTICK, L. E. & REID, I. 1989. Is structure the main control of river drainage and sedimentation in rifts? *Journal of African Earth Sciences*, **8**, 165–182.

GURNIS, M. 1985. Large scale mantle convection and the aggregation and dispersal of supercontinents. *Nature*, **332**, 695–699.

HAQ, B. U., HARDENBOL, J. & VAIL, P. R. 1987. Chronology of fluctuating sea levels since the Triassic. *Science*, **235**, 1156–1167.

——, BOYD, R. L., EXON, N. F. & VON RAD, U. 1992. Evolution of the Central Exmouth Plateau: A Post-Drilling Perspective. *Proceedings of the ODP*, **122**, 801–816.

HOLBROOK, W. S. & KELEMEN, P. B. 1993. Large igneous province on the U.S. Atlantic margin and implications for magmatism during continental breakup. *Nature*, **364**, 433–436.

HOPPER, J. R., MUTTER, J. C., LARSON, R. L., MUTTER, C. Z. & THE NORTHWEST AUSTRALIA STUDY GROUP. 1992. Magmatism and rift evolution: Evidence from northwest Australia. *Geology*, **20**, 853–857.

KARNER, G. D., DRISCOLL, N. W. & PEIRCE, J. 1991. Gravity and magnetic signature of Broken Ridge, Southeast Indian Ocean. *In*: WEISSEL, J. K., PEIRCE, J., TAYLOR, E. & ALT, J. (eds) *Proceedings of the ODP, Scientific Results*, **121**, 681–686.

——, —— & WEISSEL, J. K. 1993. Response of the lithosphere to in-plane force variations. *Earth and Planetary Science Letters*, **114**, 397–416.

——, EGAN, S. E. & WEISSEL, J. K. 1992. Modeling the tectonic development of the Tucano and Sergipe–Alagoas rift basins, Brazil. *In*: ZIEGLER, P. A. (ed.) *Geodynamics of Rifting, Volume III. Thematic Discussions*. Tectonophysics, **215**, 133–160.

——, DRISCOLL, N. W. McGINNIS, J. P., BRUMBAUGH, W. D. & CAMERON, N. 1997. Tectonic significance of syn-rift sedimentary packages across the Gabon–Cabinda continental margin. *Marine and Petroleum Geology*, **14**, 973–1000.

KEEN, C. E. & DEHLER, S. A. 1997. Extensional styles and gravity anomalies at rifted continental margins: Some North Atlantic examples. *Tectonics*, **16**, 744–754.

——, COURTNEY, R. C., DEHLER, S. A. & WILLIAMSON, M. C. 1994. Decompression melting at rifted margins: comparison of model predictions with the distribution of igneous rocks on the eastern Canadian margin. *Earth and Planetary Science Letters*, **121**, 403–416.

KELEMEN, P. B. & HOLBROOK, W. S. 1993. Origin of thick, high-velocity igneous crust along the U.S. east coast margin. *Journal of Geophysical Research*, **100**, 10 077–10 094.

KIRK, R. B. 1985. A seismic stratigraphy case history in the Eastern Barrow Sub-basin, North West Shelf, Australia. *In*; BERG, O. R. & WOLVERTON, D. G. (eds) *Seismic Stratigraphy II: An Integrated Approach to Hydrocarbon Exploration*. AAPG Memoir, **39**, 183–207.

KOOI, H. S., CLOETINGH, S. & BURRUS, J. 1992. Lithospheric necking and regional isostasy at extensional basins, 1. Subsidence and gravity modelling with an application to the Gulf of Lions margin (SE France). *Journal of Geophysical Research*, **97**, 17 553–17 571.

KOPSEN, E., & McGANN, G. 1985. A review of the hydrocarbon habitat of the Eastern and Central Barrow–Dampier Sub-Basin, Western Australia. *Australian Petroleum Exploration Association Journal*, **25**, 154–176.

KUSZNIR, N. J., MARSDEN, G. & EGAN, S. S. 1991. A flexural cantilever simple-shear/pure-shear model of continental lithosphere extension: Application to the Jeanne d'Arc basin and Viking Graben. *In*: ROBERTS, A. M., YIELDING, G. & FREEMAN, B. (eds.) *The Geometry of Normal Faults*. Geological Society, London, Special Publications, **56**, 41–60.

LEEDER, M. & GAWTHORPE, R. 1987. Sedimentary models for extensional tilt-block/half-graben basins. *In*: COWARD, M. P., DEWEY, J. F. & HANCOCK, P. L. (eds) *Continental Extensional Tectonics*. Geological Society, London, Special Publications, **28**, 139–152.

LISTER, G. S. & DAVIS, G. A. 1989. The origin of metamorphic complexes and detachment faults formed during Tertiary continental extensional in the northern Colorado River region, USA. *Journal of Structural Geology*, **11**, 65–94.

LUDDEN, J. N. & DIONNE, J. N. 1992. The geochemistry of oceanic crust at the onset of rifting in the Indian ocean. *In*: GRADSTEIN, F. M., LUDDEN, J. N. et al. (eds) *Proceedings of ODP, Scientific Results*, **123**, 791–799.

McCLAY, K. R. & WHITE, M. J. 1995. Analogue models of orthogonal and oblique rifting. *Marine Petroleum Geology*, **12**, 137–151.

McKENZIE, D. P. 1984. A possible mechanism for epeirogenic uplift. *Nature*, **307**, 616–618.

—— & BICKLE, M. J. 1988. The volume and composition of melt generated by extension of the lithosphere. *Journal of Petrology*, **29**, 625–679.

MARSDEN, G., YIELDING, G., ROBERTS, A. M. & KUSZNIR, N. J. 1990. Application of a flexural cantilever simple-shear/pure-shear model of continental lithosphere extension to the formation of the North sea. *In*: BLUNDELL, D. J. & GIBBS, A. D. (eds) *Tectonic Evolution of the North Sea Rifts*. Oxford University Press.

MUTTER, J. C., LARSON, R. L. & NORTHWEST AUSTRALIA STUDY GROUP. 1989. Extension of the Exmouth Plateau, offshore northwestern Australia: Deep seismic reflection/ refraction evidence for simple and pure shear mechanisms. *Geology*, **17**, 15–18.

——, TALWANI, M. & STOFFA, P. 1982. Origin of seaward-dipping reflectors in oceanic crust off the Norwegian margin by 'subaerial seafloor spreading'. *Geology*, **10**, 353–357.

ROBERTS, A. M., YIELDING, G., KUSZNIR, N. J. WALKER, I. & DORN-LOPEZ, D. 1993. Quantitative analysis of Triassic extension in the Northern Viking Graben. *Journal of the Geological Society, London*, **152**, 15–26.

ROSS, M. I. & VAIL, P. R. 1994. Sequence stratigraphy of the lower Neocomian Barrow Delta, Exmouth Plateau, Northwest Australia. *In*: PURCELL, P. G. & PURCELL, R. R. (eds) *The Sedimentary Basins of Western Australia*. Proceedings of the Petroleum Exploration Society of Australia Symposium, Perth, 436–447.

SAGER, W. W., FULLERTON, L. G., BUFFLER, R. T. & HANDSCHUMACHER, D. W. 1992. Argo Abyssal Plain magnetic lineations revisited: Implications for the onset of seafloor spreading and tectonic evolution of the Eastern Indian Ocean. *In*: GRADSTEIN, F. M., LUDDEN, J. N. et al. *Proceedings of the Ocean Drilling Program*. Scientific Results, Vol. **123**, 659–669.

SANDWELL, D. T. & SMITH, W. H. F. 1992. Global marine gravity from ERS-1, Geosat and Seasat reveals new tectonic fabric. *EOS*, **73**, 133.

SCHUMM, S. A. 1993. River response to baselevel change: Implications for sequence stratigraphy. *Journal of Geology*, **101**, 279–294.

STAGG, H. M. J. & COLWELL, J. B. 1994. The structural foundations of the Northern Carnarvon basin. *In*: PURCELL, P. G. & PURCELL, R. R. (eds) *The Sedimentary Basins of Western Australia*. Proceedings of the Petroleum Exploration Society of Australia Symposium, Perth, 349–354.

——, WILLCOX, J. B. & NEEDHAM, D. J. L. 1989. Werner deconvolution of magnetic data: Theoretical models and application to the Great Australian Bight. *Australian Journal of Earth Sciences*, **36**, 109–122.

STEIN, A. 1994. Rankin platform, Western Australia: Structural development and exploration potential. *In*: PURCELL, P. G. & PURCELL, R. R. (eds) *The Sedimentary Basins of Western Australia*. Proceedings of Petroleum Exploration Society of Australia Symposium, Perth, 509–523.

SYMONDS, P. A., COLLINS, C. D. N. & BRADSHAW, J. 1994. Deep structure of the Browse basin: Implications for basin development and petroleum exploration. *In*: PURCELL, P. G. & PURCELL, R. R. (eds) *The Sedimentary Basins of Western Australia*. Proceedings of the Petroleum Exploration Society of Australia Symposium, Perth, 315–331.

——, PLANKE, S., FREY, O. & SKOGSEID, J. 1998. Volcanic evolution of the Western Australian continental margin and its implications for basin development. *In*: PURCELL, P. G. & PURCELL, R. R. (eds) *The Sedimentary Basins of Western Australia 2*. Proceedings of Petroleum Exploration Society of Australia Symposium, Perth, 33–54.

TAIT, A. M. 1985. A depositional model for the Dupuy Member and the Barrow Group in the Barrow Sub-Basin, Northwestern Australia. *Australian Petroleum Exploration Association Journal*, **25**, 282–290.

TRON, V. & BRUN, J. P. 1991. Experiments on oblique rifting in brittle–ductile systems. *Tectonophysics*, **188**, 71–84.

VEENSTRA, E. 1985. Rift and drift in the Dampier Sub-Basin, a seismic and structural interpretation. *Australian Petroleum Exploration Association Journal*, **25**, 177–189.

VON RAD, U. & EXON, N. F. 1983. Mesozoic–Cenozoic sedimentary and volcanic evolution of the starved passive continental margin off northwest Australia. *In*: WATKINS, J. S. & DRAKE, C. L. (eds) *Studies in Continental Margin Geology*. *AAPG Memoir*, **34**, 253–281.

——, EXON, N. F. & HAQ, B. U. 1992. Rift-to-drift history of the Wombat Plateau, Northwest Australia: Triassic to Tertiary Leg 122 results. *In*: VON RAD, U. & HAQ, B. U. *et al. Proceedings of the Ocean Drilling Program*, Scientific Results, Vol. **122**, 765–800

WARRIS, B. J. 1993. The hydrocarbon potential of the Palaeozoic basins of Western Australia. *Australian Petroleum Exploration Association Journal*, **33**, 123–137.

WATTS, A. B. 1988. Gravity anomalies, crustal structure and flexure of the lithosphere at the Baltimore Canyon Trough. *Earth and Planetary Science Letters*, **89**, 221–238.

—— & MARR, C. 1995. Gravity anomalies and the thermal and mechanical structure of rifted continental margins. *In*: BANDA, E., TALWANI, M. & TORNÈ, M (eds) *Rifted Ocean–Continent Boundaries*. Kluwer, 65–94.

—— & FAIRHEAD, J. D. 1997. Gravity anomalies and magmatism along the western continental margin of the British Isles. *Journal of the Geological Society, London*, **154**, 523–529.

—— & STEWART, J. 1998. Gravity anomalies and segmentation of the continental margin offshore west Africa. *Earth and Planetary Science Letters*, **156**, 239–252.

WEISSEL, J. K. & KARNER, G. D. 1989. Flexural uplift of rift flanks due to mechanical unloading of the lithosphere during extension. *Journal of Geophysical Research*, **94**, 13 919–13 950.

—— & SMITH, W. H. F. 1995. New version of the generic mapping tools released. *EOS*, **76**, 329.

WERNICKE, B. 1981. Low-angle normal faulting in the Basin and Range Province: Nappe tectonics in an extending orogen. *Nature*, **291**, 645–648.

——1985. Uniform-sense normal simple shear of the continental lithosphere. *Canadian Journal of Earth Sciences*, **22**, 108–125.

—— & BURCHFIEL, B. C. 1982. Modes of extensional tectonics. *Journal of Structural Geology*, **4**, 105–115.

WHITE, R. S. & MCKENZIE, D. P. 1989. Magmatism at rift zones: the generation of volcanic continental margins and flood basalts. *Journal of Geophysical Research*, **94**, 7685–7729.

WITHJACK, M. O. & JAMISON, W. R. 1986. Deformation produced by oblique rifting. *Tectonophysics*, **126**, 99–124.

WILLCOX, J. B. & EXON, N. F. 1976. The regional geology of the Exmouth Plateau. *Australian Petroleum Exploration Association Journal*, **16**, 1–11.

WILLIAMS, A. F. & POYNTON, D. J. 1985. The geology and evolution of the South Pepper Hydrocarbon Accumulation. *Australian Petroleum Exploration Association Journal*, **25**, 235–247.

YEATES, A. N., BRADSHAW, M. T., DICKENS, J. M. *et al.* 1987. The Westralian Superbasin, an Australian link with Tethys. *In*: MCKENZIE, K. G. (ed.) *Shallow Tethys 2*. Proceedings of an International Symposium, Shallow Tethys, Wagga Wagga. Balkema, 199–213.

# Structural and magmatic segmentation of the Tertiary East Greenland Volcanic Rifted Margin

JEFFREY A. KARSON[1] & C. KENT BROOKS[2]

[1] Division of Earth & Ocean Sciences, Duke University, Durham, NC 27708-0230, USA
(email: jkarson@eos.duke.edu)
[2] Danish Lithosphere Center, 10 Øster Voldgade, 1350 Copenhagen K, Denmark

**Abstract:** The Tertiary East Greenland Volcanic Rifted Margin is characterized by massive magmatic construction that produced a distinctive crustal architecture including: (1) a thick pile of flood basalts continuing offshore as seismically imaged 'seaward-dipping reflector sequences'; (2) an extensive margin-parallel mafic dyke swarm; and (3) shallow crustal gabbroic plutons and deeper crustal 'underplated' material. These igneous units developed in the framework of an asymmetrical, crustal-scale fold, or 'flexure', that accommodated major subsidence along the continent–ocean transition. Extensive exposures along the margin reveal that the flexure and associated igneous structures define rift segments separated by various types of structural discontinuities. First-order segments occur between major triple-rift junctions as at Kangerlussuaq. At an intermediate scale, second-order accommodation zones bound margin segments *c.* 100 km in length with long-lived structural and/or magmatic expressions. Third-order discontinuities, spaced at tens of kilometres, correspond to smaller accommodation zones at abrupt along-strike changes fault or magmatic structures. Outcrop-scale transfer and transform faults occur at still smaller scales. Some of the larger accommodation zones appear to be related to pre-existing Precambrian structures and may have helped localize relatively late, post-flexure alkalic intrusions. The style of segmentation provides a link between similar segmentation patterns in continental rifts and mid-ocean ridge spreading centres that persist long after continental separation.

Volcanic rifted margins (VRMs) are generally considered to form where continental rifting occurs near a mantle hotspot (White & McKenzie 1989, 1995; Coffin & Eldholm 1994). Accordingly, VRM are characterized by a distinctive style of crustal structure created by the interplay of voluminous magmatic construction and mechanical extension. Typically, they have relatively abrupt continent–ocean transition zones, commonly only a few tens of kilometres wide where rifted continental crust cut by mafic intrusions and underplated by mafic igneous material (Kelemen & Holbrook 1995; Lizarralde & Holbrook 1997) passes laterally into 'thick oceanic crust', in which the thicknesses of both the volcanic and plutonic units are larger than those of typical oceanic crust (Mutter *et al.* 1982; Bott 1987; White *et al.* 1987; Eldholm & Grue 1994).

Although most VRMs have submerged below sea level, or been buried beneath thick lava sequences, subaerial exposures of continent–ocean transitions have been described in several locations worldwide (DuToit 1929; Wager 1934; Nielsen 1975, 1978; Brooks & Nielsen 1982*a, b*; Bohannon 1986; Geoffrey *et al.* 1998). The southern East Greenland coast provides one of the world's best exposed and most laterally continuous windows into rock units and structures near the continent–ocean transition of a VRM (Nielsen 1975, 1978; Myers 1980; Brooks & Nielsen 1982*a, b*; Larsen & Jakobsdóttir 1988; Larsen & Saunders 1999). The extensive exposures reveal not only important aspects of the continent–ocean transition, but also significant along-strike variations in structural style that are emphasized in this paper. In a previous study, Myers (1980) emphasized the along-strike variations in magmatic construction and mapped an en echelon array of overlapping and partially nested dyke swarms and related mafic plutonic centres. More recently, detailed statistical analyses of dyke-swarm profiles along the margin have refined this view (Bromann *et al.* 1996; Bromann 1999). Faulting along the margin has been described previously mainly in the Kangerlussuaq area (Nielsen 1975, 1978; Nielsen & Brooks 1981).

In this paper, the along-strike distribution of large-scale magmatic features are integrated with major structural features to define various scales of segmentation that are likely to be fundamental aspects of the architecture VRMs

*From*: MAC NIOCAILL, C. & RYAN, P. D. (eds) 1999. *Continental Tectonics*. Geological Society, London, Special Publications, **164**, 313–338. 1-86239-051-7/99/$15.00 © The Geological Society of London 1999.

worldwide. After a brief review of the regional tectonic setting, follows a description of how different crustal structures, corresponding to different crustal levels, define segments of the coastal flexure. The nature of various types of discontinuities that separate these segments, and their implications for the transition between continental rifting and seafloor spreading, are then described.

## The Tertiary East Greenland VRM

The East Greenland VRM developed as the northeast branch of the Atlantic opened between Greenland and Eurasia c. 55 Ma ago (Fig. 1). Coastal exposures of the East Greenland margin were first described by Wager (1934), but subsequent onshore and offshore investigations have greatly refined the understanding of the general structure and composition of this classic example of a VRM [Brooks & Nielsen (1982a, b), Nielsen (1975, 1978) and Larsen & Saunders (1999 and refs cited therein) provide a comprehensive review]. Large-scale structures on the East Greenland side of the margin appear to be more or less symmetrical with those of the conjugate Norwegian and British margins (Talwani & Eldholm 1977; Hinz et al. 1987; Eldholm &

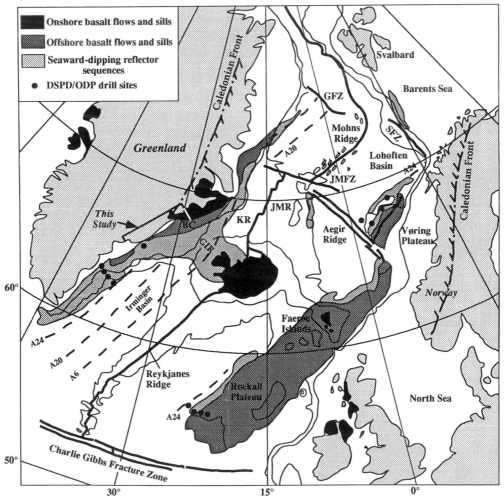

**Fig. 1.** Generalized tectonic map of the volcanic rifted margins of the northeast Atlantic Region and location of onland and offshore flood basalts, including areas underlain by seaward-dipping reflector sequences. BC, Blosseville Coast; GFZ, Greenland Fracture Zone; GIR, Greenland–Iceland Ridge; JMFZ, Jan Mayen Fracture Zone; JMR, Jan Mayan Ridge; KR, Kolbeinsey Ridge, Magnetic Anomalies A6, A20, A24. Box shows location of study area.

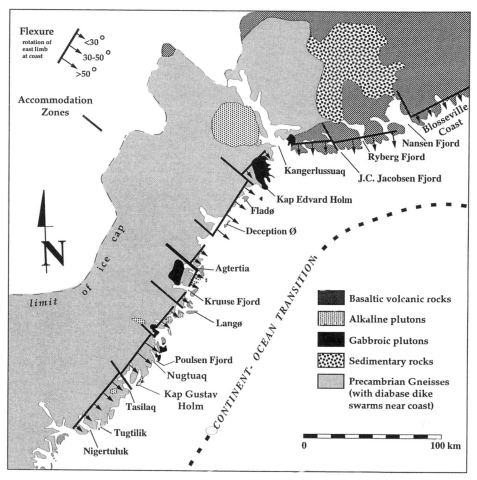

**Fig. 2.** Highly generalized geological map of the southern East Greenland VRM showing the hinge line of the coastal flexure which, together with dyke swarms and fault structures, defines rift segments that are tens of kilometres in length. The segments are separated by northwest trending accommodation zones. Kangerlussuaq is the failed arm of a triple-rift junction that formed on a domal uplift.

Grue 1994). Therefore, the onshore structures of southern East Greenland may provide information on geological structures that apply broadly to the other VRMs around the Northeast Atlantic, and perhaps to VRMs worldwide.

As at most VRMs, the transition from continental to oceanic crust is relatively narrow along the East Greenland Margin. Typically, it is only a few tens of kilometres wide (Larsen 1978; Larsen & Jakobsdóttir 1988; Larsen & Saunders 1999). Magnetic anomalies and anomalously thick oceanic crust created near the Iceland hotspot lie only a few kilometres offshore along much of the margin. Therefore, exposures along the southern East Greenland coast represent the rifted basement and the

landward portion of the continent–ocean transition zone.

The remarkable glaciated exposures along the central and southern East Greenland coast provide windows into the internal crustal structure of this margin for at least 1200 km (Fig. 1). The northern part of this exposure, conjugate to the Vøring margin, cuts north–south across Caledonian basement and Palaeozoic basins (Surlyk 1977; Larsen & Marcussen 1992; Price et al. 1997). To the south, exposures along the Blosseville Coast are dominated by flood basalts (Pedersen et al. 1997), representing the onshore 'feather-edge' of the basalts that comprise the seaward-dipping reflector sequence (SDRS) imaged offshore along much of the margin

(Fig. 1; Larsen & Jakobsdóttir 1988; Larsen & Saunders 1999). Just to the southwest, the coast curves to an east–west trend at a triple-rift junction at Kangerlussuaq before continuing to the southwest where rifted Precambrian continental basement is cut by numerous mafic intrusions. Along the East Greenland coast to the south, mafic intrusions become progressively less common over *c.* 300 km and the continent–ocean transition zone appears to be further offshore. This segment of the margin terminates at an abandoned, offshore triple-rift junction south of Greenland (Kristoffersen & Talwani 1977).

In this paper, the focus is on structures which developed along the East Greenland VRM southwest of Kangerlussuaq (Fig. 2). Along this part of the margin, rift-related faulting and magmatic structures are spectacularly exposed for *c.* 250 km, owing to uplift and deep dissection (Wager 1934; Brooks & Nielsen 1982*a*; Brooks *et al.* 1996). Exposures are limited to an across-strike width which is typically <50 km from the coast to the ice-cap. However, this is sufficiently wide to expose the landward limit of rifted crust and a portion of the continent–ocean transition zone in most areas (Fig. 2; Nielsen 1975; Larsen 1978; Brooks & Nielsen 1982*b*; Hinz *et al.* 1987; Larsen & Jakobsdóttir 1988). A regional-scale dome-like uplift centred on the Kangerlussuaq triple-rift junction is a classic example of a hot-spot-related uplift and failed arm along the rifted margin (Brooks 1973, 1979). Reconstruction of the plateau basalt pile (Pedersen *et al.* 1997), sedimentary units (Clift *et al.* 1995), fission-track thermochronology (Gleadow & Brooks 1979; Hansen 1996) and other considerations, suggest that present exposures have been exhumed from crustal levels of less than a few kilometres (Brooks 1979; Brooks & Nielsen 1982*a*).

Rifting in the Kangerlussuaq region began *c.* 100 Ma ago when basement-derived clastic sedimentary deposits and marine shales were deposited in a series of fault-bounded basins on high-grade Precambrian basement northeast of Kangerlussuaq (Higgins & Soper 1981; Nielsen *et al.* 1981; Brooks & Nielsen 1982*a*). An angular unconformity between these and overlying Tertiary units (Curewitz, 1999; Dam *et al.* 1998) may be part of a much broader unconformity recognized around the Northeast Atlantic that is related to uplift driven by the Iceland hotspot (Dam *et al.* 1998; Larsen & Saunders 1999).

By Palaeocene time, voluminous shallow-water to subaerial basaltic to intermediate lavas and pyroclastic units >2 km thick (Lower Series) covered much of the area. These pass upward into monotonous plateau lavas (Middle and Upper Series) which reach thicknesses of >7 km onshore to the northeast (Pedersen *et al.* 1997) and at least 10 km thick in a well-developed SDRS offshore (Larsen & Jakobsdóttir 1988; Larsen & Saunders 1999). This SDRS has been studied extensively in offshore areas along-strike to the southeast (Larsen *et al.* 1994; Duncan *et al.* 1996; Larsen & Saunders 1999). The area southwest of Kangerlussuaq is considered to represent the upper crustal levels of the continent–ocean transition that developed beneath the feather-edge of the seaward-thickening lava sequence exposed along the Blosseville Coast to the northeast and in the SDRS to the southeast (Fig. 1).

## Major structures of the continent–ocean transition

The growth of the seaward-thickening wedge of lavas that pass laterally into the SDRS offshore was accommodated by the subsidence and flexure of the edge of the rifted Precambrian basement and part of the Tertiary igneous assemblage (Fig. 3). The hinge line of this flexure (defined here by the landward limit of tilting) is more or less parallel to the coastline, and located in the belt of onshore exposures along this part of the margin. The geometry of the SDRS show that the hinge line migrated progressively seaward during the evolution of the margin. The geometry and kinematics of the flexure and related structures provide important clues to the early development of major structural and magmatic features of the margin. Previous studies of the coastal flexure have focused mainly on local cross-sectional structures (Wager & Deer 1938; Faller & Soper 1979; Myers 1980; Nielsen & Brooks 1981). Here, the nature of along-strike variations in the flexure and related features are examined.

Onshore, from Kangerlussuaq southwest for at least 250 km (Fig. 2), granulite to amphibolite facies Precambrian basement gneisses (Bridge-water *et al.* 1976) are intruded by laterally persistent swarms of basaltic dykes and irregularly spaced gabbroic plutons occurring at intervals of tens of kilometres (Wager & Deer 1938; Myers 1980; Nielsen 1978; Brooks & Nielsen 1982*a*; Bromann *et al.* 1996; Tegner *et al.* 1998). Only a few exposures of pre-rift sedimentary rocks and lavas remain in this area, but they provide important constraints on the development of the margin. Specifically, the tilted sedimentary rocks and lavas, and deformed and rotated dykes and plutons, define the geometry of the coastal flexure, the dominant structural feature

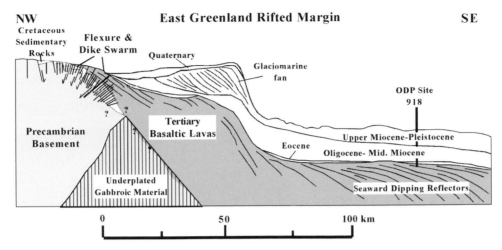

**Fig. 3** Generalized cross section of the East Greenland VRM. Note that the flexure is defined by different structures at different crustal levels.

of the margin (Fig. 3). Three major faulting events can be identified on the basis of their relationships to the flexure (Karson et al. 1996). (1) Pre-flexure faults appear to be limited to those bounding the half-graben sedimentary basins northeast of Kangerlussuaq. Major fault zones landward of the flexure also occur in areas where dykes are sparse, or absent, and lack cross-cutting relations that constrain their ages. These may also predate the flexure. (2) Intense faulting, distributed cataclastic deformation and pseudotachylyte generation attended the development of the flexure (Karson et al. 1998). Numerous normal faults are kinematically linked to minor transfer and transform faults, as well as major strike-slip to oblique-slip faults of accommodation zones that separate discrete rift segments. (3) Post-flexure faults that offset the rock units that were previously tilted in the flexure are the most obvious, but are relatively widely spaced (kilometres to tens of kilometres) and include both landward- and seaward-dipping normal faults.

The magmatic units of the margin were intruded before, during and after the development of the flexure. Cross-cutting relationships provide a relative chronology of magmatism and tectonic deformation. A growing number of radiometric dates on the magmatic units helps to place this chronology in the broader context of events along the entire VRM.

Available radiometric dates for magmatic units along the margin indicate that the development of the lava pile, dyke swarms and mafic to silicic plutons was complex in both time and space [see Saunders et al. (1997), Larsen &

Saunders (1999) and Tegner et al. (1998) for recent summaries]. The oldest dykes and lavas are dated at c. 62 Ma (Storey et al. 1996); however, the most voluminous flood basalts apparently were not erupted until 57–55 Ma, at the time of continental break-up (Larsen & Saunders 1999) which is generally considered to have occurred at chron 24r (c. 55 Ma; Saunders et al. 1997; Larsen & Saunders 1999). Several gabbroic plutons and at least part of the dyke swarm were not intruded until after c. 50 Ma (Tegner et al. 1998). Alkalic dykes and plutons were intruded along the margin at c. 35 Ma (Myers 1980; Brooks & Nielsen 1982a, b).

The oldest lavas, dykes and gabbroic plutons that were intruded before c.55 Ma are intensely deformed and appear to have predated the flexure. Plutons and dykes intruded at c. 55 Ma are variably deformed and appear to have been intruded during the development of the flexure, at about the time of continental break-up. Swarms of dykes that have not been dated and plutons intruded at <50 Ma show almost no signs of deformation or tilting, and therefore appear to postdate the flexure. Thus, the flexure is bracketed by available dates to an interval of c. 5 Ma at c. 59–54 Ma.

## Internal structure of VRM rift segments

The general structural expression of the margin described above varies significantly along the margin south of Kangerlussuaq (Fig. 2). In addition to major triple-rift junctions at Kangerlussuaq (Brooks 1973) and at the southern tip of

Greenland (Kristoffersen & Talwani 1977), many smaller discontinuities are also evident. They are expressed as abrupt changes in geologic structure between adjacent rift segments. The individual segments have roughly consistent internal structures with distinctive styles of magmatic construction and tectonic deformation.

The major structural and magmatic features that occur in the rift segments along this part of the East Greenland VRM, approximately from north to south, corresponding to progressively deeper crustal levels, are described below. The along-strike variations that occur in this region, and their implications for the large-scale architecture of the VRM, are emphasized. Firstly, a brief description of the rock units that define continuous segments of the margin and the flexure is given. This is followed by a description of the various types of discontinuities that separate these segments and their relations to older and younger structures. These features are then considered in the light of other studies of rift segmentation and continental margin evolution.

## Major rock units

*Upper and Middle Series lavas.* Perhaps the most obvious manifestation of the coastal flexure is the seaward-increasing dip of lava flows. Middle–Upper Series lavas are extensively exposed along the Blosseville Coast north of Nansen Fjord (Fig. 2), where they reach a total stratigraphic thickness of at least 7 km (Pedersen *et al.* 1997). Broad dip domains, tens of kilometres across, define major fault-bounded blocks but, except near the coast, dips seldom exceed 10°. At the coast, the dip increases to as much as 20°. Zeolite facies horizons, considered to represent burial metamorphism and hence approximately palaeohorizontal surfaces, also dip gently seaward, suggesting that they predate the flexure (Neuhoff *et al.* 1997).

The hinge line of the flexure in this area is roughly linear and parallel to the coast, i.e. east-northeast (Fig. 2). Systematic seaward-dipping lava flows, defining the hinge line and the eastern limb of the flexure, are limited to a width of *c.* 10 km along the coast. Most of the eastern limb is submerged and is probably continuous with SDRS offshore. Flexure in this

area may have been partially accommodated by faulting, but no major along-strike discontinuities are apparent (Pedersen *et al.* 1997). The relatively young Middle and Upper Series lavas in this area may have buried an older, much more complicated, margin structure.

*Lower Series Lavas and sedimentary rocks.* To the south, from Nansen Fjord to Kap Edvard Holm (Fig. 2), the Lower Series lavas include interlayered volcaniclastic units. This assemblage overlies basement and locally developed Cretaceous clastic sedimentary basins (Fig. 4b and c; Soper *et al.* 1976; Nielsen & Brooks 1981; Dam *et al.* 1998; Curewitz 1992). The volcaniclastics and older sedimentary rocks are intruded by voluminous mafic sills (Fig. 4d). Radiometric dates indicate that the sill complex north of Kangerlussuaq is *c.* 55 Ma (Tegner *et al.* 1998). In this region, all of these units are affected by the flexure within *c.* 10 km of the coast. North of Kangerlussuaq, bedding and Lower Series lava flows dip as steeply as 55° at the coast. These define a nearly east–west hinge line. To the south of Kangerlussuaq, at Kap Edvard Holm, lavas, volcaniclastic sediments and sills dip *c.* 20°SE, defining a northeast trending hinge line. The strongly tilted and densely intruded Lower Series lavas and rift-related sedimentary rocks near Kangerlussuaq contrast with the Middle to Upper Series lavas to the north and show that the flexure was well developed before the eruption of these somewhat younger units.

Farther to the south, lavas and associated sedimentary rocks crop out only in a few small exposures (Fig. 5), some of which are highly deformed and affected by local metamorphism near Tertiary plutons. Bedding typically dips moderately to the southeast. Local exposures at Dødemandspynten, just south of Agtertia Fjord, are highly deformed and have folded and boudinaged dykes. Bedding presently dips gently seaward. North of Poulsen Fjord, and at Kap Gustav Holm, sedimentary rocks dip *c.* 50° seaward (Myers *et al.* 1993). At Nugtuaq, south of Poulsen Fjord, they steepen to subvertical dips (Fig. 5).

*Lava–basement non-conformity.* At the base of the Tertiary lava pile, the contact with underlying Precambrian basement gneisses is smooth

**Fig. 4.** Lavas and sedimentary rocks in the coastal flexure. (**a**) Contact between basement and overlying lavas dips seaward on northeast side of Poulsen Fjord. (**b**) Dipping sedimentary bedding, lava flows and sills cut by sparse, steeply dipping dykes at J. C. Jacobsen Fjord. (**c**) Seaward-dipping (right) pyroclastic sedimentary rocks and lavas cut by sparse landward-dipping dykes west of Kap Edward Holm. (**d**) Strong concentration of sills in lavas and sedimentary rocks near Kap Edvard Holm. Cliffs are *c.* 500 m high in all of these photos.

and continuous. The contact is disrupted by relatively late, post-flexure, normal faults in some places, but this continuous surface can be mapped over many kilometres from the ice cap to the sea in some places. The contact is subhorizontal inland and increases in dip over c. 20 km as it approaches the coast. The contact is moderately to steeply dipping where it is exposed along the coast (Fig. 4a). At Nugtuaq, this contact and overlying bedding in lavas and volcaniclastic sediments is nearly vertical. At Kap Edvard Holm and Kap Gustav Holm, more moderate dips of c. 50 are present.

*Dyke swarms.* Along much of the coast south of Kangerlussuaq, outcrops are dominated by Precambrian basement gneisses with varying proportions of diabase dykes. Any basaltic lavas and sedimentary rocks once present here have been removed by uplift, erosion and glaciation. Fission-track thermochronology shows that no more than a few kilometres of material has been removed from this region by erosion (Hansen 1996). The dykes represent the mid-crustal feeder systems for the Tertiary basaltic lavas and are linked to them by field relations (Wager & Deer 1938; Nielsen 1978; Myers 1980; Brooks & Nielsen 1982a), geochemistry (Nielsen 1978; Hanghøj 1998), radiometric dates and palaeomagnetic studies (Faller & Soper 1979; Guenther *et al.* 1996). Reconstructions of the margin suggest that the volume of basaltic material in the dyke swarm may be at least as large as that of the previously overlying basaltic lavas (Nielsen 1978; Brooks & Nielsen 1982a).

The distribution of dykes across the exposed part of the East Greenland VRM are broadly systematic with the density of dykes increasing seaward over as much as 20–30 km (Myers 1980; Bromann 1999). Basement exposures near the ice cap and nunataks typically have very few dykes or no dykes at all, suggesting that no significant magmatic accretion occurred further landward. From near the hinge line of the flexure seaward, the density of dykes increases rapidly to c. 40% of the outcrop area. In local dense swarms, and some coastal exposures, sheeted dyke complexes with >95% dykes are present (Nielsen 1978; Myers 1980). These may represent a transition to oceanic crust just offshore (Nielsen 1978; Brooks & Nielsen 1982b). Detailed investigations of the widths

and spacings of dykes in the coastal dyke swarm reveal complex patterns of accretion across the VRM (Bromann 1999).

Significant variations in the host rock, the internal structure and the composition of the dyke swarm occur along the margin. Just north of Kangerlussuaq, the dyke swarm intrudes volcaniclastic sedimentary rocks and basalts. To the south, dykes intrude Precambrian basement rocks and therefore appear to represent a deeper crustal level (Fig. 6a). Locally, dyke swarms also intrude gabbroic plutons (Fig. 6b; Myers 1980; Tegner *et al.* 1998). In most areas the dykes are subparallel and trend in a northeasterly direction, more or less parallel to the coastline.

In detail, the internal structure of the coastal dyke swarm within the coastal flexure is complex. It appears to have been constructed by multiple generations of partially overlapping, laterally discontinuous swarms (Myers 1980). These are likely to be associated with magmatic centres marked by gabbroic plutons (Myers 1980; Bromann *et al.* 1996), as seen in other magmatic environments (Saemundsson 1978; Speight *et al.* 1982; Helgason & Zentilli 1985; Jolly & Sanderson 1995). In addition, many exposures show a systematic geometry in cross-section. In general, relatively old, more deformed and metamorphosed dykes are cut by succeeding generations of less deformed and metamorphosed dykes. The older dykes commonly dip landward with the lowest angle dykes found toward the coast. Dykes dipping as gently as 20° are common in some areas. Later, cross-cutting dykes also dip landward, but much more steeply. Thus, in cross-section, the dyke swarm has an asymmetric fanning geometry (Fig. 8). The geometry, cross-cutting relations, and variations in deformation and metamorphism of the dykes suggests that they were intruded before, during and after the development of the coastal flexure. Thus, they record a history of progressive intrusion and rotation (Fig. 8).

Radiometric dates on a limited number of dykes and gabbroic plutons that are intruded by dyke swarms (Tegner *et al.* 1998; Storey *et al.* 1996) demonstrate that there are dyke swarms of different ages that are at least, in part, superimposed on one another in this area (Nielsen 1975, 1978; Bromann *et al.* 1996). In some places, relatively young, post-flexure, dyke swarms dominate. These swarms obscure older

**Fig. 5.** Deformed (meta-) sedimentary rocks and basaltic lavas in the eastern (seaward) limb of the flexure.
(**a**) Metasediments and folded metadiabase dyke, Dødemandspynten. (**b**) Near-vertical bedding (parallel to pen) in volcaniclastic sandstones and basalts, Nugtuaq. (**c**) Detail of bedding near outcrop shown in (**b**).
(**d**) Moderately dipping bedding in deformed metasediments northeast of Poulsen Fjord. (**e**) Stretching lineation and quartz–diorite vein in basement-derived metaconglomerate, Dødemandspynten.

**Fig. 6.** Landward-dipping dykes. (**a**) Array of variably landward-dipping diabase dykes on Deception Island. Note systematic cross-cutting relations with more deformed, more gently dipping dykes cut by less deformed, more steeply dipping dykes. (**b**) Tilted gabbro pluton with landward-dipping dykes, Imilik Gabbro, northeast of Poulsen Fjord. Cliffs in both views are $c$. 300 m high.

structures, including older dyke swarms (Nielsen 1975, 1978), and indicate, at least locally, intense post-flexure intrusive activity.

*Gabbroic plutons.* Like the coastal dyke swarm, gabbroic plutons appear to have been intruded during a time interval that spans the development of the flexure. These probably define major magmatic centres along the margin (Wager & Deer 1938; Myers 1980; Brooks & Nielsen 1982a). The gabbroic bodies range in age from >55 to c. 42 Ma, and display a wide variation in the degree and style of deformation and metamorphism (Tegner *et al.* 1998). Several features distinguish the older plutons, intruded before or during the development of the flexure, from younger ones, intruded after the flexure. (1) Although they are only suggestive, the dips of the margins and igneous layering in the most deformed plutons hint that the older plutons have been rotated from original orientations. (2) Available palaeomagnetic data (Schwarz *et al.* 1979) indicate variable amounts of tilting. (3) These plutons are cut by numerous semi-ductile to cataclastic shear zones and locally have weak foliations. (4) Diabase dykes cut all the plutons but the older ones are characterized by dense dyke swarms that dip moderately landward (Fig. 6b), again qualitatively marking the flexure. The younger gabbro plutons are cut by few dykes and these tend to be very steeply dipping. Deformation fabrics have not been found in them.

## Structural style within rift segments

The structural style within individual rift segments is defined by three main elements: (1) faults; (2) flexure; and (3) dyke swarms. Each of these is now briefly described.

### Geometry and kinematics of faulting

Faulting and more distributed cataclastic deformation were most intense during the development of the flexure, and may have been more or less continuous during this phase of development of the margin. Faults that predate and postdate the flexure are also common features of the East Greenland VRM (Nielsen 1975, 1978; Pedersen *et al.* 1997; Larsen *et al.* 1994; Karson *et al.* 1996, 1998). Faults, and associated structures, that formed during this interval appear to be more or less continuous along-strike on the scale of individual rift segments (tens of kilometres). They are mainly normal faults with landward and seaward dips dominating in different

segments. Nielsen (1975) described landward-dipping faults in the Kangerlussuaq area, but in other segments to the southeast there is no preferential landward- or seaward-dip of faults. In some areas, low-angle normal faults and cataclastic shear zones are common. Crude foliations are locally developed in these areas. Regions of concentrated strain, created by fault geometry and kinematics, commonly show pervasive cataclastic deformation. Pseudotachylytes, suggesting seismic slip, are common along many of these fault zones (Karson *et al.* 1998). In addition to the dominant normal faults, oblique-slip accommodation zones and minor transfer or transform faults are also important parts of this extensional environment. These are described in more detail below.

In areas of mainly basement exposures, faults accommodate most of the extension associated with rifting; elsewhere, especially toward the coast, dykes account for most of the extension. A general lack of appropriate structural markers makes an accurate estimate of the stretching factor from faults impossible. Cross-cutting relations between dykes and faults do, however, permit determination of a relative geochronology for many fault zones. It appears that many faults cut relatively old dykes and are themselves cut by later, less deformed and metamorphosed dykes. Thus, it appears that many faults were active during the development of at least part of the dyke swarm and the flexure.

### Geometry and kinematics of the coastal flexure

The coastal flexure is a major asymmetrical fold that is an integral part of the East Greenland VRM (Wager & Deer 1938; Nielsen & Brooks 1981). In the present context, the term 'flexure' is used in a kinematic sense, rather than the dynamic sense as related to the elastic rigidity and effective elastic thickness of the lithosphere (Weissel & Karner 1989). Similar asymmetrical, crustal-scale folds develop in the hanging walls of major detachment systems in rifts (Lister & Davis 1989), continental margins (Etheridge *et al.* 1989) and mid-ocean ridge spreading centres (Pálamason 1980; Mutter & Karson 1992). A similar flexure also occurs on the West Greenland margin (Geoffroy *et al.* 1998).

The flexure is defined by originally horizontal rock units that have been progressively rotated toward seaward-dipping attitudes, locally by as much as 90° (Figs 3–5). Originally vertical structures have been rotated similarly to dip landward. The hinge line of the flexure is roughly

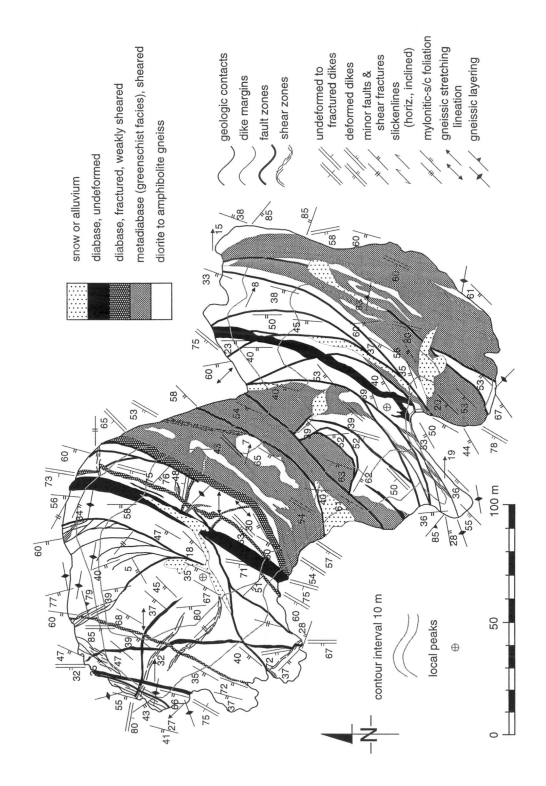

snow or alluvium

diabase, undeformed

diabase, fractured, weakly sheared

metadiabase (greenschist facies), sheared

diorite to amphibolite gneiss

geologic contacts

dike margins

fault zones

shear zones

undeformed to
fractured dikes

deformed dikes

minor faults &
shear fractures

slickenlines
(horiz., inclined)

mylonitic-s/c foliation

gneissic stretching
lineation

gneissic layering

contour interval 10 m

local peaks

100 m

50

N

0

**(b)**

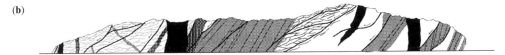

**Fig. 7.** Systematic cross-cutting relations in the coastal dyke swarm. (**a**) Map of small island at Similaq, Tasilaq Fjord, showing multiple generations of cross-cutting dykes and faults in cataclastically deformed Precambrian gneiss. (**b**) Northwest–southeast cross-section of the area shown in (a): note how younger, steeper dykes cut older, more gently dipping, sheared dykes.

parallel to the shoreline. The eastern limb of the flexure is commonly deeply dissected by fjords and locally submerged offshore. Landward exposures are dominated by Precambrian basement gneiss. Seaward, there is an increasing volume of magmatic material in the form of lavas, dykes and mafic plutons that dominate the eastern limb of the flexure. The position of the hinge line does not appear to correspond to specific magmatic structures, such as specific dyke densities or lava thickness, so it affects the

basalt pile and dyke swarm in different ways at different points along the margin.

The flexure was accommodated by widespread faulting and cataclastic deformation across the continent–ocean transition. The style of deformation varies with respect to major rock units (Fig. 9), with layer-parallel slip dominant in bedded lavas and sedimentary rocks, slip on faults at a low angle to dyke margins in areas of dense dykes and discrete, widely spaced fault zones in gneissic basement. Regions of intense

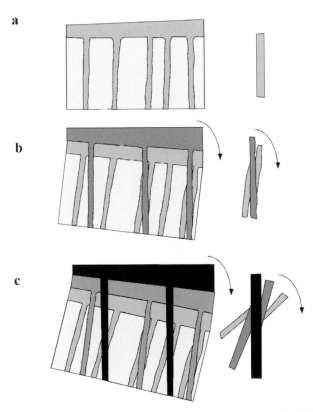

**Fig. 8.** Generalized relationship between dykes and lava flows on the East Greenland VRM. Progressive tilting and seaward subsidence of the margin accommodates the growth of the seaward-dipping wedge of lavas (seismically imaged offshore as seaward-dipping reflectors). Progressive tilting and intrusion of dykes reflects the evolution of the flexure.

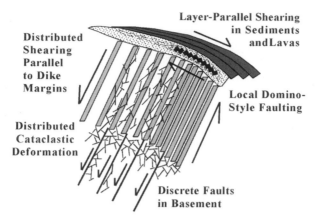

**Fig. 9.** Schematic illustration of brittle-strain accommodation in various rock units of the coastal flexure. Layer-parallel slip dominates in the deformed lavas and sedimentary rocks. Locally, layers are deformed by slip on closely spaced, steeply dipping fractures resulting in domino-style rotation. In areas dominated by diabase dykes in gneissic basement, small-scale faults are subparallel to dyke margins. In basement areas, discrete fault zones occur. In many areas, strain compatibility is maintained by deformation in areas of distributed cataclastic deformation.

cataclastic deformation, with a distinctive rose colour, appear to be developed in areas of strain compatibility, especially along accommodation zones.

Seaward subsidence was related to the growth of seaward-thickening lava units that are now mostly offshore. The overall geometry of the lava units is similar to that described for Iceland by Pálmason (1980). The dip of rock units in the seaward limb of the flexure, and the position of its hinge line, commonly vary from segment to segment along the margin. In some places the variations involve only minor lateral or vertical offsets (hundreds of metres?) of the flexure hinge line. Elsewhere, completely different parts of the flexure are developed in adjacent segments implying large displacements (kilometers) and/ or along-axis variations in rift structures.

*Internal structure of dyke swarms*

In many areas, tectonic extension recorded by dyke intrusion is much greater than that from faulting. An extreme example of this occurs in coastal exposures of sheeted dyke complexes. The dilation from dyke intrusion ranges from 20 to 60% across many parts of the margin (Nielsen 1978; Bromann *et al.* 1996; Bromann 1999); however, significant variations occur between adjacent rift segments, suggesting that magmatic extension events were confined to individual segments on the scale of tens of kilometres. The

continuity of variably rotated dyke swarms, and their interplay with fault zones, is substantially obscured by the intrusion of later, steeply dipping dyke swarms, some of which apparently cut across segment boundaries.

Assuming that individual pre- or synflexure dykes within the dyke swarm were intruded in a near-vertical orientation, they can be used, at least qualitatively, to define the coastal flexure. Although the dykes need not have been intruded vertically, several lines of evidence support this generalization. Firstly, where dykes and lava flows or sedimentary rocks are present, whether rotated by the flexure or not, the oldest dykes have a very high angle of orientation with respect to the lowest flows and/or beds (Nielsen 1975, 1978; Myers 1980). Secondly, in areas where little or no evidence of Tertiary deformation is observed, the dykes are dominantly very steeply dipping. Thirdly, preliminary palaeomagnetic data from areas with gently dipping dykes indicate that large post-cooling tectonic rotations have indeed occurred (Guenther *et al.* 1996). In some areas, especially near magmatic centres, it is likely that non-vertical cone sheets were intruded; however, the number of dykes observed in the extensive exposures along the coast and the walls of fjords, permit the identification of regional trends that are consistent for tens of kilometres and local variations within the rift segments.

Where originally horizontal markers are not present, the geometry of the dyke swarm in basement rocks is used to help define the hinge line and the magnitude of rotation in the flexure. In doing so it is assumed that early or pre-flexure dykes are present and that they record maximum rotations. This permits mapping of the general geometry of the flexure over >200 km along the margin, much more continuously than if sedimentary rocks and lavas are used alone. If the axis of the flexure hinge line were parallel to the strike of the dykes, then the dips of the dykes would define the flexure rotation axis. Locally, discrepancies between the flexure axis, defined by the dip of sedimentary rocks and lavas, and the dip of the dykes indicates that the hinge line is not always parallel to the strike of the dykes. Palaeomagnetic studies currently in progress will help further constrain the rotation axis in these areas.

## Along-strike variations in margin structures

Major geological units that are discontinuous along the East Greenland VRM have been identified in previous studies. Myers (1980) mapped individual dyke swarms in terms of domains defined by their density and structural trends. He defined several highly elongated swarms that are arranged in en echelon arrays along the coast. Bromann (1999) interprets the coastal dyke swarm in terms of three sigmoidal, en echelon segments: Nigertuluk–Agtertia; Agtertia to Kangerlussuaq; and from Kangerlussuaq eastward to Nansen Fjord (Fig. 2). These data are here incorporated with other structural considerations to outline major structural and magmatic units along the margin.

### Segment boundaries

Boundaries between discrete segments of the margin are defined by abrupt along-strike variations in rift-related rock units or structures. In some cases, these appear to reflect displacements of once continuous rift structures. Most displacements appear to have overall oblique offsets, but vertical offsets are the most obvious. In some cases, offset structures may be the result of original jogs in the margin across which somewhat different structural features have developed. The latter type are analogous to oceanic ridge–ridge transform faults across which slip is related to offset spreading centres and not to the displacement of once continuous ridge axes. In most cases, the boundary zone, or contact between adjacent segments, is not exposed, but the character of the discontinuity is inferred

from the contrast in rock units and structures across the boundary.

Vertical offsets are marked by the juxtaposition of rock units corresponding to different crustal levels. For example, across some boundaries, Precambrian gneisses cut by basaltic dykes face basaltic lavas, or dense dyke swarms abut lavas with abundant sills. These are the result of along-strike variations in processes that thicken or thin the crust, e.g. stretching, growth of lava sequences, sill injection or pluton intrusion. In most cases it is not clear if the boundary is a fault zone that has displaced a once continuous structure or if it is a boundary that resulted from local differences in crustal construction.

Lateral offsets are marked by the juxtaposition of segments of the margin that have been extended with different structural styles, magnitudes or with different stretching histories. These are commonly marked by abrupt changes across segment boundaries in the density of dykes and resulting horizontal dilation. Elsewhere, deformation styles such as fault dip-direction (landward v. seaward), fault density and integrated displacement are discontinuous. Tracing some of these boundaries landward, the densities of dykes and faults decrease, and the contrast between adjacent segments becomes indistinct along the trace of the boundary.

Although lateral offsets could be produced by a number of different mechanisms, e.g. strike-slip faulting of once continuous structures, this does not appear to be the case. Instead, along-strike variations in dyke, fault and flexure geometries occur in adjacent segments, suggesting that the boundaries are complexly faulted areas that accommodated along-strike differences in extension and flexure with overall strike-slip displacements. A component of horizontal displacement would also result from along-strike variations in subsidence along the flexure or crustal inflation.

In most of the segment boundaries identified, there are substantial variations in the rock units across them, as well as variations in the style and magnitude of faulting and the density of dykes. In addition, the dip of structures such as bedding, lava flows or foliations within the rock units can vary. Thus, it is inferred that several different mechanisms may contribute to creating the segment boundaries. These include along-strike differences in the total amount and style of crustal thickening (or thinning), differences in horizontal extension, and differences in the shape, hinge-line trend and hinge-line position relative to other structures. The variations of features across these boundaries, and their inferred structural and kinematic diversity, prompts their interpretation as accommodation zones rather than specifying them as, e.g., strike-slip or oblique-slip faults.

## Structures along accommodation zones

Most of the boundaries between segments of the margin occur at major fjords, and it is believed that glaciers have preferentially removed faulted and fractured rock along topographic steps at accommodation zones. However, no significant discontinuities are detected across many of the smaller fjords and it does not therefore appear that all fjords correspond to accommodation zones.

In a few places, accommodation zone structures have been examined directly. These occur in a few accommodation zones that are wider than the fjords, on islands and promontories extending into fjords, and in places where the fjords do not extend inland all the way to the ice-cap. In these areas a number of different types of structures are recognized, depending upon the scale of the offset across the accommodation zones.

On the outcrop scale, small strike-slip faults link terminations of dykes or dyke swarms. In some cases, these have transform fault geometries. Accordingly, they do not extend beyond the dyke margins and have horizontal displacements that are similar to the widths of the dykes. More commonly, these appear to have localized along pre-existing faults or shear zones that trend at a high angle to the margin (i.e. northwest) and are only a few metres wide (Fig. 10a and b). It is suspected that many northwest tending faults are geometrically similar to these, but have been reactivated by later movements related to the intrusion of dykes that terminate near the same surface, or block movements related to continued faulting or the flexure. Where the density of dykes is very low, the kinematics of faulting is more clear and margin-parallel normal faults (both landward and seaward dipping) commonly terminate along northwest trending strike-slip faults. These probably include simple transfer faults (i.e. jogs in normal fault systems) and more complex fault zones that have linked different normal faults to either side (e.g. Morley *et al.* 1990).

These relatively small faults and shear zones are typically 1–10 m wide. Some of the shear zones have schistose fault rocks with mixed cataclastic and crystal–plastic microstructures. Others are relatively diffuse cataclastic fault zones with anastomosing fault arrays separating lenses of cataclastically deformed basement gneisses and dykes. In some places, dykes have intruded along these fault zones. Commonly, these dykes have been deformed by subsequent fault movements (Fig. 10c and d).

Larger scale accommodation zones are up to several kilometres wide. They appear to have permitted rift segments to either side and to have been extended by different systems of normal faults and dykes. In several areas, relatively late dyke swarms transect the accommodation zones, thereby obscuring differences across them and possibly 'locking' them.

The accommodation zones include very complex arrays of anastomosing mylonites and cataclasites. Surrounding gneisses and some dykes are pervasively fractured and reduced to cataclasites. Pseudotachylyte veins and breccia dykes (Fig. 11a) are common in these fault zones (Karson *et al.* 1998), suggesting that seismic slip may have occurred preferentially along these areas relative to the margin-parallel fault zones.

In several instances, the accommodation zones appear to have exploited pre-existing Precambrian or Palaeozoic crustal heterogeneities. Some are located along northwest trending, steeply dipping schistose to mylonitic shear zones, locally >100 m wide, with amphibolite facies mineral assemblages (Fig. 11b and c). Low-temperature (greenschist facies or lower grade) cataclastic deformation overprints the older fabrics. These older mylonite zones do not cut the weakly metamorphosed Tertiary mafic dykes.

The Precambrian shear zones and Tertiary accommodation zones locally cut or lie along large (tens of metres wide), northwest trending mafic dykes (Fig. 11d). These dykes are commonly foliated and locally mylonitic. Some have amphibolite facies assemblages. These northwest trending dykes do not cut the Tertiary dykes and, like the mylonitic shear zones, appear to be much older structures. A single $^{40}Ar/^{39}Ar$ date on one of these dykes at Agtertia Fjord confirms a Precambrian age (Storey, pers. comm.).

The large northwest trending dykes and shear zones appear to have provided appropriately oriented zones of weakness that were exploited by Tertiary accommodation zones. It is noted that not all of these older structures have been reactivated. Some lie within segments between accommodation zones where they are obscured by the Tertiary dyke swarms.

Tertiary alkaline intrusions (Brooks & Nielsen 1982*a, b*) occur along the inland trace of some of the accommodation zones. These plutons appear to postdate deformation along the accommodation zones and may have preferentially intruded the crust along these discontinuities. Thus, the spacing of these bodies along the margin may be

**Fig. 10.** Structures in accommodation zones. (**a**) Pseudotachylyte breccia in west-northwest trending strike-slip fault zone at Agtertia Fjord. (**b**) Mylonitic shear zone with strike-slip kinematic indicators in Precambrian basement gneiss at Lango. The (Precambrian?) shear zone is cut by many dykes but does not deform them. (**c**) Mylonitic (Precambrian?) ultramafic dyke in basement gneiss strikes northwest and has right-lateral strike-slip kinematic indicators, Flado. (**d**) Dark-weathering Precambrian(?) metagabbro dykes. Dykes are near vertical and strike at a high angle to the rifted margin; this example is just north of Agtertia Fjord.

**Fig. 11.** Small-scale transfer or transform fault zones. (**a**) Strike-slip ductile shear zone cutting small diabase dyke at Tasilaq Fjord. (**b**) Same shear zone as in (a), transected by undeformed diabase dyke. (**c**) Strike-slip fault zone (viewed at high angle to strike) with variably deformed diabase dykes north of Nigertuluk. Different dykes are offset by different distances indicating periodic intrusion during part of the displacement history. (**d**) Detail of slickensided columnar joint surfaces in deformed fault-parallel dyke shown in (c).

a function of their preferred sites of intrusion along accommodation zones.

## Examples of margin discontinuities

The discontinuities in rift structures found along the East Greenland VRM can be grouped with respect to the scale of intervening segments, the orientations of structures and their persistence over the course of rift evolution. There follows discussions of large-scale first-order discontinuities, and smaller scale second- and third-order features in terms of rift segments and the structures that bound them.

### First-order discontinuities: triple-rift junctions

Two triple-rift junctions occur along the East Greenland VRM and mark first-order discontinuities in the margin. At the southern tip of Greenland, a triple-rift developed offshore at the intersection of the Mid-Atlantic Ridge, the Reykjanes Ridge and the Labrador Sea spreading centre. Structures related to the development of the adjacent margins are mainly submerged. At the Kangerlussuaq triple junction, the northeastern and southwestern branches of the rift have become rifted margins that pass laterally into oceanic crust (Brooks 1973). The northwestern branch is a failed rift, extending into the continental crust of the interior of Greenland. The location of this triple-rift junction approximately coincides with the point at which

Caledonian structures of northern East Greenland project to the rifted margin (Henriksen & Higgins 1976). This geometry is common along many rifted margins (Burke & Dewey 1973) and is reflected in major rift-related continental dyke swarms (Ernst *et al.* 1995).

The northwest trending failed arm lies along the Kangerlussuaq Fjord. Numerous structures to the southwest and northeast are truncated at this feature. Nevertheless, only a few faults and dykes are parallel to this trend (Brooks & Nielsen 1982*a*). This region is well known as a site of repeated magmatism during the evolution of the margin and is generally considered to mark the point at which the Iceland Plume crossed the margin, although the timing of this intersection is a matter of debate (Lawver & Müller 1994; Saunders *et al.* 1997; Larsen & Saunders 1999).

To the northeast, rift structures along the Blosseville Coast appear to curve smoothly into a relatively straight section of the VRM with few discontinuities. Most of this area is seaward-dipping lavas with the axis of the flexure trending subparallel to the coastline. Near Kangerlussuaq, uplift and erosion have been somewhat greater and the Lower Series lavas and dyke swarms are exposed. In general, it appears that the southern limb of the flexure lies well offshore, at least several kilometres, along this part of the margin.

The southwestern branch is composed of somewhat deeper crustal material with variably deformed gneisses, dyke swarms and gabbro plutons, exposed for >250 km along the coast. In general, the hinge line of the flexure lies near

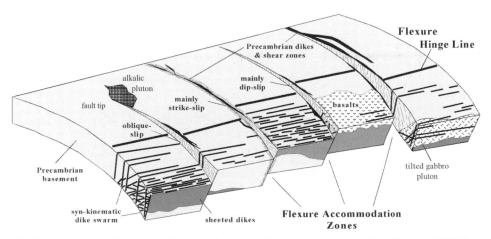

**Fig. 12.** Schematic block diagram of the style of segmentation developed on the East Greenland VRM. Accommodation zones tend to be localized along Precambrian mylonitic shear zones and large mafic dykes that strike at a high angle to the margin. Tertiary alkalic intrusions and pseudotachylyte breccias also commonly occur along these zones. Accommodation zones separate rift segments that are typically a few tens of kilometres in length and that contain different densities of dykes, fault geometries and flexure geometries.

the coast or onshore; however, it is offset from one segment to the next across accommodation zones spaced by the order of tens of kilometres along the margin (Fig. 12).

## Second-order accommodation zone at Agtertia Fjord

One of the most obvious discontinuities in the structure of the margin occurs at Agtertia Fjord, *c*. 100 km southwest of Kangerlussuaq (Fig. 2). These types of features represent second-order discontinuities in the rift structure. To the northeast of Agtertia Fjord, dyke swarms trending 30°, with varying dips, define the hinge line of the flexure. To the southwest, the hinge line of the flexure is displaced to the northwest and can be defined in coastal exposures by outcrops of lava and sedimentary rocks dipping steeply seaward and (locally) vertically. The oldest Tertiary dykes in these areas dip <20° landward and, collectively, the dyke swarms have a more northerly trend. In addition, several large gabbroic intrusions occur along this part of the margin and may be genetically related to dyke swarms (Myers 1980; Myers *et al.* 1993; Bromann *et al.* 1996). At least parts of these gabbros are highly deformed and cut by dyke swarms that have been rotated to moderately landward-dipping orientations. It is not yet known if genetically or temporally related magmatic units occur both northeast and southwest of the Agtertia Fjord accommodation zone, but ongoing geochemical studies will clarify these relationships and refine the view of the evolution of this discontinuity.

The accommodation zone itself is *c*. 20 km wide. Deformation is not pervasive across this area, but is rather focused along several northwest trending fault zones. Kinematic indicators show mainly normal, right-lateral oblique displacement. It is noted that Myers (1980) mapped a left-lateral step in dyke swarms here which could be reconciled with a transform fault structure.

Some of the fault zones overprint older, northwest trending, strike-slip, mylonitic shear zones. They also deform northwest trending, Precambrian, metadiabase to metagabbro dykes. Between some of the fault zones, basement gneisses and some Tertiary mafic dykes are intensely deformed. Pervasive cataclasis is especially evident in the gneisses where granitic and amphibolitic banding has been destroyed by closely spaced fractures and associated alteration. Mafic dykes are similarly affected. Crustal blocks between the fault zones show little

evidence of the damage that is so intense in the nearby fault zones. Although earlier Tertiary dykes are cut and rotated along these fault zones, later dykes cut across undisturbed. These relationships suggest a limited time span of activity for the accommodation zone, or at least some fault strands within it. Rock units and structures spanning much of the magmatic history of the margin are discontinuous in this area, suggesting that it was a relatively long-lived discontinuity along the VRM.

## Third-order discontinuities

The large Agtertia Fjord accommodation zone is dominated by mechanical deformation and may mark an area where both substantial vertical and lateral displacements occurred. Other smaller accommodation zones appear to have more modest displacements and some may be the result of along-strike changes of magmatic constructional style. Kinematically, these are similar to partial transfer zones along continental rifts (Morley *et al.* 1990) and they are referred to here as third-order discontinuities.

An example occurs between Kap Edvard Holm and Fladø, *c*. 30 km south of Kangerlussuaq (Fig. 2). Here, a typically well-developed dyke swarm, with *c*. 40–80% of dykes hosted by gneissic basement (Bromann 1999), strikes directly toward continuous cliff exposures inland from Kap Edward Holm (Myers 1980). To the west of the cliff exposures, inland nunataks show the lava–basement nonconformity dipping seaward below the cliff exposures, which are dominated by altered pyroclastic deposits, basaltic lava flows, mafic and silicic sills, and sparse diabase dykes. A large gabbroic intrusion occurs along the northern and eastern edges of the exposure. This dramatic along-strike variation appears to be related to both a vertical offset (Kap Edward Holm cliffs down relative to Fladø) and to a change in magmatic style (sills and gabbroic plutons to the northeast in contrast to dykes only to the southwest).

Strike-slip displacements may also occur across this boundary. The flexure to the southwest is defined only by dykes on Fladø. They decrease in dip rapidly over only a few kilometres across the island, suggesting a sharp flexure. Directly along the trend of this flexure, and for several kilometres southeastward, the lavas and sedimentary rocks of the Kap Edward Holm cliffs have a gentle and monotonous dip to the southeast in a very gentle flexure. Although the hinge lines of these two segments of the flexure are not laterally offset significantly, the different shape of the flexure across the accommodation zone suggests oblique-slip displacement.

At J. C. Jacobsen Fjord, to the northeast of Kangerlussuaq (Fig. 2), another type of accommodation zone occurs. To the southwest, moderate dyke densities occur in a thick succession of lavas, pyroclastic rocks and sills. Across the fjord, to the northeast, sills up to at least 200 m thick intrude a similar sequence and near the coast a sheeted dyke complex occurs. The preferential intrusion of dykes and sills on the northeast side of fjord suggests that both vertical and horizontal separations may have been accommodated. Along the trend of this discontinuity, to the northwest, a fault separates fractured basement rocks from Cretaceous and Palaeocene(?) sedimentary units with abundant sills (Curewitz 1999). Subsidence, related to basin formation, and localized magmatic inflation by sills would have been accommodated across this boundary. This example suggests that some northwest trending fault zones, representing third-order discontinuities along the margin, were relatively long-lived features with respect to the evolution of the margin.

A similar relationship exists along Nansen Fjord, just to the northeast. This third-order discontinuity appears to have a dominantly vertical offset and it also projects landward to the northwest along a Cretaceous basin boundary.

Similar discontinuities in structure or magmatic style mark other discontinuities along the margin (Fig. 12). Some have marked lateral offsets and others have mainly horizontal offsets. It is believed that many of the minor changes in the along-strike geology across northwest trending fjords in this region can be regarded as third-order discontinuities.

## Discussion

### Segmentation of VRM

Discontinuities that have developed along divergent plate boundaries are known to occur in a wide variety of settings, including continental rifts (Rosendahl 1987; Karson & Curtis 1989; Ebinger & Hayward 1996; Dickas & Mudrey 1997), rifted continental margins (Schouten & Klitgord 1982; Bonatti 1985; Lister et al. 1986; Etheridge et al. 1989; Uchupi 1989), and slow- and fast-spreading mid-ocean ridges (Lin et al. 1990; Sempéré et al. 1990; Macdonald et al. 1991). Although there is general agreement that 'segmentation' is typical of these environments, it appears that the segmentation is defined differently by different workers, according to their particular methodologies. For example, in the

western branch of the East African Rift, rift segments and accommodation zones are identified on the basis of half-graben defined on the basis of seismic reflection data (e.g. Rosendahl et al. 1986). Alternatively, segmentation along the East Pacific Rise is defined by topographic and magmatic associations (Macdonald et al. 1991; Sinton & Detrick 1992). In other environments, where magmatic and structural segmentation both appear to play a role, it is not clear if magmatic and structural discontinuities coincide or if they are somehow genetically related (Karson & Curtis 1989; Ebinger & Hayward 1996; Mutter & Karson 1992). Subaerially exposed VRMs provide an outstanding opportunity to examine the geometry and kinematics of rift segments and accommodation zones that developed at continent–ocean transition zones.

In this preliminary study, it is emphasized that the segmentation of the East Greenland margin includes both magmatic segmentation and mechanical segmentation. The magmatic segmentation, noted by previous workers, is especially well defined by the coast-parallel dyke swarm (Myers 1980; Bromann et al. 1996). Several discrete dyke swarms, that overlap in time and space, appear to have developed, creating a very complex array of subparallel units. At least some of these appear to be related to central mafic intrusive complexes (Myers 1980; Brooks & Nielsen 1982a, b; Myers et al. 1993; Tegner et al. 1998). Future radiometric dating, geochemical investigations and palaeomagnetic studies will help refine the internal structure of the East Greenland dyke swarm and provide additional constraints on its evolution. The dyke swarms are not only important in defining the individual segments, but also in marking accommodation zones and providing kinematic constraints on crustal displacements in these areas. The dyke swarms are also a key link between the spatially restricted exposures of flood basalts and underlying crustal structure and magmatic evolution.

Although the magmatic segmentation of this VRM has received significant attention, the mechanical segmentation has not previously been considered. This reconnaissance of the area southwest of Kangerlussuaq suggests that horizontal extension from slip along major fault zones equals, or exceeds, that of dyke dilation, especially in areas of low dyke density. Even in areas of abundant dykes, minor faults and fractures accommodate significant slip, probably mainly related to the development of the coastal flexure. The deformation associated with the dilation from the dyke swarms and the development of the flexure is far greater than the more obvious, widely spaced, post-flexure faults. Furthermore, it appears that relatively late dyke swarms, which may define magmatic segments >100 km long (Bromann et al. 1996; Bromann 1999), were superimposed on an earlier, more complex, segmented rift structure with segments only a few tens of kilometres long.

## VRM accommodation zones and later structures

Based on a synthesis of data from the East Africa Rift, Rosendahl (1987) speculated on how rift segmentation might be related to the segmentation of rifted continental margins and spreading centres as two lithosphere plates diverged. Uchupi (1989) followed this theme by comparing rift segmentation to that of the rifted margin of eastern North America. These and similar studies along other margins (e.g. Lister et al. 1986; Etheridge et al. 1989) suggested that accommodation zones and other major discontinuities along a continental rift may determine the spacing and locations of oceanic transform faults.

The East Greenland VRM provides a direct comparison to the segmentation of rift structures on Iceland and the Reykjanes Ridge, as well as in the intervening Northeast Atlantic. In Iceland, rift segments are defined on the basis of active seismicity and volcanism. On the seafloor, mid-ocean ridge morphology and magnetic anomalies define segments and discontinuities. Although the thick lavas and sedimentary cover, and paucity of magnetic reversals near the East Greenland margin, preclude any direct correlations, the spacing and orientation of segment boundaries invites comparisons.

Major discontinuities occur along the East Greenland VRM at the two triple-rift junctions. These also mark sharp kinks in the trend of the margin. Offshore, these appear to correspond to major transform faults along the spreading centre. Like the major kink in the West African margin at the failed arm occupied by the Niger River (Burke & Dewey 1973), these triple rifts mark places where spreading kinematics result in large lateral offsets that persist as transform faults.

The northwest trending accommodation zones of the East Greenland VRM are well aligned with the transforms of the Northeast Atlantic; however, about twice as many accommodation zones are recognized in the margin as there are transforms in the adjacent seafloor. It therefore appears that not all accommodation zones are destined to become transform offsets, or if they

do initially, they are eliminated by asymmetric spreading (Schouten & Klitgord 1982). It is also possible that some accommodation zones have offsets that are just too small to be resolved with the available seafloor data. It is suggested that only the largest of the accommodation zones along the margin, such as at Agtertia Fjord, are likely to persist as segment discontinuities that will be inherited as mid-ocean ridge spreading centres develop offshore. The thick volcanic crust associated with the SDRS probably buries and masks many of these features near the East Greenland margin and along the Greenland–Iceland Ridge.

The spacing of rift segments on Iceland is comparable to that of the second-order discontinuities of the East Greenland margin and therefore may resemble the surface of expression of the segmented VRM during its early evolution. Specifically, one might compare the Agtertia Fjord accommodation zone to second-order discontinuities such as the (east–west) South Iceland Seismic Zone or the on-land portion of the Tjornes Fracture Zone in northeast Iceland. Smaller rift segments with plutons and related dyke swarms are bounded by third-order discontinuities along the East Greenland VRM. These are comparable in scale to the en echelon volcanic ridges and intervening step-overs along the Reykjanes Ridge just south of Iceland (Murton & Parson 1993).

A continuing question regarding rift segmentation is the possible relationship between structural segmentation and magmatic centres. The coincidence of central volcanoes and graben or half-graben in the East African Rift suggests that the crust may be thermally weakened at spaced magmatic centres that determine the locus of extension and subsidence (Karson & Curtis 1989; Ebinger & Hayward 1996). In this view, accommodation zones would form between these centres in response to different timing or magnitude of extension. Accommodation zones would therefore be areas of less voluminous magmatism, and slip in them should generally postdate the earliest magmatism.

In an alternative view, heterogeneous extension along a rift focuses mantle upwelling and magmatism at the most highly extended areas. In this scenario, accommodation zones would be determined by the initial rift geometry and typically be located away from the volcanic centres. Faulting in the rifts and accommodation zones should predate the earliest magmatic activity.

In the East Greenland VRM, it appears that many accommodation zones are located preferentially along pre-existing zones of crustal weakness. Future dating of dyke rocks and fault zones will shed light on whether magmatic activity mainly preceded or postdated accommodation zone faulting. However, at present, it is clear that not all magmatic segment boundaries have obvious strike-slip or oblique-slip cross faults that could be identified as accommodation zones. Preliminary palaeomagnetic studies of dyke swarms show that, despite apparent continuity and parallelism, some of the swarms are composites formed during different magnetic polarity intervals. In addition, there appear to be many places where relatively late dyke swarms cut across earlier accommodation zones. It is suspected that, along the Blosseville Coast, voluminous late lavas have buried this type of earlier, segmented structure.

These characteristics suggest that only a few widely spaced (c. 100 km) accommodation zones represent second-order discontinuities in the rifted margin structure. They coincide with boundaries between magmatic segments and are persistent over the evolution of the margin. They may also help control the location of transform offsets as adjacent spreading centres develop. Smaller accommodation zones, which are identified as third-order discontinuities, spaced at tens of kilometres, have smaller and more variable vertical and lateral offsets. They have a variety of expressions with respect to magmatic constructions and appear to have been ephemeral features that developed periodically during the evolution of the margin to accommodate local-scale, along-strike, magmatic and structural variations. Magmatic construction dominated the late stages of evolution of the VRM resulting in relatively continuous dyke swarms and overlying lavas that tend to mask the earlier segmented structure.

## Conclusions

The East Greenland VRM, similar to other divergent plate boundaries, developed in a segmented fashion. The largest segments (c. 1000 km long) are separated by first-order discontinuities, exemplified by the Kangerlussuaq triple-rift junction that may be traced to the present plume centre in Iceland or major transform faults. Second-order discontinuities occur at major accommodation zones (c. 100 km spacing) that separate long-lived, regional, structural or magmatic trends. They appear to reactivate older basement structures and are persistent over the evolution of the margin. Second-order rift discontinuities may evolve into oceanic transforms that develop as seafloor spreading begins. The smallest segments, defined on the basis of consistent internal magmatic and fault structures,

can be traced for tens of kilometres along-strike. The third-order discontinuities (tens of kilometres spacing) that separate them appear to have been active for shorter intervals of time in response to along-strike variations in faulting and magmatism associated with the development of the coastal flexure. Collectively, these features result in a spatially and temporally complex assemblage of magmatic and structural features along the continent–ocean transition zone. As seafloor spreading began, this complex segmented structure was intruded and buried by voluminous magmatic construction that resulted in the offshore seaward-dipping reflector sequences. These features developed as an intermediate stage of in the evolution of a magmatic continental rift to a high-magma budget spreading centre.

This work is part of an ongoing comprehensive study of the Tertiary East Greenland Volcanic Rifted Margin conducted by the Danish Lithosphere Center (DLC), which is funded by the Danish National Research Foundation. Structural studies were supported by the DLC and National Science Foundation grant EAR-9508250 (to JAK). Discussions, in and out of the field, with members of the DLC and, in particular, M. Bromann and D. Curewitz contributed significantly to the development of the ideas in this paper. We thank C. MacNiocaill and K. Cox for their constructive reviews.

## References

BOHANNON, R. G. 1986. Tectonic configuration of the western Arabian continental margin, southern Red Sea. *Tectonics*, **5**, 277–499.

BONATTI, E. 1985. Transition from a continental to an oceanic rift: Punctiform initiation of sea floor spreading in the Red Sea. *Nature*, **316**, 33–37.

BOTT, M. H. P. 1987. The continental margin of central East Greenland in relation to North Atlantic plate tectonic evolution. *Journal of the Geological Society, London*, **144**, 561–568.

BRIDGEWATER, D., KETO, L., MCGREGOR, V. R. & MYERS, J. S. 1976. Archaean gneiss complex of Greenland. *In*: ESCHER, A. & WATT, W. S. (eds) *Geology of Greenland*. The Geological Survey of Greenland, 18–75.

BROMANN, M. 1999. *Structure of rift-related igneous systems and associated crustal flexures*. PhD dissertation, University of Copenhagen, Denmark.

——, BROOKS, C. K., NIELSEN, T. F. D., SVENNINGSEN, O. & KARSON, J. A. 1996. Infrastructure of the coast-parallel dike swarm along the rifted continental margin of southeast Greenland. *Transactions of the American Geophysical Union, EOS*, **77**, F827.

BROOKS, C. K. 1973. Rifting and doming in southern East Greenland. *Nature*, **244**, 23–25.

——1979. Geomorphological observations at Kanger-dlugssuaq, East Greenland. *Meddelelser Om Grønland Geoscience*, **1**, 3–21.

—— & NIELSEN, T. F. D. 1982a. The Phanerozoic development of the Kangerlussuaq area, East Greenland. *Meddelelser Om Grønland Geoscience*, **9**, 1–30.

—— & ——1982b. The E Greenland continental margin: a transition between oceanic and continental magmatism. *Journal of the Geological Society, London*, **139**, 265–275.

—— & FIELD PARTIES. 1996. The East Greenland oceanic rifted margin-onshore DLC fieldwork. *Bulletin Grønlands Geologiske Undersøgelse*, **172**, 95–102.

BURKE, K. & DEWEY, J. F. 1973. Plume generated triple junctions: key indicators in applying plate tectonics to old rocks. *Journal of the Geological Society, London*, **81**, 406–433.

CLIFT, P. D., TURNER, J. & OCEAN DRILLING PROGRAM LEG 152 SCIENTIFIC PARTY. 1995. Dynamic support by the Icelandic plume and vertical tectonics of the northeast Atlantic continental margins. *Journal of Geophysical Research*, **100**, 24 473–24 846.

COFFIN, M. F. & ELDHOLM, O. 1994. Large igneous provinces: crustal structure, dimensions, and external consequences. *Reviews of Geophysics*, **32**, 1–36.

CUREWITZ, D. 1999. *Investigations of faulting, magmatism and hydrothermal activity in extensional enviroments*. PhD dissertation, Duke University, Durham, USA.

DAM, G., LARSEN, M. & SØNDERHOLM, M. 1998. Sedimentary response to mantle plumes: implications from Paleocene onshore successions, West and East Greenland. *Geology*, **26**, 207–210.

DICKAS, A. B. & MUDREY, M. G., JR 1997. Segmented structure of the Middle Proterozoic Mid-continent Rift System, North America. *In*: OJAKANGAS, R. W., DICKAS, A. B. & GREEN, J. C. (eds) *Middle Proterozoic to Cambrian Rifting, Central North America*. Geological Society of America, Special Papers, **312**, 37–44.

DUNCAN, R. A., LARSEN, H. C., ALLEN, J. *et al*. 1996. *Proceedings of the Ocean Drilling Program, Initial Reports*, **163**.

DUTOIT, A. L. 1929. The volcanic belt of the Lebombo – a region of tension. *Transactions of the Royal Society of South Africa*, **18**, 189–218.

EBINGER, C. J. & HAYWARD, N. J. 1996. Soft plates and hot spots: views from afar. *Journal of Geophysical Research*, **101**, 21 859–21 876.

ELDHOLM, O. & GRUE, K. 1994. North Atlantic volcanic margins: dimensions and production rates. *Journal of Geophysical Research*, **99**, 2955–2968.

ERNST, R. E., HEAD, J. W., PARFITT, E., GROSFILS, E. & WILSON, L. 1995. Giant radiating dike swarms on Earth and Venus. *Earth Science Reviews*, **39**, 1–58.

ETHERIDGE, M. A., SYMONDS, P. A. & LISTER, G. S. 1989. Application of the detachment model to reconstruction of conjugate passive margins. *In*: TANKARD, A. J. & BALKWILL, H. R. (eds) *Extensional Tectonics and Stratigraphy of the North Atlantic Margins*. AAPG Memoir, **46**, 23–40.

FALLER, A. M. & SOPER, N. J. 1979. Paleomagnetic evidence for the origin of the coastal flexure and dyke swarm in central E. Greenland. *Journal of the Geological Society, London*, **136**, 737–744.

GEOFFROY, L., GELARD, J. P., LEPRVRIER, C. & OLIVIER, PH. 1998. The coastal flexure of Disko (West Greenland), onshore expression of the oblique reflectors. *Journal of the Geological Society, London*, **155**, 463–473.

GLEADOW, A. J. W. & BROOKS, C. K. 1979. Fission track dating, thermal histories, and tectonics of igneous intrusions in East Greenland. *Contributions to Mineralogy and Petrology*, **71**, 45–60.

GUENTHER, L. D., VEROSUB, K. L., HURST, S. D. & KARSON, J. A. 1996. Paleomagnetic constraints on the evolution of the Tertiary East Greenland coast-parallel dike swarm: tectonic rotations at a volcanic rifted margin. *Transactions of the American Geophysical Union, EOS*, **77**, F824.

HANFHØJ, K. 1998. *Magmatic evolution during continental rifting in the North Atlantic: Constraints from the geochemistry of the East Greenland coastal dyke swarm*. PhD dissertation, University of Copenhagen, Denmark.

HANSEN, K. 1996. Thermotectonic evolution of a rifted margin: fission track evidence from the Kangerlussuaq area, SE Greenland. *Terra Nova*, **8**, 458–469.

HELGASON, J. & ZENTILLI, M. 1985. Field characteristics of laterally emplaced dikes: anatomy of an exhumed Miocene dike swarm in Reydarfjördur, Eastern Iceland. *Tectonophysics*, **115**, 247–274.

HENRIKSEN, N. & HIGGINS, A. K. 1976. East Greenland Caledonian fold belt. *In*: ESCHER, A. & WATT, W. S. (eds) *Geology of Greenland*. The Geological Survey of Greenland, 182–247.

HIGGINS, A. C. & SOPER, N. J. 1981. Cretaceous–Palaeogene sub-basaltic and intrabasaltic sediments of the Kangerlussuaq area, Central East Greenland. *Geological Magazine*, **118**, 337–448.

HINZ, K., MUTTER, J. C., ZEHNDER, C. M. & GROUP, N. S. 1987. Symmetric conjugation of continent–ocean boundary structures along the Norwegian and East Greenland Margins. *Marine and Petroleum Geology*, **4**, 166–187.

JOLLY, R. J. H. & SANDERSON, D. J. 1995. Variations in the form and distribution of dykes in the Mull swarm, Scotland. *Journal of Structural Geology*, **17**, 1543–1557.

KARSON, J. A. & CURTIS, P. C. 1989. Tectonic and magmatic processes in the Eastern Branch of the East African Rift and implications for magmatically active continental rifts. *In*: ROSENDAHL, B. R., ROGERS, J. J. W. & RACH, N. M. (eds) *Rifting in Africa: Karroo to Recent*. Journal of African Earth Sciences, Special Volume, **8**, 431–453.

——, BROOKS, C. K., STOREY, M. & PRINGLE, M. 1998. Tertiary faulting and pseudotachylytes in the East Greenland volcanic rifted margin: Seismogenic faulting during magmatic construction. *Geology*, **26**, 39–42.

——, CUREWITZ, D., BROOKS, C. K., STOREY, M., LARSEN, H. C. & PRINGLE, M. S. 1996. Geometry and kinematics of faulting on the Tertiary

East Greenland volcanic rifted margin. *Transactions of the American Geophysical Union, EOS*, **77**, F839.

KELEMEN, P. B. & HOLBROOK, W. S. 1995. Origin of thick, high-velocity igneous crust along the U.S. east coast margin. *Journal of Geophysical Research*, **100**, 10007–10094.

KRISTOFFERSEN, Y. & TALWANI, M. 1977. Extinct triple junction south of Greenland and the Tertiary motion of Greenland relative to North America. *Geological Society of America*, **88**, 1037–1049.

LARSEN, H. C. 1978. Offshore continuation of East Greenland dyke swarm and North Atlantic Ocean formation. *Nature*, **274**, 220–223.

—— & JAKOBSDÓTTIR, S. 1988. Distribution, crustal properties and significance of seawards-dipping sub-basement reflectors off E Greenland. *In*: MORTON, A. C. & PARSON, L. M. (eds) *Early Tertiary Volcanism and the Opening of the NE Atlantic*. Geological Society, London, Special Publications, **39**, 95–114.

—— & MARCUSSEN, C. 1992. Sill-intrusion, flood basalt emplacement and deep crustal structure of the Scoresby Sund region, East Greenland. *In*: STOREY, B. C, ALABASTER, T. & PANKHURST, R. J. (eds) *Magmatism and the Causes of Continental Break-up*. Geological Society, London, Special Publications, **68**, 365–386.

—— & SAUNDERS, A. D. 1999. Tectonism and volcanism at the SE Greenland rifted margin: A record of plume impact and later continental rupture. *In*: SAUNDERS, A. D., LARSEN, H. C. & WISE, S. (eds) *Proceedings of the Ocean Drilling Program, Scientific Results*, in press.

——, ——, CLIFT, P. *et al.* 1994. *Proceedings of the Ocean Drilling Program, Initial Reports*, **153**.

LAWVER, L. A. & MÜLLER, R. D. 1994. Iceland hotspot track. *Geology*, **22**, 311–314.

LIN, J., PURDY, G. M., SCHOUTEN, H., SEMPÉRÉ, J. C. & ZERVAIS, C. 1990. Evidence from gravity data for focused magmatic accretion along the Mid-Atlantic Ridge. *Nature*, **334**, 627–632.

LISTER, G. S. & DAVIS, G. A. 1989. The origin of metamorphic core complexes. *Journal of Structural Geology*, **11**, 65–94.

——, ETHERIDGE, M. A. & SYMONDS, P. A. 1986. Detachment faulting and the evolution of passive continental margins. *Geology*, **14**, 246–250.

LIZARRALDE, D. & HOLBROOK, W. S. 1997. U.S. mid-Atlantic margin structure and early thermal evolution. *Journal of Geophysical Research*, **102**, 22855–22875.

MACDONALD, K. C., SCHEIRER, D. S. & CARBOTTE, S. M. 1991. Mid-ocean ridges: discontinuities, segments, and giant cracks. *Science*, **253**, 986–994.

MORLEY, C. K., NELSON, R. A., PATTON, T. L. & MUNN, S. G. 1990. Transfer zones in the East African rift system and their relevance to hydrocarbon exploration in rifts. *AAPG Bulletin*, **74**, 1234–1253.

MURTON, B. J. & PARSON, L. M. 1993. Segmentation, volcanism and deformation of oblique spreading centres: a quantitative study of the Reykjanes Ridge. *Tectonophysics*, **222**, 237–257.

MUTTER, J. C. & KARSON, J. A. 1992. Structural processes at slow-spreading ridges. *Science*, **257**, 627–634.

——, TALWANI, M. & STOFFA, P. L. 1982. Origin of seaward-dipping reflectors in oceanic crust off the Norwegian margin by 'subaerial sea-floor spreading'. *Geology*, **10**, 353–357.

MYERS, J. S. 1980. Structure of the coastal dyke swarm and associated plutonic intrusions of East Greenland. *Earth and Planetary Science Letters*, **46**, 407–418.

——, GILL, R. C. O., REX, D. C. & CHARNLEY, N. R. 1993. The Kap Gustav Holm Tertiary plutonic centre, East Greenland. *Journal of the Geological Society, London*, **150**, 259–276.

NIELSEN, T. F. D. 1975. Possible mechanism of continental breakup in the North Atlantic. *Nature*, **253**, 182–184.

——1978. The Tertiary dike swarm of the Kangerlussuaq area, East Greenland. An example of magmatic development during continental break-up. *Contributions to Mineralogy and Petrology*, **67**, 63–78.

—— & BROOKS, C. K. 1981. The East Greenland rifted continental margin: An examination of the coastal flexure. *Journal of the Geological Society, London*, **138**, 559–568.

——, SOPER, N. J., BROOKS, C. K., FALLER, A. M., HIGGINS, A. C. & MATTHEWS, D. W. 1981. The pre-basaltic sediments and the Lower Basalts at Kangerlussuaq, East Greenland: their stratigraphy, lithology, palaeomagnetism and petrology. *Meddelelser Om Grønland Geoscience*, **6**.

NEUHOFF, P. S., WATT, W. S., BIRD, D. K. & PEDERSEN, A. K. 1997. Timing and structural relations of regional zeolite zones in basalts of the East Greenland continental margin. *Geology*, **25**, 803–806.

PALMASON, G. 1980. A continuum model of crustal generation in Iceland: kinematic aspects. *Journal of Geophysics*, **47**, 7–18.

PEDERSEN, A. K., WATT, M., WATT, W. S. & LARSEN, L. M. 1997. Structure and stratigraphy of the early Tertiary basalts of the Blosseville Kyst, East Greenland. *Journal of the Geological Society, London*, **154**, 565–570.

PRICE, S., BRODIE, J., WITHAM, A. & KENT, R. W. 1997. Mid-Tertiary rifting and magmatism in the Traill Ø region, East Greenland. *Journal of the Geological Society, London*, **154**, 419–434.

ROSENDAHL, B. R. 1987. Architecture of continental rifts with special reference to East Africa. *Annual Reviews of Earth and Planetary Sciences*, **15**, 445–503.

——, REYNOLDS, D. J., LORBER, P. M. *et al.* 1986. Structural expressions of rifting. *In*: FROSTICK, L. E., RENAUT, R. W., REID, I. & TIERCELIN, J.-J. (eds) *Sedimentation in the African Rifts*. Geological Society, London, Special Publications, **25**, 29–43.

SAEMUNDSSON, K. 1978. Fissure swarms and central volcanoes of the neovolcanic zones of Iceland. *In*: BOWES, D. R. & LEAKE, B. G. (eds) *Crustal Evolution of the British Isles and Adjacent Regions*.

Geological Society, London, Special Publications, **10**, 415–432.

SAUNDERS, A. D., FITTON, J. G., KERR, A. C., NORRY, M. J. & KENT, R. W. 1997. North Atlantic Igneous Province. *In*: MAHONEY, J. J. & COFFIN, M. F. (eds) *Large Igneous Provinces*. American Geophysical Union, Geophysical Monograph, **100**, 45–93.

SCHOUTEN, H. & KLITGORD, K. D. 1982. The memory of the accreting plate boundary and the continuity of fracture zones. *Earth and Planetary Science Letters*, **59**, 255–266.

SCHWARZ, E. J., COLEMAN, L. C. & CATTROLL, H. M. 1979. Paleomagnetic results from the Skaergaard Intrusion, East Greenland. *Earth and Planetary Science Letters*, **42**, 437–443.

SEMPÉRÉ, J. C., PURDY, G. M. & SCHOUTEN, H. 1990. Segmentation of the Mid-Atlantic Ridge between 24° and 30°40′N. *Nature*, **344**, 427–431.

SINTON, J. M. & DETRICK, R. S. 1992. Mid-ocean ridge magma chambers. *Journal of Geophysical Research*, **97**, 197–216.

SPEIGHT, J. M., KELHORN, R. R., SLOAN, T. & KNAPP, R. J. 1982. The dyke swarms of Scotland. *In*: SUTHERLAND, D. S. (ed.) *Igneous Rocks of the British Isles*, Wiley, 449–459.

SOPER, N. J., HIGGINS, A. C., DOWNIE, C., MATTHEWS, D. W. & BROWN, P. E. 1976. Late Cretaceous–early Tertiary stratigraphy of the Kangerlussuaq area, east Greenland, and the age of the north-east Atlantic. *Journal of Geological Society, London*, **132**, 85–104.

STOREY, M., DUNCAN, R. A., LARSEN, H. C. *et al.* 1996. Impact and rapid flow of the Iceland plume beneath Greenland at 61 Ma. *Transactions of the American Geophysical Union, EOS*, **77**, F839.

SURLYK, F. 1977. Mesozoic faulting in East Greenland. *Geologie en Mijnbouw*, **56**, 311–327.

TALWANI, M. & ELDHOLM, O. 1977. Evolution of the Norwegian–Greenland Sea. *Geological Society of America Bulletin*, **88**, 969–999.

TEGNER, C., DUNCAN, R. A., BERNSTEIN, S., BROOKS, C. K., BIRD, D. K. & STOREY, M. 1998. $^{40}$Ar–$^{39}$Ar geochronology of Tertiary mafic intrusions along the East Greenland rifted màrgin: relation to flood basalts and the Iceland hotspot track. *Earth and Planetary Science Letters*, **156**, 75–88.

UCHUPI, E. 1989. The tectonic style of the Atlantic Mesozoic rift system. *Journal of African Earth Sciences*, **8**, 143–164.

WAGER, L. R. 1934. Geological investigations in East Greenland Part I General geology from Angmagsalik to Kap Dalton. *Meddelelser Om Grønland*, **105**, 1–46.

—— & DEER, W. A. 1938. A dyke swarm and crustal flexure in East Greenland. *Geological Magazine*, **75**, 39–46.

WEISSEL, J. K. & KARNER, G. D. 1989. Flexural uplift of rift flanks due to mechanical unloading of the lithosphere during extension. *Journal of Geophysical Research*, **94**, 13 919–13 950.

WHITE, R. S. & MCKENZIE, D. P. 1989. Magmatism of rift zones: the generation of volcanic continental margins and flood basalts. *Journal of Geophysical Research*, **94**, 7685–7729.

—— & ——1995. Mantle plumes and flood basalts. *Journal of Geophysical Research*, **100**, 17 543–17 586.

WHITE, R. S., SPENCE, G. D., FOWLER, S. R., MCKENZIE, D. P., WESTBROOK, G. K. & BOWEN, A. N. 1987. Magmatism at rifted continental margins. *Nature*, **330**, 439–444.

# Index